HANDBOOK OF APPLICABLE MATHEMATICS

Volume I: Algebra

HANDBOOK OF APPLICABLE MATHEMATICS

Chief Editor: Walter Ledermann

Editorial Board: Robert F. Churchhouse
Harvey Cohn
Peter Hilton
Emlyn Lloyd
Steven Vajda

Assistant Editor: Carol Jenkins

Volume I: ALGEBRA
Edited by Walter Ledermann, *University of Sussex*
and Steven Vajda, *University of Sussex*

Volume II: PROBABILITY
Emlyn Lloyd, *University of Lancaster*

Volume III: NUMERICAL METHODS
Edited by Robert F. Churchhouse, *University College Cardiff*

Volume IV: ANALYSIS
Edited by Walter Ledermann, *University of Sussex*
and Steven Vajda, *University of Sussex*

Volume V: GEOMETRY AND COMBINATORICS
Edited by Walter Ledermann, *University of Sussex*
and Steven Vajda, *University of Sussex*

Volume VI: STATISTICS
Edited by Emlyn Lloyd, *University of Lancaster*

HANDBOOK OF

APPLICABLE MATHEMATICS

Chief Editor: Walter Ledermann

Volume I: Algebra

Edited by

Walter Ledermann

and

Steven Vajda

University of Sussex

A Wiley–Interscience Publication

JOHN WILEY & SONS

Chichester – New York – Brisbane – Toronto

M8452
22/4/85

British Library Cataloguing in Publication Data:

Handbook of applicable mathematics.
  Vol. I: Algebra
  1. Mathematics
  I. Ledermann, Walter
  II. Vajda, Steven
  510   QA36   79-42724
  ISBN 0 471 27704 5

Printed in United States of America

Contributing Authors

Peter Hilton, Case Western Reserve University, Cleveland and Battelle Research Center, Seattle, U.S.A.

T. H. Jackson, University of York, Heslington, York, U.K.

Carol Jenkins, University of Sussex, Brighton, U.K.

W. Ledermann, University of Sussex, Brighton, U.K.

Des MacHale, University College, Cork, Ireland.

I. Stewart, University of Warwick, Coventry, U.K.

D. O. Tall, University of Warwick, Coventry, U.K.

Kathleen Trustrum, University of Sussex, Brighton, U.K.

P. J. Unsworth, University of Sussex, Brighton, U.K.

S. Vajda, University of Sussex, Brighton, U.K.

H. P. Williams, University of Edinburgh, U.K.

S. Wylie, 29 Almoners Avenue, Cambridge, U.K.

Contents

Introduction
to the
Handbook of Applicable Mathematics

Today, more than ever before, mathematics enters the lives of every one of us. Whereas, thirty years ago, it was supposed that mathematics was only needed by somebody planning to work in one of the 'hard' sciences (physics, chemistry), or to become an engineer, a professional statistician, an actuary or an accountant, it is recognized today that there are very few professions in which an understanding of mathematics is irrelevant. In the biological sciences, in the social sciences (especially economics, town planning, psychology), in medicine, mathematical methods of some sophistication are increasingly being used and practitioners in these fields are handicapped if their mathematical background does not include the requisite ideas and skills.

Yet it is a fact that there are many working in these professions who do find themselves at a disadvantage in trying to understand technical articles employing mathematical formulations, and who cannot perhaps fulfil their own potential as professionals, and advance in their professions at the rate that their talent would merit, for want of this basic understanding. Such people are rarely in a position to resume their formal education, and the study of some of the available textbooks may, at best, serve to give them some acquaintance with mathematical techniques, of a more or less formal nature, appropriate to current technology. Among such people, academic workers in disciplines which are coming increasingly to depend on mathematics constitute a very significant and important group.

Some years ago, the Editors of the present Handbook, all of them actively concerned with the teaching of mathematics with a view to its usefulness for today's and tomorrow's citizens, got together to discuss the problems faced by mature people already embarked on careers in professions which were taking on an increasingly mathematical aspect. To be sure, the discussion ranged more widely than that—the problem of 'mathematics avoidance' or 'mathematics anxiety', as it is often called today, is one of the most serious problems of modern civilization and affects, in principle, the entire community—but it was decided to concentrate on the problem as it affected professional effectiveness. There emerged from those discussions a novel format for presenting mathematics to this very specific audience. The intervening years have been spent in putting this novel conception into practice, and the result is the Handbook of Applicable Mathematics.

THE PLAN OF THE HANDBOOK

The 'Handbook' consists of two sets of books—*guide books* and *core volumes.* On the one hand, there are a number of guide books, written by experts in various fields in which mathematics is used (e.g., medicine, sociology, management, economics). These guide books are by no means comprehensive treatises; each is intended to treat a small number of particular topics within the field, employing, where appropriate, mathematical formulations and mathematical reasoning. In fact, a typical guide book consists of a discussion of a particular problem, or related set of problems, and shows how the use of mathematical models serves to solve the problem. Wherever any mathematics is used in a guide book, it is cross-referenced to an article (or articles) in the core volumes.

There are 6 core volumes devoted respectively to Algebra, Probability, Numerical Methods, Analysis, Geometry and Combinatorics, and Statistics. These volumes are texts of mathematics—but they are no ordinary mathematical texts. They have been designed specifically for the needs of the professional adult (though we believe they should be suitable for any intelligent adult!) and they stand or fall by their success in explaining the nature and importance of key mathematical ideas to those who need to grasp and to use those ideas. Either through their reading of a guide book or through their own work or outside reading, professional adults will find themselves needing to understand a particular mathematical idea (e.g., linear programming, statistical robustness, vector product, probability density, round-off error); and they will then be able to turn to the appropriate article in the core volume in question and *find out just what they want to know*—this, at any rate, is our hope and our intention.

How then do the content and style of the core volumes differ from a standard mathematical text? First, the articles are designed to be read by somebody who has been referred to a particular mathematical topic and would prefer not to have to do a great deal of preparatory reading; thus each article is, to the greatest extent possible, self-contained (though, of course, there is considerable cross-referencing within the set of core volumes). Second, the articles are designed to be read by somebody who wants to get hold of the mathematical ideas and who does not want to be submerged in difficult details of mathematical proof. Each article is followed by a bibliography indicating where the unusually assiduous reader can acquire that sort of 'study in depth'. Third, the topics in the core volumes have been chosen for their relevance to a number of different fields of application, so that the treatment of those topics is not biased in favour of a particular application. Our thought is that the reader—unlike the typical college student!—will already be motivated, through some particular problem or the study of some particular new technique, to acquire the necessary mathematical knowledge. Fourth, this is a handbook, not an encyclopedia—if we do not think that a particular aspect of a mathematical topic is likely to be useful or interesting to the kind of reader we have in mind, we have omitted it. We have not set out to include everything known on a particular topic, and we are not catering for the professional mathematician. The Handbook has been written as a contribution to the *practice* of mathematics, not to the *theory*.

The reader will readily appreciate that such a novel departure from standard textbook writing—this is neither 'pure' mathematics nor 'applied' mathematics as traditionally interpreted—was not easily achieved. Even after the basic concept of the Handbook had been formulated by the Editors, and the complicated system of cross-referencing had been developed, there was a very serious problem of finding authors who would write the sort of material we wanted. This is by no means the way in which mathematicians and experts in mathematical applications are used to writing. Thus we do not apologize for the fact that the Handbook has lain so long in the womb; we were trying to do something new and we had to try, to the best of our ability, to get it right. We are sure we have not been uniformly successful; but we can at least comfort ourselves that the result would have been much worse, and far less suitable for those whose needs we are trying to meet, had we been more hasty and less conscientious.

It is, however, not only our task which has not been easy. Mathematics itself is not easy. The reader is not to suppose that, even with his or her strong motivation and the best endeavours of the editors and authors, the mathematical material contained in the core volumes can be grasped without considerable effort. Were mathematics an elementary affair, it would not provide the key to so many problems of science, technology and human affairs. It is universal, in the sense that significant mathematical ideas and mathematical results are relevant to very different 'concrete' applications—a single algorithm serves to enable the travelling salesman to design his itinerary, and the refrigerator manufacturing company to plan a sequence of modifications of a given model; and could conceivably enable an intelligence unit to improve its techniques for decoding the secret messages of a foreign power. Given this universality, mathematics cannot be trivial. And, if it is not trivial, then some parts of mathematics are bound to be substantially more difficult than others.

This difference in level of difficulty has been faced squarely in the Handbook. The reader should not be surprised that certain articles require a great deal of effort for their comprehension and may well involve much study of related material provided in other referenced articles in the core volumes—while other articles can be digested almost effortlessly. In any case, different readers will approach the Handbook from different levels of mathematical competence and we have been very much concerned to cater for all levels.

THE REFERENCING AND CROSS-REFERENCING SYSTEM

To use the Handbook effectively, the reader will need a clear understanding of our numbering and referencing system, so we will explain it here. Important items in the core volumes or the guidebooks—such as definitions of mathematical terms or statements of key results—are assigned sets of numbers according to the following scheme. There are six categories of such mathematical items, namely:

(i) Definitions

(ii) Theorems, Propositions, Lemmas and Corollaries

(iii) Equations and other Displayed Formulae

(iv) Examples

(v) Figures

(vi) Tables

Items in any one of these six categories carry a triple designation a.b.c. of arabic numerals, where 'a' gives the *chapter* number, 'b' the *section* number, and 'c' the number of the individual *item*. Thus items belonging to a given category, for example, definitions, are numbered in sequence within a section, but the numbering is independent as between categories. For example, in Section 5 of Chapter 3 (of a given volume), we may find a displayed Formula labelled (5.3.7) and also Lemma 5.3.7. followed by Theorem 5.3.8. Even where sections are further divided into *subsections*, our numbering system is as described above, and takes no account of the particular subsection in which the item occurs.

As we have already indicated, a crucial feature of the Handbook is the comprehensive cross-referencing sytem which enables the reader of any part of any core volume or guide book to find his or her way quickly and easily to the place or places where a particular idea is introduced or discussed in detail. If, for example, reading the core volume on Statistics, the reader finds that the notion of a *matrix* is playing a vital role, and if the reader wishes to refresh his or her understanding of this concept, then it is important that an immediate reference be available to the place in the core volume on Algebra where the notion is first introduced and its basic properties and uses discussed.

Such ready access is achieved by the adoption of the following system. There are six core volumes, identified by the Roman numerals as follows:

I Algebra

II Probability

III Numerical Methods

IV Analysis

V Geometry and Combinatorics

VI Statistics

A reference to an item will appear in square brackets and will *typically* consist of a pair of entries [see A, B] where A is the volume number and B is the triple designating the item in that volume to which reference is being made. Thus '[see II, (3.4.5)]' refers to equation (3.4.5) of Volume II (Probability). There are, however, two exceptions to this rule. The first is simple a matter of economy!—if the reference is to an item in the same volume, the volume number designation (A, above) is suppressed; thus '[see Theorem 2.4.6]', appearing in Volume III, refers to Theorem 2.4.6. of Volume III. The second exception is more fundamental and, we contend, wholly natural. It may be that we feel the need to refer to a substantial discussion rather than to a single mathematical item (this could well have been the case in the reference to 'matrix', given as an example above). If we judge that such a comprehensive reference is appropriate, then the second entry B of the reference may carry only two numerals—or even, in an extreme case, only one. Thus the reference '[see I, 2.3]' refers to Section 3 of Chapter 2 of Volume I and recommends the reader to study that entire section to get a complete picture of the idea being presented.

Bibliographies are to be found at the end of each chapter of the core volumes and at the end of each guide book. References to these bibliographies appear in the text as '(Smith (1979))'.

It should perhaps be explained that, while the referencing *within* a chapter of a core volume or *within* a guide book is substantially the responsibility of the author of that part of the text, the cross-referencing has been the responsibility of the editors as a whole. Indeed, it is fair to say that it has been one of their heaviest and most exacting responsibilities. Any defects in putting the referencing principles into practice must be borne by the editors. The successes of the system must be attributed to the excellent and wholehearted work of our invaluable colleague, Carol Jenkins.

CHAPTER 1

Sets

There are at least two main strands in mathematical thought: *concepts* and *computations*. But the two are closely intertwined: blind computation without a proper conceptual grasp can easily produce nonsense (as computer scientists say, 'garbage in, garbage out'), and conceptual insight is largely useless unless backed up by computational power.

Since both kinds of mathematics are necessary, each is as 'applicable' as the other. However, applications of concepts usually take a less direct form than those of computational techniques, which latter therefore tend to be emphasized in 'applied mathematics'. Sometimes this obscures the importance of the underlying concepts.

So it is with set theory. In itself, it has few direct applications of any consequence; but it provides a sufficiently sensitive, precise, and flexible language for the accurate pinpointing of mathematical ideas; and it helps to keep separate ideas that are traditionally confused, when in fact they have important differences. In particular, set theory clarifies the concept of a 'function'—indeed it arose from a serious crisis in Fourier analysis of functions—in a way that turns out useful, later on, in complicated situations where functions of several variables take values subject to several constraints on the variables. Traditional notation under such circumstances can be very confusing.

1.1. SETS

A *set* is any collection of mathematical (or other) objects, specified in such a way that we can tell in principle whether or not any given object belongs to it. Phrases such as

'The set of all prime numbers' [see § 4.2],

'The set of points in the plane equidistant from two given points',

'The set of zeros of the cosine function' [see V, § 1.2.3 or IV, § 2.12],

'The set of natural numbers n expressible as $n = x^2 + y^2$ for natural numbers x, y' [see § 2.1]

define sets in the required manner. For example, to tell whether or not 19 belongs to the fourth set, we need only seek solutions of the equation $19 = x^2 + y^2$.

1

Since none exist with natural numbers x and y, it follows that 19 is not in the set.

The objects belonging to a set are called its *elements*, *members*, or *points*. For clarity in this chapter we shall, as far as possible, use capital letters X, Y, Z, S, T, ... for sets and lower case x, y, z, s, t, ... for elements. The notation

$$x \in X$$

means 'x is an element of X', whereas

$$x \notin X$$

means 'x is not an element of X'. With the above example for X, we saw that $19 \notin X$.

Two sets X and Y are equal, written

$$X = Y,$$

if they have precisely the same elements. If not, we write

$$X \neq Y.$$

A set is said to be *finite* or *infinite* according as it contains a finite or infinite number of elements.

The simplest notation for a set is to list its elements inside braces { }. Thus

$$X = \{3, 5, 7, 12, 23\}$$

denotes a set with five elements, the numbers 3, 5, 7, 12, and 23. Order within the list, or repetition of elements, makes no difference. For example $\{1, 2, 3\} = \{3, 1, 2\} = \{1, 3, 3, 2, 1, 1, 2, 1\}$. This notation can be adapted in various ways: for example

$$X = \{1, 2, 3, \ldots, n\}$$

means X contains the elements 1, 2, 3, ... up to and including n; and

$$X = \{1, 2, 3, \ldots\}$$

means X contains *all* whole numbers, namely 1, 2, 3, ... The use of dots is only satisfactory in unambiguous circumstances. For precision we could specify this last set as

$$X = \{x \mid x \text{ is a whole number and } x \geq 1\}.$$

In general,

$$X = \{x \mid P(x)\}$$

where $P(x)$ is some property involving x, means 'the set of all x for which the property $P(x)$ is true'.

Standard symbols for the more important sets of numbers are desirable. Almost universally adopted are the following:

$\mathbb{N}$ = the natural numbers 0, 1, 2, 3, ...
$\mathbb{Z}$ = the integers ... $-2, -1, 0, 1, 2, \ldots$
$\mathbb{Q}$ = the rational numbers p/q, where $p, q \in \mathbb{Z}$ and $q \neq 0$,
$\mathbb{R}$ = the real numbers [see § 2.6.1],
$\mathbb{C}$ = the complex numbers [see § 2.7.1].

For brevity, it is usual to write

$$\{x \in Y \mid P(x)\}$$

instead of

$$\{x \mid x \in Y \text{ and } P(x)\}.$$

EXAMPLES 1.1.1 to 1.1.4. Let
1. $X = \{x \in \mathbb{N} \mid x^2 - 5x + 6 = 0\}$,
2. $Y = \{x \in \mathbb{R} \mid x^3 = x\}$,
3. $Z = \{x \in \mathbb{Z} \mid 0 \leq z \leq 27 \text{ and } z = r^2 \text{ for some } r \in \mathbb{Z}\}$, and
4. $T = \{x \in \mathbb{R} \mid x^5 + x = 1\}$.

We discuss each of these sets in turn:

1. The solutions of the equation $x^2 - 5x + 6 = 0$ are of course $x = 2$ and $x = 3$. Since 2 and 3 belong to $\mathbb{N}$, we have more explicitly $X = \{2, 3\}$.

2. Similarly $Y = \{-1, 0, 1\}$.

3. The perfect squares between 0 and 27 are 0, 1, 4, 9, 16, 25. So $Z = \{0, 1, 4, 9, 16, 25\}$.

4. Unlike the previous three examples, there is no easy way to list the elements of T, because there is no formula for solving the quintic equation $x^5 + x = 1$. This does not prevent T from being a well defined set; for given any $\theta \in \mathbb{R}$ we can decide whether or not $\theta \in T$ by working out $\theta^5 + \theta$ and seeing whether the result equals 1.

As these examples show, the solutions (within a given set such as $\mathbb{Z}$ or $\mathbb{R}$) of a system of equations or inequalities form sets. It is unusual for such a system to have exactly one solution, and this is a natural place to make use of set theory.

1.2. SET ALGEBRA

Next we develop a kind of 'algebra' of sets—methods for combining sets to form other sets, together with some general laws that hold for these combinations.

1.2.1. Unions and Intersections

The *intersection* of two sets X and Y is the set whose elements are those objects belonging *both* to X and to Y. It is denoted

$$X \cap Y,$$

and defined formally by $X \cap Y = \{x \mid x \in X \text{ and } x \in Y\}$. For instance, if $X = \{1, 2, 3, 4, 5\}$ and $Y = \{2, 3, 5, 7, 11\}$, then $X \cap Y = \{2, 3, 5\}$.

The *empty set*, or *null set*, denoted $\varnothing$, is the set with *no* elements. Sets with at least one element are said to be *non-empty*. Formally, we may define

$$\varnothing = \{x \mid x \neq x\}.$$

The main use of $\varnothing$ is to express the non-existence of certain elements. Thus

$A \cap B = \emptyset$ means that A and B have no elements in common, or are *disjoint*. Disjointness is often important.

EXAMPLE 1.2.1.   Some solutions of geometric problems may be formulated in set-theoretic terms. Let P and Q be two points in the plane, distance 6 units apart. Find all points R which are equidistant from P and Q, and distance 5 units from P. Let

$$X = \{R \mid R \text{ is equidistant from } P \text{ and } Q\},$$
$$Y = \{R \mid R \text{ is 5 units from } P\}.$$

The points R we seek are the elements of $X \cap Y$. Now in Figure 1.2.1 X is the perpendicular bisector of the line between P and Q [see V, § 1.2.1]; and Y is the circle centre P, radius 5 [see V, § 1.1.8]. These intersect, by elementary geometry, in two points R_1, R_2. (By Pythagoras [see V, § 1.1.3], R_1 and R_2 lie 4 units either side of PQ.) Note how set-theoretic and geometric intersection correspond.

The *union* of X and Y, written

$$X \cup Y,$$

consists of the elements of X together with those of Y: thus $X \cup Y = \{x \mid x \in X \text{ or } y \in Y \text{ (or both)}\}$. With X, Y as in the numerical example above,

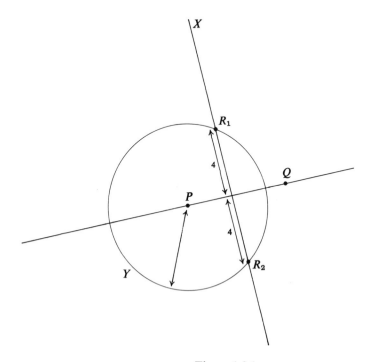

Figure 1.2.1

$X \cup Y = \{1, 2, 3, 4, 5, 7, 11\}$. In the geometric example, $X \cup Y$ is the circle Y together with the line X, 'glued together' into a single set of points.

1.2.2. Venn Diagrams

Unions, intersections, and other set-theoretic operations are often pictured using a *Venn diagram*, which represents the sets concerned as regions in the plane, chosen so that any 'general formula' about sets which is valid for the regions chosen is automatically valid in general. For a detailed and precise account of what this means, see Stewart (1975b). For X, Y as above the relevant Venn diagram takes the form of Figure 1.2.2. The elements of X are written inside a closed loop (here an ellipse [see V, § 1.3.1], though circles are common), and similarly for Y, in such a way that the elements common to both lie within the overlap, which perforce represents $X \cap Y$. The union $X \cup Y$ is the region obtained by glueing both loops together.

We may use a similar 'general' picture without any elements written in Figure 1.2.3. For formulae involving three sets we use three loops, overlapping as in Figure 1.2.4.

Four or more sets require more complicated diagrams (for example, no diagram using four circles is general enough, because not all possible combinations of overlapping and non-overlapping occur): Figure 1.2.5 shows one way

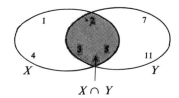

Figure 1.2.2

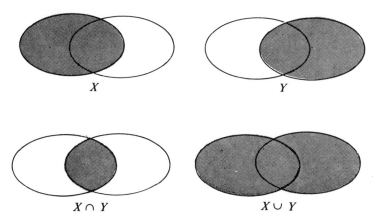

Figure 1.2.3

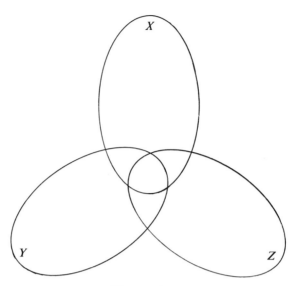

Figure 1.2.4

using ellipses. The numbers 1–16 indicate the sixteen regions corresponding to all possible combinations of membership or non-membership of the four sets A, B, C, D. For example region 13 corresponds to elements in A, B, D but not C; region 10 to elements in A, C but not B, D. (See also Stewart and Tall (1977), p. 56.)

Using Venn diagrams we can illustrate some of the 'laws' of set theory. For any three sets X, Y, Z the following equations are always valid:

Commutative law for $\cap$: $\qquad$ $X \cap Y = Y \cap X.$ $\qquad$ (1.2.1)

Associative law for $\cap$: $\qquad$ $(X \cap Y) \cap Z = X \cap (Y \cap Z).$ (1.2.2)

Commutative law for $\cup$: $\qquad$ $X \cup Y = Y \cup X.$ $\qquad$ (1.2.3)

Associative law for $\cup$: $\qquad$ $(X \cup Y) \cup Z = X \cup (Y \cup Z).$ (1.2.4)

Distributive law for $\cup$ over $\cap$: $\qquad$ $X \cup (Y \cap Z) = (X \cup Y) \cap (X \cup Z).$

$\qquad$ (1.2.5)

Distributive law for $\cap$ over $\cup$: $\qquad$ $X \cap (Y \cup Z) = (X \cap Y) \cup (X \cap Z).$

$\qquad$ (1.2.6)

For example, to check (1.2.6) we work out the two sides of the equation as in Figure 1.2.6. The results are clearly equal. (Provided a suitably 'general' Venn diagram is used, this method actually gives a rigorous proof: see Stewart (1975b).)

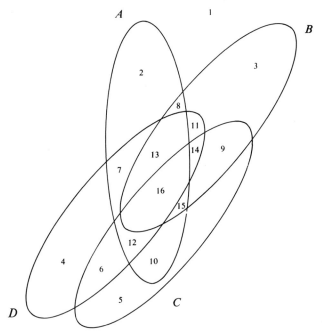

Figure 1.2.5

1.2.3.   Subsets

A set S is a *subset* of a set T, or is *contained* in T, if every element of S is an element of T. We write

$$S \subseteq T.$$

(Many authors, especially topologists, use '$\subset$' where we have written '$\subseteq$'.) For instance, $\{2, 4, 6\} \subseteq \{1, 2, 3, 4, 5, 6\}$ since each of the elements 2, 4, 6 of the first set belongs also to the second. But $\{2, 4, 6\}$ is not a subset of $\{1, 2, 3, 4, 5\}$, because one of its elements, namely 6, is not an element of the second set. We write

$$S \nsubseteq T.$$

If S is not a subset of T. By convention (or logical hairsplitting) $\varnothing \subseteq T$ for all T; and clearly $T \subseteq T$ for all T. A subset of T not equal to T itself is called a *proper* subset; for this we write

$$S \subsetneqq T.$$

This clumsy symbol avoids the ambiguity of the more attractive possibility '$\subset$'.

On a Venn diagram we illustrate the relation $S \subseteq T$ by drawing S *inside* T, as in Figure 1.2.7.

WARNING.   Do not confuse '$\in$' with '$\subseteq$'. When $T = \{1, 2, 3\}$ the *elements*

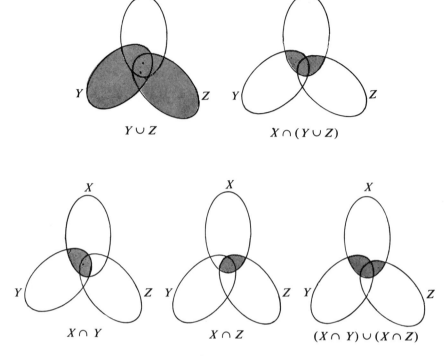

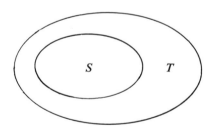

Figure 1.2.6

Figure 1.2.7

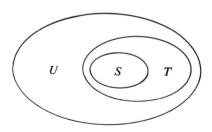

Figure 1.2.8

$x \in T$ are $x = 1, 2, 3$. The *subsets* $X \subseteq T$ are $X = \emptyset, \{1\}, \{2\}, \{3\}, \{1, 2\}, \{1, 3\}$, $\{2, 3\}, \{1, 2, 3\}$.

Also do not confuse an element x with the set $\{x\}$, known as a *singleton*, whose only element is x. For example, with T as above, $\{T\}$ is a set whose only element is T; this is quite different from T, whose elements are 1, 2, 3. As here, a set can commonly occur as an element of another set. For example if $S = \{$all lines in the plane$\}$ then each element of S is a line, that is, a *set* of points. A less mathematical example might be $\{$all Trade Unions$\}$.

The following statements are equivalent:
1. $S \subseteq T$.
2. $S \cup T = T$.
3. $S \cap T = S$.
The subset relation '$\subseteq$' is *transitive*, that is:
4. If $S \subseteq T$ and $T \subseteq U$ then $S \subseteq U$ (see Figure 1.2.8).
Finally, the following property is often useful in proving two sets equal:
5. $S = T$ if and only if $S \subseteq T$ and $T \subseteq S$.

1.2.4. Complements

The *difference*

$$A \backslash B$$

of two sets A, B is the set $\{x \in A \mid x \notin B\}$, whose Venn diagram is Figure 1.2.9. Some sample laws involving it, provable by Venn diagrams, are:

$$(A \backslash B) \cap C = (A \cap C) \backslash B. \tag{1.2.7}$$

$$(A \backslash B) \cup C = (A \cup C) \backslash ((A \cap B) \backslash C). \tag{1.2.8}$$

$$A \backslash (B \cup C) = (A \backslash B) \backslash C. \tag{1.2.9}$$

$$A \backslash (B \cap C) = (A \backslash B) \cup (A \backslash C). \tag{1.2.10}$$

$$A \backslash A = \emptyset. \tag{1.2.11}$$

$$A \backslash \emptyset = A. \tag{1.2.12}$$

$$\emptyset \backslash A = \emptyset. \tag{1.2.13}$$

$$A \backslash B \subseteq A. \tag{1.2.14}$$

In many mathematical discussions it is possible to choose a 'universe of discourse' U, such that all the sets under consideration are subsets of U. For example, if discussing sets of real numbers, the natural choice is $U = \mathbb{R}$ [see § 1.1]. This has the advantage of restricting attention to the relevant objects ('...in a discussion about dogs, if one wishes to talk about all non-sheepdogs, it is pointless to worry about camels...', Stewart (1975a), p. 57). In these circumstances we call

$$A^c = U \backslash A$$

the *complement* of A (relative to U), often suppressing the bracketed phrase. Figure 1.2.10 shows this.

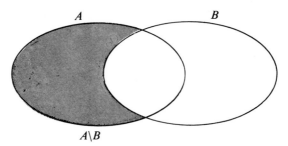

$A\backslash B$

Figure 1.2.9

It is tempting to introduce a *universe* Ω containing all possible sets as subsets, in which case A^c (relative to Ω) becomes 'everything not in A'. But such a set turns out to be 'too big'; more seriously, the assumption that Ω exists leads to logical paradoxes (see Stewart (1975a) p. 287, or Stewart and Tall (1977) p. 49). This is why more restricted universes of discourse are used, along with *relative* complements.

Two useful identities are the *DeMorgan Laws*: for all subsets A and B of U,

$$(A \cap B)^c = A^c \cup B^c. \tag{1.2.15}$$

$$(A \cup B)^c = A^c \cap B^c. \tag{1.2.16}$$

1.2.5. Membership Tables

An efficient alternative to Venn diagrams, as a method of proof of set-theoretic identities whose rigour is evident, is the use of *membership tables*. These are closely analogous to the 'truth tables' of symbolic logic [see § 16.3]: their use, to avoid the common circumlocutory reduction of a set-theoretic identity to a proposition in logic which is then proved by a truth table, was

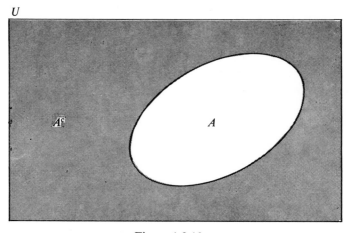

Figure 1.2.10

advocated by Ellis (1970). They seem not to be widely known (despite having been invented independently by a number of people).

We introduce two symbols I, O meaning 'in' and 'out' respectively. The following tables indicate, for each combination of 'in' and 'out' for an element with regard to A and B, whether that element is 'in' or 'out' of the particular combination concerned.

A	B	$A \cap B$	$A \cup B$	$A \backslash B$	A	A^c
I	I	I	I	O	I	O
I	O	O	I	I	O	I
O	I	O	I	O		
O	O	O	O	O		

To check an identity such as $(A \cup B) \cup C = A \cup (B \cap C)$ we draw up the membership tables for both sides:

A	B	C	$(A \cup B)$	$(A \cup B) \cup C$	$(B \cup C)$	$A \cup (B \cup C)$
I	I	I	I	I	I	I
I	I	O	I	I	I	I
I	O	I	I	I	I	I
I	O	O	I	I	O	I
O	I	I	I	I	I	I
O	I	O	I	I	I	I
O	O	I	O	I	I	I
O	O	O	O	O	O	O
				↑		↑

Since the two arrowed columns agree exactly, it follows that an element belongs to $(A \cup B) \cup C$ if and only if it belongs to $A \cup (B \cup C)$; therefore the two are equal.

Similarly we may deal with expressions of the form $X \subseteq Y$, where X and Y are combinations of other sets: the condition for validity is that whenever an I appears in the X column, an I also appears in the Y column (but opposite an O in the X column any Y entry is acceptable). For this shows that *if* an element belongs to X, *then* it belongs to Y.

1.2.6. Ordered Pairs

If a and b are two elements we can form the set $\{a, b\}$. Since

$$\{a, b\} = \{b, a\}$$

we call this set an *unordered pair*. Now it is often important to distinguish ordering in a pair: for example in coordinate geometry, where the point with coordinates

(1, 5) is quite different from (5, 1). Accordingly we introduce the notion of an *ordered pair*

$$(a, b)$$

whose crucial property is this:

If $(a, b) = (c, d)$ then $a = c$ and $b = d$, and conversely.

There is a fancy definition

$$(a, b) = \{\{a, b\}, \{a\}\},$$

designed precisely to achieve the above property. See Stewart and Tall (1977) p. 64. We call a the *first component* of (a, b) and b the *second component*.

We can then define ordered triplets, quadruplets, ..., n-tuplets by

$$(a, b, c) = ((a, b), c)$$

$$(a, b, c, d) = (((a, b), c), d)$$

$$\vdots$$

$$(a_1, \ldots, a_n) = ((a_1, a_2), \ldots, a_n)$$

with the property

If $(a_1, \ldots, a_n) = (b_1, \ldots, b_n)$ then $a_1 = b_1, \ldots, a_n = b_n$, and conversely.

$$(1.2.17)$$

The *Cartesian product* (named in honour of Descartes) of two sets A and B is

$$A \times B = \{(a, b) \mid a \in A \text{ and } b \in B\}. \qquad (1.2.18)$$

For instance if $A = \{2, 4, 6\}$ and $B = \{5, 7\}$ then $A \times B$ is

$$\{(2, 5), (4, 5), (6, 5), (2, 7), (4, 7), (6, 7)\}$$

which is shown geometrically in Figure 1.2.11. We also have repeated Cartesian products

$$A_1 \times \ldots \times A_n = \{(a_1, \ldots, a_n) \mid a_1 \in A_1, \ldots, a_n \in A_n\}.$$

The importance of Cartesian products will become apparent when we consider relations and functions later in this chapter. We note in passing that the plane of coordinate geometry, in this notation, becomes

$$\mathbb{R}^2 = \mathbb{R} \times \mathbb{R},$$

and the coordinate geometry representation of 3-dimensional space involves

$$\mathbb{R}^3 = \mathbb{R} \times \mathbb{R} \times \mathbb{R}$$

[see also § 5.2].

Diagrams like Figure 1.2.11 (see Stewart and Tall (1977) p. 67) or suitable

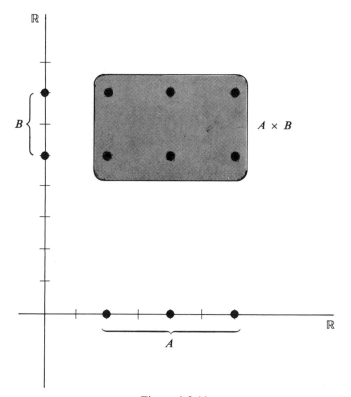

Figure 1.2.11

modifications of the membership table method, or other arguments, prove the following general identities:

$$(A \cup B) \times C = (A \times C) \cup (B \times C) \tag{1.2.19}$$

$$(A \cap B) \times C = (A \times C) \cap (B \times C). \tag{1.2.20}$$

$$A \times (B \cup C) = (A \times B) \cup (A \times C). \tag{1.2.21}$$

$$A \times (B \cap C) = (A \times B) \cap (A \times C). \tag{1.2.22}$$

$$(A \times B) \cap (C \times D) = (A \cap C) \times (B \cap D). \tag{1.2.23}$$

$$(A \times B) \cup (C \times D) \subseteq (A \cup C) \times (B \cup D). \tag{1.2.24}$$

Note in (1.2.24) that we cannot replace '$\subseteq$' by '$=$' in general.

1.2.7.   Sets of Sets

The elements of a given set are often sets in their own right. For example, given any set X we may define the *power set*

$$\mathscr{P}(X) = \{Y \mid Y \subseteq X\}.$$

By definition, every element of $\mathscr{P}(X)$ is a set—in fact a subset of X.

It is useful to use the notation

$$\{S_\alpha \mid \alpha \in A\}$$

to denote a set whose elements are sets S_α, where α runs through an *index set A*. This may be defined formally in terms of a function [see § 1.4]

$$f: A \to \mathscr{S}$$

where $\mathscr{S}$ is a set whose elements are themselves sets, by writing

$$f(\alpha) = S_\alpha.$$

For example suppose $A = \{1, 2, 3, 4\}$, and

$$S_1 = \{1, 3, 5, 7\}$$
$$S_2 = \{2, \sqrt{3}, \pi, 4, 5\}$$
$$S_3 = \{0, -1, 3, 5\}$$
$$S_4 = \{\tfrac{22}{7}, 5, 1\}$$

then

$$\mathscr{S} = \{S_1, S_2, S_3, S_4\} = \{S_\alpha \mid \alpha \in A\}.$$

Such an object we call an *indexed family of sets*. We may define

$$\bigcup_{\alpha \in A} S_\alpha = \{x \mid x \in S_\alpha \text{ for some } \alpha \in A\},$$

$$\bigcap_{\alpha \in A} S_\alpha = \{x \mid x \in S_\alpha \text{ for all } \alpha \in A\}.$$

In the example above we have

$$\bigcap_{\alpha \in A} S_\alpha = S_1 \cap S_2 \cap S_3 \cap S_4 = \{5\}$$

$$\bigcup_{\alpha \in A} S_\alpha = S_1 \cup S_2 \cup S_3 \cup S_4 = \{-1, 0, 1, 2, 3, 4, 5, 7, \sqrt{3}, \tfrac{22}{7}, \pi\}.$$

Index sets may be infinite, rather than finite as here—indeed that is when they are most useful. For example, the set of all intervals in $\mathbb{N}$,

$$\mathscr{S} = \{x \subseteq \mathbb{N} \mid x = \{m, m + 1, \ldots, m + k\} \text{ for } m, k \in \mathbb{N}\}$$

may be indexed by taking $A = \mathbb{N} \times \mathbb{N}$, and for each $\alpha = (m, k) \in A$,

$$S_{(m,k)} = \{m, m + 1, \ldots, m + k\}.$$

Then

$$\bigcup_{\alpha \in A} S_\alpha = \mathbb{N}, \text{ the natural numbers,}$$

and

$$\bigcap_{\alpha \in A} S_\alpha = \varnothing, \text{ the empty set.}$$

1.3. RELATIONS

In mathematics we encounter many different kinds of relation between mathematical objects. Between numbers there are relations such as

$=$	equal
$>$	greater than
$<$	less than
$\geq$	greater than or equal
$\leq$	less than or equal
$\neq$	unequal
$\ngtr$	not greater than
$\nless$	not less than
$\mid$	divides (between integers)

and for other objects we have such things as

$\subseteq$	subset of (between sets)
$\in$	member of (between elements and sets)
$\equiv$	congruent (between triangles) [see V, § 1.1.3]
$\parallel$	parallel (between lines) [see V, § 1.1.2].

We seek a precise set-theoretic definition of a 'relation'.

1.3.1. The Definition of a Relation

The common property of all the above relations is that, for certain specified types of mathematical object, *any* two such objects either are, or are not, related; and in principle we can tell which occurs. By considering the set A of objects involved we see that the crucial item is the set of pairs (a, b) of elements of A for which a is related to b. For this set conveys the same information as the relation: the questions 'is a related to b?' and 'is (a, b) in the given set?' are equivalent.

We can handle the set of pairs by set theory, whereas the unformalized term 'related to' is more vague. So we canonize the set of pairs in a formal definition:

DEFINITION 1.3.1.   *A relation on a set A is a subset of $A \times A$; a relation between sets A and B is a subset of $A \times B$. (For the definition of $A \times A$ and $A \times B$ see § 1.2.6.)*

It is clear that a relation *between* A and B may be viewed as a relation *on* $A \cup B$, and henceforth we consider only relations on a single set. If $R \subseteq A \times A$ is a relation, we write

$$a \, \mathrm{R} \, b$$

to mean that

$$(a, b) \in \mathrm{R}.$$

Then

$$\mathrm{R} = \{(a, b) \mid a \, \mathrm{R} \, b\}.$$

This notational trick allows us to retain traditional notation for relations. For example, let $A = \{1, 2, 3, 4\}$ and let

$$R = \{(1, 2), (1, 3), (1, 4), (2, 3), (2, 4), (3, 4)\}.$$

Clearly $(a, b) \in R$ if and only if $a < b$ (in the usual sense). But $(a, b) \in R$ means $a \, R \, b$, in the above notation; so we have $a \, R \, b$ if and only if $a < b$. So we could, without ambiguity, define $<$ to be R. While it takes a little getting used to having '$<$' as a symbol for a set, it is very convenient to do so. It allows us to define a relation R either as a subset of $A \times A$, directly; or by specifying exactly when $a \, R \, b$ is true (which is equivalent to specifying which pairs (a, b) are elements of the set R) which is often more natural.

If $R \subseteq A \times A$ is a relation, then its *restriction*

$$R|_{B \times B}$$

to a subset $B \subseteq A$ is defined to be $R \cap (B \times B)$. This is therefore the 'same' relation as R, but cut down to apply only to elements of B.

1.3.2. Order Relations

The prototype order relation is '$\leq$' on $\mathbb{Z}$ (or $\mathbb{N}$ or $\mathbb{Q}$ or $\mathbb{R}$): For the definition of these sets see § 1.1. We abstract its main properties for a general relation R as follows. A relation R on a set A is called a *total order* if, for all $a, b, c \in A$,

$$a \, R \, a. \tag{1.3.1}$$

$$\text{If } a \, R \, b \text{ and } b \, R \, c \text{ then } a \, R \, c. \tag{1.3.2}$$

$$\text{If } a \, R \, b \text{ and } b \, R \, a \text{ then } a = b. \tag{1.3.3}$$

$$\text{Either } a \, R \, b \text{ or } b \, R \, a. \tag{1.3.4}$$

Taking $A = \mathbb{Z}$, $R = \leq$, we obtain familiar properties of the relation 'less than or equal to'.

If (1.3.4) does not hold, we call R a *partial order*. For example $R = |$ ('divides') is a partial order on the set $\{1, 2, 3, 4, 5, 6, 7, 8, 9, 10, 11, 12, 13, 14, 15\}$, but not a total order since neither of $5|7$ nor $7|5$ is true.

If R is a total (partial) order on A and B is a subset of A, then the restriction $R|_{B \times B}$ is also a total (partial) order.

1.3.3. Equivalence Relations

An *equivalence relation* R on a set A is a relation which satisfies the following conditions, for all $a, b, c \in A$:
   1. (Reflexive) $a \, R \, a$.
   2. (Symmetric) If $a \, R \, b$ then $b \, R \, a$.
   3. (Transitive) If $a \, R \, b$ and $b \, R \, c$ then $a \, R \, c$.

Of the relations listed at the start of this section, only '$=$', '$\equiv$', and '$\|$' are equivalence relations. An arbitrary equivalence relation on a set A is usually denoted '$\sim$'.

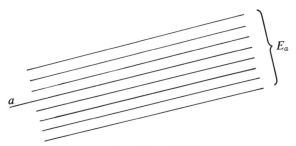

Figure 1.3.1

Given an equivalence relation $\sim$, we define, for each $a \in A$, the *equivalence class*

$$E_a = \{b \in A | a \sim b\}.$$

For example, if $\sim$ is $\parallel$ and a is some line, then the class E_a consists of all the lines parallel to a (Figure 1.3.1). (Each equivalence class therefore corresponds to a unique *direction* across the plane; indeed, this is one way to make the concept 'direction' precise.)

The three properties defining equivalence relations have an important consequence:

For a given equivalence relation, the equivalence classes are either equal or disjoint.   (1.3.5)

That is, if they overlap at all they coincide. Since every $a \in A$ belongs to its own equivalence class E_a (by law 1) the equivalence classes decompose A into disjoint pieces: we say they form a *partition* of A. This, by definition, is a set $\mathscr{P}$ of subsets of A, such that
1. Every $a \in A$ belongs to some $P \in \mathscr{P}$.
2. If $P, Q \in \mathscr{P}$ and $P \neq Q$ then $P \cap Q = \varnothing$.
Figure 1.3.2 shows a typical partition, for which

$$\mathscr{P} = \{P_1, P_2, \ldots, P_{13}\}$$

where $P_i \cap P_j = \varnothing$ for $i \neq j$, and $A = P_1 \cup P_2 \cup \ldots \cup P_{13}$.

Conversely, given any partition $\mathscr{P}$ of A, we can define a relation $\sim$ by $a \sim b$ if and only if a and b belong to the same element of $\mathscr{P}$. Then $\sim$ is an equivalence relation, and its equivalence classes define the same partition as $\mathscr{P}$.

EXAMPLE 1.3.1. $A = \mathbb{Z}$, and $a \sim b$ if and only if $a - b$ is a multiple of 5. It is not hard to check the conditions 1–3, to show that $\sim$ is an equivalence relation. The equivalence classes are

$$E_0 = \{\ldots -10, -5, 0, 5, 10, \ldots\}$$
$$E_1 = \{\ldots -9, -4, 1, 6, 11, \ldots\}$$
$$E_2 = \{\ldots -8, -3, 2, 7, 12, \ldots\}$$
$$E_3 = \{\ldots -7, -2, 3, 8, 13, \ldots\}$$
$$E_4 = \{\ldots -6, -1, 4, 9, 14, \ldots\}.$$

Note that there are no others: for instance $E_5 = E_0$, $E_6 = E_1$, and in general $E_n = E_r$ where r ($=0, 1, 2, 3,$ or 4) is the remainder on dividing n by 5.

Therefore $\sim$ partitions $\mathbb{Z}$ into five disjoint pieces. Similar effects result from replacing 5 by any other integer n. This idea is very useful, and we return to it in section 1.3.4.

Of course, if $\sim$ is not an equivalence relation we may still define E_a by the above formula; but we no longer get a partition. For example, if $A = \{1, 2, 3, 4, 5, 6, 7, 8\}$ and $\sim$ is 'divides' (as in § 1.3.2), then

$$E_1 = \{1, 2, 3, 4, 5, 6, 7, 8\}$$
$$E_2 = \{2, 4, 6, 8\}$$
$$E_3 = \{3, 6\}$$
$$E_4 = \{4, 8\}$$
$$E_5 = \{5\}$$
$$E_6 = \{6\}$$
$$E_7 = \{7\}$$
$$E_8 = \{8\}$$

and the pieces overlap in all kinds of ways (Figure 1.3.3).

1.3.4. The Integers Modulo n

Given an integer n we define a relation $\equiv_n$ on $\mathbb{Z}$ by

$$a \equiv_n b \text{ if and only if } n \text{ divides } a - b.$$

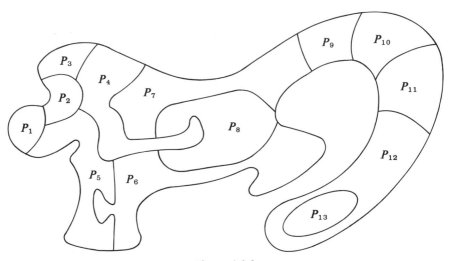

Figure 1.3.2

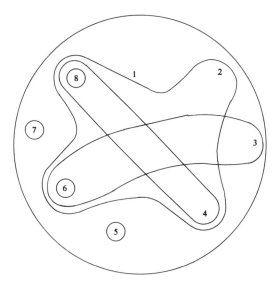

Figure 1.3.3

This is an equivalence relation; and two numbers are in the same equivalence class if and only if they leave the same remainder on division by n. There are n equivalence classes $E_0, E_1, \ldots, E_{n-1}$; thereafter $E_n = E_0$ and everything repeats.

It is convenient to change notation and write m_n in place of E_m, and we do so. We denote the set of equivalence classes $\{0_n, 1_n, \ldots, (n-1)_n\}$ by $\mathbb{Z}_n$. The relation $\equiv_n$ is called *congruence modulo n*, and the classes are *congruence classes* [see § 4.3.1].

It seems to be Gauss who first noticed that it is possible to do arithmetic with congruence classes. All we do is define

$$m_n + k_n = (m + k)_n$$
$$m_n k_n = (mk)_n.$$

(Certain technical problems arise with this definition, but may be resolved: see Stewart and Tall (1977) p. 76). This 'arithmetic' closely resembles ordinary arithmetic of integers, except that we may throw away multiples of n in any calculation. For example,

$$5_7 + 3_7 = 8_7 = 1_7$$

and we never need anything other than $0_7, 1_7, \ldots, 6_7$. This algebraic system is called the *integers modulo n* and denoted $\mathbb{Z}_n$ [see § 4.3.2].

It is highly useful in situations that repeat after some interval n; for example in crystallography rotations through angles $2\pi/n$ compose according to addition in $\mathbb{Z}_n$, and the multiplication of complex nth roots of unity does likewise. Phenomena that repeat across a lattice in space also tend to involve these ideas in their mathematical description.

1.4. FUNCTIONS

One of the problems that gave rise to set theory (though not as directly as one might imagine, see Kline (1972) p. 970) was making precise the meaning of 'function'. An early definition, due to Euler, is 'a curve drawn by freely leading the hand', which is not definite enough for modern tastes; while the 18th and 19th century idea of 'something defined by a formula' foundered on the question of what kinds of formula were permissible.

We encounter many functions in mathematics: the square x^2, the cube x^3, ..., the nth power x^n [see § 3.1]; the reciprocal $1/x$ [see § 3.2]; the square root $\sqrt{x}$, the cube root $\sqrt[3]{x}$, ..., the nth root $\sqrt[n]{x}$ [see § 3.3]; the logarithm to base 10, $\log_{10} x$ [see § 3.6], the natural logarithm $\log x$ and the exponential e^x [see IV, § 2.11]; the sine $\sin x$, cosine $\cos x$, tangent $\tan x$, secant $\sec x$, cosecant $\operatorname{cosec} x$, cotangent $\cot x$ [see IV, § 2.12]; the factorial $x!$ [see (3.7.1) and IV, § 10.2.1]; the absolute value $|x|$ [see, for example, (2.6.5)] ... Beyond these are hyperbolic functions [see § 2.13], Bessel functions [see IV, § 10.4], error functions [see, for example, IV, Theorem 3.6.2(ii)], and the like: a whole host of special functions, each with its own particular applications and uses.

In calculus we consider more complicated, compound functions, such as

$$\sqrt{\frac{x^3 + \tan\,(7x\,.\cos\,(2x - 5))}{\log x}}.$$

All of this 'x-temporization' with ever more complicated formulae may be very well, but it does not tell us what a function is in general.

1.4.1. The Definition of a Function

A classical approach to the problem is to start with an 'independent variable' x, and say that a 'dependent variable' y is a function of x if, whatever value x takes, we can compute y by means of some formula involving x. Now things like the sine are only functions if we know how to calculate them, so our concept of functional dependence is limited to what we can calculate, and depends on just what we permit. Again, a function like $|x|$ seems to require two different formulae:

$$|x| = \begin{cases} x & \text{if } x \geq 0 \\ -x & \text{if } x < 0. \end{cases}$$

Is this legitimate? (It is certainly a problem for many students of calculus, who have trouble integrating it!) Nonetheless the classical concept contains the germ of the modern.

As a starting-point, let us say that $f(x)$ is a function of x if, given any x, we know *in principle* how to calculate $f(x)$. (We may not actually be able to perform the calculation. Nobody knows $\sin 1°$ to a billion decimal places; but we do know what steps are necessary to find it, even if in practice the computation would take too long.)

This definition leaves much to be desired. First, it is not really $f(x)$ that is the function: it is more arguably the symbol f. That is, 'log', rather than 'log x', is the *function*; log x is its *value* at x. Next, it does not tell us what manner of creature x is. Not only do we need x to be a 'variable'—whatever that may be—but we also need to know over what objects x can vary. Finally, it would be nice to know what sort of object $f(x)$ might be.

Set theory suggests we replace the 'variable' x by a *set* of possible values that x may take; this in turn suggests a second set for the values $f(x)$. Note that while a set of values may be more congenial to the modern mind, it plays exactly the same role as the classical 'variable'. E. C. Zeeman refers to this process as the 'nounization of verbs': it seems to be rather common. We therefore are led to formulate a revised definition as follows:

DEFINITION 1.4.1.   Let D and C be any two sets. We say that f is a *function* from D to C, or that f is a *map* of D into C, written

$$f: D \to C,$$

if for each $x \in D$ there is a uniquely defined element $f(x) \in C$. We call D the *domain* of f and C the *codomain*.

Grown men—engineers who happily calculate bridges taking thousands of tons strain—have been known to pale at the sight of this functional arrow. Perhaps they imagine that the tip is poisoned. We shall add a little set-theoretic polish to the above definition in a moment (we still haven't said what a function *is*, only what it *does*—though this is arguably more important anyway). But first a few examples.

EXAMPLE 1.4.1. $f: \mathbb{R} \to \mathbb{R}, f(x) = x^2$. The domain is $\mathbb{R}$ [see § 2.6.1], and so is the codomain. Since x^2 is *uniquely* defined for *each* $x \in \mathbb{R}$, this is a function in the above sense.

EXAMPLE 1.4.2.   Let $D = \{x \in \mathbb{R} \mid x \neq 0\}; f: D \to \mathbb{R}, f(x) = 1/x$. Again $f(x)$ is a uniquely defined element of the codomain for each element x of the domain D, so we have a function. Note how 0 has to be excluded from the domain, because $1/0$ is not defined.

EXAMPLE 1.4.3.   Let $D = \{x \in \mathbb{R} \mid x \geq 0\}, f: D \to \mathbb{R}, f(x) = \sqrt{x}$ where this denotes the *positive* square root. This is a function, but only because we have specified which square root we mean. (We could, of course, have specified the negative square root, or even positive for some x and negative for the rest—anything goes as long as we do not have two possibilities for the same x.) We do *not* allow functions to be 'multivalued' at this stage of the game; and by the time (in complex analysis) that we do, it will turn out that we are talking about a rather different thing. There is nothing especially *wrong* with the concept of a 'multivalued function' (which we will see boils down to the same thing as a relation, in point of fact), but it requires very cautious handling and we prefer to exclude it at this stage.

Note that D is chosen as it is because negative reals do not have (real) square roots.

EXAMPLE 1.4.4.   Let $f: D \to \mathbb{R}$, where

$$f(x) = \left(\frac{(\cos x) + e^x}{(e - e^x)(x^2 - 5x + 6)} \right)^3.$$

We have not specified D: what should it be?

Evading the issue entirely we could take, say, $D = \{0\}$, but then our function would only be defined for $x = 0$, with value $f(0) = 1/27(e - 1)^3$. Not a very interesting function. Clearly we want D to be as large as possible, consistent with good mathematical sense. Now $f(x)$ will be defined uniquely as a real number provided the denominator of the fraction is not zero. The first term $(e - e^x)$ vanishes only for $x = 1$; the second $(x^2 - 5x + 6)$ only for $x = 2, 3$. So we should take $D = \mathbb{R}\backslash\{1, 2, 3\}$.

Often this largest 'natural' domain for some function defined by a formula is referred to as *the* domain of the formula. While logically this puts the cart before the horse, it is psychologically appealing.

With more general sets for C and D, we can concoct all sorts of functions:

EXAMPLE 1.4.5.   Let $D = \{$all circles in the plane$\}$, $C = \mathbb{R}$ and $f(x) = $ the area of x. This is a function: for example, if x is the circle centre $(17, 99.23)$ and radius 5, then $f(x) = 25\pi$ [see V (1.1.1)].

EXAMPLE 1.4.6. $D = \{$all straight lines in the plane$\}$, $C = \{$all angles$\}$, $f(x) = $ the angle between x and the y-axis.

EXAMPLE 1.4.7. $D = \{$all smooth curves starting at $(1, 1)$ and ending at $(3, 7)\}$, $C = \mathbb{R}$, $f(x) = $ the length of x.

EXAMPLE 1.4.8. $D = \{$all subsets of $\{1, 2, 3, 4\}\}$, $C = \mathbb{N}$, $f(x) = $ the number of elements in the set x.

EXAMPLE 1.4.9. $D = \{$all functions $g: \{x \in \mathbb{R} \mid 0 \le x \le 1\} \to \mathbb{R}\}$, $C = \mathbb{R}$, and

$$f(g) = \int_0^1 g(x)dx$$

if the integral exists see IV, §4.1, $f(g) = 0$ if not.

These examples only scratch the surface of what is possible—and commonly encountered in the mathematical literature.

Finally we turn to the set-theoretic polish. The trick is to consider (much as we did for relations) not the rather vague creature f, but the set of all ordered pairs $(x, f(x))$ as x runs through D. Provided this has certain properties, to be listed in a moment, it is just as useful as f: namely, to compute $f(x)$, we look for a pair whose first term is x, and then its second term gives the desired value of $f(x)$. Thus we may make the following formal

DEFINITION 1.4.2.   A *function* $f: D \to C$ is a subset f of $D \times C$, such that
(1) For each $d \in D$ there exists $c \in C$ such that $(d, c) \in f$.
(2) For each $d \in D$, c_1 and $c_2 \in C$, if both $(d, c_1) \in f$ and $(d, c_2) \in f$, then $c_1 = c_2$.
(1) says that $f(d)$ is defined for all $d \in D$; (2) that it is uniquely defined. Given a function f defined in this way, then for any $x \in D$ we define $f(x)$ to be the unique element of C such that $(x, f(x)) \in f$. This recovers our usual notation for a function from the set-theoretic formulation (much as occurred for relations).

In this strict sense a function is defined as something rather like a table of values. For example if we define $f: \{1, 2, 3, 4, 5\} \to \mathbb{Z}$ by $f(x) = x^2$, then the corresponding set f is given by

$$f = \{(1, 1), (2, 4), (3, 9), (4, 16), (5, 25)\}.$$

The ordered pairs amount to a table of $f(x)$ given x.

Geometrically we can interpret this another way. For a function $f: \mathbb{R} \to \mathbb{R}$, we have $f = \{(x, f(x)) \mid x \in \mathbb{R}\}$. But it is clear (Figure 1.4.1) that this is the *graph* of f in the usual sense of the term (see Stewart and Tall (1977) p. 92). Thus the set-theoretic approach identifies a function with its graph.

If $f: D \to C$ is a function, then the set

$$f(D) = \{f(d) \mid d \in D\} \subseteq C \tag{1.4.1}$$

is called the *image* (or *range*) of f. If $X \subseteq D$, we write

$$f(X) = \{f(x) \mid x \in X\} \tag{1.4.2}$$

and call this the *image of X under f*.

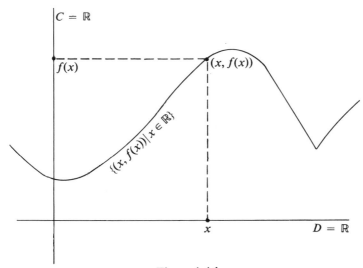

Figure 1.4.1

We remark in passing that 'the' codomain of a function is not uniquely determined; in fact any set containing the image will serve as codomain. It is usual to choose whichever codomain is most convenient: the exact image is often complicated and not the best choice.

1.4.2. Fundamental Properties of Functions

In this section we introduce a number of terms which describe basic properties of functions.

If $f: A \to B$ and $g: C \to D$ are functions, and the image of f is a subset of C, then we may *compose* f and g to obtain a function

$$g \circ f: A \to D \tag{1.4.3}$$

such that

$$g \circ f(a) = g(f(a))$$

for $a \in A$ (Figure 1.4.2). The restriction that $f(A) \subseteq C$ is necessary: for example

$$\log: \{x \in \mathbb{R} \mid x > 0\} \to \mathbb{R}$$

$$\sin: \mathbb{R} \to \mathbb{R},$$

and the image of sin is $\{x \in \mathbb{R} \mid -1 \le x \le 1\}$ [see IV, § 2.14] which is not contained in the domain of log [see § 3.6]. Therefore we cannot define

$$\log \sin x$$

for all $x \in \mathbb{R}$. We get round this by cutting down the domain of sin to those x for which $\sin x > 0$, namely the set

$$\{x \in \mathbb{R} \mid 2n\pi < x < (2n + 1)\pi \text{ for some } n \in \mathbb{Z}\};$$

then the image of *this* is $\{x \in \mathbb{R} \mid 0 < x < 1\}$ which *is* contained in the domain of log; so log sin x is defined for all x in this set.

Composition of functions is (trivially) associative; that is if $f: A \to B$, $g: C \to D$, $h: E \to F$, with $f(A) \subseteq C$ and $g(C) \subseteq E$, then

$$h \circ (g \circ f) = (h \circ g) \circ f \tag{1.4.4}$$

as functions from A to F.

The *identity function* $i_A: A \to A$, for any set A, is defined by

$$i_A(a) = a \tag{1.4.5}$$

for all $a \in A$. If $f: A \to B$, then $f \circ i_A = f = i_B \circ f$.

We say that $f: A \to B$ is *injective* or *one-one*, or is an *injection*, if for all $a, b \in A$, $f(a) = f(b)$ implies $a = b$. In some situations an injective function is

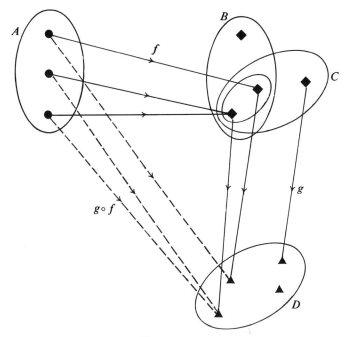

Figure 1.4.2
Stewart and Tall (1977) p. 97

called an *embedding*; that is, *f* maps different elements of *A* to different elements of *B*. Likewise *f* is *surjective*, or *onto*, or a *surjection*, if $f(A) = B$. (Notice that this depends on specifying a particular codomain *B*. Every function is a surjection if we take its image as codomain, but often we do not.) If *f* is both injective and surjective we say that it is *bijective*, or a *bijection*, or a *one-to-one correspondence*.

For example $\exp: \mathbb{R} \to \mathbb{R}$, $\exp(x) = e^x$ [see IV, § 2.11] is injective but not surjective; $f: \mathbb{R} \to \{x \in \mathbb{R} \mid x \geq 0\}, f(x) = x^2$ is surjective but not injective; and $f: \mathbb{R} \to \mathbb{R}, f(x) = x^3$ is bijective.

Given $f: A \to B$, a function $g: B \to A$ is called a *left inverse* for *f* if $g \circ f = i_A$; a *right inverse* for *f* if $f \circ g = i_B$, and an *inverse* for *f* if *both* these conditions hold. It may be shown (Stewart and Tall (1977) p. 99) that *f* has a left inverse if and only if it is injective; a right inverse if and only if it is surjective; and an inverse if and only if it is bijective.

EXAMPLE 1.4.10. $f: \mathbb{R} \to \mathbb{R}$, $f(x) = x^3$. This is bijective, and has inverse $g: \mathbb{R} \to \mathbb{R}$, $g(x) = \sqrt[3]{x}$.

EXAMPLE 1.4.11. $f: \mathbb{R} \to \{x \in \mathbb{R} \mid x \geq 0\}, f(x) = x^2$. This is surjective. The function $g: \{x \in \mathbb{R} \mid x \geq 0\} \to \mathbb{R}$ given by $g(x) = \sqrt{x}$ (positive root) is a right inverse, since $(\sqrt{x})^2 = x$. But it is *not* a left inverse, for

$$\sqrt{x^2} = \begin{cases} x & \text{if } x \geq 0 \\ -x & \text{if } x < 0. \end{cases}$$

Consider $f: \mathbb{R} \to \mathbb{R}, f(x) = \sin x$. This is neither injective nor surjective, so has no inverse, left or right. But what about the 'inverse trigonometric function' $\sin^{-1} x$ (or arc sin x), [see IV, § 2.14]?

If we choose a more sensible codomain, $\{x \in \mathbb{R} \mid -1 \leq x \leq 1\}$, then we obtain a *right* inverse $\sin^{-1} x$, defined to be the unique y with $-\pi/2 \leq y \leq \pi/2$ for which $\sin y = x$. But this is *not* a left inverse, for $\sin^{-1} \sin 10\pi = \sin^{-1} 0 = 0 \neq 10\pi$.

In many ways the most satisfactory procedure is to restrict the domain, too. Then

$$\sin: \{x \in \mathbb{R} \mid -\pi/2 \leq x \in \pi/2\} \to \{x \in \mathbb{R} \mid -1 \leq x \leq 1\}$$

is a bijection, and has an inverse $\sin^{-1}$ defined as before.

This is not mere hairsplitting: it is well known that inverse trigonometric functions must be handled with caution, and for precisely the above reasons. Of course, if we allow 'multivalued functions' we can use a different approach; but it runs into the same problem as soon as we try to specify *which* of the multiple values we mean. (And this is a serious problem: in evaluating definite integrals, say, by a formula involving $\sin^{-1} x$, choosing the wrong value from among those possible gives us the wrong answer [IV, ex 4.3.2(v)].)

If $f: \mathbb{R} \to \mathbb{R}$ is a bijection, then the graph of f^{-1} is obtained by reflecting the graph of f about the diagonal line $y = x$ (Figure 1.4.3). If $A, B \subseteq \mathbb{R}$ the same goes for a bijection $f: A \to B$ [see V, § 3.4.5].

When an inverse function for $f: A \to B$ exists (that is, when f is a bijection) we use the notation

$$f^{-1}: B \to A \qquad (1.4.6)$$

for its inverse function. Clearly $(f^{-1})^{-1} = f$.

There is another use of the symbol f^{-1} which is not restricted to bijections. For *any* $f: A \to B$ and any subset $X \subseteq B$, we put

$$f^{-1}(X) = \{a \in A \mid f(a) \in X\}. \qquad (1.4.7)$$

This is called the *inverse image* or *pullback* of X under f.

If $f: A \to B$ is a function, and $X \subseteq A$, we define the *restriction*

$$f|_X: A \to B \qquad (1.4.8)$$

of f to X by

$$f|_X(x) = f(x)$$

for $x \in X$. It is therefore just like f, but with its domain cut down to X. This apparently innocuous procedure is in fact quite useful in more advanced work.

In analysis we encounter functions of several variables, such as

$$f(x, y) = (\cos x^2) - 2 \tan xy.$$

We can avoid separate consideration of these by noting that they are simply

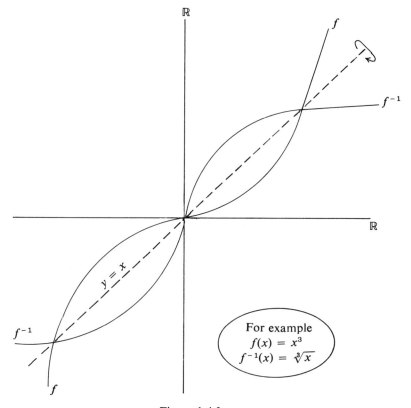

Figure 1.4.3

functions, in the usual sense, defined on a set of ordered pairs (x, y). So a function of two variables is a function

$$f: A \times B \to C$$

for sets A, B; and a function of n variables is

$$f: A_1 \times \ldots \times A_n \to C.$$

1.5. BINARY OPERATIONS

A very large part of abstract algebra is currently approached through the idea of a *binary operation*. This assigns to pairs of elements x, y of a set S another element, say $x \circ y$, of the same set. Formally, therefore, a binary operation is defined to be a function

$$f: S \times S \to S$$

where S is a set. The symbol $\circ$ used above plays the role of f, but instead of $f(x, y)$ we write $x \circ y$.

For example, if $S = \mathbb{Z}$, the integers, and $\circ$ is the operation of addition, we define f by

$$f(x, y) = x + y.$$

If $\circ$ is multiplication, we set

$$f(x, y) = xy.$$

Such examples are the reason why the notation $x \circ y$ is preferred to $f(x, y)$—or indeed to $\circ(x, y)$, which would be perfectly reasonable. However, $+(x, y)$ is not an appealing notation for the sum!

Note that the definition of a binary operation *implies* that for all x and y in S, the 'product' $x \circ y$ is *also* in S. This property is often expressed as 'S is *closed* under $\circ$'. The definition also requires $x \circ y$ to be defined for *all* $x, y \in S$; so for example we cannot define a binary operation on $\mathbb{Z}$ by putting $x \circ y = x/y$, because division by zero is not possible in $\mathbb{Z}$.

Numerous axiomatic systems in abstract algebra involve sets S on which are defined various binary operations, subject to certain rules. For example, a binary operation $\circ$ on a set S is said to be:

Commutative if $x \circ y = y \circ x$ for all $x, y \in S$;   (1.5.1)

Associative if $(x \circ y) \circ z = x \circ (y \circ z)$ for all $x, y, z \in S$.   (1.5.2)

If $*$ is a second binary operation on S, then we say that $\circ$ is *distributive* over $*$ if $x \circ (y*z) = (x \circ y)*(x \circ z)$, and $(x*y) \circ z = (x \circ z)*(y \circ z)$, for all $x, y, z \in S$.   (1.5.3)

Thus $+$ and $\times$ are commutative and associative binary operations on $\mathbb{Z}$, and $\times$ is distributive over $+$. However, $-$ is not commutative or associative on z, because in general

$$x - y \neq y - x$$

$$(x - y) - z \neq x - (y - z).$$

And $+$ is not distributive over $\times$, because

$$x + (y \times z) \neq (x + y) \times (x + z).$$

As well as binary operations, we may define n-ary operations for any natural number n, to be functions $S \times S \times \ldots \times S \rightarrow S$, where on the left there are n copies of S. (By convention, 0-ary operations are 'constants', such as 0 or 1 in $\mathbb{Z}$; 1-ary operations are things like the negative $-x$ of x, or the reciprocal x^{-1}.) Ternary operations, however, can generally be expressed as repeated binary operations (possibly defined on a bigger set) and are seldom encountered: this accounts for the predominance of binary operations in the literature.

If $\circ$ is a binary operation on a set S, and if $\sim$ is an equivalence relation on S, we say that $\sim$ is a *congruence relation* (relative to $\circ$) if it satisfies the condition:

$$x \sim x' \quad \text{and} \quad y \sim y' \quad \text{implies} \quad x \circ y \sim x' \circ y'. \qquad (1.5.4)$$

For example, congruence modulo n, for fixed n, is a congruence relation on $\mathbb{Z}$ (relative to $+$ or to $\times$) in this sense. On the other hand 'having the same sign' is not a congruence relation relative to subtraction, because $-7 \sim -2$, and $-3 \sim -14$, but

$$-4 = -7 - (-3) \nsim -2 - (-14) = 12.$$

The importance of congruence relations is that the operation $\circ$ can be defined on equivalence classes under $\sim$, by setting

$$E_x \circ E_y = E_{x \circ y}.$$

This operation is well defined only when $\circ$ is a congruence relation.

1.6.   CARDINAL NUMBERS

Let A be a set with finitely many elements. We write

$$|A|,$$

called the *cardinality* of A, to denote the number of elements in A.

Writing B^A to denote the set of all functions $f: A \to B$, the following equations hold:

$$|A \cup B| = |A| + |B| \quad \text{if } A \cap B = \varnothing. \tag{1.6.1}$$

$$|A \cup B| = |A| + |B| - |A \cap B|. \tag{1.6.2}$$

$$|A \times B| = |A| |B|. \tag{1.6.3}$$

$$|B^A| = |B|^{|A|}. \tag{1.6.4}$$

$$|\mathscr{P}(A)| = 2^{|A|}. \tag{1.6.5}$$

Formula (1.6.2) may be extended to three or more sets; for example

$$|A \cup B \cup C| = |A| + |B| + |C| - |A \cap B| - |B \cap C|$$
$$- |A \cap C| + |A \cap B \cap C|; \tag{1.6.6}$$

and the extension to four or more sets is obvious.

These ideas, as Cantor discovered, may be generalized to infinite sets. The resulting transfinite numbers (or *infinite cardinals*) possess many, but not all, of the familiar properties of arithmetic; they form a hierarchy of 'different types of infinity'. While they do not have many practical applications, one distinction that they draw is crucial in probability theory, so we shall briefly indicate the main ideas. (For more details, see Stewart and Tall (1977) Chapter 12.)

The basic assumption is that two sets A and B 'have the same number of elements' if there is a *bijection* $f: A \to B$. This defines an equivalence relation; to each equivalence class we assign an object, the *cardinal* of each element in the class. Arithmetic of cardinals is defined by means of equations (1.6.1), (1.6.3) and (1.6.4) above.

A set A is finite, with cardinal n, if and only if there is a bijection $f: A \to \{1, 2, \ldots, n\}$. If no such bijection exists, it is an infinite set. If there is a bijection $f: A \to \mathbb{N}$, we say that A has cardinal $\aleph_0$. This is the smallest infinite cardinal. A set which is finite or has cardinal $\aleph_0$ is called *countable*.

Any union of countably many countable sets is countable. The sets $\mathbb{N}$, $\mathbb{Z}$, $\mathbb{Q}$ are all countable. But the set $\mathbb{R}$ of real numbers is *uncountable*, by a famous proof due to Cantor (Stewart and Tall (1977) p. 237).

This is a fundamental distinction. For many purposes, while countable sets can be handled, uncountable sets are 'too large'. An important case is probability theory, where the probability of a *countable* union of disjoint events is the sum (a convergent series in the sense of IV, § 1.7) of the separate probabilities. It is impossible to deal with an uncountable union this way, because there is no sensible definition of the sum of uncountably many real numbers, except in the trivial case where all but a countable number of them are zero.

I.S.

REFERENCES

Ellis, F. (1970). 'Venn vill they ever learn ?', *Manifold*, **6**, 44–47.

Kline, M. (1972). *Mathematical Thought from Ancient to Modern Times*, Oxford University Press.

Stewart, I. N. (1975a). *Concepts of Modern Mathematics*, Penguin Books, Harmondsworth, Middlesex.

Stewart, I. N. (1975b). The truth about Venn diagrams, *Math. Gazette*.

Stewart, I. N. and Tall, D. O. (1977). *The Foundations of Mathematics*, Oxford University Press.

CHAPTER 2

Numbers

2.1. THE NATURAL NUMBERS

The first numbers which we meet as children are whole numbers 1, 2, 3, ... or, as they are usually called, *natural numbers*. The young child has actually absorbed quite a lot about the nature of numbers even before being formally introduced to them; since right from its first days it is surrounded by the notions of plurality, and takes in such everyday facts as that people have *one* nose, *two* eyes, and cats, dogs and cows have *four* legs. The idea of number is a fundamental one which people grow up with from their earliest beginnings.

Numbers are properties of the physical objects in the universe rather like hardness or colour. The essential difference between the concept of number and most other properties is that it says something about a *collection* or *set* of objects [see § 1.1]. For example the collections shown in Figure 2.1.1 all share the property of 'threeness'. Numerical properties are rather like electrical properties of substances, since although it is difficult to say what electricity *is*, we can work with it and use it to our advantage by knowing how it behaves. Similarly we can say how numbers behave and how they mirror certain aspects of the world around us. We shall let $\mathbb{N}$ be the name of the collection, or set, of all natural numbers, and write

$$n \in \mathbb{N}, \tag{2.1.1}$$

(to be read as 'n is a member of $\mathbb{N}$', or 'n is an element of $\mathbb{N}$') meaning that n is a natural number.

The simplest facts to observe about natural numbers are:

$\mathbb{N}$A1 Any two natural numbers a and b can be combined (by addition) to produce a unique third number $a + b$ called their sum;

$\mathbb{N}$M1 Any two natural numbers a and b can be combined (by multiplication) to produce a unique third number ab or $a . b$ called their product;

$\mathbb{N}$O1 Every natural number has a *size* associated with it; so that given two numbers a and b say, we have just one of the following possibilities; either they are the same, or a is larger than b (written $a > b$), or b is larger than a ($b > a$ or $a < b$).

Furthermore

$\mathbb{N}$O2 $a < a + 1$ for every a.

31

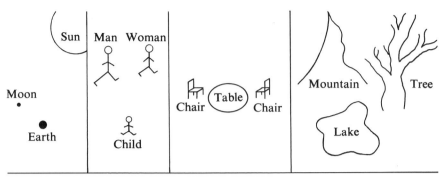

Figure 2.1.1

Property A1 is a reflection of the observed fact that if we have 2 chairs in one room, 3 chairs in another, and we bring them together, the total will be the same number no matter how many times we repeat the experiment. Property M1 corresponds to the fact that if we have 3 rows of objects, each row containing two objects, then if we count the total number of objects present the answer is always 6 no matter what sort of objects we use. Property O1 says that if two rooms, labelled a and b, each contain some people, then they either contain the same number of people, or a contains more people than b, or b contains more people than a.

These properties of natural numbers are quite simple and familiar but it is helpful to set them down explicitly; and once we have done so we can observe further laws that these two operations of addition and multiplication, and the relation of order (O1) satisfy. If a, b, c are natural numbers then we have [cf. § 16.1],

NA2 $a + b = b + a$ (the commutative law of addition).

NA3 $a + (b + c) = (a + b) + c$ (the associative law of addition).

NA4 If $a + c = b + c$ then $a = b$ (the cancellation law of addition).

NM2 $a.1 = a$ (1 is the identity of multiplication).

NM3 $ab = ba$ (commutative law of multiplication).

NM4 $a(bc) = (ab)c$ (associative law of multiplication).

NM5 If $ac = bc$ then $a = b$ (cancellation law of multiplication).

NO3 If $a < b$ and $b < c$ then $a < c$ (the transitive property of order).

EXAMPLE 2.1.1

    (i) $2 + 3 = 5$ and also $3 + 2 = 5$.

    (ii) $1 + (2 + 3) = (1 + 2) + 3$.

    (iii) If $a + 3 = 7$ and also $b + 3 = 7$, then necessarily $a = b = 4$.

    (iv) $3.1 = 3$.

    (v) $5.4 = 4.5$.

    (vi) $3(5.7) = (3.5).7$.

    (vii) If $3a = 3b = 18$ then necessarily $a = b = 6$.

    (viii) $2 < 2 + 1 = 3$ and $3 < 3 + 1 = 4$ so $2 < 4$.

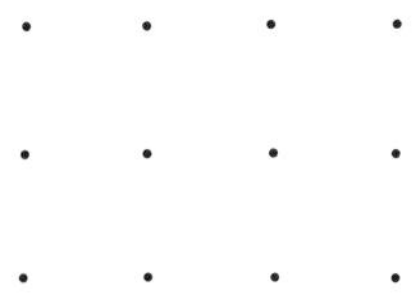

. . . .

. . . .

. . . .

Figure 2.1.2

The above formal statements of these laws may look strange and slightly forbidding to anyone unused to symbolic notation, but they are just statements of common facts which we all use, perhaps unconsciously, in everyday life. For example the statement of the commutativity of multiplication (M3) says that if we have several rows of dots, each containing the same number, such as shown in Figure 2.1.2, then we obtain the same total number of dots whether we regard them as being arranged in 3 rows of 4 each or whether we turn the page on its side and regard them as being arranged in 4 rows of 3 each.

We also have three laws which connect addition, multiplication and order, namely for any natural numbers $a, b, c,$

NAM $a(b + c) = ab + ac$ (distributive law).

NAO If $a < b$ then $a + c < b + c$.

NMO If $a < b$ then $ac < bc$.

EXAMPLES 2.1.2

(i) If we want to calculate $3(5 + 7)$, we may either first add the 5 and 7 to obtain

$$3(5 + 7) = 3.12 = 36,$$

or we may write

$$3(5 + 7) = 3.5 + 3.7 = 15 + 21 = 36.$$

The distributive law is just a general statement of this fact—that we may always take the multiplier inside the brackets.

(ii) It is easy to deduce by repeated applications of O2 and O3 that

$$3 < 3 + 1 = 4 < 4 + 1 = 5 < 5 + 1 = 6 < 6 + 1 = 7,$$

so we know, without using O2 and O3 again, that

$$8 = 3 + 5 < 7 + 5 = 12.$$

(iii) Also since $3 < 7$, we can easily conclude from MO that

$$30 = 3.10 < 7.10 = 70$$

Finally we have two laws which help to distinguish the natural numbers from other types of numbers which we shall meet later. The first concerns the order relation.

NO4 *The well-ordering principle* Any non-empty set of natural numbers contains a smallest member.

For example, suppose we have a room containing several boxes, each containing a different number. Then we can be certain that one of the boxes contains a smaller number than any of the other boxes. It is important to realize that the principle remains true even if we allow our room to contain infinitely many natural numbers. The same thing is not true of most other number systems.

Our last law really concerns addition, but it is important enough to deserve a separate name.

$\mathbb{N}$A5 *The principle of induction*: If a collection of natural numbers contains 1 and contains $n + 1$ whenever it contains the number n, then it must contain every natural number.

This principle specifies more clearly the additive structure of $\mathbb{N}$, since it says that each natural number can be obtained by just adding 1's together. It is a way of stating that nothing is to be included as a natural number except $1, 1 + 1, 1 + 1 + 1,\ldots$.

EXAMPLE 2.1.3.   Suppose we want to use the principle of induction to show that the sum of the numbers between 1 and n inclusive is always $n(n + 1)/2$, which we write as

$$\sum_{i=1}^{n} i = \frac{n(n + 1)}{2}. \tag{2.1.2}$$

We first let T be the collection of numbers n for which the formula (2.1.2) is true. Then $1 \in T$, or in other words (2.1.2) is true when $n = 1$; for if 1 is the only number to be counted, the sum on the left is certainly 1, and when $n = 1$ the expression on the right of (2.1.2) is also $1 . 2/2 = 1$. Next we must see whether T necessarily contains $n + 1$ whenever it contains n. If $n \in T$ then (2.1.2) holds for that n, so for $n + 1$,

$$\sum_{i=1}^{n+1} i = \sum_{i=1}^{n} i + (n + 1)$$

$$= \frac{n(n + 1)}{2} + (n + 1) \quad \text{(from the truth of (2.1.2) for } n)$$

$$= \frac{(n + 1)(n + 2)}{2}.$$

This last expression is exactly what the right-hand side of (2.1.2) becomes if n is replaced by $n + 1$. So $n \in T$ does indeed imply $n + 1 \in T$, and since $1 \in T$ the principle of induction assures us that $T = \mathbb{N}$, or in other words that (2.1.2) is true for every natural number n.

There is another form of induction, called the principle of complete induction, which is equivalent to our previous statement but which can sometimes be used in cases where the first principle is hard to apply directly.

The principle of complete induction: If a collection of natural numbers contains 1 and contains $n + 1$ whenever it contains the number n and all numbers less

symbolically the result of walking 2 steps to the right followed by 2 steps to the left. We shall let 0, called zero, stand for this notion of no movement. The set $\{\ldots -3, -2, -1, 0, 1, 2, 3, \ldots\}$ of all the negative numbers, the natural numbers, and zero, is called the set $\mathbb{Z}$ (from the German 'zahl', meaning 'number') of *integers* or whole numbers, and the natural numbers are usually called *positive integers*. We also write

$$a \in \mathbb{Z} \tag{2.2.1}$$

to mean that a is an integer.

We could of course just specify how far the walker was from the starting point by giving the number of steps without regard to their direction. If the position reached was denoted by the integer a, then the number of steps involved is a itself if a is a natural number or zero, and $-a$ if $a = -k$ is negative (whereby $-(-k)$ we just mean k). In other words this quantity, denoted by $|a|$ and called the *modulus* or *absolute value* of a, is a measure of the size of a. Formally we put

$$|a| = \begin{cases} a & \text{if } a \geq 0 \\ -a & \text{if } a < 0. \end{cases} \tag{2.2.2}$$

If a and b are both integers we always have

$$|ab| = |a| \cdot |b|, \tag{2.2.3}$$

no matter what the signs of a and b are. We do not always have $|a + b| = |a| + |b|$, since this is only true if a and b have the same sign or if one of them is zero. If one is positive and the other negative we have $|a + b| < |a| + |b|$, so at least we can always say

$$|a + b| \leq |a| + |b|. \tag{2.2.4}$$

Another useful inequality which always holds irrespective of the signs of a and b is

$$|a - b| \geq ||a| - |b||. \tag{2.2.5}$$

EXAMPLES 2.2.1
  (i) $|a| = |-a|$ for any a.
  (ii) $|5| = 5, |-3| = 3$ and $|0| = 0$.
  (iii) $|(-3)(-5)| = |3 \cdot (-5)| = |(-3) \cdot 5| = |3.5| = 15$.
  (iv) $|5 - 3| = 2$ but $|5| + |-3| = 5 + 3 = 8$.

If we now examine the properties of our addition of integers we find that it is quite similar to the addition of natural numbers. For, looking at our previous list of properties and using the same names, we have

ZA1 Any two integers can be combined by addition to produce a third integer called their sum.

ZA2 Addition is commutative: if a and b are any two integers, $a + b = b + a$.

ZA3 Addition is associative: $a + (b + c) = (a + b) + c$ for any $a, b, c \in \mathbb{Z}$.

ZA4 The cancellation law holds ($a + c = b + c$ implies $a = b$).

The counterpart of A5, the principle of induction, does not hold for integers since no negative integer, nor zero, can be obtained by adding 1s together. This corresponds to the obvious fact that if we start from 1 and keep walking to the right we shall never arrive at the origin nor anywhere to the left of it. However to make up for this loss we have two properties which together are stronger than the cancellation law.

$\mathbb{Z}$A4 (i) There is a special integer 0 such that for any integer a, $a + 0 = a$; in other words 0 is the 'identity' of addition.

$\mathbb{Z}$A4 (ii) Corresponding to each integer a, there is an integer a' (depending upon a) such that $a + a' = 0$.

We can obtain a more uniform nomenclature if for example we regard the '−' in −2 not as meaning 'move to the left instead of the right as specified by 2' but as meaning 'move in the opposite direction to that specified by 2'. Then the a' in A4(ii) can be written as $-a$ even if a is negative. In order to see that A4(i) and A4(ii) give the cancellation law, we suppose that

$$a + c = b + c,$$

and add $-c$ to both sides

$$(a + c) + (-c) = (b + c) + (-c).$$

Then using commutativity and associativity we get

$$a + (c + (-c)) = b + (c + (-c))$$

whence $a = a + 0 = b + 0 = b$.

One of the nice properties of the addition of integers is that if a and b are any two integers there is always an integer x such that $a + x = b$. Indeed x is given by $b + (-a)$, which is often written as $b - a$. Adding $-a$ to b is called 'subtracting a from b'. If we are working entirely with natural numbers then, given two natural numbers a and b, the corresponding assertion would be that either $a = b$, or there is a natural number x such that $a + x = b$, or there is a natural number y such that $a = b + y$.

The additive properties $\mathbb{Z}$A1–$\mathbb{Z}$A4(ii) of the integers can be conveniently summarized by saying that $\mathbb{Z}$, with the operation of addition, forms an Abelian group [see Definition 8.2.2].

2.2.2. Further Properties of Integers

So far we cannot multiply arbitrary pairs of integers, but we can see how to make sensible definitions of products involving negative integers if we look again at Figure 2.2.2 and re-examine the idea of multiplication of natural numbers. If a and b are natural numbers then, in terms of our picture, $a.b$ means 'move b steps to the right, and do this a times'. If we try to extend this idea as far as possible then $a.0$ should mean 'stay where you are, and do this a times'. Clearly no resulting displacement is produced by these instructions and so $a.0 = 0$. Indeed if multiplication of integers is still to be distributive with respect to addition we have to have $a.0 = 0$ for any intger a, since $a.0 = a(0 + 0) = a.0 + a.0$ whence subtracting $a.0$ from both sides we get $0 = a.0$. Also for natural numbers $a, b, a(-b)$ should mean 'move b steps to the left, and do this a times'. This means that we take the same total number of steps as when we were calculating $a.b$, but all to the left; so that if we move along ab steps and then $a(-b)$ steps we should get back to our starting place. That is, $a(-b) = -a.b$. This again fits in with the requirements of the distributive law since for any integers a and b we must have $ab + a(-b) = a(b + (-b)) = a.0 = 0$. Similarly if multiplication of integers is to be commutative we must have $(-a)b = b(-a) = -(ba) = -ab$ for any integers a and b. These two requirements, that $a.0 = 0.a = 0$ for any integer a, and that $a(-b) = (-a)b = -ab$ for any integers a and b, then define the product of every pair of integers so that $\mathbb{Z}$M1, the analogue of $\mathbb{N}$M1, holds. Further we also have

$\mathbb{Z}$M2 1 is the identity of multiplication.

$\mathbb{Z}$M3 Multiplication is commutative.

$\mathbb{Z}$M4 Multiplication is associative.

$\mathbb{Z}$M5 A cancellation law holds: if a, b, c are integers with $c \neq 0$, $ac = bc$ implies $a = b$.

$\mathbb{Z}$AM Multiplication is distributive with respect to addition.

EXAMPLES 2.2.2

(i) For any integer a, $a(-1) = -(a.1) = -a$.

(ii) For any two integers a and b, $(-a)(-b) = -((-a)b) = -(-ab) = ab$.

It is easy to extend the order relation to include the integers, because in terms of our description as points on a line we say that $a < b$ if b is further to the right than a. This is actually the same as saying that $a < b$ if there is a *natural number* n with $a + n = b$. We still have almost all our original laws [cf. § 1.3.2].

$\mathbb{Z}$O1 For any two integers a and b, either $a = b$, or $a < b$, or $b < a$.

$\mathbb{Z}$O2 $a < a + 1$ for every a.

$\mathbb{Z}$O3 Order is a transitive relation [see § 1.3.3].

The original well-ordering principle for natural numbers no longer holds for the integers since for example the whole set of integers has no smallest member. However an amended version holds.

$\mathbb{Z}$O4 If a set S of integers is not empty and if there is an integer A which is less than every member of S, then S contains a smallest member.

We have a similar connection between addition and order to the original one since $\mathbb{N}$AO becomes $\mathbb{Z}$AO if a, b, c are allowed to be any integers. However the connection between multiplication and order has to be altered slightly as follows;

$\mathbb{Z}$MO Suppose a and b are integers with $a < b$, then

$$\text{if } c \text{ is a natural number, } ac < bc;$$

$$\text{if } c \text{ is a negative integer, } ac > bc.$$

EXAMPLES 2.2.3
  (i) $2 < 5$ whence $(-3).2 > (-3).5$;
  (ii) $-3 < 7$ whence $-12 = 4(-3) < 4.7 = 28$.

EXAMPLE 2.2.4. The Fibonacci sequence $u_1, u_2, \ldots$ is the sequence of natural numbers specified by

$$u_1 = 1, u_2 = 1, \text{ and for } n \geq 2, u_{n+1} = u_n + u_{n-1}; \qquad (2.2.6)$$

or, in other words, each term after the second is the sum of the preceding two terms. We shall prove that for each $k \geq 2$

$$u_{k-1}u_{k+1} - u_k{}^2 = (-1)^k. \qquad (2.2.7)$$

We do this by appealing to a form of the principle of induction; and so first note that (2.2.7) is true if k is the first natural number for which it makes sense, namely 2. For then the left-hand side of (2.2.7) is $u_1u_3 - u_2{}^2 = 1.2 - 1^2 = 1$, and the right side is also $(-1)^2 = 1$. Next if (2.2.7) holds for $k = n$, then for $k = n + 1$ the left side of (2.2.7) is

$$u_n u_{n+2} - u_{n+1}^2 = u_n(u_n + u_{n+1}) - u_{n+1}(u_{n-1} + u_n)$$

$$= u_n{}^2 - u_{n-1}u_{n+1}$$

$$= -(u_{n-1}u_{n+1} - u_n{}^2)$$

$$= -(-1)^n \quad \text{(from the truth of (2.2.7) for } k = n)$$

$$= (-1)^{n+1}.$$

So the truth of (2.2.7) for a certain integer k implies its truth for the next integer, and as it is true for $k = 2$, it therefore holds for all integers greater than 2.

  Note that the natural numbers are precisely those integers which are each greater than zero. So the well-ordering principle for natural numbers can be regarded as the special case of $\mathbb{Z}O4$ in which $A = 0$.

EXAMPLE 2.2.5. Suppose that a certain gambling club allows its members to run up debts provided they do not exceed a fixed limit. Then if everyone has won some money we can represent their winnings by natural numbers and conclude from the well-ordering principle that there is someone whose winnings are less than or equal to those of anyone else. If there are some people with debts, we can use integers to represent each person's winnings, since if somebody owes 2 units of money we say that they have won -2 units. The well-ordering principle for integers then assures us that there is someone with a debt which is larger than, or perhaps equal to anyone else's. Of course these observations are simple and obviously true in ordinary circumstances but the well-ordering principle allows us to draw the same conclusions about the hypothetical situation in which there are an infinite number of players.

2.3. RINGS

  We know from § 8.2.1 that the word 'group' is just the general term for any set with an operation which possesses some properties akin to those of addition of integers. Frequently mathematicians have to work with sets possessing operations which have even more properties in common with the integers, and they have given such systems a special name. A *ring* is a set R of things, on which has been specified two different operations, or ways of combining pairs of elements. We shall denote these operations by '$\oplus$' and '$\odot$' and they must satisfy the following conditions

R1 Any two elements can be combined by $\oplus$ to produce a third element.

R2 $\oplus$ is commutative: if a, b are any two elements,
$$a \oplus b = b \oplus a.$$

R3 $\oplus$ is associative: if a, b, c are any three elements,
$$a \oplus (b \oplus c) = (a \oplus b) \oplus c.$$

R4 There is a special element 0 such that for any element a,
$$a \oplus 0 = a.$$

R5 Corresponding to each element a, there is an element a' (depending upon a) such that
$$a \oplus a' = 0.$$

R6 Any two elements can be combined by $\odot$ to produce a third element.

R7 $\odot$ is associative: if a, b, c are any three elements,
$$a \odot (b \odot c) = (a \odot b) \odot c.$$

R8 $\odot$ is distributive with respect to $\oplus$: if a, b, c are any three elements,
$$a \odot (b \oplus c) = (a \odot b) \oplus (a \odot c) \text{ and } (a \oplus b) \odot c = (a \odot c) \oplus (b \odot c).$$

In the above eight laws we have enclosed the usual signs for multiplication and addition by small circles to emphasize that even if the elements of the ring are ordinary numbers, the ways of combining them may not be ordinary addition and multiplication. However the above laws look more natural for addition and multiplication so we shall sometimes drop the circles if no confusion is likely to result. Notice that we could use the terminology of Chapter 8 to give a shorter description of a ring since the conditions $R1-R5$ just say that R is an Abelian group with respect to the operation $\oplus$ [see Definition § 8.2.2] and $R6$, $R7$ say that R is a semigroup with respect to the operation $\odot$ [see Definition 8.4.2]. The set of integers with the usual operations of addition and multiplication is of course an example of a ring but as we know $\mathbb{Z}$ satisfies several extra properties which may not be true for other rings. For instance $\mathbb{Z}$ has a multiplicative identity 1, with the usual meaning of 'identity' in a semigroup; but the set E of all even integers (those which are divisible by 2), together with the usual operations of addition and multiplication, is a ring which has no identity of multiplication. In fact, following the discussion in § 8.2.4 it is sensible to say that E is a *subring* of $\mathbb{Z}$. Later we shall meet other special types of rings; and in § 6.3 we shall see examples of rings of *matrices* in which the operation $\odot$ is not commutative and in which the cancellation laws for $\odot$ do not hold.

If the ring R has an identity (of multiplication), if also the operation $\odot$ is commutative, and if the cancellative laws for $\odot$ do hold, the ring has all the properties of the integers except, possibly, those connected with order, and is appropriately called an *integral domain*.

It should be noted that the failure of the cancellation laws for $\odot$ is equivalent to the existence of *zero divisors* in R: these are non-zero elements of R, say a and b, with the property $a \odot b = 0$, even though neither a nor b are themselves zero.

EXAMPLE 2.3.1.   The ring $\mathbb{Z}$ of all integers is of course an example of an integral domain, but the subring of even integers is not an integral domain since it does not have a multiplicative identity.

EXAMPLE 2.3.2.   The set of all polynomials in a single unknown X, and with integer coefficients [see § 14.1], forms an integral domain. The set of polynomials with integer coefficients in any number of variables $X_1, \ldots, X_n$ [see § 14.15] also forms an integral domain.

In § 2.7 we shall meet the Gaussian integers which form another common example of an integral domain.

2.4. RATIONAL NUMBERS AND FIELDS

2.4.1. Definition and Properties of the Rational Numbers

Sometimes, when using the natural numbers to count objects, we deal with things that can themselves be divided into smaller parts. We can deal with the possibility of counting just some of the parts by introducing the notation 'm/n', where m and n are natural numbers to mean 'm equal parts, n of which make a whole'. An expression m/n is called a *rational (or fractional) number* with m the *numerator* and n the *denominator* of the rational number.

EXAMPLE 2.4.1.   The rational number $1/2$ is the same as $2/4$, and in general for any natural numbers a, b and k, ka/kb is the same as a/b.

A rational number a/b can be of even greater use if we allow the numerator a to be any integer. For instance $0/2$ could stand for 'no parts, 2 of which make a whole'—in other words, zero. Similarly, for any n, $0/n$ is the same as 0. Following our ideas about negative integers a sensible interpretation of say $-1/3$ would be to regard it as meaning 'one part out of 3 of a whole which we owe to someone else'. This would be the same as $-(1/3)$. Another point to realize is that each integer a is a special type of rational number, since a is the same as $a/1$.

We can see how to form the sum and product of two rational numbers if we consider some particular examples. For instance

$$2/3 + 1/5 = 2.5/3.5 + 3.1/3.5,$$

and this means 'take 10 parts out of 15 and then add a further 3 parts out of 15'. So we get a total of 13 parts out of 15, whence

$$2.5/3.5 + 3.1/3.5 = (10 + 3)/15.$$

Analogously, for any integers a, c and natural numbers b, d, we have

$$a/b + c/d = (ad + bc)/bd. \qquad (2.4.1)$$

Also a meaningful interpretation of say $2/3(2/5)$, which fits in with our previous notions, would be to divide $2/5$ into 3 parts and take 2 of them. We can do this if we first write $2/5$ as $6/15$ and then we see that one part out of three of $6/15$ is $2/15$ whence $2/3(2/5)$ is $2.2/15 = 4/15$. Similarly for any integers a, c and $b, d \in \mathbb{N}$ we have

$$a/b . c/d = ac/bd. \qquad (2.4.2)$$

Moreover if we want to compare the sizes of say $2/3$ and $4/5$, we can proceed easily if we first write them with the same denominator, as $2.5/15$ and $4.3/15$. This means 10 pieces of a certain kind compared with 12 pieces of the same kind. Since in $\mathbb{N}$, $10 < 12$, it is then natural to say that $10/15 < 12/15$. Similarly for the rational numbers a/b, c/d we say that

$$a/b < c/d \quad \text{if} \quad ad < bc \text{ in } \mathbb{Z}. \qquad (2.4.3)$$

The letter $\mathbb{Q}$ is reserved for the set or collection of all rational numbers, so that as before we can write $r \in \mathbb{Q}$ meaning that r is a rational number. Most of the properties of $\mathbb{Q}$ will now be quite familiar. For instance $\mathbb{Q}$ is an Abelian group with respect to addition [see Definition 8.2.2]:

QA1 Any two elements of $\mathbb{Q}$ can be combined by addition to produce a third element of $\mathbb{Q}$.

QA2 Addition is associative in $\mathbb{Q}$: given any three members r, s, t of $\mathbb{Q}$, $r + (s + t) = (r + s) + t$.

QA3 Addition is commutative in $\mathbb{Q}$: if r, s are any two rationals, $r + s = s + r$.

QA4 There is a special number $0 \in \mathbb{Q}$, the identity of addition, such that for any rational r, $r + 0 = r$.

QA5 Corresponding to each rational number r, there is a rational $-r$ such that $r - r = r + (-r) = 0$.

The same argument as we used for $\mathbb{Z}$ now shows that the cancellation laws hold for addition in $\mathbb{Q}$. Indeed they hold for any group, whether it is Abelian or not [see Proposition 8.2.1]: If G is a group whose operation is written as $\oplus$ say, then for group elements a, b, c: $a \oplus c = b \oplus c$ implies $a = b$, and $c \oplus a = c \oplus b$ implies $a = b$. Also as in the case of $\mathbb{Z}$ we can solve any equation $r + x = s$ where r, s are rationals; for x is given by $s - r$. Similarly, in any group with operation $\oplus$, we can always find a group element x such that $x \oplus a = b$, for given group elements a, b [again, see Proposition 8.2.1].

As far as multiplication in $\mathbb{Q}$ is concerned we have even stronger properties than those we are accustomed to for $\mathbb{Z}$.

QM1 Any two rational numbers can be combined by multiplication to produce a third rational called their product.

QM2 Multiplication in $\mathbb{Q}$ is associative: given any three members r, s, t of $\mathbb{Q}$, $r(st) = (rs)t$.

QM3 Multiplication in $\mathbb{Q}$ is commutative: if r, s are two rationals $rs = sr$.

QM4 1 is the identity of multiplication: if r is any rational, $r \cdot 1 = 1 \cdot r = r$.

QM5 Corresponding to each rational r, apart from zero, there is a rational number r' (depending upon r), such that $rr' = r'r = 1$.

If $r = a/b$ where a is a natural number, then r' in QM5 equals b/a (see (2.4.2.) above). If a is a negative integer then $-a \in \mathbb{N}$ and $r' = -b/-a$. If $r \in \mathbb{N}$ then r' is the rational $1/r$, and similarly for any rational $r \neq 0$ we write $1/r$ for r' and call it the inverse or *reciprocal* of r. Note that in QM5 we must exclude $r = 0$, because (2.4.2) shows that $0 \cdot s = 0$ for each rational s so 0 cannot possibly have a reciprocal.

The above properties of multiplication stipulate that the set $\mathbb{Q}^*$ of all non-zero rational numbers is an Abelian group with respect to multiplication [see Definition 8.2.2]. The cancellation laws for multiplication by non-zero numbers then hold. That is, if $t \neq 0$ and $rt = st$ then $r = s$, and also $tr = ts$ implies $r = s$. We cannot say that $r \cdot 0 = s \cdot 0$ implies $r = s$ since for example $1 \cdot 0 = 2 \cdot 0 = 0$. Also we can solve any equation $rx = s$ where $r, s, \in \mathbb{Q}$ and $r \neq 0$; for x is given by $s(1/r) = s/r$. Multiplying by $1/r$ is called 'dividing by r'. Again as in $\mathbb{Z}$ a distributive property holds:

QAM Multiplication is distributive with respect to addition: if r, s, t are any rational numbers, $r(s + t) = rs + rt$.

2.4.2. Fields

The set $\mathbb{Q}$ with its addition and multiplication is one of the main examples of an algebraic system called a *field* since a set F together with two operations, say $\oplus$ and $\odot$, between pairs of its elements, is a field if the following conditions hold.

F1 F contains at least two elements.

F2 F is an Abelian group with respect to the operation $\oplus$ [see Definition 8.2.2].

F3 If 0 is the identity of $\oplus$, then all the elements of F, except 0, form an Abelian group with respect to $\odot$.

F4 The operation $\odot$ is distributive with respect to the operation $\oplus$.

The conditions $F2$, $F3$ and $F4$ are just the analogues, written in condensed form, of QA1–QA5, QM1–QM5, and QAM. As in the case of rings we shall sometimes call the field operations '$+$' and '$\cdot$' (or 'juxtaposition') even if they are not ordinary addition and multiplication, to emphasize the similarity between the operations in an arbitrary field and addition and multiplication of rational numbers. When the field operations are written as $+$ and juxtaposition then just as for $\mathbb{Q}$, we can solve any equation of the form $ax + b = 0$ in an arbitrary field. An equation such as $ax + b = 0$ is actually the most general sort of linear equation; one that is in which the unknown number x does not appear combined with itself, as x^2 for instance.

EXAMPLE 2.4.2. Which rational number x satisfies $2x + 3 = 0$? Since $2x + 3 = 0$, $2x$ must be -3 and so x is $-3/2$.

EXAMPLE 2.4.3. (i) Find the rational x such that $3x - 1 = 5/7$. This can be written as $3x - 1 - 5/7 = 0$, or $3x - 12/7 = 0$. So $3x = 12/7$ and $x = 4/7$.

(ii) Find the rational number x such that $3x + 1 = 5/7$. This is the same as $3x + (1 - 5/7) = 0$ or $3x + 2/7 = 0$. Hence $3x = -2/7$ and so $x = -2/21$.

EXAMPLE 2.4.4. Are there any rational numbers x which satisfy $x^2 + 2x - 3 = 0$? We have $(x^2 + 2x + 1) - 4 = 0$ or $x^2 + 2x + 1 = (x + 1)^2 = 4$. Now $2^2 = (-2)^2 = 4$ so the equation is satisfied by $x + 1 = 2$ and $x + 1 = -2$, or $x = 1$ and $x = -3$.

EXAMPLE 2.4.5. Are there any rational numbers x such that $x^2 = 2$? Suppose there is such an $x = a/b$ with $a \in \mathbb{Z}$, $b \in \mathbb{N}$. If a and b have a common divisor $d > 0$ then $a = da_1$, $b = db_1$, say, so that $x = da_1/db_1 = a_1/b_1$. We may thus 'divide out' all the common positive divisors of a and b (apart from 1) and obtain $x = a_1/b_1$ where a_1, b_1 only have 1 as a common positive divisor. Hence $x^2 = a_1^2/b_1^2 = 2$, so that $a_1^2 = 2b_1^2$. This means that a_1^2 is even whence a_1 must be even; so if $a_1 = 2k$ for some integer k then $(2k)^2 = 4k^2 = 2b_1^2$ and $b_1^2 = 2k^2$. Thus b_1 is also even, but this contradicts the fact that a_1, b_1 have no common divisor apart from 1. So there is no rational number x with $x^2 = 2$.

The last two examples show that in the field $\mathbb{Q}$, equations involving x^2 may not always have solutions. We shall later meet examples of fields in which more polynomial equations of degree greater than 1 (and sometimes all such equations) have solutions. The fields which we shall encounter later in this chapter (see § 2.6 and § 2.7) each have the property that they contain $\mathbb{Q}$. This means that the set of elements of each of them contains the set of elements of $\mathbb{Q}$. Also if two rationals are added or multiplied in the larger field, the result is the same as if they were combined according to the field operations in $\mathbb{Q}$. When a field F_1 contains a field F_2 in this sense, we say that F_2 is a *subfield* of F_1, or that F_1 is an *extension field* of F_2. Every field is in particular a ring, (see § 2.3) indeed an integral domain, so it is not surprising that there are not as many examples of fields as there are of rings or, even more so, of groups.

In the case of $\mathbb{Q}$, the relation of order satisfies properties which are familiar from our study of $\mathbb{Z}$.

QO1 For any two rational numbers r and s, either $r = s$, or $r < s$, or $s < r$.

QO2 $r < r + 1$ for every $r \in \mathbb{Q}$.

QO3 Order is a transitive relation.

Property QO2 actually follows from the other order axioms. It should be mentioned here that any field upon which an order relation [see § 1.3.2] has been defined is called an *ordered* field.

There is no counterpart of the well-ordering principle here, for consider the set $S = \{1/n\}$ of all reciprocals of natural numbers. All the elements of S are positive (greater than zero) but for each n we have $1/(n + 1) < 1/n$ so there is no smallest element of S. However we have familiar connections with addition and multiplication.

QAO If $r < s$ then $r + t < s + t$ for $r, s, t \in \mathbb{Q}$,

QMO Suppose r, s, t are rational numbers with $r < s$, then
  if $t > 0$, $rt < st$;
  if $t < 0$, $rt > st$.
  Of course, if $t = 0$ then $rt = st = 0$.
One nice feature of the notation a/b is that it shows clearly that each rational number is just a pair of integers. We could indeed make this more explicit by writing the rational number a/b as (a, b). We would then add two of these pairs of integers by

$$(a, b) + (c, d) = (ad + bc, bd); \tag{2.4.4}$$

multiply them by writing

$$(a, b).(c, d) = (ac, bd); \tag{2.4.5}$$

and order them by

$$(a, b) < (c, d) \quad \text{if } ad < bc. \tag{2.4.6}$$

These statements are just the translations of (2.4.1), (2.4.2) and (2.4.3) into this alternative notation. The only pairs which are not allowable are those in which the second entry is zero or a ngative integer. The latter kind can be allowed if we agree that $(-a, -b)$ is to be the same as (a, b), (in other words $-a/-b = a/b$). The exclusion of zero as a second entry is a necessary restriction however, corresponding to the fact that we cannot divide by zero in the more usual a/b notation.

Note that we must think of the pairs $(1, 2), (2, 4), (3, 6), \ldots$ as being equivalent, or identical, since they each represent the same rational number. More formally we have to put

$$(a, b) = (c, d) \quad \text{whenever } ad = bc; \tag{2.4.7}$$

or in other words

$$a/b = c/d \quad \text{whenever } ad = bc. \tag{2.4.8}$$

In practice this causes no inconvenience since we know that if we use $3/6$ in a calculation we shall get a result equivalent to that obtained if we had used $1/2$. In general, using equivalent fractions leads to equivalent results.

EXAMPLE 2.4.6.
  (i) $3/6 + 4/10 = (3.10 + 4.6)/6.10 = 54/60$; and if we add the fractions $1/2$ and $2/5$ (equivalent respectively to $3/6$ and $4/10$) we get $(1.5 + 2.2)/2.5 = 9/10$, which is equivalent to $6.9/6.10 = 54/60$.
  (ii) $8/14(3/9 - 2/15) = 8/14(3.15 - 2.9)/9.15 = 8.27/9.14.15 = 216/1890$; whereas using the equivalent fractions $4/7$ and $1/3$ instead of $8/14$, $3/9$, we have $4/7(1/3 - 2/15) = 4/7.9/45 = 4/7.1/5 = 4/35 = 4.54/35.54 = 216/1890$.

2.5. DECIMAL EXPANSIONS

A disadvantage of the a/b notation is that it can involve doing a lot of arithmetic in order to manipulate rational numbers. For example, it is not immediately apparent that $8/3 < 19/7$; we have to calculate 8.7 and 3.19 in order to verify this. Again it takes a few seconds to see that $5/12 + 9/14 = 89/84$. The difficulty in these cases is that we are dealing with different denominators: it is easy to add $3/16$ and $5/16$ for instance, or to see which one is the smaller. So it would be helpful if we could write all rational numbers with the same denominator, or failing this, with a simple sequence of denominators. We could not use a denominator such as 144 to represent each rational, since $1/145$ cannot be written in the form $a/144$ with an integral a. In the same way there is no number N which could be used as a denominator for every rational, since $1/(N + 1)$ cannot be written with denominator N. It is possible though to write each rational number as a sum of fractions, with a pre-arranged sequence of increasing denominators. The sequence most commonly used is the one which is usually used to write the natural numbers, namely the sequence $1, 10 = 10^1, 100 = 10^2$, $1000 = 10^3, \ldots$ of 1 and positive powers of 10.

EXAMPLES 2.5.1.
   (i) $1/2 = 5/10$;
   (ii) $3/20 = 15/100 = 1/10 + 5/100$;
   (iii) $-7/25 = -28/100 = -(2/10 + 8/100)$.
In order to avoid the possibility of numbers having two or more representations, as in (ii) and (iii) above, we shall restrict each fraction with denominator 10 or more to have numerator between 0 and 9 inclusive. This means that in (ii) the standard expansion of $3/20$ will be $1/10 + 5/100$.

If we are given a rational in the form a/b we can easily obtain an expansion such as in the above examples by putting $a/b = a.10^n/b.10^n$ where we choose 10^n, if possible, so that $a.10^n/b$ is an integer K say. Then $a/b = K/10^n$. However, there are many fractions a/b where we can multiply by larger and larger powers of 10 without $a.10^n/b$ ever being an integer. An example of this behaviour is given by $2/3$, where we have

$$\frac{2}{3} = \frac{\dfrac{2.10}{3}}{10} = \frac{6\frac{2}{3}}{10}$$

$$= \frac{\dfrac{200}{3}}{100} = \frac{66\frac{2}{3}}{100}$$

$$= \frac{\dfrac{2000}{3}}{1000} = \frac{666\frac{2}{3}}{1000}.$$

The last equality tells us that the difference between $\frac{2}{3}$ and $\frac{666}{1000}$ is less than $\frac{1}{1000}$. Similarly $\frac{2}{3} - \frac{6666}{10,000} < \frac{1}{10,000}$, and for any natural number n,

$$\frac{2}{3} - \frac{66\ldots6}{10^n} < \frac{1}{10^n} \quad \text{where there are } n \text{ 6's.}$$

So the sums $\frac{6}{10}$, $\frac{6}{10} + \frac{6}{100}$, ..., approach $\frac{2}{3}$ more and more closely, and make it seem reasonable to write

$$\frac{2}{3} = \frac{6}{10} + \frac{6}{10^2} + \frac{6}{10^3} + \ldots + \frac{6}{10^n} + \ldots . \tag{2.5.1}$$

A discussion of such infinite sums can be made completely watertight and mathematically sound, but for the present we shall merely note that no contradiction is caused by writing expansions such as in (2.5.1) or by working with them like finite expansions. Note that 1 can be written as $\frac{1}{1}$ or as

$$1 = \frac{9}{10} + \frac{9}{100} + \frac{9}{1000} + \ldots . \tag{2.5.2}$$

This can be seen by an argument like that used to give (2.5.1) or by multiplying both sides of (2.5.1) by $\frac{3}{2}$ since

$$\frac{3}{2} \cdot \frac{2}{3} = 1 \quad \text{and} \quad \frac{3}{2}\left(\frac{6}{10^n}\right) = \frac{9}{10^n}.$$

We may similarly write any integer N as

$$N - 1 + \frac{9}{10} + \frac{9}{10^2} + \ldots,$$

but we shall use N itself as the standard representation of N. Actually as the following examples show, we usually write integers themselves by expressing them as sums of multiples of 1 and of the positive powers of 10.

EXAMPLES 2.5.2.
   (i) 1234 is shorthand for $1(10^3) + 2.10^2 + 3.10 + 4.1$.
   (ii) -617 means $-(6.10^2 + 10 + 7.1)$.
   (iii) 409 stands for $4.10^2 + 9.1$.
When we see an integer like 123 we automatically understand that the digit at the extreme right denotes the number of units or 1's in the given integer; the next digit to the left is the number of 10's (the *coefficient* of 10), and so on. This simple device of having a place notation is quite easy to work with and saves writing a great many needless symbols. It would be very helpful if we had a similar quick way of writing something like

$$\frac{4}{10} + \frac{5}{10^2} + \frac{6}{10^3}.$$

We could not write 456 for that would risk confusion with an integer; so we use an extra sign to show that these digits do not denote an integer. The sign always used is '·'—the decimal point—and it is placed immediately to the left of the coefficient of $\frac{1}{10}$ in our number. Thus 0·456 denotes

$$\frac{4}{10} + \frac{5}{10^2} + \frac{6}{10^3}.$$

This notation can be combined with the notation for integers as in the following examples.

EXAMPLES 2.5.3
  (i) 123·456 denotes

$$1.10^2 + 2.10 + 3.1 + \frac{4}{10} + \frac{5}{10^2} + \frac{6}{10^3}.$$

(ii) 0·0216 stands for

$$\frac{2}{10^2} + \frac{1}{10^3} + \frac{6}{10^4}.$$

(iii) $-23·2$ is short for

$$-\left(2.10 + 3.1 + \frac{2}{10}\right).$$

The fractions

$$\frac{1}{10}, \frac{1}{10^2}, \frac{1}{10^3}, \cdots,$$

are called negative powers of 10 [see § 3.2] and we write

$$10^{-1} = \frac{1}{10}, \qquad 10^{-2} = \frac{1}{10^2},$$

and for each natural number n,

$$10^{-n} = \frac{1}{10^n}.$$

We also write $10^0 = 1$. So we can express every number as a sum of multiples of powers (positive, negative and zero) of 10, in the form

$$a_1 a_2, \ldots, a_n \cdot b_1 b_2 \ldots \tag{2.5.3}$$

where each of the a's and b's lies between 0 and 9 inclusive and there are possibly an infinite number of b's. The last remark is to include such cases as in (2.5.1); but instead of writing $\frac{2}{3} = 0·666\ldots$, we actually write $\frac{2}{3} = 0·\dot{6}$ where the dot over the 6 shows that it is indefinitely repeated. There are also some cases such as $\frac{1}{7} = 0.142857142857\ldots$ where several digits are repeated together. For brevity we indicate such a situation by putting a dot over the first and last digits to be repeated, as in $\frac{1}{7} = 0·\dot{1}4285\dot{7}$.

EXAMPLES 2.5.4
   (i) $3\frac{1}{16} = 3{\cdot}0625$.
   (ii) $\frac{25}{3} = 8\frac{1}{3} = 8{\cdot}\dot{3}$.
   (iii) $\frac{5}{6} = 0{\cdot}833\ldots = 0{\cdot}8\dot{3}$.
   (iv) $-\frac{3}{7} = -0{\cdot}\dot{4}2857\dot{1}$.

EXAMPLE 2.5.5.   A curious property of the repeating period of $\frac{1}{7}$ is that the integer 142857 can be multiplied by 1, 2, 3, 4, 5 or 6 by choosing a suitable cyclic permutation of its digits [see Definition 8.1.3]. For 2(142857) = 285714, 3(142857) = 428571, 4(142857) = 571428, 5(142857) = 714285, 6(142857) = 857142; but 7(142857) = 999999(!)

EXAMPLES 2.5.6.   Obtain the decimal expansions of (i) 27/5, (ii) 7/33, (iii) 5/14. The easiest way to proceed (in the absence of a pocket calculator) is to divide the denominator into the numerator (which we suppose has an infinite number of zeros after its decimal point). The necessary long divisions are shown below

$$
\begin{array}{lll}
\quad\ 5{\cdot}4 & \quad\ 0{\cdot}\dot{2}\dot{1} & \quad\ 0{\cdot}3\dot{5}71428 \\
\text{(i) } 5)\overline{27{\cdot}0} & \text{(ii) } 33)\overline{7{\cdot}00} & \text{(iii) } 14)\overline{5{\cdot}0000000} \\
\quad\ 25 & \quad\ \nearrow 6\,6 & \quad\ 4\,2 \\
\quad\ \overline{20} & \quad\ \overline{40} & \quad\ \rightarrow\overline{80} \\
\quad\ 20 & \quad\ 33 & \quad\ 70 \\
\quad\ \overline{} & \quad\ \rightarrow\overline{7} & \quad\ \overline{100} \\
& & \quad\ 98 \\
& & \quad\ \overline{20} \\
& & \quad\ 14 \\
& & \quad\ \overline{60} \\
& & \quad\ 56 \\
& & \quad\ \overline{40} \\
& & \quad\ 28 \\
& & \quad\ \overline{120} \\
& & \quad\ 112 \\
& & \quad\ \rightarrow\overline{8}
\end{array}
$$

We place the decimal point of the quotient above the decimal point of the numerator, but otherwise we ignore the decimal point in performing the division. We continue the division until we get an exact quotient, in which case the rational number has a finite decimal expansion; or until we find the first remainder which is the same as a previous one—which occurred at or after the decimal point, in which case the rational has an infinite expansion and the repeating period is the part of the quotient found between the common remainders (as indicated in cases (ii) and (iii) above). Expansions such as (2.5.3) are called *decimal expansions* and expansions $\cdot b_1 b_2 \ldots$ are *decimal fractions*, with the spaces where the a's and b's are being the *decimal places*.

Two finite decimal expansions can be added very easily by just adding the digits in the corresponding decimal places, and if the sum of the two digits is 10 or more we 'carry 1' to the next place on the left. So that for instance $0.5 + 0.7 = 1.2$, corresponding to the fact that

$$\frac{5}{10} + \frac{7}{10} = \frac{12}{10} = 1\frac{2}{10}.$$

They can also be multiplied by temporarily forgetting about the decimal points and multiplying the two numbers as if they were integers. Then, if there are s decimal places to the right of the decimal point in the first number and t places to the right of the decimal point in the second number, we insert the decimal point in the product so that there are $s + t$ places to the right of it. For example, to multiply 3.14 by 2.72 we multiply 314 by 272 obtaining 85408, and insert the decimal point so that there are four places to the right of it: thus $(3.14).(2.72) = 8.5408$. It isn't always so easy to multiply, or even to add decimal expansions if they are not finite. Suppose however that we are given two infinite decimal expansions. (A finite fractional expansion may be considered as an infinite one by adjoining an infinite number of zeros to the right of it.) For each one consider an associated finite expansion obtained by writing down those digits which occur to the left of its decimal point and only the first s digits which occur to the right of it. Then the sum of the two finite expansions differs from the true sum by at most a 2 in the sth decimal place. Also if n is the greatest number of digits to the left of the decimal point in any of the numbers then the product of the finite expansions differs from the true product by at most a 2 in the $(s - n)$th place to the right of the decimal point. Thus even in the case of multiplication we can get as much accuracy as we wish, by taking s large enough. Usually in practical applications sufficient accuracy is obtained by working with a fixed number (often 4 or 5) of the first few digits to the right of the decimal point.

It is easy to tell which of two decimal expansions is the greater (whether they are infinite or not). We may assume that both numbers are positive, since if one is positive and the other negative the result is obvious—likewise if one of them is zero—and if they are both negative we replace the word 'greater' by 'smaller' at each occurrence in the procedure below. If one of them has more digits to the left of the decimal point then it is the greater. Otherwise the two numbers must differ in at least one decimal place. We look at the first place (reading from the left) in which they differ, and whichever number has the larger digit there is the greater number.

EXAMPLES 2.5.7
  (i) $2.573 + 0.617 = 3.190 = 3.19$.
  (ii) $\frac{1}{9} + \frac{2}{9} = \frac{1}{3}$ so $0.\dot{1} + 0.\dot{2} = 0.\dot{3}$, whereas $0.111 + 0.222 = 0.333$ which differs from $\frac{1}{3}$ by less than $\frac{1}{1000} = 0.001$.
  (iii) $(\frac{25}{3})(\frac{192}{7}) = \frac{1600}{7} = 228\frac{4}{7} = 228.\dot{5}7142\dot{8}$; but $\frac{25}{3} = 8.\dot{3}$, $\frac{192}{7} = 27.\dot{4}2857\dot{1}$ so $(8.\dot{3}).(27.\dot{4}2857\dot{1}) = 228.\dot{5}7142\dot{8}$. If instead of the true values of $\frac{25}{3}$ and $\frac{192}{7}$ we consider the approximations 8.333 and 27.428, we find that

(8·333). (27·428) = 228·557524; and this does indeed differ from the true value of the product $(\frac{25}{3})(\frac{192}{7})$ by less than 2 (indeed by less than 1) in the (3 − 2)th = first decimal place to the right of the decimal point.

(iv) 97·829 < 120·6.

(v) 67·354 < 68·243.

(vi) −3·467̇6̇ < −3·46̇.

Multiplying two numbers expressed in decimal notation is not any easier than when they are expressed in the form a/b; but it is the great advantage of the decimal notation when dealing with addition and order that makes its use so widespread.

2.6.   REAL NUMBERS

2.6.1.   Definition of the Real Numbers

We have already seen examples of rationals whose decimal expansions terminate or are finite, and some with infinite periodic expansions; but what we have not mentioned is that every rational behaves in one of these two ways. That is, if a rational number has an infinite decimal expansion then its expansion is periodic from some point onwards. Conversely, any infinite expansion which is periodic from some point onwards is the expansion of a rational number. This raises the question of whether there are numbers having infinite decimal expansions which are not periodic, but whose digits can all be specified by some rule. Numbers like this do exist, for such a rule makes them as accessible and meaningful as infinite periodic expansions.

EXAMPLE 2.6.1.   Consider the number 0·1010010001... where there is a zero between the first 1 and the second, two zeros before the next 1, three zeros before the next, and in general r zeros between the rth and $(r + 1)$th 1's. The 1's occur in the places numbered 1, 3, 6, ..., $\frac{1}{2}n(n + 1)$, ... to the right of the decimal point, and there are zeros everywhere else. So we can easily see whether any particular place has a 1 or a 0 in it. It is not hard to show that the expansion is not periodic; but we now know as much about its decimal expansion as about an infinite periodic expansion.

EXAMPLE 2.6.2.   Consider the number 0·363636... where the digits in the odd-numbered places after the decimal point are 3's, and all the others are 6's. This rule defines the number unambiguously just as in Example 2.6.1. It is evidently a periodic decimal and so is a rational number, but the way in which it is specified puts it on exactly the same footing as the number in the previous example.

EXAMPLE 2.6.3.   Consider any number whose fractional part is formed by writing, immediately after the decimal point, all the positive integers, one after

another, in their natural order (and in decimal notation!) as $0.123456789101112\ldots$.
This rule again defines the number completely since we can clearly write down
as much of the expansion as we like and so discover which digit is in any particular decimal place. It is not perhaps immediately obvious that it is not periodic;
but this is so, and hence such a number cannot be rational.

Suppose we are given an infinite *purely periodic* decimal fraction x; that is to
say, its repeating period starts immediately after the decimal point and if

$$x = .\dot{b}_1\ldots\dot{b}_n$$

the integer n is the *length* of the period of x. We may write

$$x = .\dot{b}_1\ldots b_n\dot{b}_1\ldots b_n$$

$$= .\dot{b}_1\ldots b_n + .00\ldots0\dot{b}_1\ldots\dot{b}_n$$

$$= .\dot{b}_1\ldots b_n + .\dot{b}_1\ldots\dot{b}_n/10^n$$

$$= .\dot{b}_1\ldots b_n + x/10^n.$$

Hence $x(1 - 1/10^n) = .\dot{b}_1\ldots\dot{b}_n = b_1\ldots b_n/10^n$,   so   that $x(10^n - 1)/10^n =$
$b_1\ldots b_n/10^n$ and

$$x = b_1\ldots b_n/(10^n - 1). \tag{2.6.1}$$

If we have a number $y = .a_1\ldots a_k\dot{b}_1\ldots\dot{b}_n$ whose period does not start immediately after the decimal point, we can write

$$y = .a_1\ldots a_k + .0\ldots0\dot{b}_1\ldots\dot{b}_n$$

$$= a_1\ldots a_k/10^k + \dot{b}_1\ldots\dot{b}_n/10^k$$

$$= a_1\ldots a_k/10^k + b_1\ldots b_n/10^k(10^n - 1) \quad \text{from (2.6.1.)}$$

$$= [(a_1\ldots a_k)(10^n - 1) + b_1\ldots b_n]/10^k(10^n - 1). \tag{2.6.2}$$

EXAMPLES 2.6.4

(i) The period of the number $x = 0.\dot{3}\dot{6}$ of Example 2.6.2 is of length 2, so
from (2.6.1)

$$x = \frac{36}{(10^2 - 1)} = 36/99 = 4/11.$$

(ii) It is easy to express 5.76 as a quotient of two integers since it is $576/100 =$
$144/25$.

(iii) $3.6\dot{3}2\dot{4} = 3.6 + 0.0\dot{3}2\dot{4} = 36/10 + 0.\dot{3}2\dot{4}/10 = 36/10 + 324/10(999)$
$= 36/10 + 12/10.37 = (18.37 + 6)/5.37 = 672/185.$

Numbers with infinite decimal expansions which are not eventually periodic are
called *irrational* numbers, and the collection of all rational and irrational
numbers is the set $\mathbb{R}$ of *real numbers*. Examples 2.6.1 and 2.6.3 above can be
varied to give many other irrational numbers and, in a sense explained in the
chapter on sets (§ 1.6), there are far more irrational than rational numbers.

2.6.2. Properties of Real Numbers

We can add and multiply any two real numbers by the method previously described for operating with infinite decimal expansions. The procedure used— of choosing a natural number s, cutting off the infinite expansions after the first s places to the right of the decimal point, and working with the finite expansions so obtained—is in effect a rule which enables us to specify the digits in any finite number of places of the resulting sum or product. Also the rule for specifying which of two different real numbers is the greater is exactly the same as that previously described for decimal expansions. It can be shown that with this addition and multiplication $\mathbb{R}$ is a field. That is, it satisfies the laws $F1 - F4$ of § 2.4.2 if we read $\mathbb{R}$, $+$, $.$, instead of F, $\oplus$, $\odot$ each time. So we can always divide by real numbers, apart from zero and we can solve any linear equations with real coefficients.

EXAMPLE 2.6.5.   We can describe the process of division by real numbers most easily if we number the decimal places so that place 0 is immediately to the left of the decimal point, place 1 is next to the left and we continue with the positively numbered places to the left of that; then place -1 is immediately to the right of the decimal point, place -2 on the right of that, and so on. Suppose now that r_1, r_2 are two real numbers, and that (reading from the left) the first non-zero digit of r_i ($i = 1, 2$) occurs in place n_i. We first consider the case of r_1 having a finite decimal expansion. Then we can use long division (as in Example 2.5.6) to divide the digits of r_1 into r_2, and we insert the decimal point in the quotient so that if $r_2 \geq 10^{n_2 - n_1} r_1$ the first non-zero digit in the quotient occurs in place $n_2 - n_1$; and if $r_2 < 10^{n_2 - n_1} r_1$ the first non-zero digit in the quotient occurs in place $n_2 - n_1 - 1$. When r_1 has an infinite, perhaps non-periodic, decimal expansion, we choose a natural number s and truncate r_1 by writing down any digits to the left of its decimal point and only the first s digits to the right of the decimal point. Then the quotient we obtain on dividing r_2 by the truncated expansion will differ from r_2/r_1 by at most a 1 in place $(1 + n_2 - 2n_1) - s$. So we can achieve any desired accuracy in our division by simply taking sufficiently good approximations to the divisor.

The *order relation* in $\mathbb{R}$ is quite similar to that for $\mathbb{Q}$ since we obtain $\mathbb{R}O1$, $\mathbb{R}O2$, $\mathbb{R}O3$, $\mathbb{R}AO$ and $\mathbb{R}MO$ simply by reading the corresponding order axioms for $\mathbb{Q}$ in § 2.4.2 and letting r, s, t be real numbers each time. Again there is no well-ordering principle as otherwise it would hold for $\mathbb{Q}$, since rationals are particular instances of reals. The order relation in $\mathbb{R}$ does have an important property called '*completeness*', which $\mathbb{Q}$ does not possess, but this really belongs in the realm of Analysis [see IV, § 11.3] so we shall not attempt to explain it here. Further light can be thrown on the relation between $\mathbb{R}$ and $\mathbb{Q}$ if we observe that in $\mathbb{R}$ the rationals and irrationals are interleaved; in the sense that between any two distinct rationals q_1 and q_2, say $q_1 < q_2$, there is always an irrational r, so that $q_1 < r < q_2$. There are indeed infinitely many irrationals between q_1 and q_2. Also there are always infinitely many rational numbers between any two given irrationals.

EXAMPLE 2.6.6.   Find an irrational number lying between 0·375 and 0·376. The number 0·375101001..., formed by writing the digits of the number in Example 2.6.1 immediately after 0.375, is a real number which lies between 0·375 and 0·376. Also the digits 101001... are not eventually periodic, so the digits 375101001... cannot be, and hence the number 0·375101001... is irrational. Notice that if the two given rational numbers had been 0·375 and 0·3751 we could have just altered our example to 0·3750101001...; and this again is an irrational number lying between the given rationals.

EXAMPLE 2.6.7.   Let us show that for a given real number $h \geq -1$, and any natural number n, we always have [see (3.10.2)]

$$(1 + h)^n \geq 1 + nh, \tag{2.6.3}$$

where '$\geq$' means 'greater than or equal to'. We shall use the method of induction [see § 2.1]; so we first show that the result is true for $n = 1$. This is because, when $n = 1$, $(1 + h)^1 = 1 + h = 1 + 1.h$, so the two sides of (2.6.3) are actually equal. Next we must show that the truth of (2.6.3) for a certain n implies its truth for $n + 1$. Now $h \geq -1$ means $1 + h \geq 0$, so that when we consider $n + 1$

$$\begin{aligned}
(1 + h)^{n+1} &= (1 + h)(1 + h)^n \geq (1 + h)(1 + nh) \quad &\text{(from } \mathbb{R}\text{MO and} \\
&= 1 + (n + 1)h + nh^2 &\text{the truth of} \\
&\geq 1 + (n + 1)h. &\text{(2.6.3) for } n)
\end{aligned}$$

So the truth of (2.6.3) for n does imply its truth for $n + 1$, and since it is true for $n = 1$ the principle of induction implies that it holds for every natural number n. The inequality (2.6.3) in fact holds for every $h \geq -2$, and each n, as we can see by using a different induction argument. We need only consider $-2 \leq h < -1$, and this means $-1 \leq 1 + h < 0$. Under these conditions we can now establish

$$-1 \leq (1 + h)^n \leq 1 \tag{2.6.4}$$

for any natural number n. Because, from $-1 \leq (1 + h) < 0$, (2.6.4) certainly holds for $n = 1$; and if it is true for a certain n then for $n + 1$ we have (remembering that $1 + h$ is negative),

$$-1 \leq 1 + h \leq (1 + h)^{n+1} \leq -(1 + h) \leq 1.$$

So the induction principle implies that (2.6.4) holds for every natural number n when $-2 \leq h < -1$. In order to complete the proof of (2.6.3) we note that it holds with equality when $n = 1$, and when $n \geq 2$

$$\begin{aligned}
1 + nh &< 1 - n \quad &\text{since } h < -1 \\
&\leq -1 &\text{since } n \geq 2 \\
&\leq (1 + h)^n &\text{from (2.6.4).}
\end{aligned}$$

Another helpful way to look at the order relation in $\mathbb{R}$ is to represent each real number by a point on a straight line; which we suppose to be continued indefinitely in both directions as in Figure 2.6.1. A number r is represented in the

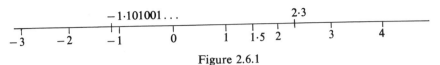

Figure 2.6.1

picture by a point which is to the left of all points representing numbers larger than r. Such a linear portrayal of the real numbers is commonly called the *real line*. It is of course an extension of the linear representation of the integers used in Figure 2.2.2.

We used our linear picture of the integers to suggest the definition of the absolute value of an integer, and we can similarly define the *modulus or absolute value* $|r|$ of a real number r by

$$|r| = \begin{cases} r & \text{if } r \geq 0 \\ -r & \text{if } r < 0. \end{cases} \qquad (2.6.5)$$

For any real numbers r_1, r_2, we have properties corresponding to (2.2.3), (2.2.4) and (2.2.5)

$$|r_1 r_2| = |r_1| \cdot |r_2|. \qquad (2.6.6)$$

$$|r_1 + r_2| \leq |r_1| + |r_2|, \qquad (2.6.7)$$

and

$$|r_1 - r_2| \geq |\,|r_1| - |r_2|\,|. \qquad (2.6.8)$$

EXAMPLES 2.6.8

(i) $|1 \cdot 732 - 2 \cdot 6| = |-0 \cdot 868| = 0 \cdot 868 < |1 \cdot 732| + |-2 \cdot 6| = 4 \cdot 332$
and
$|1 \cdot 732 - 2 \cdot 6| = |\,|1 \cdot 732| - |2 \cdot 6|\,| = 2 \cdot 6 - 1 \cdot 732 = 0 \cdot 868.$

(ii) $|1 \cdot 732 + 2 \cdot 6| = |1 \cdot 732| + |2 \cdot 6| = 4 \cdot 332,$
and
$|1 \cdot 732 + 2 \cdot 6| = |1 \cdot 732 - (-2 \cdot 6)| > |\,|1 \cdot 732| - |-2 \cdot 6|\,| = 2 \cdot 6 - 1 \cdot 732$
$= 0 \cdot 868.$

2.6.3. Sets of Real Numbers

Suppose we want to investigate a particular set of real numbers, such as the set of real numbers x which satisfy the inequality

$$x^4 - 5x^3 + 5x^2 + 5x - 6 = (x - 1)(x + 1)(x - 2)(x - 3) \leq 0. \quad (2.6.9)$$

A useful first step is to say where its members lie in relation to the other numbers on the real line. One of the easiest ways to do this is to be able to say that its members are all less than or equal to some known number. In the present instance, it is true that they are all less than 10; but it is also true, and gives us more information, to say that all such numbers are less than or equal to 3. In general if S is a set of real numbers all of whose members are less than or equal to some number A we say that S is *bounded above* (by A) and that A is an *upper bound* for S. Any number bigger than such an A will also be an upper bound, so any set which is bounded above has an infinite number of upper bounds. However there will always be one and only one smallest possible upper bound, called the *least upper bound* or *supremum* of S (abbreviated to l.u.b. or sup). In our initial example, 10 is an upper bound for the set of numbers satisfying (2.6.9) and 3 is actually the least upper bound.

Even if a set of real numbers is not bounded above its members might still all be greater than or equal to a number B. If this is the case, we say that the set is *bounded below* and that B is a *lower bound*. For example the set of natural numbers 1, 2, 3, ... is not bounded above, but it is bounded below since all its members are greater than 0. There will be infinitely many lower bounds if there are any, and the unique largest one of the lower bounds is called the *greatest lower bound* (g.l.b.) or *infimum* (inf) of the set.

EXAMPLES 2.6.9
(i) The set of natural numbers has g.l.b. 1 since 1 is less than or equal to each natural number, and no number larger than 1 is a lower bound.
(ii) The set of negative real numbers is not bounded below, but it is bounded above and has l.u.b. 0.
(iii) The set of numbers satisfying (2.6.9) is bounded below and its g.l.b. is -1.
(iv) The set of rational numbers such that the square of each is less than 2 has l.u.b. $\sqrt{2}$ and g.l.b. $-\sqrt{2}$.

If a set of numbers is bounded above and below we simply say that it is *bounded*. This is equivalent to the statement that the modulus of each member of the given set is less than, or equal to, some fixed positive number M.

EXAMPLES 2.6.10
(i) The set of numbers satisfying (2.6.9) is bounded, and a suitable M is 3.
(ii) The set of numbers in Example 2.6.9(iv) is bounded, and we can take M to be $\sqrt{2}$.

Some of the simplest bounded sets of numbers are those which contain all the real numbers between their g.l.b. and the l.u.b.; for example $\{x \mid 3 < x < 7\}$. Sets of this form are called *intervals*, and we distinguish different types. Intervals which contain both their g.l.b. and their l.u.b. (as well as all the numbers between) are called *closed intervals*, and are of the form

$$\{x \mid a \le x \le b\} \qquad (2.6.10)$$

for some real numbers a (the g.l.b. of the set) and b (the l.u.b.) where $a \le b$. We denote the set (2.6.10) by $[a, b]$. Intervals which contain neither their g.l.b. nor their l.u.b. are called *open intervals* and are of the form

$$\{x \mid a < x < b\}, \qquad (2.6.11)$$

where the g.l.b. a and l.u.b. b are real numbers with $a < b$. We denote the set (2.6.11) by (a, b). The remaining type of interval contains either its g.l.b. or l.u.b.

but not both. They are called *semi-open intervals* and are of one of the two forms

$$\{x \mid a < x \le b\} \quad \text{and} \quad \{x \mid a \le x < b\}, \qquad (2.6.12)$$

denoted by $(a, b]$ and $[a, b)$ respectively.

EXAMPLES 2.6.11
   (i) The set $\{x \mid 3 < x < 7\}$ is the open interval $(3, 7)$.
   (ii) The non-negative numbers whose squares are each less than 2 form a semi-open interval $[0, \sqrt{2})$.
   (iii) The set of real numbers x which satisfy

$$(x + 1)(x - 2) = x^2 - x - 2 \le 0$$

is the closed interval $[-1, 2]$.
   (iv) The set of numbers in Example 2.6.9(iv) is not an interval since it does not contain any of the irrational numbers between $-\sqrt{2}$ and $\sqrt{2}$ (such as $\frac{1}{2}\sqrt{2}$).
   (v) The set of numbers which satisfy (2.6.9) is not an interval, since its g.l.b. is -1 and its l.u.b. is 3 but it does not contain $\frac{3}{2}$ for instance. However it is the union of two closed intervals $[-1, 1]$ and $[2, 3]$.

We can now extend our terminology to speak meaningfully of *infinite intervals* in cases where the set in question is not bounded above (or below). Infinite open intervals are those of the form

$$\{x \mid x > a\} \quad \text{and} \quad \{x \mid x < b\} \qquad (2.6.13)$$

which are denoted for convenience by $(a, +\infty)$ and $(-\infty, b)$ respectively. Here a and b are ordinary real numbers; but, as far as we are concerned at the moment, the symbol ∞ is meaningless by itself and is only to be used as part of a composite symbol, as shorthand for expressions like those in (2.6.13). Similarly, the infinite closed intervals are those of the form

$$\{x \mid x \ge a\} \quad \text{and} \quad \{x \mid x \le b\}, \qquad (2.6.14)$$

which are denoted by $[a, +\infty)$ and $(-\infty, b]$. Here again a and b are real numbers and we must be careful not to use the symbol ∞ by itself. Sometimes the whole real line is called the open interval $(-\infty, +\infty)$.

EXAMPLES 2.6.12
   (i) The positive real numbers constitute the open interval $(0, +\infty)$.
   (ii) The set of real numbers which satisfy

$$x^3 + 3x^2 + 3x + 2 = (x + 2)(x^2 + x + 1) \le 0$$

form the closed interval $(-\infty, -2]$.
   (iii) The set of real numbers which satisfy

$$x^2 - x - 2 = (x + 1)(x - 2) > 0$$

is the union of the infinite open intervals $(-\infty, -1)$ and $(2, +\infty)$.

2.6.4.   Roots of Real Numbers

One of the most important facts about real numbers is that every positive real number has *nth roots*. That is, if r is a positive real number and n is a natural number then there is real number $l > 0$ such that

$$l^n = \underbrace{l \ldots l}_{n\ ls} = r.$$

(2.6.15)

The number l is called an *n*th root of r, and we write $l = r^{1/n}$, or $l = \sqrt[n]{r}$ if n is at least 2 (when $n = 2$ we write $l = \sqrt{r}$ and call l the *square root* of r) [see also, § 3.3].

EXAMPLE 2.6.13.   Find a number $l = 5^{1/3}$. We call such an l a *cube root* of 5. Firstly $1 = 1^3 < l^3 < 2^3 = 8$ so $1 < l < 2$. Next we find that $1 \cdot 7^3 = 4 \cdot 913$ whereas $1 \cdot 8^3 = 5 \cdot 832$, so $1 \cdot 7 < l < 1 \cdot 8$. Also $1 \cdot 71^3 = 5 \cdot 000211 > l^3$ so, $1 \cdot 70 < l < 1 \cdot 71$. Thus $l = 1 \cdot 70 \ldots$, and we see that this trial and error procedure is in effect a rule for prescribing any number of decimal places of l.

The field $\mathbb{Q}$ has not this property, as we have seen there is no rational number whose square is 2. However there is a real number $\sqrt{2} = 1 \cdot 414 \ldots$ whose square is 2. As usual in mathematics $\sqrt{r}$ (where $r > 0$ is real) means the positive real number whose square is r. The existence of such roots now means that we can solve many more polynomial equations than before; for example all those of the form $x^n - r = 0$ where r is positive, although we should note that there are still some relatively simple equations such as $x^2 + 1 = 0$ which have no real solutions.

Almost all the numbers connected with the physical world are rational,* but the introduction of the real numbers enables us to give numerical representation to some otherwise anomalous situations. For example if we draw a right angled triangle with two sides of length 1, the theorem of Pythagoras [see V, § 1.1.3] tells us that the length l of the hypotenuse will be such that $l^2 = 1^2 + 1^2 = 2$, so that $l = \sqrt{2}$ as in Figure 2.6.2.

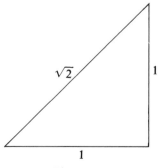

Figure 2.6.2

* Indeed the mathematician L. Kronecker (1823–91) is reputed to have said 'God made the integers, everything else is the work of man'.

2.7. COMPLEX NUMBERS

2.7.1. The Field of Complex Numbers

The desire to be able to solve all quadratic equations, and hopefully all polynomial equations, led mathematicians to introduce a new field of numbers containing the real numbers. The idea is to follow the examples of our previous model of the rationals as ordered pairs, [see § 2.4.2]. This time we consider pairs (x_1, x_2) of real numbers where two pairs are only to be considered the same if their respective first entries are equal and also their second entries are equal; and if $x = (x_1, x_2)$, $y = (y_1, y_2)$ are two pairs of real numbers we add them by

$$x + y = (x_1 + y_1, x_2 + y_2), \tag{2.7.1}$$

and multiply them by

$$x . y = (x_1 y_1 - x_2 y_2, x_1 y_2 + x_2 y_1). \tag{2.7.2}$$

The set of all possible pairs of real numbers, with the above ways of combining them (which we are certainly free to call 'addition of real pairs' and 'multiplication of real pairs'), is the set $\mathbb{C}$ of *complex numbers*. The first coordinate x_1 of (x_1, x_2) is called the *real part* of the complex number, the second coordinate x_2 is the *imaginary part*. The perhaps unfortunate terms 'complex' and 'imaginary' should not be taken to imply that these numbers are unusually difficult or that any part is illusory.

EXAMPLES 2.7.1
  (i) $(-2, 3) + (2, 7) = (0, 10)$,
  (ii) $(8/7, -5/3) + (-8/7, 5/3) = (0, 0)$,
  (iii) $(1, 1) . (1, -1) = (2, 0)$,
  (iv) $(6, 0) . (0, 1) = (0, 6)$,
  (v) for any real x_1, x_2, $(1, 0)(x_1, x_2) = (x_1, x_2)(1, 0) = (x_1, x_2)$.
  It is not hard to check that the set $\mathbb{C}$ together with the above operations of addition and multiplication forms a field [see § 2.4.2]. For example the additive identity is $(0, 0)$ and the additive inverse of (x_1, x_2) is $(-x_1, -x_2)$. From (v) above the multiplicative identity is $(1, 0)$, and if $(x_1, x_2) \neq (0, 0)$, the multiplicative inverse of (x_1, x_2) is

$$\left(\frac{x_1}{x_1^2 + x_2^2}, \frac{-x_2}{x_1^2 + x_2^2} \right),$$

since

$$(x_1, x_2) \left(\frac{x_1}{x_1^2 + x_2^2}, \frac{-x_2}{x_1^2 + x_2^2} \right) = \left(\frac{x_1^2 + x_2^2}{x_1^2 + x_2^2}, 0 \right) = (1, 0).$$

EXAMPLE 2.7.2. Solve the equation $(1, 2)(x_1, x_2) = (1, 1)$. Multiplying both sides by the multiplicative inverse of $(1, 2)$, or in other words dividing both sides by $(1, 2)$, we obtain

$$\left(\frac{1}{5}, \frac{-2}{5}\right)(1, 2)(x_1, x_2) = (1, 0)(x_1, x_2) = \left(\frac{1}{5}, \frac{-2}{5}\right)(1, 1),$$

whence

$$(x_1, x_2) = (\tfrac{1}{5}, -\tfrac{2}{5})(1, 1) = (\tfrac{3}{5}, -\tfrac{1}{5}).$$

One of the most obvious differences between the field $\mathbb{C}$ and the previous number systems we have considered is that there is no order relation in $\mathbb{C}$. That is to say, no possible relationship between complex numbers could simultaneously satisfy the counterparts of QO1—QO3, QA0 and QM0. We shall see this later, but note now that for complex numbers z_1, z_2 we never say 'z_1 is less than z_2', nor write '$z_1 < z_2$'.

We get a clearer view of the links between real numbers and complex numbers if we notice how complex numbers with zero imaginary part combine. When x_1, y_1 are real numbers,

$$(x_1, 0) + (y_1, 0) = (x_1 + y_1, 0),$$
$$(x_1, 0).(y_1, 0) = (x_1 y_1, 0),$$
$$(x_1, 0) - (y_1, 0) = (x_1 - y_1, 0),$$

and if $y_1 \neq 0$,

$$(x_1, 0) \text{ divided by } (y_1, 0) \text{ is } (x_1/y_1, 0).$$

Thus we can perform all arithmetic operations on complex numbers with zero imaginary part by performing the usual real operations on the real parts alone. So the set of complex numbers with zero imaginary part is a subfield of $\mathbb{C}$; and we can regard this subfield as *being* the field of real numbers called by other names. (The famous mathematician H. Poincaré (1854–1912) once said that mathematics is the art of calling different things by the same name.) For each real x we identify x with the pair $(x, 0)$ and so think of real numbers as being particular complex numbers. Using the definitions of addition and multiplication in (2.7.1) and (2.7.2), and our agreement about real numbers, we can now write

$$(x_1, x_2) = (x_1, 0) + (0, x_2)$$
$$= (x_1, 0) + (x_2, 0)(0, 1)$$
$$= x_1 + x_2(0, 1), \tag{2.7.3}$$

(since we are identifying x_1, x_2 with $(x_1, 0)$, $(x_2, 0)$). For convenience we denote $(0, 1)$ by the special symbol i (engineers often use j instead of i) and so write the complex number (x_1, x_2) as $x_1 + ix_2$. The sum and product of two complex numbers are then written as

$$(x_1 + ix_2) + (y_1 + iy_2) = (x_1 + y_1) + i(x_2 + y_2), \tag{2.7.4}$$
$$(x_1 + ix_2).(y_1 + iy_2) = (x_1 y_1 - x_2 y_2) + i(x_1 y_2 + x_2 y_1). \tag{2.7.5}$$

One of the most familiar properties of complex numbers must be the fact that

$$i^2 = -1. \qquad (2.7.6)$$

This is because $i^2 = (0, 1)(0, 1) = (-1, 0) = -1$. Although (2.7.6) may seem a strange result at first it should not be seen as in any way contradicting previous experience since i is a new number outside $\mathbb{R}$. It is rather like the fact that there is no rational number whose square is 2, but if we go outside the rationals we can find a real number whose square is 2.

We can now see from (2.7.6) that all real numbers possess square roots (which may be complex). If the real number r is positive, it has two real square roots $\sqrt{r}$ and $-\sqrt{r}$. The number zero has only itself as a square root, and the negative real number $-k$ has two purely imaginary square roots $i\sqrt{k}$ and $-i\sqrt{k}$. Even more remarkable is the fact that *every complex number a + ib has (complex) square roots*. For if $(x + iy)^2 = a + ib$, where x, y, a, b are real, then expanding the left-hand side gives

$$x^2 - y^2 + 2ixy = a + ib.$$

By the remarks about equality of complex numbers just before (2.7.1), we can equate real and imaginary parts here and obtain $x^2 - y^2 = a$ and $2xy = b$. The solutions to these equations are given by

$$x = \pm\sqrt{\tfrac{1}{2}(a + \sqrt{a^2 + b^2})}, \quad y = \pm\sqrt{\tfrac{1}{2}(\sqrt{a^2 + b^2} - a)},$$

where the initial signs are chosen to the same if b is positive, and different if b is *negative*. Thus $a + ib$ has two square roots given by

$$\pm\sqrt{\tfrac{1}{2}(a + \sqrt{a^2 + b^2})} \pm i\sqrt{\tfrac{1}{2}(\sqrt{a^2 + b^2} - a)}. \qquad (2.7.7)$$

EXAMPLES 2.7.3

(i) Find the square root of i. Here $i = 0 + 1.i$, and with $a = 0$, $b = 1$, (2.7.7) gives $\sqrt{i} = \pm(\sqrt{\tfrac{1}{2}} + i\sqrt{\tfrac{1}{2}}) = \pm(1 + i)/\sqrt{2}$. We can easily verify this since $[\pm(1 + i)/\sqrt{2}]^2 = (1^2 + i^2 + 2i)/2 = 2i/2 = i$.

(ii) Find the square root of $3 + 4i$. Here we have

$$\sqrt{3 + 4i} = \pm\left[\sqrt{\tfrac{1}{2}(3 + \sqrt{3^2 + 4^2})} + i\sqrt{\tfrac{1}{2}(\sqrt{3^2 + 4^2} - 3)}\right]$$

$$= \pm[\sqrt{\tfrac{1}{2}.8} + i\sqrt{\tfrac{1}{2}.2}] = \pm(2 + i),$$

and indeed $(2 + i)^2 = [-(2 + i)]^2 = 4 + i^2 + 4i = 3 + 4i$.

Since we can extract square roots with impunity we can solve any quadratic equation, even one with complex coefficients. For suppose we have an equation such as

$$ax^2 + bx + c = 0 \qquad (2.7.8)$$

where $a, b, c \in \mathbb{C}$, and we suppose $a \neq 0$. The equation is the same as $(2ax + b)^2 + 4ac - b^2 = 0$, or $(2ax + b)^2 = d$ where $d = b^2 - 4ac$. Therefore $2ax + b = \pm\sqrt{d}$ or

$$x = (-b \pm \sqrt{d})/2a. \qquad (2.7.9)$$

If $d = 0$ the only solution is $x = -b/2a$. If $d \neq 0$ we know that the complex number d has two complex square roots so (2.7.9) gives two distinct complex solutions to the equation (2.7.8).

EXAMPLES 2.7.4

(i) Solve the quadratic equation $2x^2 + x + 1 = 0$. Here $16x^2 + 8x + 8 = 0$ or $(4x + 1)^2 + 7 = 0$. So $4x + 1 = \sqrt{-7}$ and $x = (-1 + \sqrt{-7})/4$. The equation has no real solutions but two complex solutions $(-1 + i\sqrt{7})/4$ and $(-1 - i\sqrt{7})/4$.

(ii) Solve the equation $x^2 + 3x + (3/2 - i) = 0$. We have from (2.7.9) that $x = (-3 \pm \sqrt{9 - 4(3/2 - i)})/2 = (-3 \pm \sqrt{3 + 4i})/2$. We know from the last example that the two square roots of $3 + 4i$ are $2 + i$ and $-(2 + i)$, so the two solutions of the equation are $(-3 + (2 + i))/2 = (-1 + i)/2$ and $(-3 - (2 + i))/2 = (-5 - i)/2$.

EXAMPLE 2.7.5. Suppose it were possible to order the complex numbers so that all our accustomed properties of order held. Then i or $-i$ would be positive. If $i > 0$ then $i.i > 0.i$, or $i^2 > 0$. Similarly if $-i > 0$, $(-i)^2 > 0$. In either case $i^2 = (-i)^2 = -1 > 0$. Also $1^2 = 1 > 0$; whence adding the last two inequalities would give $1 + (-1) = 0 > 0$, which cannot hold at the same time as $0 = 0$.

EXAMPLE 2.7.6. For any real x_1, x_2

$$(x_1 + ix_2)(x_1 - ix_2) = x_1^2 + x_2^2. \qquad (2.7.10)$$

The number $x_1 - ix_2$ is called the *conjugate complex number*, or simply the conjugate, of $x_1 + ix_2$; and if $x_1 + ix_2$ is written as z, the conjugate is denoted by $\bar{z}$.

EXAMPLE 2.7.7. If $z = x_1 + ix_2$ and $w = y_1 + iy_2$ then

$$\overline{z + w} = \bar{z} + \bar{w} \quad \text{and} \quad \overline{zw} = \bar{z}.\bar{w}. \qquad (2.7.11)$$

That is, the conjugate of a sum is formed by taking the sum of the conjugates, and the conjugate of a product is formed by taking the product of the conjugates. For the left-hand side of the first equality in (2.7.11) is

$$\overline{(x_1 + ix_2) + (y_1 + iy_2)} = \overline{(x_1 + y_1) + i(x_2 + y_2)}$$

$$= (x_1 + y_1) - i(x_2 + y_2)$$

$$= (x_1 - ix_2) + (y_1 - iy_2) = \bar{z} + \bar{w}.$$

The left-hand side of the second part of (2.7.11) is

$$\overline{(x_1 + ix_2)(y_1 + iy_2)} = \overline{(x_1y_1 - x_2y_2) + i(x_1y_2 + x_2y_1)}$$

$$= (x_1y_1 - x_2y_2) - i(x_1y_2 + x_2y_1)$$

and the right-hand side is $(x_1 - ix_2)(y_1 - iy_2)$, which is also $(x_1y_1 - x_2y_2) - i(x_1y_2 + x_2y_1)$.

EXAMPLES 2.7.8
  (i) $\bar{3} = 3$, $\overline{-7} = -7$ and for any real r, $\bar{r} = r$.
  (ii) $\bar{i} = -i$.
  (iii) $\overline{2 + 3i} = 2 - 3i$, $\overline{1 + i} = 1 - i$, and in accordance with (2.7.10) the products $(2 + 3i)(2 - 3i) = 13$, $(1 + i)(1 - i) = 2$ are real numbers. Also $\overline{(2 + 3i) + (1 + i)} = \overline{3 + 4i} = 3 - 4i$ and $(2 - 3i) + (1 - i) = 3 - 4i$; while $\overline{(2 + 3i)(1 + i)} = \overline{-1 + 5i} = -1 - 5i$, and $(2 - 3i)(1 - i) = -1 - 5i$.

EXAMPLE 2.7.9.   We can deduce from (2.7.10) that complex numbers can be used to give results about real numbers alone. Suppose that x_1, x_2, y_1, y_2 are real numbers. Then

$$(x_1{}^2 + x_2{}^2)(y_1{}^2 + y_2{}^2) = (x_1 + ix_2)(x_1 - ix_2)(y_1 + iy_2)(y_1 - iy_2)$$
$$= [(x_1 + ix_2)(y_1 + iy_2)][(x_1 - ix_2)(y_1 - iy_2)]$$
$$= [(x_1y_1 - x_2y_2) + i(x_2y_1 + x_1y_2)]$$
$$\times [(x_1y_1 - x_2y_2) - i(x_2y_1 + x_1y_2)].$$

Now the two numbers in square brackets are conjugates, so, following (2.7.10), their product is $(x_1y_1 - x_2y_2)^2 + (x_2y_1 + x_1y_2)^2$. Therefore

$$(x_1{}^2 + x_2{}^2)(y_1{}^2 + y_2{}^2) = (x_1y_1 - x_2y_2)^2 + (x_2y_1 + x_1y_2)^2. \quad (2.7.12)$$

The identity (2.7.12) involves only real numbers and could of course be verified by simply multiplying out the brackets on each side; although it would have been more difficult to find without the aid of complex numbers. If x_1, x_2, y_1, y_2 are integers then (2.7.12) says that if two numbers are each sums of two integer squares then their product is also such a sum. For example $13 = 3^2 + 2^2$ and $17 = 4^2 + 1^2$, so the product $13.17 = 221$ should also be a sum of two squares; and indeed from (2.7.12)

$$13.17 = (3^2 + 2^2)(4^2 + 1^2) = (3.4 - 2.1)^2 + (2.4 + 3.1)^2 = 10^2 + 11^2.$$

EXAMPLE 2.7.10.   The set of complex numbers such as $1 - i$, 0, $7 + 19i$, $\ldots$ whose real and imaginary parts are both *integers* (instead of arbitrary real numbers) forms an integral domain [see § 2.3] called the domain of *Gaussian integers*.

2.7.2. Geometrical Representation of Complex Numbers

In Figure 2.6.1. we used a geometrical picture of $\mathbb{R}$ to gain greater insight into real numbers. It is similarly often helpful to represent complex numbers geometrically. We can do this by using two straight lines at right angles as in Figure 2.7.1. The horizontal line is marked with the numbers of zero imaginary part—the

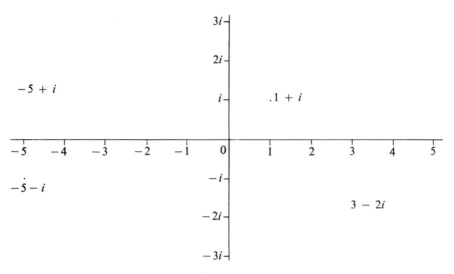

Figure 2.7.1

real numbers, and the vertical line (or axis) measures the numbers with zero real part—the imaginary numbers. Each complex number $x_1 + ix_2$ then corresponds uniquely to the point in the plane obtained by measuring x_1 units along the horizontal axis and then x_2 units parallel to the vertical axis. In Figure 2.7.1 the positions of $1 + i$, $3 - 2i$, $-5 - i$ and $-5 + i$ have been marked. This geometrical representation was originally conceived by Gauss in 1799 and Argand in 1806, and is commonly known as the *Argand Diagram*.

We defined the modulus of a real number as its distance from the origin of the real line. In exactly the same way we define the modulus $|z|$ of a complex number z as the distance from the point representing z to the origin of the Argand Diagram. If $z = x + iy$ we can see, as in Figure 2.7.2, that

$$|z| = \sqrt{x^2 + y^2} \qquad (2.7.13)$$

(cf. Pythagoras' theorem; V, § 1.1.3).

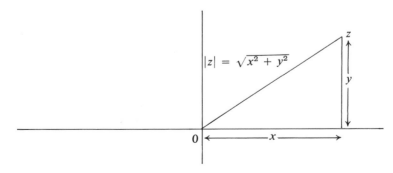

Figure 2.7.2

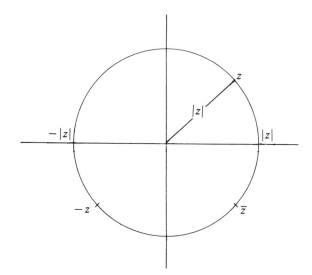

Figure 2.7.3

In Figure 2.7.1 we have $|1 + i| = \sqrt{1^2 + 1^2} = \sqrt{2}$, $|3 - 2i| = \sqrt{9 + 4} = \sqrt{13}$ and $|-5 + i| = |-5 - i| = \sqrt{26}$. These last two numbers $-5 + i$ and $-5 - i$ are conjugates, and in general for any z,

$$|-z| = |z| = |\bar{z}|. \tag{2.7.14}$$

Of course, $-z$ and $\bar{z}$ are not the only numbers which have the same modulus as z. All the infinitely many complex numbers which lie on a circle centre the origin and radius $|z|$ share the same modulus. This is illustrated in Figure 2.7.3 for $z = 1 + i$ and $|z| = \sqrt{2}$. We can also rewrite (2.7.10) more succinctly as

$$z.\bar{z} = |z|^2, \tag{2.7.15}$$

and for two complex numbers z_1, z_2 (which could be the same), we have the familiar properties

$$|z_1 z_2| = |z_1||z_2|, \tag{2.7.16}$$

$$|z_1 + z_2| \leq |z_1| + |z_2|, \tag{2.7.17}$$

and

$$|z_1 - z_2| \geq ||z_1| - |z_2||. \tag{2.7.18}$$

We shall see shortly why inequality (2.7.17) is commonly known as the '*triangle inequality*'.

EXAMPLES 2.7.11

(i) $|i| = 1$, $|-5| = 5$, and if $z = x + 0.i$ is real, $|z| = |x|$ is the usual absolute value of the real number x.

(ii) For $z_1 = 1 - i$, $z_2 = 3 + i$ we have $|z_1| = \sqrt{2}$, $|z_2| = \sqrt{10}$, whereas $z_1 z_2 = 4 - 2i$ and $|z_1 z_2| = \sqrt{16 + 4} = \sqrt{20} = |z_1| \cdot |z_2|$.

(iii) $(2 + i) + (3 - 2i) = 5 - i$ so $|(2 + i) + (3 - 2i)| = \sqrt{26} = 5.099\ldots$, and this is less than $|2 + i| + |3 - 2i| = \sqrt{5} + \sqrt{13} = 5.84\ldots$. Also $|(2 + i) + (3 - 2i)| = |(2 + i) - (2i - 3)| > ||2 + i| - |2i - 3|| = |\sqrt{5} - \sqrt{13}| = \sqrt{13} - \sqrt{5} = 1.369\ldots$.

(iv) $2(-1 + 3i) = -2 + 6i$ and $|-2 + 6i| = \sqrt{4 + 36} = \sqrt{40} = 2\sqrt{10} = 2|-1 + 3i|$.

EXAMPLES 2.7.12

(i) If $z = 5 - 3i$, $\bar{z} = 5 + 3i$ and $z\bar{z} = (5 - 3i)(5 + 3i) = 5^2 - (3i)^2 = 34 = |z|^2$. This means that $(5 - 3i)[(5 + 3i)/34] = 1$ so that the reciprocal of $5 - 3i$ is $(5 + 3i)/34$; and in general the inverse or reciprocal z^{-1} of a complex number z is given by

$$z^{-1} = \bar{z}/|z|^2. \tag{2.7.19}$$

(ii) Simplify the fraction $(-2 + 7i)/(4 - i)$. We deal with fractions like this by multiplying both the top and bottom of the expression by the conjugate of the denominator, and obtain here

$$\frac{(-2 + 7i)}{(4 - i)} = \frac{(-2 + 7i)(4 + i)}{(4 - i)(4 + i)} = \frac{(-15 + 26i)}{|4 - i|^2} = \frac{(-15 + 26i)}{17}.$$

We can also represent the algebraic operations of complex addition and multiplication in this geometrical picture. To add the complex numbers $z = x_1 + ix_2$ and $w = y_1 + iy_2$ we first mark the positions of z and w in the Argand diagram. Then from z we draw a line in the same direction as, and equal in length to, the line from 0 to w. This new line joins z to the point representing $z + w$. The process is illustrated by the continuous lines of Figure 2.7.4 for $z = \frac{7}{2} + \frac{1}{2}i$ and $w = \frac{1}{2} + 2i$, whose sum is $4 + (5/2)i$.

We can now see why the inequality $|z + w| \leq |z| + |w|$ is known as the triangle inequality—because it corresponds to the geometrical fact that any one side of a triangle is smaller than the sum of the other two. Equality is only

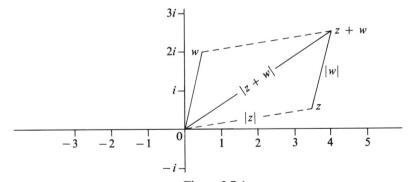

Figure 2.7.4

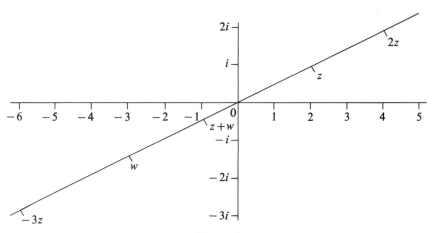

Figure 2.7.5

possible if the triangle is squashed flat so that z and w are in the same direction.

Evaluating $z + w$ is the same as finding $w + z$, so we could equally well have drawn a line from w in the same direction as, and equal in length to, the line from 0 to z. This alternative would have produced the dashed lines in Figure 2.7.4. The four outside lines in the figure form a parallelogram so the method is often known as the *parallelogram law*. The method always applies, even in cases such as $z = 2 + i$, $w = -3 - 3/2i$, illustrated in Figure 2.7.5. If we add z to itself we obtain $2z$, which therefore lies on the extended straight line from 0 to z. In general if k is a positive real number we obtain kz by drawing a line from 0 which is k times as long, and in the same direction, as the line from 0 to z. If $-k$ is a negative real number we obtain $-kz$ by drawing a line from 0 which is k times as long as, but in the opposite direction to, the line from 0 to z. See Figure 2.7.5 for the cases $k = 2$, and -3. It is also easy to multiply a given complex number z by i for it can be shown that this just amounts to rotating the line from 0 to z through a right angle in an anticlockwise direction. In Figure

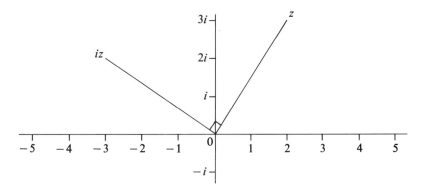

Figure 2.7.6

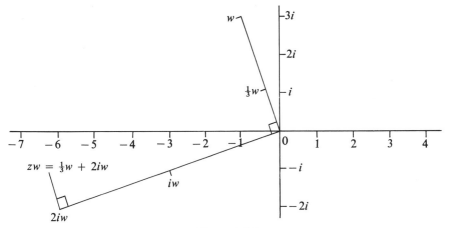

Figure 2.7.7

2.7.6 this is demonstrated for $z = 2 + 3i$ where $iz = i(2 + 3i) = 2i + 3i^2 = -3 + 2i$. If we were given two complex numbers $z = x_1 + ix_2$ and w, we could write $zw = x_1w + i(x_2w)$, and so use the above methods to find zw entirely by geometrical construction.

EXAMPLE 2.7.13. Construct zw, where $z = \frac{1}{3} + 2i$ and w is marked in Figure 2.7.7. We have $zw = \frac{1}{3}w + 2(iw)$, and the necessary steps are shown in Figure 2.7.7.

EXAMPLE 2.7.14. If $z = -1 - \frac{1}{2}i$ and $w = 1 + 2i$, zw is constructed in Figure 2.7.8.

A particularly instructive example is given by representing the product of i with itself geometrically, as in Figure 2.7.9, for we obtain another way of interpreting $i^2 = -1$.

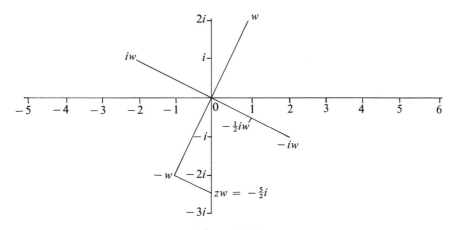

Figure 2.7.8

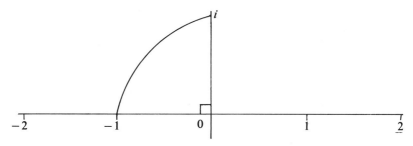

Figure 2.7.9

2.7.3.   Polar Coordinates of Complex Numbers

It is very convenient and fruitful to represent complex numbers by points in a plane, but so far we have always specified the point which represents the complex number $z = x + iy$ by giving its *Cartesian coordinates* x and y [see V, § 1.2.1]. It is equally possible to specify z by giving its *polar coordinates* [see V, § 1.2.6]: $r = |z|$, the modulus of z; and θ the angle which the line from the origin to z makes with the real axis. In Figure 2.7.10 the values of r and θ have been indicated for $z = 6, 1 + i, -3, -1 - i$ and $-i$. The value of θ is always given in radians [see § 17.2.2], and from Figure 2.7.10 we see that for points $x + iy$ with $y < 0$, the angle θ is negative and strictly between $-\pi$ and 0; while for $y \geq 0$, θ is between 0 and π inclusive. So every complex number z can be specified by an r and θ which satisfy

$$r \geq 0 \quad \text{and} \quad -\pi < \theta \leq \pi. \tag{2.7.20}$$

The number $z = 0$ is the only complex number whose modulus is zero and it does not have an associated angle θ. We gave the polar coordinates of $1 + i$ as $(\sqrt{2}, \pi/4)$, but we could just as well have used the angles $2\pi + \pi/4, 4\pi + \pi/4,$

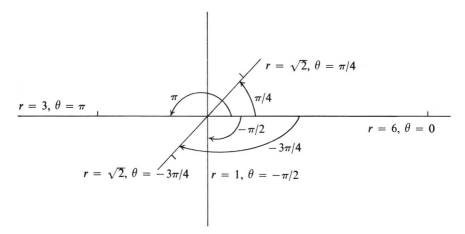

Figure 2.7.10

$-2\pi + \pi/4$, or $\pi/4 + 2k\pi$ for any integer k. Similarly any $z \neq 0$ can be specified by $|z|$ and any one of an infinity of angles which differ from each other by multiples of 2π. Whenever possible though we shall agree to stick to the angle θ which satisfies (2.7.20) and call this the *argument*, arg z, of z. The number $z = 0$ does not have an argument.

When the polar coordinates (r, θ) of z are given, the Cartesian coordinates of z are determined by

$$x = r \cos \theta, \qquad y = r \sin \theta, \tag{2.7.21}$$

and then

$$z = r \cos \theta + ir \sin \theta = r(\cos \theta + i \sin \theta). \tag{2.7.22}$$

On the other hand when z is given as $z + iy$, we have

$$r = |z| = \sqrt{x^2 + y^2} \tag{2.7.23}$$

and then, from (2.7.21) θ is given by $\cos \theta = x/r$, $\sin \theta = y/r$. Notice that these relations imply

$$\tan \theta = y/x,$$

but this is not enough by itself to specify θ completely, since for instance $\tan \theta = 1$ does not distinguish between $\theta = \pi/4$ and $\theta = -3\pi/4$.

EXAMPLES 2.7.15

    (i) $|i| = 1$, arg $i = \pi/2$.

    (ii) $|1 + \sqrt{3}i| = \sqrt{1 + 3} = 2$, so from (2.7.21) $\cos \theta = \frac{1}{2}$, $\sin \theta = \sqrt{3}/2$ whence $\theta = \pi/3$ and $1 + i\sqrt{3} = 2(\cos \pi/3 + i \sin \pi/3)$.

    (iii) $|1 - i\sqrt{3}| = 2$, so $\cos \theta = 1/2$, $\sin \theta = -\sqrt{3}/2$ whence $\theta = -\pi/3$ and $1 - i\sqrt{3} = 2(\cos(-\pi/3) + i \sin(-\pi/3)) = 2(\cos(\pi/3) - i \sin(\pi/3))$.

    (iv) When $(r, \theta) = (3, 3\pi/4)$, $z = 3(\cos 3\pi/4 + i \sin 3\pi/4) = 3(-1/\sqrt{2} + i/\sqrt{2})$.

It is very difficult to add or subtract complex numbers in polar form (the easiest way being to convert them into Cartesian form first!) but it is very easy to take conjugates or multiply or divide in polar notation. If $z = r(\cos \theta + i \sin \theta)$ then

$$\bar{z} = r(\cos \theta - i \sin \theta) = r(\cos(-\theta) + i \sin(-\theta)) \tag{2.7.24}$$

and from (2.7.19), $z^{-1} = \bar{z}/r^2$ so

$$z^{-1} = 1/r(\cos(-\theta) + i \sin(-\theta)). \tag{2.7.25}$$

This means that $|\bar{z}| = |z|$, $|z^{-1}| = 1/|z|$ and, since $-\pi < \theta < \pi$ implies $-\pi < -\theta < \pi$, we have arg $\bar{z} = $ arg $z^{-1} = -$arg z unless arg $z = \pi$. When arg $z = \pi$, or in other words $z = -r$ is a negative real number, $\bar{z} = -r$ and $z^{-1} = -1/r$ are also negative real numbers and arg $\bar{z} = $ arg $z^{-1} = $ arg $z = \pi$.

If $z_1 = r_1(\cos \theta_1 + i \sin \theta_1)$ and $z_2 = r_2(\cos \theta_2 + i \sin \theta_2)$ the product $z_1 z_2$ is given very simply by

$$z_1 z_2 = r_1 r_2(\cos(\theta_1 + \theta_2) + i \sin(\theta_1 + \theta_2)). \tag{2.7.26}$$

We already know that $|z_1 z_2| = r_1 r_2 = |z_1| \cdot |z_2|$; but $\arg z_1 z_2$ might not be given by $\theta_1 + \theta_2 = \arg z_1 + \arg z_2$ for $\theta_1 + \theta_2$ may not lie in the range (2.7.20) even though θ_1 and θ_2 both do. What we can say is that

$$\arg z_1 z_2 = \arg z_1 + \arg z_2, \qquad \arg z_1 + \arg z_2 - 2\pi \quad \text{or} \quad \arg z_1 + \arg z_2 + 2\pi,$$

depending on which of the three possibilities satisfies (2.7.20). We can also combine (2.7.25), (2.7.26) to deduce

$$z_1 z_2^{-1} = [r_1(\cos \theta_1 + i \sin \theta_1)][1/r_2(\cos (-\theta_2) + i \sin (-\theta_2))]$$

$$= r_1/r_2(\cos (\theta_1 - \theta_2) + i \sin (\theta_1 - \theta_2)). \tag{2.7.27}$$

EXAMPLES 2.7.16

   (i) Multiply the complex number $z = r(\cos \theta + i \sin \theta)$ by i. Using $i = \cos \pi/2 + i \sin \pi/2$ and (2.7.26) we have

$$iz = r(\cos (\theta + \pi/2) + i \sin (\theta + \pi/2)),$$

the result of which is to rotate z through a right angle as we have already seen.

  (ii) Multiply $z = r(\cos \theta + i \sin \theta)$ by $\cos \alpha + i \sin \alpha$. Here we have

$$r(\cos \theta + i \sin \theta)(\cos \alpha + i \sin \alpha) = r(\cos (\theta + \alpha) + i \sin (\theta + \alpha))$$

so the effect is to rotate z through an angle α.

 (iii) Simplify $(2 + 2i)/(1 - i\sqrt{3})$. Here we may either multiply both numerator and denominator by $1 + i\sqrt{3}$ and obtain $[(1 - \sqrt{3}) + (1 + \sqrt{3})i]/2 \simeq -0.366 + 1.366i$; or we may argue that

$$2 + 2i = 2\sqrt{2}(\cos (\pi/4) + i \sin (\pi/4)),$$

$$1 - i\sqrt{3} = 2(\cos (-\pi/3) + i \sin (-\pi/3)),$$

so $(2 + 2i)/(1 - i\sqrt{3}) = 2\sqrt{2}/2[\cos (\pi/4 + \pi/3) + i \sin (\pi/4 + \pi/3)] = \sqrt{2}(\cos 7\pi/12 + i \sin 7\pi/12) \simeq -0.366 + 1.366i$.

When $z_1 = z_2 = z = r(\cos \theta + i \sin \theta)$, the product formula (2.7.26) gives

$$z^2 = r^2(\cos 2\theta + i \sin 2\theta).$$

If we now multiply by z again we get

$$z^3 = r^3(\cos \theta + i \sin \theta)(\cos 2\theta + i \sin 2\theta)$$

$$= r^3(\cos 3\theta + i \sin 3\theta) \quad \text{from (2.7.26)}$$

and for any integer n,

$$z^n = [r(\cos \theta + i \sin \theta)]^n = r^n(\cos n\theta + i \sin n\theta). \tag{2.7.28}$$

This formula actually holds for n positive, negative or zero if as customary we interpret z^0 as 1 and z^{-n} as meaning $1/z^n$. A particularly important case is when $r = |z| = 1$ and (2.7.28) becomes

$$(\cos \theta + i \sin \theta)^n = \cos n\theta + i \sin n\theta. \tag{2.7.29}$$

This result is known as *de Moivre's Theorem*, after A. de Moivre (1667–1754).

EXAMPLE 2.7.17. Find the cube of $\frac{1}{2} + i\sqrt{3}/2$. We could of course evaluate $(\frac{1}{2} + i\sqrt{3}/2)^3$ by straightforward multiplication keeping the Cartesian coordinates. Instead we may write $\frac{1}{2} + i\sqrt{3}/2 = \cos \pi/3 + i \sin \pi/3$, and have from (2.7.29)

$$(\tfrac{1}{2} + i\sqrt{3}/2)^3 = (\cos \pi/3 + i \sin \pi/3)^3 = \cos \pi + i \sin \pi = -1.$$

A rather striking application of de Moivre's Theorem is that it can be used to get formulae, which only involve real numbers, for cosines and sines of multiple angles. For example let us deduce the known trigonometric results $\cos 2\theta = \cos^2 \theta - \sin^2 \theta$ and $\sin 2\theta = 2 \sin \theta \cos \theta$ for any angle θ. We use (2.7.29) with $n = 2$

$$(\cos \theta + i \sin \theta)^2 = \cos 2\theta + i \sin 2\theta,$$

and then expand the left-hand side by the binomial theorem [see (3.10.1)] to obtain

$$\cos^2 \theta + i^2 \sin^2 \theta + 2i \cos \theta \sin \theta = \cos 2\theta + i \sin 2\theta.$$

Equating real and imaginary parts we have $\cos^2 \theta - \sin^2 \theta = \cos 2\theta$, $2 \cos \theta \sin \theta = \sin 2\theta$. Similarly, by using de Moivre's result with $n = 3$

$$(\cos \theta + i \sin \theta)^3 = \cos 3\theta + i \sin 3\theta,$$

and expanding the left-hand side, we find, after equating real and imaginary parts, that

$$\cos 3\theta = \cos^3 \theta - 3 \cos \theta \sin^2 \theta$$

and

$$\sin 3\theta = 3 \cos^2 \theta \sin \theta - \sin^3 \theta.$$

2.7.4. Roots of Complex Numbers

We saw in equation (2.7.7) that every complex number has square roots. The formula (2.7.28) tells us that each complex number has nth roots for any natural number n. If $\alpha = r(\cos \theta + i \sin \theta)$, then $w = r^{1/n}(\cos \theta/n + i \sin \theta/n)$ is an nth root of α, because from (2.7.28)

$$w^n = (r^{1/n})^n(\cos n.\theta/n + i \sin n\theta/n) = r(\cos \theta + i \sin \theta) = \alpha. \quad (2.7.30)$$

Actually each α has n nth roots (two square roots, three cube roots, and so on) and with the above notation they are all of the form $\omega_0 w, \omega_1 w, \ldots, \omega_{n-1} w$ where each ω_k satisfies

$$\omega_k^n = 1. \quad (2.7.31)$$

In other words $\omega_0, \ldots, \omega_{n-1}$ are the n *nth roots of* 1. These nth roots of 1 (or nth roots of unity) can be given explicitly by the formulae

$$\omega_k = \cos 2\pi k/n + i \sin 2\pi k/n, \quad \text{for } k = 0, 1, \ldots, n - 1. \quad (2.7.32)$$

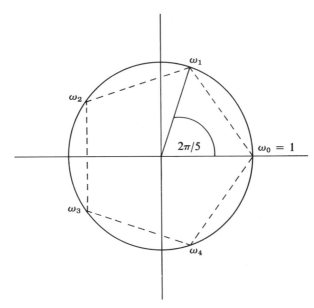

Figure 2.7.11

It is particularly interesting to plot their positions in the Argand diagram, and this has been done in Figure 2.7.11 for $n = 5$. Each of the nth roots of 1 has modulus 1 and so they all lie on the circle centre the origin and radius 1. Their most remarkable property is that they form the vertices of a regular n-gon [see V, § 1.1.4] inscribed in this circle. One vertex is always at $1 = \omega_0$ and the angle between adjacent vertices is $2\pi/n$.

EXAMPLE 2.7.18.   Find the three cube roots of 1. From (2.7.32) the cube roots are

$$\omega_0 = \cos 0 + i \sin 0 = 1,$$

$$\omega_1 = \cos 2\pi/3 + i \sin 2\pi/3 = -\tfrac{1}{2} + i\sqrt{3}/2,$$

$$\omega_2 = \cos 4\pi/3 + i \sin 4\pi/3 = -\tfrac{1}{2} - i\sqrt{3}/2.$$

In the above example we note that $\omega_2 = \omega_1^2$. This is a particular case of the fact that for an arbitrary n

$$\omega_1^k = (\cos 2\pi/n + i \sin 2\pi/n)^k = \cos 2\pi k/n + i \sin 2\pi k/n = \omega_k. \qquad (2.7.33)$$

If we put $\omega = \omega_1$ we may express all the nth roots of 1 as $1, \omega, \omega^2, \ldots, \omega^{n-1}$ and because of this property ω is called a *primitive nth root*. The nth roots of the complex number α in (2.7.30) can now be written as $w, \omega w, \ldots, \omega^{n-1}w$.

We can now solve all quadratic equations with complex coefficients and certain others of the form $z^n - \alpha = 0$. It is a famous theorem, 'The Fundamental Theorem of Algebra' [see Theorem 14.5.2], that any polynomial equation, $f(x) = 0$, (even one with complex coefficients) has at least one complex root:

that is, a complex number z (which may be real) such that $f(z) = 0$. A consequence is that all roots of such an equation are complex. This fact alone, that
complex numbers suffice to provide solutions for all polynomial equations, would
make them very important. As we have seen, they also have very close connections with geometry, and this makes their further properties very useful in
problems which can be posed geometrically.

2.8. QUATERNIONS

Since complex numbers (pairs of real numbers) can be used so successfully to
deal with the geometry of the plane, mathematicians tried to find a similar algebra
for three dimensional vectors (triples of real numbers) which would do the same
thing for the geometry of space. This is not possible for triples but it is possible
for quadruples of reals—or what is the same thing, pairs of complex numbers.
Unfortunately as we shall see below we are forced to give up yet one more of the
properties which make life so (relatively) straightforward in the case of real and
rational numbers. Pairs (α, β) of complex numbers α, β are called *quaternions* and,
mirroring the role of i in the complex case, it is convenient to introduce a new
place symbol j so that we can write (α, β) as $\alpha + \beta j$. If $X = \alpha + \beta j$ and $Y = \gamma + \delta j$ are two quaternions then we define their sum and product by

$$X + Y = (\alpha + \gamma) + (\beta + \delta)j, \qquad (2.8.1)$$

and

$$X . Y = (\alpha\gamma - \beta\bar{\delta}) + (\alpha\delta + \beta\bar{\gamma})j, \qquad (2.8.2)$$

where the bar above denotes complex conjugation. If

$$\alpha = x_1 + ix_2, \qquad \beta = x_3 + ix_4, \qquad \gamma = y_1 + iy_2, \qquad \delta = y_3 + iy_4$$

then

$$X = x_1 + ix_2 + jx_3 + kx_4, \qquad Y = y_1 + iy_2 + jy_3 + ky_4,$$

where $k = ij$. This means that the only things which differ from one quaternion
to another are the real coefficients—the x's and y's. So we could just as well
regard quaternions as being quadruples of real numbers: $X = (x_1, x_2, x_3, x_4)$,
$Y = (y_1, y_2, y_3, y_4)$. From this viewpoint the particular quaternions $(0, 1, 0, 0)$,
$(0, 0, 1, 0)$ and $(0, 0, 0, 1)$ represent i, j and $k = ij$ respectively. Also (2.8.1) and
(2.8.2) become

$$X + Y = (x_1, x_2, x_3, x_4) + (y_1, y_2, y_3, y_4)$$
$$= (x_1 + y_1, x_2 + y_2, x_3 + y_3, x_4 + y_4), \qquad (2.8.3)$$

and

$$X . Y = (x_1, x_2, x_3, x_4) . (y_1, y_2, y_3, y_4)$$
$$= (x_1 y_1 - x_2 y_2 - x_3 y_3 - x_4 y_4, x_1 y_2 + x_2 y_1 + x_3 y_4 - x_4 y_3,$$
$$x_1 y_3 - x_2 y_4 + x_3 y_1 + x_4 y_2, x_1 y_4 + x_2 y_3 - x_3 y_2 + x_4 y_1).$$
$$(2.8.4)$$

For instance the first coordinate of the right-hand side of (2.8.4) is just the real part of $\alpha\gamma - \beta\delta$; but (2.8.2) is much easier to write, and to remember, than (2.8.4) so we shall continue to use the notation of (2.8.1) and (2.8.2). The set of all quaternions is denoted* by $\mathbb{H}$, and it satisfies all the field axioms $F1$–$F4$ of section 6, *except* for the commutativity of multiplication. This can be seen by taking $\alpha = \delta = 0$, $\beta = 1$ and $\gamma = i$ in (2.8.2), when $X = j$, $Y = i$, and (2.8.2) implies

$$ji = -ij. \tag{2.8.5}$$

So we can add, subtract and multiply quaternions, and divide by non-zero ones, just as with real or complex numbers, but we have to be careful when writing the product of two quaternions A and B since AB may not be the same as BA. However, if b is real, we always have $Ab = bA$.

If we take $i = (0, 1, 0, 0)$, $j = (0, 0, 1, 0)$, $k = (0, 0, 0, 1)$ as before, we could write the quaternion $X = (x_1, x_2, x_3, x_4)$ as $x_1 + x_2 i + x_3 j + x_4 k$ and we have the relationships

$$i^2 = j^2 = k^2 = -1 \tag{2.8.6}$$

and

$$ij = -ji = k, \quad jk = -kj = i, \quad ki = -ik = j. \tag{2.8.7}$$

We could then, if desired, use just (2.8.6) and (2.8.7) together with the distributive laws to find any product $(x_1 + x_2 i + x_3 j + x_4 k)(y_1 + y_2 i + y_3 j + y_4 k)$.

The quaternion $\bar{X} = x_1 - x_2 i - x_3 j - x_4 k$ is called the *conjugate quaternion* to $X = x_1 + x_2 i + x_3 j + x_4 k$, and as in (2.7.10) the product $X\bar{X}$ is always real, for we have

$$X\bar{X} = x_1^2 + x_2^2 + x_3^2 + x_4^2. \tag{2.8.8}$$

For two quaternions X, Y we also have a statement analogous to (2.7.11):

$$\overline{X + Y} = \bar{X} + \bar{Y} \quad \text{and} \quad \overline{XY} = \bar{Y}\bar{X}. \tag{2.8.9}$$

EXAMPLE 2.8.1.   We could not write $\overline{XY} = \bar{X}\bar{Y}$ always, instead of as in (2.8.9), because consider $X = j$, $Y = i$. Then $XY = -ij$, from (2.8.5), so that $\bar{X}\bar{Y} = +ij$; but $\bar{X} = -j$, $\bar{Y} = -i$ whence $\bar{X}\bar{Y} = ji = -\overline{XY}$.

We can now use our knowledge about quaternions to deduce a relationship to do with real numbers alone. For if X and Y are as in (2.8.3) and (2.8.4) then

$$(x_1^2 + x_2^2 + x_3^2 + x_4^2)(y_1^2 + y_2^2 + y_3^2 + y_4^2)$$

$$= X\bar{X}(y_1^2 + y_2^2 + y_3^2 + y_4^2)$$

$$= X(y_1^2 + y_2^2 + y_3^2 + y_4^2)\bar{X}$$

$$= X(Y\bar{Y})\bar{X}$$

$$= (XY)(\bar{Y}\bar{X})$$

$$= (XY)(\overline{XY}) \quad \text{from (2.8.9)}$$

* After W. R. Hamilton (1805–65) who first discovered them.

Equation (2.8.8) tells us that the product of a quaternion and its conjugate is the sum of the squares of its coordinates. So $(XY)\,(\overline{XY})$ will be the sum of the squares of the coordinates of XY, and these are displayed in (2.8.4). Therefore for any real numbers $x_1,\ldots,x_4,y_1,\ldots,y_4$, the product $(x_1^2 + x_2^2 + x_3^2 + x_4^2)$. $(y_1^2 + y_2^2 + y_3^2 + y_4^2)$ is always again a sum of four squares, namely the squares of the expressions in (2.8.4).

Quaternions are not so widely used as the other number systems we have been considering, but they show once again that when we are looking at numbers in one class it is often helpful to be able to regard them as special cases of numbers of a more general kind.

T.H.J.

BIBLIOGRAPHY

Cohen, L. W. and Ehrlich, G. (1977). *The Structure of the Real Number System* (rev. ed.), Krieger.

Ledermann, W. (1967). *Complex Numbers*, Routledge & Kegan Paul.

Parker, F. D. (1966). *The Structure of Number Systems*, Prentice-Hall.

CHAPTER 3

Arithmetic

In this chapter we shall first concentrate our attention on powers of real numbers [see § 2.6.1] and logarithms. This will entail defining a^b for real numbers a, b (where $a > 0$) beginning with a positive integer b [see § 2.2.1] and then considering more general cases. If $a^b = c$, then b is logarithm to the base a of c (written $b = \log_a c$) and we shall deduce the usual properties of logarithms. The second part of the chapter will consider combinatorial properties of permutations and combinations.

3.1. POWERS

For any real number a we define

$$a^2 = a \times a,$$

$$a^3 = a \times a \times a,$$

and for any positive integer n, we define

$$a^n = \underbrace{a \times \ldots \times a}_{n \text{ factors}}.$$

This can be put in a more refined way by stating

$$a^1 = a$$

and, having found a^n, then the next power a^{n+1} is given by

$$a^{n+1} = a^n a.$$

By counting up the number of factors, clearly

$$a^{m+n} = a^m a^n \tag{3.1.1}$$

for positive integers m, n. (A mathematician with an eye on neatness and a little time on his hands might prove this by induction on n [see § 2.1, NA5]. For $n = 1$ it is the basic definition, and supposing $a^{m+n} = a^m a^n$ is proved for some

79

value of $n = k$, we can deduce it for $n = k + 1$ by taking $a^{m+k} = a^m a^k$, multiplying by a to get $a^{m+k}a = a^m a^k a$, whence $a^{m+k+1} = a^m a^{k+1}$.)

In the same way we also have:

$$(a^m)^n = a^{mn} \tag{3.1.2}$$

and

$$(ab)^m = a^m b^m. \tag{3.1.3}$$

If $a \neq 0$, then

$$(-a)^2 = (-1)^2 a^2 = a^2 > 0 \tag{3.1.4}$$

and so the square of a non-zero real number a is always positive. The cube a^3, however, has the same sign as a. In general even powers of a non-zero real number a are positive, whilst odd powers of a have the same sign as a.

In the expression a^n we call a the *base* and n the *exponent* or *power*. We may ask what happens to the order relation between real numbers [see § 2.6.2] on taking powers. To simplify matters we will restrict ourselves to a *positive* base a (the case of negative a follows easily if required by noting that $(-a)^{2n} = a^{2n}$ and $(-a)^{2n+1} = -a^{2n+1}$). First, varying the base and fixing the power we find

$$a < b \quad \text{if and only if} \quad a^n < b^n \quad (a, b > 0). \tag{3.1.5}$$

This follows because a product of n positive factors a is less than the product of n positive factors b if and only if a is less than b.

It means that for a fixed positive integer n the function $f(x) = x^n$ is increasing for $x > 0$ [see Figure 3.1.1 and IV, Definition 2.7.1]. (For $x < 0$ the function increases when n is odd and decreases when n is even.)

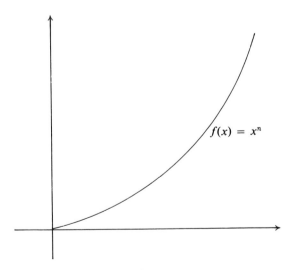

$$f(x) = x^n$$

Figure 3.1.1

Now we fix the base a and vary the exponent. When $0 < a < 1$, then multiplying the inequality by a (which is positive) we get

$$0 < a^2 < a.$$

Similarly multiplying again by a

$$0 < a^3 < a^2,$$

so in general

$$0 < a < 1 \quad \text{implies} \quad 0 < a^n < \ldots < a^3 < a^2 < a < 1. \qquad (3.1.6)$$

Similarly for $1 < a$, multiplying by a, we find

$$a < a^2$$

so

$$1 < a \quad \text{implies} \quad a < a^2 < a^3 < \ldots < a^n < \ldots. \qquad (3.1.7)$$

In particular, from (3.1.6), (3.1.7) we deduce that if $a > 0$ and $a \neq 1$, then

$$m \neq n \quad \text{implies} \quad a^m \neq a^n. \qquad (3.1.8)$$

3.2. INVERSES

Suppose that we try to define a^k for $k = 0$ or a negative integer k, in such a way that the basic rules (3.1.1), (3.1.2), (3.1.3) continue to hold. Then

$$a^0 a^n = a^{0+n} = a^n.$$

For $a \neq 0$ we have $a^n \neq 0$, so dividing through by a^n we must have

$$a^0 = 1 \quad (a \neq 0). \qquad (3.2.1)$$

We may just as easily deduce that if we wish to define a^k for k negative, say $k = -n$ where n is positive, then using (3.1.1) we must have

$$a^{-n} a^n = a^{(-n)+n} = a^0 = 1, \quad (a \neq 0)$$

so

$$a^{-n} = 1/a^n. \qquad (3.2.2)$$

Hence

$$a^{-n} = 1/\underbrace{(a \times \ldots \times a)}_{n \text{ factors}}.$$

By cancellation we find

$$a^m a^{-n} = a^m/a^n = a^{m-n}$$

for all positive integers m, n. Using this result we can soon extend equation (3.1.1)

$$a^{m+n} = a^m a^n$$

to all integers, positive, negative or zero. It is known true for $m, n > 0$ and is clearly true if either or both is zero. Suppose $m > 0$, $n < 0$, then write $n = -r$ where $r > 0$; we have

$$a^{m+n} = a^{m-r} = a^m a^{-r} = a^m a^n.$$

Finally if $m < 0$, $n < 0$, writing $m = -s$, $n = -r$, then

$$a^{m+n} = a^{-r-s} = 1/a^{r+s} = 1/a^r a^s$$

$$= (1/a^r)(1/a^s) = a^{-r} a^{-s}$$

$$= a^m a^n.$$

This shows (3.1.1) holds for all integers and similar manipulations show (3.1.2), (3.1.3) remain true for all integers as well.

3.3.   ROOTS

If $c^n = a$ where a, c are real and n is a positive integer we say that c is an *nth root* of a. For $n = 2, 3$ we also use the terms *square root, cube root*. For instance $2^2 = (-2)^2 = 4$, so $2, -2$ are both square roots of 4. Now for a non-zero real number a we always have $a^2 = (-a)^2$ is positive. So only positive numbers have (real) square roots. (Of course if we allow complex numbers, then negative numbers have *complex* square roots [see § 2.7.1], but for this chapter our discussion is restricted to real numbers.) On the other hand, if $c^3 = a$, then $(-c)^3 = -a$, so the cube root of $-a$ is $-c$.

The same phenomena occur with *n*th roots in general, depending whether n

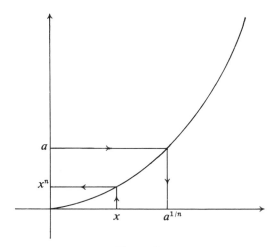

Figure 3.3.1

is even or odd. It is helpful initially to restrict our discussion to nth roots of *positive* numbers. First we demonstrate that every positive real number has a unique nth root. This is easiest done graphically by considering the graph of $f(x) = x^n$ ($x \geq 0$) (shown in Figure 3.3.1). The function $f(x) = x^n$ is strictly increasing for $x \geq 0$, so, just as we can read off nth powers from the graph in the usual way, we can reverse the procedure to determine $a^{1/n}$ for any $a > 0$. (Of course, $0^{1/n} = 0$.) Each positive real number a has a *unique* real positive nth root. For $a > 0$, and a positive integer n we define

$$\sqrt[n]{a} = c \quad \text{if} \quad c^n = a \quad \text{and} \quad c > 0. \tag{3.3.1}$$

We call $\sqrt[n]{a}$ the (*positive*) *nth root* of a. (If n is *even* then a also has a negative nth root $-\sqrt[n]{a}$ and $-a$ has no nth root. If n is *odd* then $\sqrt[n]{a}$ is the only real nth root of a and $\sqrt[n]{-a} = -\sqrt[n]{a}$.) The square root $\sqrt[2]{a}$ is usually abbreviated to $\sqrt{a}$.

Continuing from $\sqrt[n]{a} = c$, we can go on to extract an mth root of c to find

$$\sqrt[m]{c} = d \quad \text{if} \quad d^m = c \quad \text{and} \quad d > 0.$$

Hence

$$d^{mn} = c^n = a,$$

which means $d = \sqrt[mn]{a}$, and substituting $d = \sqrt[m]{c} = \sqrt[m]{\sqrt[n]{a}}$, we get

$$\sqrt[m]{\sqrt[n]{a}} = \sqrt[mn]{a}. \tag{3.3.2}$$

EXAMPLE 3.3.1. $\sqrt[2]{\sqrt[3]{64}} = \sqrt[2]{4} = 2 = \sqrt[6]{64}$.

Replacing a by a^{rn} in (3.3.2) we get

$$\sqrt[mn]{a^{rn}} = \sqrt[m]{\sqrt[n]{a^{rn}}} = \sqrt[m]{a^r}. \tag{3.3.3}$$

If $c^n = a$, $d^n = b$, then $c = \sqrt[n]{a}$, $d = \sqrt[n]{b}$ and from

$$ab = c^n d^n = (cd)^n$$

we get

$$\sqrt[n]{ab} = cd.$$

Hence

$$\sqrt[n]{ab} = \sqrt[n]{a}\sqrt[n]{b}. \tag{3.3.4}$$

EXAMPLE 3.3.2. $\sqrt{12}\sqrt{3} = \sqrt{36} = 6$.

Equation (3.3.4) can be extended to a product of r positive terms by induction on r [see § 2.1, NA5] to get

$$\sqrt[n]{(a_1 a_2 \ldots a_r)} = \sqrt[n]{a_1}\sqrt[n]{a_2}\ldots\sqrt[n]{a_r}. \tag{3.3.5}$$

We now have enough technique to describe a^k for any rational exponent $k = m/n$ where m, n are integers and $n > 0$ [see § 2.4.1].

Suppose $a^{m/n} = c$ and the basic rules (3.1.1), (3.1.2), (3.1.3) continue to hold for rational exponents, then using (3.1.2) we have

$$c^n = (a^{m/n})^n = a^m$$

so c is the nth root of a^m, i.e.

$$a^{m/n} = \sqrt[n]{(a^m)}. \tag{3.3.6}$$

If we write

$$m/n = (rm)/(rn) \quad (r > 0),$$

then we get

$$a^{(rm)/(rn)} = \sqrt[rn]{(a^{rm})}$$

$$= \sqrt[n]{(a^m)}$$

by (3.3.3) on interchanging (r, m, n) with (m, n, r),

$$= a^{m/n}.$$

Thus the definition of $a^{m/n}$ is independent of the way we write m/n (provided $n > 0$).

We can rewrite $\sqrt[n]{(a^m)}$ for $m > 0$ by using (3.3.5) with $r = m$, $a_1 = \ldots = a_m = a$, whence

$$\sqrt[n]{(a^m)} = (\sqrt[n]{a})^m.$$

This clearly holds for $m = 0$; for $m < 0$, say $m = -r$, then

$$\sqrt[n]{(a^m)} = \sqrt[n]{(a^{-r})} = \sqrt[n]{(1/a^r)} = 1/\sqrt[n]{(a^r)}$$

$$= 1/(\sqrt[n]{a})^r = (\sqrt[n]{a})^{-r} = (\sqrt[n]{a})^m.$$

Thus for all integers m, n where n is positive, we have

$$a^{m/n} = \sqrt[n]{(a^m)} = (\sqrt[n]{a})^m. \tag{3.3.7}$$

It is now a simple matter to check that with this definition of a^k for rational k, then the basic rules (3.1.1), (3.1.2), (3.1.3) hold for all *rational* values of m, n. For instance, in (3.1.1), if r, s, m, n are integers and s, n are positive, then

$$a^{r/s + m/n} = a^{(rn + ms)/sn}$$

$$= \sqrt[sn]{(a^{rn + ms})}$$

$$= \sqrt[sn]{(a^{rn} a^{ms})} \qquad \text{by (3.1.1) for integers}$$

$$= (\sqrt[sn]{(a^{rn})})(\sqrt[sn]{(a^{ms})}) \quad \text{by (3.3.4)}$$

$$= \sqrt[s]{(a^r)}\sqrt[n]{(a^m)} \qquad \text{by (3.3.3)}$$

$$= a^{r/s} a^{m/n}.$$

Cases (3.1.2), (3.1.3) are equally straightforward.

It is also important to check inequalities between positive real numbers raised to a rational exponent. First by putting $a = c^{1/n}$, $b = d^{1/n}$ in (3.1.5) we deduce for positive c, d that

$$c^{1/n} < d^{1/n} \text{ if and only if } c < d.$$

Hence replacing c, d by a, b, we find $a < b$ implies $a^{1/n} < b^{1/n}$ and by (3.1.5), for $m > 0$ this in turn implies

$$(a^{1/n})^m < (b^{1/n})^m.$$

Thus for real numbers a, $b > 0$ and a positive rational number $k = m/n$, we find

$$0 < a < b \quad \text{implies} \quad 0 < a^k < b^k. \tag{3.3.8}$$

This implies that for a positive rational k the function $f(x) = x^k$ $(x > 0)$ is an increasing function.

If we fix the base a and vary the exponent k we also get results corresponding to (3.1.6), (3.1.7). For instance if $a > 1$, and k, s are rationals with $k > s$, then $k - s$ is positive and (3.3.8) gives

$$a^{k-s} > 1^{k-s} = 1,$$

whence, multiplying by a^s, we get

$$a^k > a^s.$$

Thus

$$a > 1, \quad k > s \quad \text{implies} \quad a^k > a^s. \tag{3.3.9}$$

A similar calculation shows that

$$0 < a < 1, \quad k > s \quad \text{implies} \quad a^k < a^s. \tag{3.3.10}$$

3.4. DEFINITION OF a^b FOR REAL $a > 0$ AND ANY REAL b

Suppose we wish to give a meaning to an expression like 3^π where the exponent π is not rational [see IV, § 2.12]. We can consider the decimal expansion

$$\pi = 3\cdot14159\ldots$$

and take an approximation to a certain number of decimal places, say $3\cdot14159 = 314159/100000$, then the latter is rational [see § 2.6.1], so

$$\alpha = 3^{314159/100000}$$

is defined as in the last section. Of course this does not help us to *compute* α, but fortunately hand calculators are available to compute a^b where a, b are given to a certain number of decimal places. Using such a calculator, the decimal expansion of $3^{3\cdot14159}$ to the first five places is $31\cdot54418$.

Now if $k = 3\cdot14159$ we have

$$k < \pi < k + (1/10^5).$$

The expression 3^x is strictly increasing for *rational* x by (3.3.8), so if we wish it to be strictly increasing for all x then we would require [see IV, Definition 2.7.1],

$$3^k < 3^\pi < 3^{k+(1/10^5)}.$$

Using a hand calculator,

$$3^k = 31{\cdot}54418\ldots$$

$$3^{k+(1/10^5)} = 31{\cdot}54453\ldots$$

to five places where neither expansion has the last digit rounded up. Hence

$$31{\cdot}54418 < 3^\pi < 31{\cdot}54454$$

taking into account the rounding error. In this way we can handle a^b for any real $a > 0$ and any real b. We simply write $b = b_0{\cdot}b_1 b_2 \ldots b_n \ldots$ as a decimal expansion and sandwich a^b between $a^{b_0 \cdot b_1 \cdots b_n}$ and $a^{b_0 \cdot b_1 \cdots b_n + (1/10^n)}$. By taking a sufficient number of places in the decimal expansion of b we will end up with as close an approximation to a^b as desired. For practical purposes an 8 digit calculator gives a more than adequate approximation. (Interestingly an 8 digit approximation to a real number is *rational* [see § 2.6.1], so for practical purposes we only need to calculate a^b where a, b are both rational anyway!)

For any real number $a > 1$ we therefore can define a^b for any real b such that

$$a > 1, b < c \quad \text{implies} \quad a^b < a^c. \tag{3.4.1}$$

3.5.   THE FUNCTION $f(x) = a^x$ FOR $a > 0$

Using the theory of § 3.4 we can define a function $f(x) = a^x$ for all real x [see § 1.4.1] so that, for $a > 1$, f is strictly increasing. We note that a^x increases without limit, for if $a > 1$, let $a = 1 + b$, then

$$a^2 = 1 + 2b + b^2 > 1 + 2b$$

and by induction on n [see § 2.1, NA5]

$$a^n > 1 + nb.$$

Given a real number K, we can be certain a^x exceeds K by taking N such that $N > (K - 1)/b$, then $x > N$ implies $a^x > a^N$ by (3.4.1), and

$$a^N > 1 + Nb > 1 + \frac{(K - 1)b}{b} = K.$$

Hence

$$x > N \quad \text{implies} \quad a^x > K.$$

If $K > 0$, then

$$a^x > K$$

implies

$$a^{-x} < 1/K,$$

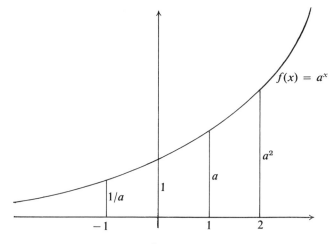

Figure 3.5.1

so a^{-x} takes arbitrarily small positive values and by the intermediate value theorem [see IV, Theorem 2.6.1], $f(x) = a^x$ takes on all values between. In set-theoretic notation [see Definition 1.4.1] this means that $f(x) = a^x$ (Figure 3.5.1) gives a strictly increasing function

$$f: \mathbb{R} \to \mathbb{R}^+$$

where $\mathbb{R}^+$ is the set of positive reals and f takes on all positive real values.

If $a = 1$, then $a^x = 1$ for all x and if $0 < a < 1$ then we find

$$f(x) = a^x \quad (0 < a < 1)$$

is strictly *decreasing* [see IV, Definition 2.7.1] taking on all positive real values. (Just put $a = 1/b$, then $a^x = 1/b^x$ where $b > 1$ and b^x takes on all positive real values.)

The reader may wish to sketch his own picture in this case.

3.6. LOGARITHMS

If a, b are real numbers where $a > 0$ and

$$a^b = c$$

then b is called the *logarithm* to base a of c, and we write

$$b = \log_a c.$$

EXAMPLE 3.6.1. $2^3 = 8$, so $\log_2 8 = 3$.

If $a \neq 1$, every positive number c has a unique logarithm to base a. We shall consider the case $a > 1$. Here $f: \mathbb{R} \to \mathbb{R}^+$ given by $f(x) = a^x$ is strictly increasing, taking on all real values (see previous section).

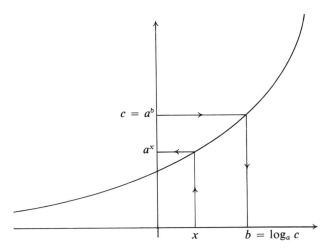

Figure 3.6.1

The inverse function [see § 1.4.2]

$$\log_a : \mathbb{R}^+ \to \mathbb{R}$$

is given by

$$\log_a c = b \quad \text{if and only if} \quad f(b) = a^b = c.$$

It can be obtained graphically by reading the graph of $f(x) = a^x$ in reverse (see

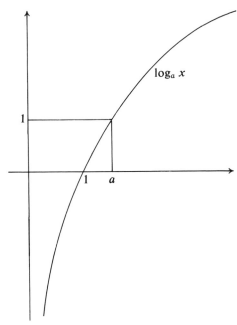

Figure 3.6.2

Figure 3.6.1). This gives the graph of the logarithm (Figure 3.6.2) by inter-changing axes [see V, § 3.4.5]. We have [see also IV, (2.11.8)]:

$$\log_a c = b \quad \text{if and only if} \quad a^b = c. \tag{3.6.1}$$

If

$$a^{b_1} = c_1, \qquad a^{b_2} = c_2,$$

then

$$c_1 c_2 = a^{b_1} a^{b_2} = a^{b_1 + b_2},$$

so

$$\log_a c_1 c_2 = b_1 + b_2.$$

Hence we get the fundamental rule for logarithms:

$$\log_a c_1 c_2 = \log_a c_1 + \log_a c_2. \tag{3.6.2}$$

In the same way, if

$$a^b = c \quad \text{or} \quad b = \log_a c,$$

then

$$a^{bk} = c^k,$$

so

$$\log_a (c^k) = bk = k \log_a c,$$

giving the second rule for logarithms:

$$\log_a (c^k) = k \log_a c. \tag{3.6.3}$$

Finally we have a remarkably simple rule for changing one base to another: suppose $c^k = x$, then $k = \log_c x$ and (3.6.3) becomes

$$\log_a x = \log_c x \log_a c. \tag{3.6.4}$$

This is most useful for computing logarithms to various bases. For instance two useful bases are $a = 10$ and $c = e$ where

$$e = 1 + 1/1! + 1/2! + \ldots + 1/n! + \ldots$$
$$= 2 \cdot 718281828 \ldots$$

[see IV, (2.11.3)]. To transform from one to the other we have

$$\log_{10} x = \log_e x \log_{10} e$$

and, using a hand calculator,

$$\log_{10} e = 0 \cdot 434294481 \ldots.$$

Knowing $\log_e x$, we can therefore find $\log_{10} x$ by multiplying by $0 \cdot 434294481 \ldots.$
Putting $x = a$ in (3.6.4) gives

$$1 = \log_a a = \log_c a \log_a c.$$

Hence

$$\log_c a = 1/(\log_a c). \tag{3.6.5}$$

This last computation can be illustrated by taking $a = 2$, $b = 16$, then $2^4 = 16$, so

$$\log_2 16 = 4$$

whilst

$$16^{1/4} = 2,$$

so

$$\log_{16} 2 = 1/4.$$

3.7. PERMUTATIONS

We now turn to combinatorial problems. First we use subtle counting methods to specify the number of different ways choices may be made in a specified order. For instance given a choice of 4 different routes from town A to town B and 3 routes from town B to town C, we may wish to compute the number of different routes from A to C. It is clear that for each of the 4 routes from A to B there are 3 from B to C, so the total number from A to C is

$$4 \times 3 = 12.$$

More generally, if we make two successive choices, the first in r_1 ways, the second in r_2 ways, then the number of different ways of making the two choices is $r_1 \times r_2$. This rule holds good even if the first choice affects the objects left for the second choice, provided that having made the first choice there are always r_2 ways of making the second. For instance if we are making two selections in order from, say, five objects A, B, C, D, E, then having made the first choice, in each case a different set of objects is left, but there always remain four objects from which to make the second choice. The total number of different ways of choosing the first then the second is then

$$5 \times 4 = 20.$$

In a simple case like this we can enumerate the various ways to check the calculation. If the first choice is A, the second may be B, C, D or E, giving the choices AB, AC, AD, AE. Likewise the choices with B first are BA, BC, BD, BE; with C first: CA, CB, CD, CE; with D first: DA, DB, DC, DE: with E first: EA, EB, EC, ED. These make 5 lots of 4, or 20 in all, as expected.

The same method generalizes to a succession of choices to give: *The Fundamental Principle of Choice*: If a first choice may be made in r_1 ways, then for each of these ways a second choice may be made in r_2 ways, then a third in r_3 ways and so on, up to an mth choice being made in r_m ways, then the total number of ways the choices may be made in the given order is

$$r_1 r_2 r_3 \ldots r_m.$$

This principle is useful in cases where the numbers involved are so large we would never dream of enumerating all the cases:

EXAMPLE 3.7.1. Five teams consisting of 7, 11, 10, 21 and 8 members respectively must offer one member from each team for a competition. The number of different ways in which these members may be selected is

$$7 \times 11 \times 10 \times 21 \times 8 = 129{,}360.$$

A special case of the fundamental principle occurs when we make successive choices from a single set of n objects. The first choice may be made in n ways, and for each of these ways there are $n - 1$ for the second choice, then $n - 2$ for the third, and so on. We find the number of ways we can select r objects, in order, from n objects is

$$n(n - 1)(n - 2)\ldots(n - r + 1).$$

It must be emphasized that this computation is made on the assumption that the elements are taken in a specific order. A set of objects taken in a specific order is called a *permutation* of that set [see § 8.1.1]. For instance AB and BA are two different permutations of the letters A, B.

We can simplify the notation being used (though not the computation) by using the factorial symbol '!':

$$n! = 1 \times 2 \times 3 \times \ldots \times n. \tag{3.7.1}$$

This is read as 'n factorial', and the first few factorials are:

$$1! = 1$$
$$2! = 1 \times 2 = 2$$
$$3! = 1 \times 2 \times 3 = 6$$
$$4! = 1 \times 2 \times 3 \times 4 = 24$$
$$5! = 1 \times 2 \times 3 \times 4 \times 5 = 120$$
$$6! = 1 \times 2 \times 3 \times 4 \times 5 \times 6 = 720.$$

Using this notation we find

$$n(n - 1)(n - 2)\ldots(n - r + 1) = n!/(n - r)!$$

This expression is usually denoted by $_nP_r$. We have:

The Permutation Principle: The total number of permutations of r objects selected from n distinct objects is

$$_nP_r = n!/(n - r)!. \tag{3.7.2}$$

An interesting special case arises when $n = r$. Using the original version of the formula, we find

$$_nP_r = n(n - 1)(n - 2)\ldots(n - r + 1),$$

so

$$_nP_n = n(n - 1)(n - 2)\ldots1$$
$$= n!.$$

According to the permutation principle we have

$$_nP_n = \frac{n!}{(n-n)!} = \frac{n!}{0!}.$$

By defining

$$0! = 1,$$

we find the formula suits this particular case as well. In fact it is worthy of separate mention:

The Order Principle: The number of different permutations of n objects is $n!$.

3.8. COMBINATIONS

In contrast to a permutation which involves a specified order of a set of objects, a *combination* is a collection of objects in no specified order. For instance three objects A, B, C have $3! = 6$ different permutations which we may enumerate:

ABC, ACB, BAC, BCA, CAB, CBA.

These are all regarded as the same combination.

Matters become interesting when we select r objects from n objects, now without regard to the order of selection. For instance we may wish to choose a team of three competitors out of five possible candidates A, B, C, D, E, it being understood that the order of the team does not matter.

To compute the possible number of teams, let us first of all select them *in order*. According to the permutation principle there are

$$_5P_3 = 5!/2! = 60$$

different permutations. However, amongst these permutations are, for example,

ABC, ACB, BAC, BCA, CAB, CBA

which are the $3! = 6$ different permutations of the single combination A, B, C. In the same way we get 6 different permutations of A, B, D, and so on. We simply partition the 60 permutations into sublists where each sublist contains 6 different permutations of the same three competitors. There are then

$$60 \div 6 = 10$$

distinct sublists, each yielding a different combination. In this simple case we can write down the 10 combinations:

$$A, B, C; \ A, B, D; \ A, B, E; \ A, C, D; \ A, C, E; \ A, D, E;$$

$$B, C, D; \ B, C, E; \ B, D, E; \ C, D, E.$$

However it does not require a much more complicated example before enumeration of the specific combinations becomes impractical. Fortunately the general principles outlined in this example still hold good.

If we wish to find the number of possible combinations of r objects selected from n distinct objects, we first note that there are

$$n!/(n - r)!$$

different permutations. We imagine that the complete list of permutations is partitioned into sublists, where each sublist enumerates all the different possible permutations of some particular choice of r objects. Each of these sublists will contain $r!$ different permutations. The total number of sublists, and hence the total number of combinations, is

$$\frac{n!}{(n - r)!} \div r! = \frac{n!}{(n - r)!\, r!}.$$

This number is usually denoted by $_nC_r$ or $\binom{n}{r}$. We therefore have:

The Combination Principle: The total number of different combinations of r objects selected from n distinct objects is

$$_nC_r = \frac{n!}{(n - r)!\, r!}. \tag{3.8.1}$$

EXAMPLE 3.8.1. The number of different poker hands (five cards) selected from a pack of 52 cards is:

$$_{52}C_5 = \frac{52!}{47!\, 5!} = \frac{52 \times 51 \times 50 \times 49 \times 48}{2 \times 3 \times 4 \times 5} = 2{,}598{,}960.$$

3.9. PARTITIONS AND MORE GENERAL COMPUTATIONS

The methods of the previous sections may be extended to further cases with a little ingenuity. For instance we may consider the number of ways of partitioning a set of n distinct objects into subsets where the first subset contains r_1 objects, the second r_2, and so on, with the last subset containing r_k where

$$r_1 + r_2 + \ldots + r_k = n.$$

The first subset involves the choice of r_1 objects from n objects, which may be performed in

$$\frac{n!}{(n - r_1)!\, r_1!}$$

ways, then for each of these combinations, there remain $n - r_1$ objects from which to choose r_2 objects, which can be done in

$$\frac{(n - r_1)!}{(n - r_1 - r_2)!\, r_2!}$$

ways, and so on.

By the fundamental principle, the total number of ways of effecting the partition is

$$\frac{n!}{(n-r_1)!\,r_1!} \times \frac{(n-r_1)!}{(n-r_1-r_2)!\,r_2!} \times \ldots \times \frac{(n-r_1-r_2-\ldots-r_{k-1})!}{(n-r_1-r_2-\ldots-r_k)!\,r_k!}$$

which simplifies to

$$\frac{n!}{r_1!\,r_2!\,\ldots\,r_k!}. \tag{3.9.1}$$

EXAMPLE 3.9.1.   How many ways can 13 players be partitioned into a first team of 5 players, a second team of 5 players and a reserve pool of 3 reserves?

$$\text{Solution:}\ \frac{13!}{5!\,5!\,3!}.$$

A totally different problem, which ends up with the same numerical solution arises from permuting n objects, some of which are not distinct.

EXAMPLE 3.9.2.   In how many ways can the letters of the word RUNNING be permuted?

The problem here is that there are three N's which are indistinguishable. If we temporarily affix suffixes N_1, N_2, N_3 so that we have 7 distinct symbols: $RUN_1N_2IN_3G$, then by the order principle, there are 7! different permutations. Now select any one at random, say

$$UN_1N_3IRN_2G.$$

Here the three N's are in positions 2, 3, 6 in order $N_1\ N_3\ N_2$. There are $3! = 6$ different ways of permuting N_1, N_2, N_3 and placing these different permutations in the second, third and fifth place in the full permutation

$$U--IR-G$$

give six different full permutations. When we remove the suffixes, all six full permutations yield the same permutation

$$UNNIRNG$$

of RUNNING. This argument works for all permutations of RUNNING, so the number of different permutations in total is

$$7! \div 3! = 840$$

when we no longer distinguish between the N's.

A generalization of this gives the following:

THEOREM 3.9.1.   *If there are a total of n objects of which r_1 are of one kind,*

r_2 *of a second kind, and so on, up to r_k of a final kind, then the number of different permutations of the n objects is*

$$\frac{n!}{r_1!\, r_2! \,\ldots\, r_k!}.$$

EXAMPLE 3.9.3.   How many permutations are there of the eleven letters of the word 'Mississippi'?

There are four i's, four s's, two p's, so the total number of permutations is

$$\frac{11!}{4!\,4!\,2!} = 34{,}650.$$

Other uses of the basic principles are possible. For instance:

EXAMPLE 3.9.4.   How many four letter sequences can be made up using the 26 letters of the alphabet, allowing repeats of letters.

The terms of this example mean that repeats are allowed, but order matters. So we are allowed $AABA$ as a possible four letter sequence, but this is regarded as different from $ABAA$.

The first letter may be chosen in 26 ways, the second in 26, then the third and fourth in the same manner. By the fundamental principle there are

$$26 \times 26 \times 26 \times 26 = 26^4$$

four letter sequences.

This form of choosing objects in a sequence is sometimes referred to as a selection 'with replacement' because, having made each successive choice, the object just chosen is replaced in the set before making the next choice. It is easy to generalize the above example to the following:

Choice with Replacement: The number of ways of selecting r objects in order from n objects with replacement for each choice is

$$n^r.$$

This, and other refinements to the theory, follows by application of the basic principles developed earlier.

3.10.   GENERATING FUNCTIONS AND RELATIONS BETWEEN BINOMIAL COEFFICIENTS

The *Binomial Theorem* states that, for a positive integer n,

$$(x + a)^n = x^n + \binom{n}{1} x^{n-1} a + \ldots + \binom{n}{r} x^{n-r} a^r + \ldots + a^n, \quad (3.10.1)$$

where the coefficients $\binom{n}{r}$ are given in (3.8.1). This may be verified easily by

considering the product
$$(x + a_1)(x + a_2)\ldots(x + a_n) = x^n + s_1 x^{n-1} + s_2 x^{n-2} + \ldots + s_n$$
where [see (14.16.2)]
$$s_1 = a_1 + a_2 + \ldots + a_n$$
$$s_2 = a_1 a_2 + a_1 a_3 + \ldots + a_1 a_n + a_2 a_3 + \ldots$$
(which is the sum of the products $a_q a_r (q \neq s)$)
$$s_3 = a_1 a_2 a_3 + \ldots$$
(the sum of products $a_q a_r a_s$ $(q, r, s$ all different))

$$\ldots$$

$$s_n = a_1 a_2 a_3 \ldots a_n.$$

The number of monomials $a_{q_1} a_{q_2} \ldots a_{q_r}$ in the sum s_r is the number of ways of selecting m different elements $q_1, q_2, \ldots, q_r$ from the set $\{1, 2, \ldots, n\}$ of n elements, so is $\binom{n}{r}$ [see § 14.16].

Putting
$$a_1 = a_2 = \ldots = a_n = a$$
gives
$$(x + a)^n = x^n + nax^{n-1} + \ldots + \binom{n}{r} a^r x^{n-r} + \ldots + a^n,$$
as stated earlier.

If we put $x = 1$, $a = t$, then we get
$$(1 + t)^n = 1 + nt + \ldots + \binom{n}{r} t^r + \ldots + t^n. \tag{3.10.2}$$

More generally, for *any* real number k, it may be proved that
$$(1 + t)^k = 1 + kt + \ldots + \binom{k}{r} t^r + \ldots \tag{3.10.3}$$
where
$$\binom{k}{r} = k(k - 1)\ldots(k - r + 1)/r!$$
and the power series is valid for $|t| < 1$ [see IV, § 1.10].

Given any sequence $a_0, a_1, \ldots, a_n, \ldots$, [see IV, § 1.1], the power series
$$f(t) = a_0 + a_1 t + \ldots + a_n t^n + \ldots \tag{3.10.4}$$
is called the *generating function* of the series. In this sense, we find that the function
$$f(t) = (1 + t)^n$$
is the generating function for the series
$$\binom{n}{0}, \binom{n}{1}, \binom{n}{2}, \ldots, \binom{n}{r}, \ldots$$
of binomial coefficients.

The object of this new idea is to simplify computations with the binomial coefficients. For instance, we have

$$(1 + t)(1 + t)^n = (1 + t)^{n+1}$$

so that

$$(1 + t)\left(1 + \binom{n}{1}t + \ldots + \binom{n}{r}t^r + \ldots\right)$$

$$= 1 + \binom{n + 1}{1}t + \ldots + \binom{n + 1}{r}t^r + \ldots$$

On comparing coefficients of t^r we have the identity

$$\binom{n}{r} + \binom{n}{r - 1} = \binom{n + 1}{r}. \tag{3.10.5}$$

Other identities which may be found by similar methods is the sum of the binomial coefficients. Putting $t = 1$ in (3.10.2) gives

$$2^n = 1 + n + \binom{n}{2} + \ldots + \binom{n}{n}$$

so

$$\sum_{r=0}^{n} \binom{n}{r} = 2^n \quad \text{for a positive integer } n. \tag{3.10.6}$$

Putting $t = -1$ gives

$$1 - n + \binom{n}{2} - \ldots + (-1)^n\binom{n}{n} = 0$$

for a positive integer n.

A more interesting identity found by considering the coefficient of t^r in

$$(1 + t)^m(1 + t)^n = (1 + t)^{m+n}$$

is *Vandermonde's theorem*, for positive integers m, n:

$$\binom{m}{r} + \binom{m}{r - 1}\binom{n}{1} + \binom{m}{r - 2}\binom{n}{2} + \ldots + \binom{n}{r} = \binom{m + n}{r}. \tag{3.10.7}$$

3.11. USE OF GENERATING FUNCTIONS IN ENUMERATION PROBLEMS

Recall from § 3.10 that the coefficient of t^r in the product

$$(1 + a_1t)(1 + a_2t)\ldots(1 + a_nt)$$

is the sum s_r of all products of $a_1, a_2, \ldots, a_n$, taken r at a time. For instance if $n = 3, r = 2$, then

$$s_2 = a_1a_2 + a_1a_3 + a_2a_3.$$

Suppose that we consider a slightly different product, say

$$(1 + at + a^2t^2)(1 + bt)(1 + ct).$$

The coefficient of t^2 is

$$a^2 + ab + ac + bc.$$

The monomials [see (14.15.2)] which occur in this sum are all combinations of a, b, c, where repetitions of a are allowed, but not of b, c.

Considering the coefficient of t^3, we get

$$a^2b + a^2c + abc$$

which includes all combinations of a, b, c where a may appear 0, 1 or 2 times, but b or c appear once or not at all.

This method may be generalized. If we wish to consider combinations of different letters $a, b, c, \ldots$ then for each letter, say a, we consider a factor

$$\lambda_0 + \lambda_1 at + \lambda_2 a^2t^2 + \ldots + \lambda_k a^k t^k + \ldots$$

where we put

$$\left.\begin{array}{ll} \lambda_k = 1 & \text{if } a \text{ may occur } k \text{ times} \\ \lambda_k = 0 & \text{if } a \text{ may not occur } k \text{ times} \end{array}\right\}. \qquad (3.11.1)$$

Multiplying the factors together, the coefficient of t^r in the resultant product is the number of combinations of $a, b, c, \ldots$ taken r at a time under the restrictions given by (3.11.1) above.

EXAMPLE 3.11.1.   What combinations of a, b, c, are possible, taken 4 at a time, when a may occur 0, 1, or 2 times, b may occur 2 or 3 times, c may occur 0 or 1 times.

Solution.   The combinations occur as terms in the coefficient of t^4 in

$$(1 + at + a^2t^2)(b^2t^2 + b^3t^3)(1 + ct),$$

which is

$$b^3c + ab^2c + ab^3 + a^2b^2.$$

The number of such combinations is found by putting $a = b = c = 1$, when the coefficient of t^4 is the number of required combinations, namely 4.

EXAMPLE 3.11.2.   How many combinations of a, b can occur, taken 5 at a time, when a may occur any number of times, but b may only occur 2 or 3 times?

Solution.   We are only asked for the number of such combinations, so we do not need to specify them. The number is the coefficient of t^5 in

$$(1 + t + t^2 + t^3 + t^5)(t^2 + t^3)$$

which is 2. (Of course the combinations are *aaabb, aabbb*!)

EXAMPLE 3.11.3.   Given n objects $a, b, c, \ldots$, what is the number of possible combinations taken r at a time, with no restriction on the number of repetitions of each object.

Solution.   With no restriction on the number of repetitions we take a factor of the form

$$1 + at + a^2 t^2 + \ldots + a^k t^k + \ldots$$

for each object $a, b, c, \ldots$. Taking a product of n such factors and putting $a = b = c = \ldots = 1$, the number of combinations taken r at a time is the coefficient of t^r in

$$(1 + t + t^2 + \ldots + t^k + \ldots)^n = \frac{1}{(1 - t)^n}.$$

Using the binomial expansion (3.10.3) for $(1 - t)^{-n}$, we find the coefficient of t^r is

$$\binom{-n}{r}(-1)^n = \frac{(-n)(-n - 1)\ldots(-n - r + 1)(-1)^n}{r!}$$

$$= \frac{n(n + 1)\ldots(n + r - 1)}{r!}$$

$$= \binom{n + r - 1}{r}. \qquad (3.11.2)$$

Thus the number of combinations of n objects taken k at a time but allowing any number of repetitions of each is the same as the number of combinations (without repetition) of $n + r - 1$ objects taken r at a time.

In this last example the specific coefficient of t^r in

$$(1 + at + a^2 t^2 + \ldots)(1 + bt + b^2 t^2 + \ldots)(1 + ct + c^2 t^2 + \ldots)\ldots$$

consists of the sum of all monomials [see (14.15.2)]

$$a^p b^q c^k \ldots$$

where

$$p + q + k + \ldots = r.$$

The monomials of this kind are called the *homogeneous products of dimension* r formed from the letters $a, b, c, \ldots$. The number of distinct homogeneous products of dimension r which can be formed from n letters is denoted by

$$_nH_r.$$

From the last example we have

$$_nH_r = \,_{n+r-1}C_r = \binom{n + r - 1}{r}. \qquad (3.11.3)$$

D.O.T.

BIBLIOGRAPHY

A. Theory

Stewart, I. and Tall, D. (1977). *The Foundations of Mathematics*, Oxford University Press.

Swierczkowski, S. (1972). *Sets and Numbers*, Routledge and Kegan Paul.

B. Permutations, etc.

Arthurs, A. M. (1965). *Probability Theory*, Routledge and Kegan Paul.

Feller, W. (1950). *An Introduction to Probability Theory and Its Applications I*, John Wiley and Sons.

CHAPTER 4

Number Theory

4.1. DIVISIBILITY

4.1.1. Division in Rings

When discussing rational numbers in § 2.4.1 we saw that each rational $r \neq 0$ has a reciprocal rational number r' (or $1/r$) such that $r . r' = 1$. This means that for any rationals, $r \neq 0$ and s, we can always find a rational number x so that $rx = s$. We say that the quotient $x = s/r$ is the result of dividing s by r. Similarly if $r \neq 0$ and s are real numbers we can divide s by r and obtain a unique third number which will again be real [see § 2.6.2]. In general in a field we can always divide any element by another, apart from zero, and get a member of the same field [see § 2.4.2]. However if we have two integers, say $a \neq 0$ and b, we will not always be able to produce another *integer* q with $b = aq$. If we do have $b = aq$ for an integer q we say that a divides, or is a divisor of, b. When this is so we write '$a|b$', and if a does not divide b we write $a \nmid b$. In any ring R [see § 2.3] we shall say that a divides b (within the ring) if there is an element q of R with $b = aq$, and in most rings a situation like that of the integers, $\mathbb{Z}$, will apply: a given non-zero element will divide some elements but not others—with a quotient still inside the ring. It is precisely this fact, that division is sometimes possible and sometimes not, which makes its study interesting in rings.

EXAMPLES 4.1.1.   In the ring $\mathbb{Z}$
   (i) 1 and -1 divide every integer;
   (ii) 2 divides 6, but $3 \nmid 5$.

EXAMPLE 4.1.2.   In the ring of all polynomials with integer coefficients $x^2 + 1$ divides $x^4 - 1$ since $x^4 - 1 = (x^2 + 1)(x^2 - 1)$; but $x^2 + 1$ does not for instance divide $x^4 + 1$.

In any ring R (in particular in $\mathbb{Z}$) the following facts about divisibility are always true:

$$\text{if } a|b \text{ then } -a|b, \text{ and } a|-b; \qquad (4.1.1)$$

$$\text{if } a|b \text{ and } b|c \text{ then } a|c; \qquad (4.1.2)$$

$$\text{if } a|b \text{ and } a|c \text{ then } a|(bx + cy); \qquad (4.1.3)$$

101

for a, b, c, x, $y \in R$. The second property says that the idea of 'being a divisor of' is a *transitive notion* [see also § 1.3.3]. The third says that any common divisor of two numbers must also divide a sum of multiples of them. For example if we are dealing with integers, property (4.1.3) implies the facts that if an integer a divides both b and c then it must divide their sum and difference; for the first fact comes from choosing $x = y = 1$, and the second from choosing $x = 1$, $y = -1$.

Theorems about division in rings are greatly simplified, and are most useful, when the rings concerned are integral domains, that is, have all the additive and multiplicative properties of the integers [see § 2.3]. In future when we mention division in rings other than $\mathbb{Z}$, we shall confine the discussion to integral domains.

4.1.2. Euclidean Domains

The study of divisibility among integers is especially interesting because they have an order relation which is nicely linked to the division [see § 2.2.2]. If $|a|$ denotes the modulus of a as in § 2.2.1, then for integers a and b we have in addition to (4.1.1) to (4.1.3) above

$$\text{if } a|b \text{ either } b = 0 \text{ or } |a| \le |b|; \qquad (4.1.4)$$

and if a and b are any integers with $a \ne 0$ there are integers q and r so that

$$b = aq + r \text{ with either } r = 0 \text{ or } |r| < |a|. \qquad (4.1.5)$$

Property (4.1.5) is the next best thing to always being able to divide one number by another; because it says that a 'nearly divides' b in the sense that although b need not be equal to a multiple of a, it differs from a multiple of a by a remainder r which is smaller than a in absolute value.

EXAMPLES 4.1.3
  (i) $15 = 3.5 + 0$, that is, $3|5$;
  (ii) $5 = 7.0 + 5$ and the remainder 5 is less than the divisor 7;
  (iii) $11 = (-5).(-2) + 1$ with $1 < |-5|$;
  (iv) $17 = 3.5 + 2$ with $2 < 3$, and $17 = 3.6 - 1$ with $|-1| < 3$.

EXAMPLE 4.1.4.   For given integers a and b, with $a \ne 0$, there will be many different ways of expressing b as a multiple of a plus a remainder, but in most of these ways the modulus of the remainder will be greater than that of a. For instance $17 = 2.5 + 7$, or $13 = 3.6 - 5$ if $a = 3$. If a divides b, there will be just one choice of q which will satisfy (4.1.5) (with $r = 0$); and if a does not divide b there will be exactly two suitable choices of q, one of them giving a positive remainder satisfying (4.1.5) and one giving a negative remainder.

Notice that if we used a^2 instead of $|a|$ as a measure of the size of the integer a we would still have properties similar to (4.1.4) and (4.1.5) since

$$\text{if } a|b \text{ either } b = 0 \text{ or } a^2 \le b^2; \qquad (4.1.6)$$

and if $a \neq 0$ does not divide b, there are integers q and r so that

$$b = aq + r \text{ with } r^2 < a^2. \tag{4.1.7}$$

So there is nothing special about the modulus here, and, as far as investigations of factorization properties are concerned, properties (4.1.6) and (4.1.7) are just as useful as (4.1.4) and (4.1.5). In practice (4.1.4) and (4.1.5) are easier to work with because, apart from $a = 0$, $+1$ or -1, $|a|$ is always smaller than a^2. In some integral domains, other than $\mathbb{Z}$, it is possible to find a number $m(a)$ for each element a, which can be thought of as the size of a, and such that these sizes are related to the notion of divisibility in the same way as the modulus is in (4.1.4) and (4.1.5). The advantage of this is that in such domains we can prove quite general results about the possible factors of elements by exactly the same methods as we use for integers. In more detail we should like each non-zero element a of a given domain D to have ascribed to it a measure $m(a)$ so that

if $a \in D$ and $a \neq 0$ then $m(a)$ is a non-negative integer;   (4.1.8)

if $a|b$ either $b = 0$ or $m(a) \leq m(b)$;   (4.1.9)

and for any elements a, b in D with $a \neq 0$, whether a divides b or not there are elements q, r of D which satisfy the *Euclidean division property*:

$$b = aq + r \text{ with either } r = 0 \text{ or } m(r) < m(a). \tag{4.1.10}$$

Integral domains whose elements have sizes satisfying (4.1.8), (4.1.9) and (4.1.10) are called *Euclidean domains*.

EXAMPLE 4.1.5.   The ring $\mathbb{Z}$ of integers is a Euclidean domain.

EXAMPLE 4.1.6.   The domain of Gaussian integers [see Example 2.7.10] is a Euclidean domain if we take the size $m(a + ib)$ of $a + ib$ to be the natural number $a^2 + b^2$. This (Euclidean) measure $m(a + ib)$ is simply the square of the modulus of the complex number $a + ib$ as in (2.7.13). For example $1 + 3i$ does not divide $5 + 7i$ in this ring, but we have

$$5 + 7i = (1 + 3i)(3 - i) + (-1 - i)$$

and $m(-1 - i) = 2 < m(1 + 3i) = 10$. Here the *partial quotient* $3 - i$ is obtained by finding the quotient $(5 + 7i)/(1 + 3i) = \frac{13}{5} - \frac{4}{5}i$ in the complex field $\mathbb{C}$ [see § 2.7.1], and taking the Gaussian integer q which is nearest to $Q = \frac{13}{5} - \frac{4}{5}i$ in the Argand diagram [see § 2.7.2]; as in Figure 4.1.1. If there are two (or possibly four) nearest Gaussian integers to the true quotient, it doesn't matter which one we take as the partial quotient.

EXAMPLE 4.1.7.   The ring of all polynomials in a single variable x with rational coefficients is a Euclidean domain if for each polynomial $f(x)$ we set $m(f(x)) = \text{degree } (f(x))$ [see § 14.1]. For example $x + 1$ divides $x^2 - 1$ and, as

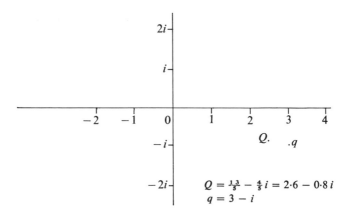

Figure 4.1.1

required $m(x + 1) = 1 < m(x^2 - 1) = 2$. Again, $(2x^2 - 1) \nmid (3x^5 + 1)$ but we have

$$3x^5 + 1 = (2x^2 - 1)(\tfrac{3}{2}x^3 + \tfrac{3}{4}x) + (\tfrac{3}{4}x + 1)$$

where $m(\tfrac{3}{4}x + 1) = 1 < m(2x^2 - 1) = 2$. The ring of polynomials in x with real (or complex) coefficients is similarly a Euclidean domain, with exactly the same definition of the size of a polynomial. However the integral domain of polynomials with integer coefficients is not a Euclidean domain.

4.1.3.   Common Divisors

Notice that when we are looking for all the divisors of an integer b we need only find the positive divisors, since the negative divisors will just be the same numbers—with 'minus signs' attached. Property (4.1.1) also assures us that the divisors of $-b$ will be the same as those of b, so if b is negative the same results will be obtained by considering $-b$ (which is positive) instead. We may thus drop all minus signs and concentrate on finding the positive divisors of positive integers. Even with these restrictions it can be quite difficult to find all the divisors of a given positive integer, especially if it is large. It may at first seem even more difficult to find the divisors common to two given natural numbers a and b; but happily there is a systematic procedure which always produces the largest number which is a common divisor of the two integers. This largest positive divisor is known as the *greatest common divisor* (g.c.d.), or *highest common factor* (h.c.f.) of a and b. Moreover the common divisors of a and b turn out to be precisely the divisors of h.c.f. (a, b). The situation is easy to deal with if the smaller of the two numbers, say a, divides the other b. For then a itself is the g.c.d. of a and b, and the common divisors are the divisors of a. In particular, if $a = 1$ it will be the greatest (and only) *positive* common divisor no matter what b is.

EXAMPLE 4.1.8.   Find the common divisors of 6 and 24. Here 6 divides 24 so the greatest common divisor is 6. The common divisors are all the divisors of 6. So the positive common divisors are 1, 2, 3, 6; and the negative common divisors are $-1, -2, -3, -6$. Note that the common divisors of 6, 24 are exactly the same as the common divisors of $-6, 24$; $6, -24$; and $-6, -24$.

When a does not divide b we can use division with remainder as in (4.1.5), for if $b = aq + r$ it happens that the common divisors of b and a are exactly the same as the common divisors of a and r, and these are respectively less than the original numbers.

EXAMPLE 4.1.9. Find the g.c.d. of 20 and 6. Here $20 = 3.6 + 2$, so the common divisors of 20 and 6 are the same as those of 6 and 2. Since 2 divides 6 the desired greatest common divisor is 2 itself.

EXAMPLE 4.1.10. Find the common divisors of 29 and 7. We have $29 = 4.7 + 1$, so the common divisors of 29 and 7 are the same as the common divisors of 7 and 1. These are the integer divisors of 1, which are $+1$ and -1.

When we express b as a multiple of a plus a remainder, it often happens that the remainder r does not divide a—in which case we still do not immediately know the g.c.d. of a and b, or equivalently, of a and r. However we can then begin again by applying division with remainder to the pair a, r; and we can continue as long as necessary until we reach a pair of numbers the smaller of which does divide the larger. The g.c.d. of this last pair of numbers (which is the smaller of the two numbers) is then the same as the g.c.d. of the original two numbers a and b.

EXAMPLE 4.1.11. Find the g.c.d. of 34 and 10. Here $34 = 3.10 + 4$ and the g.c.d. of 34 and 10 is the same as that of 10 and 4. Then $10 = 2.4 + 2$, so the g.c.d. of 10 and 4 is the same as that of 4 and 2, which is 2.

EXAMPLE 4.1.12. Find the g.c.d. of 30 and 103. This time when we apply division with remainder successively we see that

$$103 = 3.30 + 13, \tag{4.1.11}$$

$$30 = 2.13 + 4, \tag{4.1.12}$$

$$13 = 3.4 + 1. \tag{4.1.13}$$

The first line tells us that g.c.d. $(103, 30) =$ g.c.d. $(30, 13)$; the second line that g.c.d. $(30; 13) =$ g.c.d. $(13, 4)$; and the third that g.c.d. $(13, 4) =$ g.c.d. $(4, 1)$. Since 1 divides 4, g.c.d. $(103, 30) =$ g.c.d. $(4.1) = 1$.

The method of finding the greatest common divisor by repeatedly using division with remainder is called *Euclid's Algorithm*. The work involved can sometimes be reduced by choosing a negative remainder at any given stage, if there is one whose modulus is smaller than a corresponding positive remainder.

For example consider the problem of finding the common divisors of 60 and 22. We could use the following equations

$$60 = 2.22 + 16, \tag{4.1.14}$$

$$22 = 1.16 + 6, \tag{4.1.15}$$

$$16 = 2.6 + 4, \tag{4.1.16}$$

$$6 = 1.4 + 2, \tag{4.1.17}$$

$$4 = 2.2.$$

From these we see that g.c.d. $(60, 22) =$ g.c.d. $(22, 16) =$ g.c.d. $(16, 6) =$ g.c.d. $(6, 4) =$ g.c.d. $(4, 2) = 2$. However it is quicker to replace the first equation by $60 = 3.22 - 6$, for this tells us that g.c.d. $(60, 22) =$ g.c.d. $(22, 6)$ so that we have reduced the size of our first number from 60 to 6 instead of just from 60 to 16. Following through this example using our new idea we have

$$60 = 3.22 - 6, \tag{4.1.18}$$

$$22 = 4.6 - 2, \tag{4.1.19}$$

$$6 = 3.2,$$

giving g.c.d. $(60, 22) =$ g.c.d. $(22, 6) =$ g.c.d. $(6, 2) = 2$ in three steps instead of five.

It is a remarkable fact that the g.c.d. of two integers a and b can always be written as an integer multiple of a plus an integer multiple of b. Moreover we can always find one such representation by working backwards through the equations of Euclid's Algorithm. For instance we have seen that g.c.d. $(60, 22) = 2$, and equation (4.1.19) gives $2 = 4.6 - 22$. From (4.1.18) we have $6 = 3.22 - 60$, and substituting this in the previous equation produces $2 = 4(3.22 - 60) - 22 = 11(22) - 4(60)$. Exactly the same result would be given by using the equations (4.1.17) to (4.1.14) (in that order). From the fact that 2 can be expressed as a multiple of 22 plus a multiple of 60 we can see further that every even number can be so expressed. For instance, from $2 = 11(22) - 4(60)$ we have $10 = 5.2 = 55(22) - 20(60)$.

EXAMPLE 4.1.13.   Find the g.c.d. of 7 and 5, and express the g.c.d. as a multiple of 7 plus a multiple of 5. Here

$$7 = 5 + 2,$$

$$5 = 2.2 + 1,$$

so g.c.d. $(7, 5) =$ g.c.d. $(5, 2) =$ g.c.d. $(2, 1) = 1$. Moreover the second equation, followed by the first, gives $1 = 5 - 2.2 = 5 - 2(7 - 5) = 3.5 - 2.7$.

Numbers, such as 5 and 7, whose g.c.d. is 1, are called *relatively prime* or *co-prime* and have the remarkable property that every integer can be written as a sum of multiples of them. If, for example, we want to write 11 as a multiple of 5 plus a multiple of 7, we argue that $11 = 11(1) = 11(3.5 - 2.7) = 33.5 - 22.7$.

EXAMPLE 4.1.14.   Suppose that two small-scale potato merchants each sell their potatoes in 5 kilo and 7 kilo bags, and suppose that they have an agreement to sell each other potatoes whenever needed. Then they can exchange any whole number of kilos. For if merchant A wants to send 1 kilo to merchant B; all he does is to send three 5 kilos bags, and B returns two 7 kilo bags. This corresponds exactly to the fact that $1 = 3.5 - 2.7$.

The idea of Euclid's Algorithm can also be used in any other Euclidean domain to find the g.c.d. of two given numbers. For example suppose we want to find the g.c.d. of $-19 + 82i$ and $8 + 21i$ in the domain of Gaussian integers. We have

$$-19 + 82i = (8 + 21i)(3 + 2i) + (-1 + 3i),$$

$$8 + 21i = (-1 + 3i)(5 - 4i) + (1 + 2i),$$

$$-1 + 3i = (1 + 2i)(1 + i)$$

where in each case the remainder is of smaller modulus than the divisor. Then going through the equations in turn we see that g.c.d. $(-19 + 82i, 8 + 21i) =$ g.c.d. $(8 + 21i, -1 + 3i) =$ g.c.d. $(-1 + 3i, 1 + 2i) = 1 + 2i$.

The same ideas can be used to investigate the common divisors, and in particular the greatest common divisor of more than two numbers. We proceed by finding the g.c.d., say g_1 of the first two numbers; then the g.c.d. g_2 of g_1 and the next number; and each time we apply Euclid's Algorithm to the current g.c.d. and the next number in our list. The final g.c.d. which we calculate is the desired greatest common divisor of the original set of numbers.

EXAMPLE 4.1.15.   Find the common divisors of 60, 24 and 104. Here

$$60 = 2.24 + 12,$$

and $12|24$ so g.c.d. $(60, 24) = 12$. Next we examine g.c.d. $(12, 104)$ and find that $104 = 9.12 - 4$. Since $4|12$ we have g.c.d. $(60, 24, 104) =$ g.c.d. $(12, 104) = 4$. The positive common divisors are the positive divisors of 4, namely 1, 2 and 4; and the negative divisors are -1, -2 and -4.

4.2.   PRIMES AND PRIME DECOMPOSITION

The method of Euclid's Algorithm always produces the greatest common divisor of two or more numbers, and all the common divisors are precisely the divisors of this g.c.d. So the problem of finding the common divisors of several numbers can always be reduced to the problem of finding all the divisors of one number. Unfortunately there is no rule which is universally helpful here; but a useful approach to easily being able to find the factors of any given number is to try to classify numbers according to their types of divisors. We shall just do this for integers although similar schemes can be introduced in other Euclidean domains. Also, as we have observed before, we need only consider positive divisors of positive integers. The number 1 is the only number with a single positive divisor, namely 1; and it is called a *unit*. Any integer $n > 1$ has at least two positive divisors, 1 and n, and some numbers such as 2, 3, 5, 7, 11, only have two. A number with exactly two positive divisors is called a *prime*; so that, apart from being divisible by 1, a prime number has no divisors smaller than itself. A *composite* number m is one which has at least one divisor a intermediate in size between 1 and m, that is $1 < a < m$. Every number greater than 1 is either prime or composite.

EXAMPLES 4.2.1
  (i) 13 is prime and so is 79;
  (ii) 12 is composite since 3 divides 12 and $1 < 3 < 12$;
  (iii) 35 is composite since $5 | 35$ and $1 < 5 < 35$.

Although composite numbers are a lot more prolific than prime numbers, there are an infinite number of primes. In other words given any positive integer, such as 1000, we can be certain that there are primes, such as 1093, greater than 1000. Then given another larger bound, say 1,000,000, there will be still larger primes, for instance 1,269,043, and so on.

Primes are very important numbers since all other numbers can be constructed by multiplying suitable prime numbers together. Another way of saying this is that *a number greater than 1 is either prime or a product of primes*. We give a proof of this as an illustration of the method of complete induction mentioned in § 2.1. First we observe that the statement is true of the first number to which it applies, namely 2, which is a prime. Next we suppose that for some integer $n \geq 2$ the statement has been verified for all natural numbers between 2 and n inclusive, and we consider $n + 1$. The statement is certainly true for the number $n + 1$ if it is prime. If $n + 1$ is composite it must be divisible by a smaller number a, so we shall have $n + 1 = ab$ where the integers a and b are each between 2 and n. Since we are supposing that the statement has already been established for numbers up to and including n, it must be true of a and b in particular. Thus $a = a_1 \ldots a_s$, $b = b_1 \ldots b_t$ where each of $a_1, \ldots, a_s, b_1, \ldots, b_t$ is prime— and s, t just tell us the number of terms. So $n + 1 = a_1 \ldots a_s . b_1 \ldots b_t$ is a product of primes, (the conclusion is the same if one or both of a or b is prime). The principle of complete induction then assures us that the statement is true for all natural numbers.

EXAMPLES 4.2.2
  (i) $12 = 2.2.3$ and $35 = 5.7$ where each of the factors are prime;
  (ii) 101 is prime;
  (iii) $1001 = 7.11.13$.

Notice that it is impossible to write 35 as a product of primes one of which is say,

3, or any prime other than 5 or 7. This fact, that *the expression for a composite number as a product of primes is unique* (apart from writing the factors in a different order), is usually called the *Fundamental Theorem of Arithmetic*. It may happen that when a particular number is expressed as a product of primes some primes occur more than once, as in the case of 12 above. If the prime p occurs k times, then instead of writing $p.p...p$ with k terms, we abbreviate this to p^k (to be read as 'p to the power k'). The k just tells us the number of p's, so it can be helpful to write p itself as p^1. In some situations it may even be convenient to write p^0 to mean that there are no p's at all. A neat device which allows us to write p^0 in a product and still regard it as a legitimate product, is to agree to put $p^0 = 1$ for every p.

EXAMPLES 4.2.3
  (i) $3^0 = 1$, $3^1 = 3$, $3^2 = 9$, $3^3 = 27$, $3^4 = 81$;
  (ii) $12 = 2^2.3 = 2^2.3.5^0$;
  (iii) $17{,}640 = 2^3.3^2.5.7^2$;
  (iv) $120 = 2^3.3.5$ and $364 = 2^2.7.13$, but these can also be written as products which both involve the same primes, namely $120 = 2^3.3^1.5^1.7^0.13^0$ and $364 = 2^2.3^0.5^0.7^1.13^1$.

Every number greater than 1 is divisible by one or more primes, so in order to test whether a number is composite we need only see whether the number has a prime divisor less than itself. In fact it is sufficient to see whether the number has a prime divisor less than, or equal to, its square root [see § 3.3]. D. N. Lehmer's 'Factor Table for the First Ten Millions' (Carnegie Institute publication no. 105) gives the smallest prime factor, other than 2, 3, 5 or 7 of all numbers up to 10,017,000. If we do find a prime divisor of a number n, we can divide by it and then test the quotient—continuing until we know all the prime divisors of n, or in other words the *prime decomposition* of n as in (4.2.1) below

$$n = p_1{}^{a_1}.p_2{}^{a_2}...p_r{}^{a_r}. \tag{4.2.1}$$

We can insist that in (4.2.1) each power, or *exponent*, a_i is at least 1, but it may sometimes be convenient to allow some of the a's to be 0. A knowledge of the prime factorization (4.2.1) is very helpful since it enables us to deduce a lot about the divisors of n without any further work. Each divisor of n must be of the form

$$p_1{}^{\alpha_1}p_2{}^{\alpha_2}...p_r{}^{\alpha_r} \tag{4.2.2}$$

with each α_i less than or equal to a_i. That is, each divisor of n must be a product of the same primes as in (4.2.1) and each prime must occur no more times than it did originally. Also the number of divisors of n, including 1 and n itself, is given by the product

$$(a_1 + 1)(a_2 + 1)...(a_r + 1). \tag{4.2.3}$$

EXAMPLE 4.2.4.   Consider the number $144 = 2^4.3^2$. It has $(4 + 1)(2 + 1) = 15$ positive divisors and these are $1 = 2^0.3^0$, $2 = 2^1.3^0$, $4 = 2^2.3^0$, $8 = 2^3.3^0$,

$16 = 2^4.3^0$, $3 = 2^0.3^1$, $6 = 2^1.3^1$, $12 = 2^2.3^1$, $24 = 2^3.3^1$, $48 = 2^4.3^1$, $9 = 2^0.3^2$, $18 = 2^1.3^2$, $36 = 2^2.3^2$, $72 = 2^3.3^2$, and $144 = 2^4.3^2$.

The sum of the divisors of the number n in (4.2.1) is given by

$$\frac{(p_1^{a_1+1} - 1)}{(p_1 - 1)} \cdot \frac{(p_2^{a_2+1} - 1)}{(p_2 - 1)} \cdots \frac{(p_r^{a_r+1} - 1)}{(p_r - 1)}. \qquad (4.2.4)$$

So the sum of the 15 displayed divisors of 144 is

$$\frac{(2^5 - 1)}{(2 - 1)} \cdot \frac{(3^3 - 1)}{(3 - 1)} = 31.13 = 403.$$

Suppose we have another number m whose prime decomposition is

$$m = p_1^{b_1}\ldots p_r^{b_r}, \qquad (4.2.5)$$

where the primes occurring in (4.2.5) are exactly the same as those occurring in (4.2.1). (We can always use just one set of primes to write the prime factorizations of any two numbers, by making some of the exponents zero if necessary). Then the greatest common divisor of m and n can be written down immediately:

$$\text{g.c.d. } (m, n) = p_1^{c_1}\ldots p_r^{c_r}, \qquad (4.2.6)$$

where each c_i is the smaller of the two powers a_i, b_i (or their common value if $a_i = b_i$).

EXAMPLE 4.2.5.   Consider the problem of finding the g.c.d. of 1960 and 4620. We could use Euclid's Algorithm or factorize them as $1960 = 2^3.5.7^2$ and $4620 = 2^2.3.5.7.11$. We write the factorizations with the same set of primes in each case: $1960 = 2^3.3^0.5^1.7^2.11^0$ and $4620 = 2^2.3.5.7.11$. Then we can conclude at once that g.c.d. $(1960, 4620) = 2^2.3^0.5^1.7^1.11^0 = 2^2.5.7 = 140$.

4.3.   CONGRUENCES

4.3.1.   The Algebra of Congruences

In the case of the integers there is another interesting way of looking at the division with remainder property (4.1.5). For the equation $b = aq + r$ says that from the point of view of a the remainder r characterizes b in the sense of specifying how far b is from a multiple of a. If, for instance, $b = 17$ and $a = 7$ then

$$17 = 2.7 + 3.$$

So 17 and 3 only differ by a multiple of 7, and it can often be useful to regard them as being the same with respect to the number (or '*modulus*') 7. Similarly all the numbers $\ldots -11, -4, 3, 10, 17, 24, \ldots$ only differ from each other by multiples of 7 and so, as far as 7 is concerned, any two of them can be regarded as 'the same'. We actually use the technical phrase 'are congruent modulo 7' to mean 'differ by a multiple of 7'. In general suppose we are interested in classifying numbers by their remainders when divided by the integer $m > 0$. Then if b_1, b_2

leave the same remainder when divided by m, or, what is the same thing, differ by a multiple of m; we say that b_1, b_2 are congruent modulo m. If so we write

$$b_1 \equiv b_2 \,(\text{mod } m), \qquad (4.3.1)$$

to be read as 'b_1 *is congruent to* b_2 *modulo* m'. It is important to realize that the statement '$b_1 \equiv b_2 \,(\text{mod } m)$' means just the same as the statement '$b_1 = b_2 + km$ for some integer k'; or alternatively, the same as the statement that m divides the difference $b_1 - b_2$, ('$m|b_1 - b_2$').

EXAMPLES 4.3.1.
   (i) $9 \equiv -2 \,(\text{mod } 11)$;
   (ii) $m \equiv 0 \,(\text{mod } m)$ for any m;
   (iii) $365 \equiv 1 \,(\text{mod } 7)$;
   (iv) $42 \equiv 2 \,(\text{mod } 8)$;
   (v) $80 \equiv 0 \,(\text{mod } 16)$.

Examples of congruences can be found in many common situations, since they occur whenever cyclic processes are at work. Suppose that we have a clock whose face shows the hours 1 to 12. It counts hours modulo 12, because it does not for instance tell us how many hours have passed since noon yesterday. It tells us the remainder modulo 12 of the number of hours.

The congruence notation '$\equiv$' was originally designed (by C. F. Gauss, 1777–1855) deliberately to resemble the equality sign, since congruences with the same modulus can be added, subtracted and multiplied just like ordinary equations [see § 1.3.4].

EXAMPLE 4.3.2.   The congruences

$$7 \equiv 17 \,(\text{mod } 10) \text{ and } 4 \equiv 24 \,(\text{mod } 10)$$

can be added to give

$$11 \equiv 41 \,(\text{mod } 10);$$

and subtracted to give

$$3 \equiv -7 \,(\text{mod } 10).$$

EXAMPLE 4.3.3.   The terms of the congruences $7 \equiv 13 \,(\text{mod } 6)$ and $8 \equiv 14 \,(\text{mod } 6)$ can be multiplied to give the two valid congruences

$$56 = 7.8 \equiv 13.14 = 182 \,(\text{mod } 6)$$

and

$$98 = 7.14 \equiv 8.13 = 104 \,(\text{mod } 6).$$

We can express these facts symbolically by saying that for any modulus m with $a \equiv b \,(\text{mod } m)$ and $c \equiv d \,(\text{mod } m)$, we can conclude that

$$a + c \equiv b + d \,(\text{mod } m), \qquad (4.3.2)$$

$$a - c \equiv b - d \,(\text{mod } m), \qquad (4.3.3)$$

and

$$ac \equiv bd \,(\text{mod } m). \qquad (4.3.4)$$

In particular we may multiply both terms of a congruence by the same integer and obtain another true congruence. If we multiply both sides of the congruence $3 \equiv 19 \pmod 8$ by 5 we obtain the valid congruence $15 \equiv 95 \pmod 8$; and if we multiply both sides by -1 we obtain $-3 \equiv -19 \pmod 8$. In symbols,

$$\text{if } a \equiv b \pmod m \text{ then } ka \equiv kb \pmod m \text{ for any integer } k. \qquad (4.3.5)$$

If both sides of a congruence are divisible by a common factor d we can divide by that factor, but we have to be careful, for we also have to divide the modulus m by as much of d as possible. That is, we must divide m by g.c.d. (d, m).

EXAMPLE 4.3.4.   In the congruence $6 \equiv 16 \pmod{10}$ we cannot divide 6 and 16 by their common factor 2 without altering the modulus 10; for we would obtain the incorrect conclusion $3 \equiv 8 \pmod{10}$. We must also divide the modulus by 2 and obtain the valid congruence $3 \equiv 8 \pmod 5$.

EXAMPLE 4.3.5.   In the congruence $88 \equiv 4 \pmod{21}$ the common factor 4, of 88 and 4, is relatively prime to 21. So we may conclude legitimately that $22 \equiv 1 \pmod{21}$.

EXAMPLE 4.3.6.   Consider the congruence $54 \equiv 120 \pmod{33}$. Here g.c.d. $(54, 120) = 6$ and g.c.d. $(6, 33) = 3$. So we can divide each of 54, 120 by 6 if at the same time we divide the modulus by 3; and we obtain the valid congruence $9 \equiv 20 \pmod{11}$.

EXAMPLE 4.3.7.   The first day of January 1979 was a Monday. Which day of the week will 4 March 1983 fall on? Since 1980 is a leap year there are $4(365) + 1$ days till 1 January 1983; whence $4.365 + 64$ days up to 4 March 1983. Now $365 \equiv 1 \pmod 7$ and $64 \equiv 1 \pmod 7$, so

$$4.365 + 64 \equiv 4.1 + 1 = 5 \pmod 7.$$

We therefore have to count 5 days past Monday and conclude that 4 March 1983 is a Saturday.

We can also use congruences to provide simple tests for divisibility by various small numbers. This is because asking whether m divides N is the same as asking whether $N \equiv 0 \pmod m$. So let us suppose that N is a natural number [see § 2.1] which can be written in ordinary decimal notation as $a_n a_{n-1} \ldots a_0$. In other words [see § 2.5]:

$$N = a_n . 10^n + a_{n-1} . 10^{n-1} + \ldots + a_1 . 10 + a_0.$$

Since $10 \equiv 0 \pmod 2$, all higher powers of 10 will also be congruent to zero modulo 2, giving

$$N \equiv a_0 \pmod 2.$$

So N will be divisible by 2 whenever its last digit is divisible by 2, and only then. Similarly, because $10 \equiv 0 \pmod 5$, N will be divisible by 5 exactly when its last digit is divisible by 5. For the number 4 we have $10^2 \equiv 0 \pmod 4$, so that 10^3, 10^4, and all subsequent powers of 10 will be congruent to zero modulo 4. Therefore

$$N \equiv 10a_1 + a_0 \pmod 4,$$

so that N will be divisible by 4 when and only when the number formed by its last two digits is divisible by 4. The case of 9 is more interesting because $10 \equiv 1 \pmod 9$. Hence $10^2 = 10 \cdot 10 \equiv 10 \cdot 1 \equiv 1 \cdot 1 = 1 \pmod 9$, and likewise every power of 10 is congruent to 1 modulo 9. Thus

$$N = a_n \cdot 10^n + a_{n-1} \cdot 10^{n-1} + \ldots + a_0 \equiv a_n + a_{n-1} + \ldots + a_0 \pmod 9.$$

This means that N is divisible by 9 (or $N \equiv 0 \pmod 9$) if and only if the sum of its digits is divisible by 9. We also have $10 \equiv 1 \pmod 3$, so there is an analogous test for divisibility by 3: a natural number is divisible by 3 if and only if the sum of its decimal digits is divisible by 3.

4.3.2.   Finite Rings

As well as being useful in ordinary arithmetical situations congruences can provide us with interesting examples of rings (and fields) which contain only a finite number of elements [see § 2.3]. Consider for example the six numbers 0, 1, 2, 3, 4, 5. Let us use a new operation to combine two of these numbers a, b, by first taking their ordinary sum $a + b$ and then choosing the number s, between 0 and 5, with $s \equiv a + b \pmod 6$. We shall call this operation 'addition modulo 6', and write $s = a +_6 b$. Thus

$$3 + 4 = 7 \text{ and } 7 \equiv 1 \pmod 6,$$

so

$$3 +_6 4 = 1.$$

Similarly $2 +_6 4 = 0$, because $6 \equiv 0 \pmod 6$; and $3 +_6 5 = 2$. We can now construct Table 4.3.1 showing all the possible sums modulo 6. The set of numbers 0, 1, 2, 3, 4, 5 with the operation of addition modulo 6 is another example of an Abelian group [see Definition 8.2.2 and Example 8.2.7].

$+_6$	0	1	2	3	4	5
0	0	1	2	3	4	5
1	1	2	3	4	5	0
2	2	3	4	5	0	1
3	3	4	5	0	1	2
4	4	5	0	1	2	3
5	5	0	1	2	3	4

Table 4.3.1: Addition modulo 6

We can similarly define an operation of 'multiplication modulo 6' between two of the numbers 0, 1, 2, 3, 4, 5. We first form the ordinary product, ab, of the two numbers a, b, and then choose the number t, between 0 and 5, with $t \equiv ab \pmod 6$. We write $t = a \times_6 b$. So $3 \times_6 4 = 0$ because $3.4 \equiv 0 \pmod 6$, and $2 \times_6 5 = 4$. Table 4.3.2 shows all the possible products modulo 6. The numbers 0, 1, 2, 3, 4, 5, with these operations of addition modulo 6 and multiplication modulo 6, form an example of a ring with six elements and we denote it by $\mathbb{Z}_6$ (in § 2.3, read '$+_6$' for '$\oplus$' and '$\times_6$' for '$\odot$').

$\times_6$	0	1	2	3	4	5
0	0	0	0	0	0	0
1	0	1	2	3	4	5
2	0	2	4	0	2	4
3	0	3	0	3	0	3
4	0	4	2	0	4	2
5	0	5	4	3	2	1

Table 4.3.2: Multiplication modulo 6

For any natural number n, we can similarly construct a ring, called $\mathbb{Z}_n$, containing just the elements $0, 1, \ldots, n - 1$. Two numbers a, b of $\mathbb{Z}_n$ are combined by the operations of *addition modulo n* and *multiplication modulo n* to give $a +_n b$ and $a \times_n b$. As in the example of $\mathbb{Z}_6$, these operations result from taking the remainders between 0 and $n - 1$, called the *residues modulo n*, of the ordinary sum and product, $a + b$ and ab, on division by n. So

$$a +_n b \equiv a + b \pmod n$$

$$a \times_n b \equiv ab \pmod n,$$

and

$$0 \le a +_n b \le n - 1, \qquad 0 \le a \times_n b \le n - 1.$$

This process of taking the remainder after division by n is called *reduction modulo n*. Table 4.3.3 shows the results of the operations of addition modulo 5 and multiplication modulo 5 in the ring $\mathbb{Z}_5$.

$+_5$	0	1	2	3	4		$\times_5$	0	1	2	3	4
0	0	1	2	3	4		0	0	0	0	0	0
1	1	2	3	4	0		1	0	1	2	3	4
2	2	3	4	0	1		2	0	2	4	1	3
3	3	4	0	1	2		3	0	3	1	4	2
4	4	0	1	2	3		4	0	4	3	2	1

Table 4.3.3: Addition and Multiplication modulo 5

4.3.3. Finite Fields

Some of these finite rings we have constructed possess additional properties which make them fields, as in § 2.4.2. The ring $\mathbb{Z}_5$, for instance, is a field because its two operations satisfy all the requirements of *F1–F4* of § 2.4.2 if we read '$+_5$' for '$\oplus$' and '$\times_5$' for '$\odot$' each time. A particular property of a field is that division by non-zero elements is always possible: or in other words, apart from zero, each element possesses a reciprocal with respect to the operation which is the analogue of multiplication (here $\times_n$). So the ring $\mathbb{Z}_6$ cannot be a field as Table 4.3.2 shows that there is no element of $\mathbb{Z}_6$ whose product (modulo 6) with 3 is 1.

It can be shown that *for a given number n, the ring $\mathbb{Z}_n$ is a field if and only if n is a prime number*. On the other hand if p is prime, $\mathbb{Z}_p$ is essentially the only field with p elements (apart from the possibility of calling its elements and operations by different names). There are finite fields which cannot be represented as $\mathbb{Z}_n$, for any n; because the number of elements in a finite field does not have to be a prime, but *it does have to be a prime power*. That is, *if a finite field contains n elements, we must have*

$$n = p^k$$

for some prime p and natural number k. Conversely, for any prime p and natural number k, there is (essentially only one) finite field with p^k elements, and it contains a subfield of p elements which is a copy of $\mathbb{Z}_p$. The finite field with p^k elements is often called the *Galois field* with p^k elements, denoted by GF(p^k), after Évariste Galois (1811–32) who initiated the study of finite fields. By considering the prime factorization of each natural number in turn, we can see that if the number of elements of a finite field is less than 100, it must be given by one of the following 35 possibilities:

2, 3, 4, 5, 7, 8, 9, 11, 13, 16, 17, 19, 23, 25, 27, 29, 31, 32, 37, 41, 43, 47, 49, 53, 59, 61, 64, 67, 71, 73, 79, 81, 83, 89, 97.

There are many other uses for primes and prime decomposition in mathematics, but our final example is from a different area.

EXAMPLE 4.3.8. A radio signal is received from a distant star and consists of groups of dashes separated by intervals of silence. The numbers of dashes in successive groups are 1, 2, 3, 5, 7, 11, 13, 17, ... until we receive the first 1000 (or perhaps the first million) primes. We could then be fairly certain that the signals were from an intelligent source.

Numbers are some of the basic constituents of our universe, and our awareness of number patterns can help in many situations to increase our understanding of that universe.

T.H.J.

BIBLIOGRAPHY

Dudley, U. (1978). *Elementary Number Theory* (2nd. ed.), W. H. Freeman.

Hardy, G. H. and Wright, E. M. (1980). *An Introduction to the Theory of Numbers* (5th ed.), Oxford University Press.

Jackson, T. H. (1975). *Number Theory*, Routledge & Kegan Paul.

CHAPTER 5

Linear Algebra

5.1. DISPLACEMENTS

The concept of a *vector* is derived from the idea of a displacement in space, which we shall briefly describe by way of motivation.

If a particle moves from A to A', it undergoes a displacement which may be denoted by

$$\mathbf{a} = \overrightarrow{AA'},$$

the arrow indicating the sense of direction. However, as a rule we are only interested in the direction and magnitude of the displacement, irrespective of the starting point. For example, in the plane, a displacement might be described by the instruction 'move two miles north-east whatever your initial position'. Hence if the directed segments $\overrightarrow{AA'}$ and $\overrightarrow{BB'}$ are parallel and of the same length and direction, they represent the same displacement (Figure 5.1.1).

In the present context a real number will be called a *scalar*, because it can be interpreted as a point on a scale ranging from $-\infty$ to ∞ [see § 2.6.1].

We shall now turn our attention to the algebraic properties of displacements. Two modes of composition are involved: (i) addition of displacements and (ii) multiplication by a scalar.

(i) The addition of two displacements is defined by the *parallelogram law* exhibited in Figure 5.1.2. Thus the 'sum' of $\mathbf{a}$ and $\mathbf{b}$ is given by the diagonal which starts at the initial point of $\mathbf{a}$ and ends at the terminal point of $\mathbf{b}$. Alternatively, we could say that this diagonal starts at the initial point of $\mathbf{b}$ and ends at

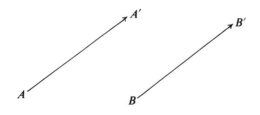

Figure 5.1.1

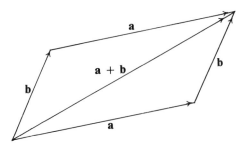

Figure 5.1.2

the terminal point of **a**. This fact, which is geometrically evident, is expressed by the equation

$$\mathbf{a} + \mathbf{b} = \mathbf{b} + \mathbf{a}.$$

The 'displacement' which signifies no movement, that is, for which starting point and end point coincide, is called the *zero displacement*. It will be denoted by **0**. Of course, it would be futile to associate a direction with the zero displacement. To every displacement **a** there corresponds a unique displacement $-\mathbf{a}$ such that $\mathbf{a} + (-\mathbf{a}) = \mathbf{0}$ (Figure 5.1.3).

(ii) The interaction between scalars and displacements is described as follows: when λ is a positive scalar, the displacement $\lambda\mathbf{a}$ has the same direction as **a** and λ times its length. When λ is negative, $\lambda\mathbf{a}$ has the opposite direction to **a** and $(-\lambda)$ times its length (see Figure 5.1.4).

When $\lambda = 0$ or $\lambda = 1$ we have that

$$0\mathbf{a} = \mathbf{0} \tag{5.1.1}$$

and

$$1\mathbf{a} = \mathbf{a}. \tag{5.1.2}$$

Finally, we observe that

$$(-1)\mathbf{a} = -\mathbf{a}.$$

The composition of displacements and scalars obeys the following general laws:

I $\mathbf{a} + \mathbf{b} = \mathbf{b} + \mathbf{a}$ (commutative law)
II $(\mathbf{a} + \mathbf{b}) + \mathbf{c} = \mathbf{a} + (\mathbf{b} + \mathbf{c})$ (associative law)
III $\mathbf{a} + \mathbf{0} = \mathbf{a}$
IV   For every **a** there exists $-\mathbf{a}$ such that $\mathbf{a} + (-\mathbf{a}) = \mathbf{0}$.
V $\lambda(\mathbf{a} + \mathbf{b}) = \lambda\mathbf{a} + \lambda\mathbf{b}$ ⎱ (distributive laws)
VI $(\lambda + \mu)\mathbf{a} = \lambda\mathbf{a} + \mu\mathbf{a}$ ⎰
VII $(\lambda\mu)\mathbf{a} = \lambda(\mu\mathbf{a})$
VIII $1\mathbf{a} = \mathbf{a}$
IX $0\mathbf{a} = \mathbf{0}$.

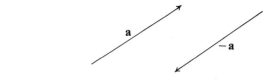

Figure 5.1.3

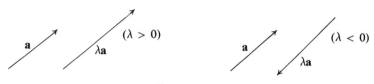

Figure 5.1.4

5.2. VECTOR SPACES

It often happens that a set of rules which are evidently true for a particular class of mathematical objects also applies to quite different situations when suitably interpreted.

Thus we shall regard a displacement as a prototype for a *vector*. However, before embarking on the abstract definition of a vector space we need to explain what we shall mean by a scalar. In most applications the scalars are the set $\mathbb{R}$ of all real numbers, as was the case with the geometrical displacements [see § 5.1]. But it is convenient to allow the 'scalars' to range over any field $\mathbb{F}$ [see § 2.4.2]; for example we may put $\mathbb{F} = \mathbb{C}$, the field of complex numbers [see § 2.7]; in the latter case the term 'scalar' is therefore no longer literally appropriate but it will still be used. We speak about a real vector space or a complex vector space or generally about a vector space over a field $\mathbb{F}$ according as the '*ground field*' that is, the field of scalars is $\mathbb{R}$, $\mathbb{C}$ or any field $\mathbb{F}$. We can now enunciate the

DEFINITION 5.2.1. A *vector space V over the field* $\mathbb{F}$ consists of a set of objects called vectors with the following properties: (i) any two vectors **a** and **b** have a sum **a** + **b**, which is a uniquely defined vector; (ii) if **a** is a vector and λ is a scalar, then a unique vector λ**a** (or **a**λ) is defined. The compositions referred to in (i) and (ii) satisfy the laws I to IX listed in § 5.1.

At first sight the reader might be put off by the abstract nature of this definition. But it will soon become apparent that vector spaces abound in mathematics. The few examples of vector spaces we are about to give here only provide a glimpse of this ubiquitous phenomenon.

EXAMPLE 5.2.1. The displacements form a vector space over the real numbers with respect to the rules for composition defined in § 5.1.

EXAMPLE 5.2.2.   Let n be a positive number. An n-*tuple* over a field $\mathbb{F}$ is a sequence

$$\mathbf{a} = (a_1, a_2, \ldots, a_n),$$

whose components $a_1, a_2, \ldots, a_n$ range independently over the whole of $\mathbb{F}$. If

$$\mathbf{b} = (b_1, b_2, \ldots, b_n)$$

is another n-tuple, we define

$$\mathbf{a} + \mathbf{b} = (a_1 + b_1, a_2 + b_2, \ldots, a_n + b_n),$$

$$k\mathbf{a} = (ka_1, ka_2, \ldots, ka_n) \quad (k \in \mathbb{F})$$

$$\mathbf{0} = (0, 0, \ldots, 0).$$

It is not hard to verify that, with these rules for composition, the n-tuples form a vector space over $\mathbb{F}$ for each value of n, which we denote by $\mathbb{F}^n$. Of particular interest are the vector spaces $\mathbb{R}^n$, called the *Euclidean Space* of dimension n, and $\mathbb{C}^n$, the n-dimensional *complex vector space*.

EXAMPLE 5.2.3.   The set of all polynomials over a field $\mathbb{F}$ of degree not exceeding n [see § 14.1] forms a vector space (over $\mathbb{F}$), where the addition of two polynomials and multiplication of a polynomial by a scalar is defined in the usual way [see § 14.2]. Thus if

$$p(x) = a_0 + a_1 x + a_2 x^2 + \ldots + a_n x^n$$

and

$$q(x) = b_0 + b_1 x + b_2 x^2 + \ldots + b_n x^n$$

then

$$p(x) + q(x) = (a_0 + b_0) + (a_1 + b_1)x + (a_2 + b_2)x^2 + \ldots + (a_n + b_n)x^n,$$

$$kp(x) = ka_0 + ka_1 x + ka_2 x^2 + \ldots + ka_n x^n.$$

This space is denoted by $P_n(\mathbb{F})$. Note that in this context we do not envisage multiplying one polynomial by another.

EXAMPLE 5.2.4.   The collection of all n times differentiable real-valued functions $f(x)$ defined over an interval $[a, b]$ [see IV, Definition 2.10.1], forms a vector space over $\mathbb{R}$. For if $f(x)$ and $g(x)$ are n times differentiable functions, so are $f(x) + g(x)$ and $kf(x)$, where k is a real number. We denote this space by C^n.

EXAMPLE 5.2.5.   All m by n matrices over a field $\mathbb{F}$ [see § 6.2] form a vector space with respect to the composition $\mathbf{A} + \mathbf{B}$ and $k\mathbf{A}$. Again, matrix multiplication does not enter into this consideration.

5.3. LINEAR DEPENDENCE

This is the most important notion in the theory of vector spaces. We begin with a geometric illustration. Consider three vectors (displacements) **a, b, c** in three dimensional space, and for simplicity let all three vectors start at the same point 0 (Figure 5.3.1). Then in general, the vectors will point into space like ribs of an open umbrella. But in exceptional circumstances they might lie flat in a single plane, and it is this case we shall consider in more detail. Assuming that **a** and **b** are not collinear, the vector **c** will then lie in the plane defined by **a** and **b**. Now every vector in this plane is of the form $\lambda\mathbf{a} + \mu\mathbf{b}$ (Figure 5.3.2). Hence, in particular

$$\mathbf{c} = \lambda\mathbf{a} + \mu\mathbf{b}.$$

It is, however, preferable to treat the three vectors on the same footing and to say that there exists a relation of the form

$$\alpha\mathbf{a} + \beta\mathbf{b} + \gamma\mathbf{c} = \mathbf{0}, \tag{5.3.1}$$

where the scalars α, β, γ are not all zero. The equation (5.3.1) then implies that we can express one of the vectors in terms of the other two.

We generalize this idea for the case of an arbitrary number of vectors in any vector space.

DEFINITION 5.3.1. The vectors $\mathbf{a}_1, \mathbf{a}_2, \ldots, \mathbf{a}_m$ are said to be *linearly dependent* if there exist scalars $\alpha_1, \alpha_2, \ldots, \alpha_m$, not all zero, such that

$$\alpha_1\mathbf{a}_1 + \alpha_2\mathbf{a}_2 + \ldots + \alpha_m\mathbf{a}_m = \mathbf{0}.$$

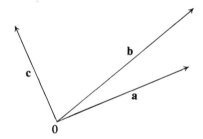

Figure 5.3.1

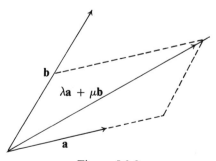

Figure 5.3.2

Again, this means that one of the vectors can be expressed as a *linear combination* of the others; for example, if $\alpha_m \neq 0$, we can write

$$\mathbf{a}_m = \lambda_1 \mathbf{a}_1 + \lambda_2 \mathbf{a}_2 + \ldots + \lambda_{m-1} \mathbf{a}_{m-1},$$

where $\lambda_i = -\alpha_i / \alpha_m$ ($i = 1, 2, \ldots, m - 1$).

If a collection of vectors is not linearly dependent, it is said to be *linearly independent*. It is useful to give a formal definition of this concept.

DEFINITION 5.3.2.   The vectors $\mathbf{a}_1, \mathbf{a}_2, \ldots, \mathbf{a}_m$ are *linearly independent* if an equation of the form

$$\alpha_1 \mathbf{a}_1 + \alpha_2 \mathbf{a}_2 + \ldots + \alpha_m \mathbf{a}_m = \mathbf{0}$$

implies that $\alpha_1 = \alpha_2 = \ldots = \alpha_m = 0$. Thus when $m > 2$ this means that linear independent vectors do not lie in the same hyperplane [see V, § 2.5].

Both definitions 5.3.1 and 5.3.2 are unsatisfactory in so far as they fail to provide a method for deciding whether or not a given collection of vectors is linearly dependent. The answer to this question is facilitated by the observation that the following operations do not interfere with linear dependence or independence. Thus in order to examine the linear dependence of $\mathbf{a}_1, \mathbf{a}_2, \ldots, \mathbf{a}_m$ we may

(α) permute the vectors in any manner,

(β) replace any one of the vectors, say $\mathbf{a}_i$, by $\lambda \mathbf{a}_i$ where λ is an arbitrary non-zero scalar,

(γ) replace $\mathbf{a}_i$ by $\mathbf{a}_i + \lambda \mathbf{a}_j$, where $j \neq i$.

The operations (α), (β) and (γ) may be repeated any number of times; for example, $\mathbf{a}_m$ may be replaced by

$$\mathbf{a}_m + \lambda_1 \mathbf{a}_1 + \ldots + \lambda_{m-1} \mathbf{a}_{m-1}$$

without altering the problem.

Any collection that contains the zero vector, say $\mathbf{a}_1 = \mathbf{0}$, is linearly dependent, because we have the non-trivial relation

$$1\mathbf{0} + 0\mathbf{a}_2 + \ldots + 0\mathbf{a}_m = \mathbf{0}.$$

In particular, a collection which consists of a single vector, is linearly dependent if and only if that vector is zero.

Two vectors $\mathbf{a}_1, \mathbf{a}_2$ are linearly dependent if and only if they are proportional that is $\mathbf{a}_1 = \lambda \mathbf{a}_2$ or $\mathbf{a}_2 = \mu \mathbf{a}_1$, where λ and μ are scalars.

When we are dealing with vectors of types $\mathbb{R}^n$ or $\mathbb{C}^n$, the question of linear dependence can be decided by a systematic procedure known as *pivotal condensation*. [Detailed information about this method is given in III, § 3.3]

Suppose we wish to examine the set of n-tuples

$$A: \mathbf{a}_1, \mathbf{a}_2, \ldots, \mathbf{a}_m$$

for linear dependence. The idea is to replace A by a set of n-tuples

$$B: \mathbf{b}_1, \mathbf{b}_2, \ldots, \mathbf{b}_m,$$

which behaves like A with regard to linear dependence, that is, A is linearly dependent if and only if B is linearly dependent. The transition from A to B is carried out by repeatedly performing the operations (α), (β) and (γ) until a system B is reached for which the problem is solved at a glance.

The method is best explained by working through a simple, but typical, example.

EXAMPLE 5.3.1. Determine whether the four 4-tuples

$$\mathbf{a} = (1, 2, -1, 0), \qquad \mathbf{b} = (2, 3, 2, 1)$$

$$\mathbf{c} = (1, 3, -2, 2), \qquad \mathbf{d} = (-1, -2, -2, -3)$$

are linearly dependent.

As a preliminary step we rearrange the vectors, if necessary, in such a way that the first component of the first vector, say $\mathbf{a}$, is non-zero and preferably equal to unity. To ensure the latter property we may replace $\mathbf{a}$ by $\mathbf{a}/a_1$ where a_1 is the first component of $\mathbf{a}$ (operation (β)). If the first components of all the n-tuples in the collection are zero, we delete the first components and proceed to examine the resulting $(n-1)$-tuples for linear dependence.

In the present example, this preliminary step is unnecessary, as the first component of $\mathbf{a}$ happens to be equal to unity. The vectors are arranged in the scheme shown in Table 5.3.1. The first row, labelled $\mathbf{a}$, contains the components

$\mathbf{a}$	①	2	-1	0 ‖	2
$\mathbf{b}$	2	3	2	1 ‖	8
$\mathbf{c}$	1	3	-2	2 ‖	4
$\mathbf{d}$	-1	-2	-2	-3 ‖	-8

Table 5.3.1

of $\mathbf{a}$ followed by their sum $(= 2)$, known as the *check sum*, which is entered after the double line. Similarly, the other rows correspond to the vectors $\mathbf{b}$, $\mathbf{c}$ and $\mathbf{d}$, each furnished with their check sum. We have ringed the first component of $\mathbf{a}$, which has been chosen as *pivot*. The process of pivotal condensation consists in adding a scalar multiple of the first row in turn to each of the subsequent rows so as to make their first entries equal to zero. For example, we replace $\mathbf{b}$ by $\mathbf{b} - 2\mathbf{a}$, as shown in Table 5.3.2. The check sums take part in this procedure: thus the 8 in the last column of Table 5.3.1 is replaced by $8 - 2 \times 2 = 4$. The calculation can be checked by verifying that this last entry is also equal to the sum of the components of the vector $\mathbf{b} - 2\mathbf{a}$; indeed,

$$0 + (-1) + 4 + 1 = 4.$$

Similarly, we construct the rows $c - a$ and $d + a$ of Table 5.3.2. In each case the

a	1	2	−1	0	‖	2
b − 2a	0	−1	4	1	‖	4
c − a	0	1	−1	2	‖	2
d + a	0	0	−3	−3	‖	−6

Table 5.3.2

first entry is zero and the last entry should be equal to the sum of the other entries in the same row. The first row plays no further role in the procedure. Next, we select a pivot in the second column, disregarding the first row, that is the entry 2. Hence either −1 or 1 can be chosen. Since we prefer the pivot to be unity, we multiply the second row throughout by −1 (operation (β)); the second row now reads:

$$2a - b : 0 \; ① - 4 - 1 \; \| \; 4.$$

Its pivot is used to reduce all entries below it to zero. Hence $c - a$ is replaced by $(c - a) - (2a - b) = -3a + b + c$, and we obtain the scheme

a	1	2	−1	0	‖	2
2a − b	0	1	−4	−1	‖	−4
−3a + b + c	0	0	3	3	‖	6
d + a	0	0	−3	−3	‖	−6.

There is now an obvious relation between the last two lines, namely;

$$-3a + b + c = -(d + a),$$

that is

$$2a - b - c - d = 0,$$

which proves that the vectors are linearly dependent.

Let us return to the general case of any collection of n-tuples. We use a cross $\times$ to denote an arbirary scalar, whilst $*$ stands for any non-zero scalar. It can then be shown that the n-tuples sketched in the scheme of Table 5.3.3 are always linearly independent. It is more difficult to say what will happen if some of the

*	×	×	...	×
0	*	×	...	×
0	0	*	...	×
⋮	⋮	⋮		⋮
0	0	0	...*...	×

Table 5.3.3

diagonal entries in Table 5.3.3 are zero. But the following special case is important: when Table 5.3.3 is a square, that is when we are concerned with n n-tuples
such that all entries below the diagonal are zero, we call Table 5.3.3 an *upper
diagonal scheme*, and we have the

PROPOSITION 5.3.1.   *The rows of an upper diagonal (square) scheme are
linearly independent if and only if all diagonal entries are non-zero.*

5.4.   BASIS AND DIMENSION

In order to specify the position of a point in three-dimensional space it is
convenient to choose a frame of reference. This consists of any set of three
linearly independent vectors $\mathbf{u}_1, \mathbf{u}_2, \mathbf{u}_3$ (Figure 5.4.1). An arbitrary vector $\mathbf{x}$ can
then be uniquely expressed in the form

$$\mathbf{x} = x_1\mathbf{u}_1 + x_2\mathbf{u}_2 + x_3\mathbf{u}_3, \tag{5.4.1}$$

where x_1, x_2, x_3 are scalars.

We mention in passing that it is not necessary to assume that the reference
vectors are of unit length and are mutually perpendicular. These concepts will
be discussed elsewhere [see § 10.2] and the notion of an orthogonal basis will
then be introduced.

The fact that a spatial frame of reference consists of precisely three vectors is
equivalent to the statement that the space, in which we live, is three-dimensional.

These ideas are carried over to a vector space of 'higher dimensions', a term
which we are now going to define. We shall only consider finite dimensions.

DEFINITION 5.4.1.   A vector space V over a field $\mathbb{F}$ of scalars is said to be of
dimension n, if there exist n vectors

$$\mathbf{u}_1, \mathbf{u}_2, \ldots, \mathbf{u}_n \tag{5.4.2}$$

such that every vector $\mathbf{x}$ of V can be uniquely expressed as

$$\mathbf{x} = x_1\mathbf{u}_1 + x_2\mathbf{u}_2 + \ldots + x_n\mathbf{u}_n, \tag{5.4.3}$$

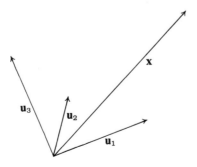

Figure 5.4.1

where $x_1, x_2, \ldots, x_n$ are suitable scalars, called the *components* or *coordinates* of $\mathbf{x}$. We say the vectors (5.4.2) form a *basis* of V and we write

$$V = [\mathbf{u}_1, \mathbf{u}_2, \ldots, \mathbf{u}_n].$$

Some remarks are called for in order to elucidate and indeed justify this definition:

(i) The word 'basis' is synonymous with the geometric terms 'frame of reference' or 'system of coordinate axes'.

(ii) A set of basic vectors is always linearly independent.

(iii) Basic vectors can be chosen in many ways; but if

$$V = [\mathbf{u}_1, \mathbf{u}_2, \ldots, \mathbf{u}_n] = [\mathbf{w}_1, \mathbf{w}_2, \ldots, \mathbf{w}_m],$$

are two bases, then $n = m$. Thus for a given vector space the number of basic vectors, if finite, is always the same. This number is called the *dimension* of V, and we write

$$n = \dim V.$$

The following proposition is of fundamental importance. It gets closest to describing in algebraic terms what one intuitively means by dimension.

PROPOSITION 5.4.1. *If V is a vector space of dimension n, then any set comprising more than n vectors of V is linearly dependent.*

EXAMPLE 5.4.1. Since $\mathbb{R}^3$ is of dimension 3, the four 3-tuples

$$\mathbf{a} = (a_1, a_2, a_3), \qquad \mathbf{b} = (b_1, b_2, b_3), \qquad \mathbf{c} = (c_1, c_2, c_3), \qquad \mathbf{d} = (d_1, d_2, d_3)$$

are linearly dependent, whatever the values of the components of the vectors.

We shall now study the relationship between two different sets of basic vectors. Suppose that

$$U: \mathbf{u}_1, \mathbf{u}_2, \ldots, \mathbf{u}_n$$

and

$$W: \mathbf{w}_1, \mathbf{w}_2, \ldots, \mathbf{w}_n$$

are two bases of V. Then each member of W can be expressed linearly in terms of U, that is we have equations

$$\mathbf{w}_i = \sum_{j=1}^{n} a_{ij}\mathbf{u}_j \quad (i = 1, 2, \ldots, n), \tag{5.4.4}$$

which describe the transition from U to W. The system (5.4.4) is specified by an n by n matrix [see § 6.1. and § 6.2]

$$\mathbf{A} = (a_{ij}).$$

Conversely, there are equations of the form

$$\mathbf{u}_j = \sum_{k=1}^{n} b_{jk}\mathbf{w}_k \quad (k = 1, 2, \ldots, n), \tag{5.4.5}$$

involving a matrix

$$\mathbf{B} = (b_{jk}).$$

On eliminating one or the other basic set from (5.4.4) and (5.4.5) and using the linear independence of basic vectors we find that

$$\mathbf{AB} = \mathbf{BA} = \mathbf{I}.$$

Hence the matrices $\mathbf{A}$ and $\mathbf{B}$ are inverse to each other [see Definition 6.4.1]. Thus *a change of basis is described by an invertible matrix* [see Theorem 6.4.2]. *Conversely, every invertible matrix can be employed to define a change of basis.*

EXAMPLE 5.4.2.   When $V = \mathbb{F}^n$, the space of n-tuples

$$(x_1, x_2, \ldots, x_n) \quad (x_i \in \mathbb{F})$$

[see Example 5.2.2], then a particularly simple basis is furnished by the set of vectors:

$$\mathbf{e}_1 = (1, 0, 0, \ldots, 0)$$
$$\mathbf{e}_2 = (0, 1, 0, \ldots, 0)$$
$$\vdots \qquad \vdots \qquad \vdots$$
$$\mathbf{e}_n = (0, 0, 0, \ldots, 1).$$

This set is sometimes referred to as the *natural basis* or the *standard basis* for $\mathbb{F}^n$.

EXAMPLE 5.4.3.   A basis for $P_n(\mathbb{F})$ is $\{1, x, \ldots, x^n\}$.

5.5. SUBSPACES

Let V be a vector space over the field $\mathbb{F}$. We often have occasion to consider certain (non-empty) subsets of vectors which we term subspaces of V.

In three-dimensional space one type of subspace consists of all vectors lying in a fixed plane through the origin 0 (see Figure 5.3.2). As we have already observed, if $\mathbf{a}$ and $\mathbf{b}$ lie in that plane, so does every vector of the form

$$\mathbf{c} = \lambda\mathbf{a} + \mu\mathbf{b},$$

where λ and μ are arbitrary constants. Another kind of subspace comprises all vectors which lie in a fixed line through 0 (Figure 5.5.1). Let $\mathbf{e}$ be a fixed non-zero vector in this line. If $\mathbf{a}$ and $\mathbf{b}$ are any two vectors in the line, we can write

$$\mathbf{a} = \alpha\mathbf{e}, \qquad \mathbf{b} = \beta\mathbf{e},$$

where α and β are suitable scalars. Again, we observe that

$$\lambda\mathbf{a} + \mu\mathbf{b} = (\lambda\alpha + \mu\beta)\mathbf{e}$$

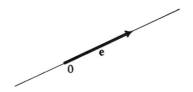

Figure 5.5.1

lies in the subspace. These considerations lead to the following definition, which applies to all vector spaces, including those of infinite dimension.

DEFINITION 5.5.1. A non-empty subset S of a vector space V is called a *subspace* of V, if $\lambda\mathbf{a} + \mu\mathbf{b}$ belongs to S whenever $\mathbf{a}$ and $\mathbf{b}$ belong to S, where λ and μ are arbitrary scalars. In particular, $\mathbf{a} + \mathbf{b}$ and $\lambda\mathbf{a}$ belong to S.

All vector spaces have the trivial subspaces $S = V$ and $S = \{\mathbf{0}\}$, the latter consisting of the zero vector only.

EXAMPLE 5.5.1. Let V be the space of real 4-tuples $\mathbf{x} = (x_1, x_2, x_3, x_4)$. Then the subset S consisting of those 4-tuples satisfying

$$x_1 - x_2 + 2x_3 + 5x_4 = 0 \qquad\qquad (5.5.1)$$

forms a subspace of V. For suppose that $\mathbf{a} = (a_1, a_2, a_3, a_4)$ and $\mathbf{b} = (b_1, b_2, b_3, b_4)$ belong to S, so that

$$a_1 - a_2 + 2a_3 + 5a_4 = 0$$

and

$$b_1 - b_2 + 2b_3 + 5b_4 = 0.$$

Then by multiplying the last two equations respectively by λ and μ and adding, it is seen that the 4-tuple $\lambda\mathbf{a} + \mu\mathbf{b}$ also satisfies (5.5.1). Thus S is a subspace of V.

On the other hand, if (5.5.1) were replaced by

$$x_1 - x_2 + 2x_3 + 5x_4 = 1,$$

then the set S would not form a subspace because if $\mathbf{x} \in S$, then $\lambda\mathbf{x} \notin S$ when $\lambda \neq 1$. Likewise, the set S of 4-tuples satisfying

$$x_1{}^2 + x_2{}^2 + x_3{}^2 - x_4{}^2 = 0$$

does not define a subspace; for if $\mathbf{x}, \mathbf{y} \in S$, it is not true in general that $\mathbf{x} + \mathbf{y} \in S$.

EXAMPLE 5.5.2. Subspaces frequently arise in connection with linear homogeneous equations of various kinds [see § 5.9]. These problems are described by equations in which the unknown quantities occur to the first degree only, there being no constant terms other than zero.

Thus if $\mathbf{A}$ is a given matrix [see § 6.2] and $\mathbf{x}$ an unknown vector, then the solutions of

$$\mathbf{Ax} = \mathbf{0} \qquad\qquad (5.5.2)$$

form a subspace of the vector space on which $\mathbf{A}$ operates. For if $\mathbf{u}$ and $\mathbf{v}$ are solutions of (5.5.2) that is

$$\mathbf{Au} = \mathbf{0}, \qquad \mathbf{Av} = \mathbf{0},$$

then

$$\mathbf{A}(\lambda\mathbf{u} + \mu\mathbf{v}) = \mathbf{0},$$

for all scalars λ, μ.

EXAMPLE 5.5.3.   Real-valued functions $f(x)$ which are twice differentiable form a vector space C^2 over the real numbers [see Example 5.2.4]. Those functions which satisfy the differential equation [see IV, § 7.3.2]

$$f''(x) + af'(x) + bf(x) = 0 \qquad (5.5.3)$$

form a subspace of this vector space. For if $u(x)$ and $v(x)$ are two solutions of (5.5.3), so is the function $\lambda u(x) + \mu v(x)$, where λ and μ are arbitrary scalars.

There is a general method for constructing subspaces of a given vector space V. Let S be any set of vectors; for simplicity we shall assume that S is a finite set, say

$$S: \mathbf{a}_1, \mathbf{a}_2, \ldots, \mathbf{a}_k. \qquad (5.5.4)$$

Then consider the collection U of all vectors of the form

$$\mathbf{u} = \xi_1 \mathbf{a}_1 + \xi_2 \mathbf{a}_2 + \ldots + \xi_k \mathbf{a}_k, \qquad (5.5.5)$$

where $\xi_1, \xi_2, \ldots, \xi_k$ range over all scalars, that is, U consists of all linear combinations of the members of S. It is easy to verify that U is a subspace of V. We write

$$U = \{\mathbf{a}_1, \mathbf{a}_2, \ldots, \mathbf{a}_k\}, \qquad (5.5.6)$$

and we say that U is *generated* (or *spanned*) by $\mathbf{a}_1, \mathbf{a}_2, \ldots, \mathbf{a}_k$. For example, let $V = \mathbb{R}^3$ and

$$\mathbf{a} = (1, 1, 0), \qquad \mathbf{b} = (0, 1, 1).$$

Then $U = \{\mathbf{a}, \mathbf{b}\}$ consist of all vectors of the form

$$\xi(1, 1, 0) + \eta(0, 1, 1) = (\xi, \xi + \eta, \eta),$$

where $\xi, \eta \in \mathbb{R}$. Alternatively, U is the collection of all triplets (x_1, x_2, x_3) satisfying the equation

$$x_1 - x_2 + x_3 = 0.$$

From now on we shall confine ourselves to finite-dimensional vector spaces. Let V be a vector space of dimension n and let U be a subspace of V. Then the following facts can be established:
   (i) U is finite-dimensional and, if dim $U = k$, we have that $0 \leq k \leq n$.
   (ii) If $\mathbf{u}_1, \mathbf{u}_2, \ldots, \mathbf{u}_k$ is a basis of U and $k < n$, then $n - k$ vectors

$$\mathbf{w}_1, \mathbf{w}_2, \ldots, \mathbf{w}_{n-k} \qquad (5.5.7)$$

can be found in V such that

$$\mathbf{u}_1, \ldots, \mathbf{u}_k, \qquad \mathbf{w}_1, \ldots, \mathbf{w}_{n-k} \qquad (5.5.8)$$

is a basis of V. We say that the basis (5.5.8) of V has been *adapted* to the subspace U; in other words, we can always construct a basis of V which incorporates a given basis of a subspace. However, it should be observed that the vectors (5.5.7) are not uniquely determined. The integer

$$\dim V - \dim U = n - k \tag{5.5.9}$$

is called the *co-dimension* of U in V, and the vectors (5.5.7) are called a *cobasis* of U in V.

(iii) Given a subspace U of V of dimension k, where $k < n$, there exists a subspace W of dimension $n - k$ such that

$$V = U + W \tag{5.5.10}$$

in the sense that every vector $\mathbf{v}$ of V can be uniquely expressed as

$$\mathbf{v} = \mathbf{u} + \mathbf{w},$$

where $\mathbf{u} \in U$ and $\mathbf{w} \in W$. We say that V is the *direct sum* of U and W and that W is a *complement* of U in V.

Indeed, referring to (5.5.8) we can put

$$W = \{\mathbf{w}_1, \mathbf{w}_2, \ldots, \mathbf{w}_{n-k}\}.$$

However, it should be pointed out that W is not uniquely determined, that is U has more than one complement in V.

5.6. RANK

In Example 5.3.1 we explained a method for deciding whether or not a system

$$S: \mathbf{a}_1, \mathbf{a}_2, \ldots, \mathbf{a}_k$$

is linearly dependent. Suppose that

$$\lambda_1 \mathbf{a}_1 + \lambda_2 \mathbf{a}_2 + \ldots + \lambda_k \mathbf{a}_k = 0$$

is a non-trivial relation between the members of S. Allowing for a renumbering, if necessary, we may assume that $\lambda_k \neq 0$. Hence

$$\mathbf{a}_k = (-\lambda_1/\lambda_k)\mathbf{a}_1 + (-\lambda_2/\lambda_k)\mathbf{a}_2 + \ldots + (-\lambda_{k-1}/\lambda_k)\mathbf{a}_{k-1},$$

that is, $\mathbf{a}_k$ is expressible as a linear combination of the other vectors of S; for short, we shall say that $\mathbf{a}_k$ is *redundant*. If this is the case we remove $\mathbf{a}_k$ and consider the smaller system

$$S': \mathbf{a}_1, \mathbf{a}_2, \ldots, \mathbf{a}_{k-1}.$$

If the vectors of S' are linearly independent, the process terminates, otherwise we remove a redundant vector which, after a suitable renumbering, may be called $\mathbf{a}_{k-1}$. Evidently, this procedure must come to an end after a finite number of steps. Suppose that when all redundant vectors have been removed, we arrive at a system $\bar{S}$ comprising r vectors of S which in the original mode of labelling are:

$$\bar{S}: \mathbf{a}_{i_1}, \mathbf{a}_{i_2}, \ldots, \mathbf{a}_{i_r}. \tag{5.6.1}$$

The number r is called the *rank* of S, and we write

$$r = r(S).$$

The concept of rank is of fundamental importance. We can describe it in a slightly different way as follows: the system S is of rank r if it contains at least one set of r linearly independent vectors, while all subsets of $r + 1$ vectors (if any) selected from S are linearly dependent. More briefly, r is the size of the largest subset of linearly independent vectors contained in S.

Yet another way to define the rank is to introduce the subspace U generated by the members of S, thus

$$U = \{\mathbf{a}_1, \mathbf{a}_2, \ldots, \mathbf{a}_k\}.$$

Then

$$r(S) = \dim U.$$

Indeed the vectors listed in (5.6.1) form a basis of U.

EXAMPLE 5.6.1.   Let

$$\mathbf{a} = (1, 0, 1, 0), \qquad \mathbf{b} = (0, 1, 1, 1),$$
$$\mathbf{c} = (1, 1, 2, 1), \qquad \mathbf{d} = (1, -1, 0, -1).$$

Then the system

$$S: \mathbf{a}, \mathbf{b}, \mathbf{c}, \mathbf{d}$$

is of rank 2. For $\mathbf{c}$ and $\mathbf{d}$ are redundant, since

$$\mathbf{c} = \mathbf{a} + \mathbf{b}, \qquad \mathbf{d} = \mathbf{a} - \mathbf{b}$$

while $\mathbf{a}$ and $\mathbf{b}$ are linearly independent. In fact

$$\{\mathbf{a}, \mathbf{b}, \mathbf{c}, \mathbf{d}\} = [\mathbf{a}, \mathbf{b}].$$

The notion of rank will now be extended to matrices. Let

$$\mathbf{A} = \begin{pmatrix} a_{11} & a_{12} & \cdots & a_{1n} \\ a_{21} & a_{22} & \cdots & a_{2n} \\ \vdots & \vdots & & \vdots \\ a_{m1} & a_{m2} & \cdots & a_{mn} \end{pmatrix}$$

be an m by n matrix. There are two ways of viewing $\mathbf{A}$ as a collection of vectors [see § 6.2.(iv)]: (i) we may think of A as a collection of n column-vectors, each being an m-tuple, thus

$$\mathbf{A} = (\mathbf{a}_1, \mathbf{a}_2, \ldots, \mathbf{a}_n), \tag{5.6.2}$$

where

$$\mathbf{a}_j = \begin{pmatrix} a_{1j} \\ a_{2j} \\ \vdots \\ a_{mj} \end{pmatrix} \quad (j = 1, 2, \ldots, n)$$

or (ii) we may resolve **A** into m row-vectors, each being an n-tuple, that is

$$\mathbf{A} = \begin{pmatrix} \mathbf{a}^{(1)} \\ \mathbf{a}^{(2)} \\ \vdots \\ \mathbf{a}^{(m)} \end{pmatrix}, \tag{5.6.3}$$

where

$$\mathbf{a}^{(i)} = (a_{i1}, a_{i2}, \ldots, a_{in}) \qquad (i = 1, 2, \ldots, m).$$

The foregoing discussion about the rank of a collection of vectors therefore leads naturally to the introduction of the *column rank* of **A** as the maximal number of linearly independent columns and of the *row rank* as the maximal number of linearly independent rows. Fortunately, it is a remarkable fact that, for every matrix **A**,

$$\text{column rank of } \mathbf{A} = \text{row rank of } \mathbf{A}. \tag{5.6.4}$$

The common value of column rank and row rank is simply referred to as the *rank* of the matrix.

In a concrete situation the rank of a matrix may be determined by subjecting either the rows or the columns to pivotal condensation [see Example 5.3.1]. This will lead to the discovery of linear relations and redundancies between the rows or columns.

EXAMPLE 5.6.2.   The matrix

$$\mathbf{A} = \begin{pmatrix} 1 & 0 & 2 & 3 & -1 \\ 2 & 1 & 0 & 2 & 0 \\ -4 & -3 & 4 & 0 & -2 \\ 1 & 1 & -2 & -1 & 1 \end{pmatrix} = \begin{pmatrix} \mathbf{a}^{(1)} \\ \mathbf{a}^{(2)} \\ \mathbf{a}^{(3)} \\ \mathbf{a}^{(4)} \end{pmatrix}$$

is of rank 2. For we observe that $\mathbf{a}^{(1)}$ and $\mathbf{a}^{(2)}$ are linearly independent (not proportional) while $\mathbf{a}^{(3)}$ and $\mathbf{a}^{(4)}$ satisfy the relations

$$\mathbf{a}^{(3)} = 2\mathbf{a}^{(1)} - 3\mathbf{a}^{(2)},$$

$$\mathbf{a}^{(4)} = -\mathbf{a}_2^{(1)} + \mathbf{a}^{(2)} \tag{5.6.5}$$

and are therefore redundant. Alternatively, but less conveniently, we may resolve **A** into columns, thus

$$\mathbf{A} = (\mathbf{a}_1, \mathbf{a}_2, \mathbf{a}_3, \mathbf{a}_4, \mathbf{a}_5).$$

We note that $\mathbf{a}_1$ and $\mathbf{a}_2$ are linearly independent, while the relations

$$\mathbf{a}_3 = 2\mathbf{a}_1 - 4\mathbf{a}_2, \qquad \mathbf{a}_4 = 3\mathbf{a}_1 - 4\mathbf{a}_2, \qquad \mathbf{a}_5 = -\mathbf{a}_1 + 2\mathbf{a}_2$$

show that $\mathbf{a}_3$, $\mathbf{a}_4$ and $\mathbf{a}_5$ are redundant. Thus, as expected, the column rank of $\mathbf{A}$ is equal to two, as is the row rank.

We append some obvious remarks about the rank:
(1) The rank of $\mathbf{A}$ is zero if and only if $\mathbf{A}$ is the zero matrix [see § 6.2(iii)].
(2) If $\mathbf{A}$ is an m by n matrix of rank r, then

$$0 \le r \le m, \qquad 0 \le r \le n.$$

(3) The most general matrix of rank unity is of the form

$$\mathbf{A} = \begin{pmatrix} \alpha_1\beta_1 & \alpha_1\beta_2 & \cdots & \alpha_1\beta_n \\ \alpha_2\beta_1 & \alpha_2\beta_2 & \cdots & \alpha_2\beta_n \\ \vdots & \vdots & & \vdots \\ \alpha_m\beta_1 & \alpha_m\beta_2 & \cdots & \alpha_m\beta_n \end{pmatrix}, \tag{5.6.6}$$

where the $m + n$ numbers $\alpha_1, \ldots, \alpha_m, \beta_1, \ldots, \beta_n$ are arbitrary except that not all $\alpha_1, \ldots, \alpha_m$ are zero and not all $\beta_1, \ldots, \beta_n$ are zero.

(4) $\qquad r(\mathbf{AB}) \le r(\mathbf{A}), \qquad r(\mathbf{AB}) \le r(\mathbf{B}) \quad$ [see § 6.2(vi)]. $\qquad$ (5.6.7)

(5) If $\mathbf{P}$ and $\mathbf{Q}$ are invertible matrices [see Theorem 6.4.2], then

$$r(\mathbf{PAQ}) = r(\mathbf{A}). \tag{5.6.8}$$

(6) $\qquad r\begin{pmatrix} \mathbf{X} & \mathbf{0} \\ \mathbf{0} & \mathbf{Y} \end{pmatrix} = r(\mathbf{X}) + r(\mathbf{Y}), \qquad$ (5.6.9)

[see § 6.6 for the definition of a partitioned matrix].

5.7. LINEAR EQUATIONS—PRELIMINARIES

An algebraic equation in n unknowns $x_1, x_2, \ldots, x_n$ is said to be *linear* if it is of the form

$$a_1 x_1 + a_2 x_2 + \ldots + a_n x_n = h, \tag{5.7.1}$$

where $a_1, a_2, \ldots, a_n, h$ are given quantities. The term is derived from the fact that, when $n = 2$, the equation

$$a_1 x_1 + a_2 x_2 = h$$

represents a straight line in the (x_1, x_2) plane provided that a_1 and a_2 are not both zero [see V (1.2.1)].

Our problem is to solve simultaneously a system of linear equations. Thus a system involving n equations and n unknowns can be written as follows:

$$\begin{aligned} a_{11}x_1 + a_{12}x_2 + \ldots + a_{1n}x_n &= h_1 \\ a_{21}x_1 + a_{22}x_2 + \ldots + a_{2n}x_n &= h_2 \\ &\cdots\cdots\cdots\cdots\cdots\cdots\cdots\cdots \\ a_{m1}x_1 + a_{m2}x_2 + \ldots + a_{mn}x_n &= h_m, \end{aligned} \tag{5.7.2}$$

where the quantities a_{ij} and h_i are given. The handling of linear equations is greatly facilitated by the use of matrix algebra. To express (5.7.2) succinctly we introduce the m by n matrix, called the *coefficient matrix* of (5.7.2),

$$\mathbf{A} = \begin{pmatrix} a_{11} & a_{12} & \cdots & a_{1n} \\ a_{21} & a_{22} & \cdots & a_{2n} \\ \vdots & \vdots & & \vdots \\ a_{m1} & a_{m2} & \cdots & a_{mn} \end{pmatrix}, \tag{5.7.3}$$

and we gather the unknowns and the right-hand members of (5.7.2) into column vectors, that is, n by 1 and m by 1 matrices respectively. Thus

$$\mathbf{x} = \begin{pmatrix} x_1 \\ x_2 \\ \vdots \\ x_n \end{pmatrix}, \qquad \mathbf{h} = \begin{pmatrix} h_1 \\ h_2 \\ \vdots \\ h_m \end{pmatrix}.$$

Then the m equations (5.7.2) reduce to the single matrix equation

$$\mathbf{Ax} = \mathbf{h}, \tag{5.7.4}$$

which presents the problem in a more attractive and suggestive manner. No significance is attached to the fact that the unknowns are arranged in a column rather than in a row. Indeed, on transposing (5.7.4) [see § 6.5] we obtain the equivalent equation

$$\mathbf{x}'\mathbf{A}' = \mathbf{h}',$$

which involves the row vector $\mathbf{x}' = (x_1, x_2, \ldots, x_n)$.

We shall see that the nature of the solution depends critically on the rank of $\mathbf{A}$. By way of introduction we present three very simple systems, which already display the germ of the general theory:

(i) $x_1 + x_2 = h_1$ (ii) $x_1 + 2x_2 = 1$ (iii) $x_1 + 2x_2 = 1$

$\quad\;\; x_1 - x_2 = h_2$ $\qquad 3x_1 + 6x_2 = 4$ $\qquad 3x_1 + 6x_2 = 3$.

The system (i) has the unique solution $x_1 = \frac{1}{2}(h_1 + h_2)$, $x_2 = \frac{1}{2}(h_1 - h_2)$ whatever the values of h_1 and h_2. The system (ii) has no solution, because on subtracting 3 times the first equation from the second equation we should arrive at the impossible deduction that $0 = 1$. In the system (iii) the two equations give the same information, and the system has infinitely many solutions $x_1 = 1 - 2t$, $x_2 = t$, where t remains arbitrary.

5.8. LINEAR EQUATIONS—THE REGULAR CASE

An n by n matrix $\mathbf{A}$ is said to be *regular* if its n columns (or rows) are linearly independent, that is if

$$n = r, \tag{5.8.1}$$

where r is the rank of $\mathbf{A}$. Such a matrix $\mathbf{A}$ is also said to be of *full rank*. Likewise the system of equations (5.7.2) is called *regular* (or *independent* or *non-singular*) if is coefficient matrix is regular; otherwise the system is called *singular* or *dependent*.

We mention two tests of regularity:

I. The square matrix $\mathbf{A}$ is regular, if and only if the system $\mathbf{Ax} = \mathbf{0}$ has only the solution $\mathbf{x} = \mathbf{0}$, that is, if and only if it is non-singular [see Definition 6.4.2].

The second test uses a result from the theory of determinants [see § 6.12]:

II. The square matrix $\mathbf{A}$ is regular if and only if det $\mathbf{A} \neq 0$.

The main result about regular systems is summarized in the following.

THEOREM 5.8.1.   *If* $\mathbf{A}$ *is a regular matrix then the system*

$$\mathbf{Ax} = \mathbf{h} \tag{5.8.2}$$

has the unique solution

$$\mathbf{x} = \mathbf{A}^{-1}\mathbf{h}. \tag{5.8.3}$$

An alternative way of writing (5.8.3) is furnished by Cramer's formulae [see (6.12.2)] which employ determinants. The result (5.8.3) is, however, chiefly of theoretical interest. In a practical situation it is advisable to proceed by pivotal condensation. Since the subject is fully treated elsewhere [see § 6.4(i)] we content ourselves with a short illustration:

EXAMPLE 5.8.1.   Solve the system

$$-x_1 + x_2 + x_3 = h_1$$

$$x_1 - x_2 + x_3 = h_2$$

$$x_1 + x_2 - x_3 = h_3,$$

where h_1, h_2 and h_3 are given.

The process of pivotal condensation will automatically check whether the matrix A is regular, since it is only in this case that a unique solution will be obtained. The rows of Table 5.8.1 are labelled R_1, R_2, ... and the entries in the first column indicate how each row has been constructed from preceding rows.

	x_1	x_2	x_3		C.S.
R_1	-1	1	1	h_1	$h_1 + 1$
R_2	1	-1	1	h_2	$h_2 + 1$
R_3	1	1	-1	h_3	$h_3 + 1$
$R_4 = R_1$	-1	1	1	h_1	$h_1 + 1$
$R_5 = R_2 + R_1$	0	0	2	$h_1 + h_2$	$h_1 + h_2 + 2$
$R_6 = R_3 + R_1$	0	2	0	$h_1 + h_3$	$h_1 + h_3 + 2$

Table 5.8.1

The last column contains the check sums, that is, the sum of each row; this number can be obtained in two ways: first, as part of the operation, for example,

$$R_5 = R_2 + R_1 \tag{5.8.4}$$

and secondly, by computing the sum of R_5 after the operation has been carried out. We now deduce from R_5 that $x_3 = \frac{1}{2}(h_1 + h_2)$ and from R_6 that $x_2 = \frac{1}{2}(h_1 + h_3)$. Substituting in R_1 we find that $x_1 = x_2 + x_3 - h_1 = \frac{1}{2}(h_2 + h_3)$. Thus a unique solution is obtained whatever the values of h_1, h_2, h_3, and this confirms that the system is regular.

5.9. LINEAR EQUATIONS—HOMOGENEOUS SYSTEMS

We consider systems of the form

$$\mathbf{Ay} = 0, \tag{5.9.1}$$

where $\mathbf{A}$ is an m by n matrix. Explicitly written, the system becomes

$$a_{11}y_1 + a_{12}y_2 + \ldots + a_{1n}y_n = 0$$

$$a_{21}y_1 + a_{22}y_2 + \ldots + a_{2n}y_n = 0$$

$$\ldots\ldots\ldots\ldots\ldots\ldots\ldots\ldots\ldots\ldots\ldots\ldots \tag{5.9.2}$$

$$a_{m1}y_1 + a_{m2}y_2 + \ldots + a_{mn}y_n = 0.$$

The equations are called *homogeneous* because if $\mathbf{y} = \mathbf{c}$ is a solution, so is $\lambda \mathbf{c}$, where λ is any scalar. In a homogeneous equation all non-zero terms are of the same degree in the unknowns, this degree being equal to unity in our case.

A system of equations which are both linear and homogeneous possesses the following important property, which in physics is called the principle of superposition: if $\mathbf{u}$ and $\mathbf{v}$ are solutions of (5.9.1), so is every vector of the form $\lambda \mathbf{u} + \mu \mathbf{v}$, where λ and μ are arbitrary scalars. We shall therefore seek only linearly independent solutions.

Every system of homogeneous equations has the *trivial* solution

$$y_1 = y_2 = \ldots = y_n = 0,$$

and if the system has one non-trivial solution, it has infinitely many. Some of the information contained in (5.9.2) may be redundant. For example, if one particular equation occurs twice, the duplicate may be omitted without affecting the solutions. Accordingly, it is not suprising that what matters is not the number of rows of $\mathbf{A}$ but the rank of $\mathbf{A}$. The main result, which is of fundamental importance, can be formulated as follows:

THEOREM 5.9.1.   *If $\mathbf{A}$ is an m by n matrix of rank r, then the homogeneous system*

$$\mathbf{Ay} = 0 \tag{5.9.3}$$

has $n - r$ *linearly independent solutions, say*

$$\mathbf{c}_1, \mathbf{c}_2, \ldots, \mathbf{c}_{n-r}, \tag{5.9.4}$$

which are called a set of basic *solutions of* (5.9.1). *Every solution is uniquely expressed as*

$$\lambda_1 \mathbf{c}_1 + \lambda_2 \mathbf{c}_2 + \ldots + \lambda_{n-r} \mathbf{c}_{n-r}.$$

The integer $n - r$ is sometimes called the *nullity* of $\mathbf{A}$. The greater the nullity the more comprehensive is the set of solutions.

It is sometimes convenient to adopt a more geometric terminology. The principle of superposition amounts to the statement that the solutions of (5.9.1) form a subspace of the space of all n-tuples described in Example 5.2.2. We may therefore speak of the *solution space* of the homogeneous system, and we have the result that

$$\text{dimension of solution space} = \text{nullity of } \mathbf{A}. \tag{5.9.5}$$

EXAMPLE 5.9.1.   Using the matrix $\mathbf{A}$ of rank 2 mentioned in Example 5.6.2, we consider the homogeneous system

$$
\begin{aligned}
y_1 \quad\quad + 2y_3 + 3y_4 - \ y_5 &= 0 \\
2y_1 + \ y_2 \quad\quad + 2y_4 \quad\quad &= 0 \\
-4y_1 - 3y_2 + 4y_3 \quad\quad - 2y_5 &= 0 \\
y_1 + \ y_2 - 2y_3 - \ y_4 + \ y_5 &= 0.
\end{aligned}
\tag{5.9.6}
$$

The equations are in one-to-one correspondence with the rows of $\mathbf{A}$. Accordingly the relations (5.6.5) imply that the last two equations in the system (5.9.6) are *redundant*. Thus we have only to satisfy the first two equations and we have five unknowns. Three of the unknowns remain indeterminate or *disposable*. In the example we may choose y_3, y_4 and y_5 as disposable unknowns and put

$$y_3 = t_1, \quad\quad y_4 = t_2, \quad\quad y_5 = t_3,$$

where t_1, t_2, t_3 are arbitrary, and it follows that

$$y_1 = -2t_1 - 3t_2 + t_3, \tag{5.9.7}$$

$$y_2 = 4t_1 + 4t_2 - 2t_3. \tag{5.9.8}$$

5.10.   LINEAR EQUATIONS—NON-HOMOGENEOUS SYSTEMS

We now consider the most general system of linear equations. This may be expressed by the matrix equation

$$\mathbf{A}\mathbf{x} = \mathbf{h}, \tag{5.10.1}$$

where $\mathbf{A}$ is an m by n matrix of rank r, [see § 5.7]. The example (ii) mentioned in § 5.7 shows that such a system need not be soluble, because the equations may be inconsistent. Therefore our first task will be to obtain a criterion which tells us whether or not a solution exists. We shall in fact mention two criteria which serve this purpose.

The matrix A has n columns, explicitly written as [see (5.6.2)]

$$A = (a_1, a_2, \ldots, a_n).$$

We now construct what is known as the *augmented matrix* by affixing h as an additional $(n + 1)$st column; thus

$$(A\ h) = (a_1, a_2, \ldots, a_n, h) \tag{5.10.2}$$

is an m by $n + 1$ matrix, and the question about solubility is answered by the following statement.

First Criterion of Solubility

The system $Ax = h$ is soluble if and only if the matrices A and $(A\ h)$ have the same rank.

For example, the systems (ii) and (iii) in § 5.7 both have the matrix

$$A = \begin{pmatrix} 1 & 2 \\ 3 & 6 \end{pmatrix},$$

which is of rank unity, since the rows are proportional [see (5.6.6)]. In (ii) the augmented matrix is

$$(A\ h) = \begin{pmatrix} 1 & 2 & 1 \\ 3 & 6 & 4 \end{pmatrix},$$

which is of rank two, since its rows are not proportional. Thus the ranks are unequal and the system is insoluble. On the other hand, in the system (iii) the augmented matrix is

$$(A\ h) = \begin{pmatrix} 1 & 2 & 1 \\ 3 & 6 & 3 \end{pmatrix},$$

whose rank is equal to unity, as is the rank of A. Accordingly the system (iii) is soluble.

Next, suppose that (5.10.1) is a soluble system, and let p be a particular solution obtained in some way. Thus we assume that

$$Ap = h.$$

If x is another solution of (5.10.1) we have that

$$A(x - p) = 0.$$

Hence $x - p$ is a solution of the associated homogeneous system

$$Ay = 0.$$

The solutions of the latter system can be found by the method previously expounded and their structure is described in Theorem 5.9.1. Thus we derive the following result:

THEOREM 5.10.1.   *If the system $Ax = h$ is soluble and if* **p** *is a particular solution, then the general solution is of the form*

$$x = p + \lambda_1 c_1 + \lambda_2 c_2 + \ldots + \lambda_{n-r} c_{n-r},$$

where $c_1, c_2, \ldots, c_{n-r}$ *is a basic set of solutions of the associated homogeneous system and* $\lambda_1, \lambda_2, \ldots, \lambda_{n-r}$ *are arbitrary scalars.*

The content of this important theorem is summarized in the statement: *the general solution of the non-homogeneous system is equal to a particular solution plus the general solution of the associated homogeneous system.*

We observe that the general solution of a soluble system involves $n - r$ arbitrary constants (parameters).

COROLLARY 5.10.2.   *The solution of (5.10.1) is unique if and only if $n = r$.*

COROLLARY 5.10.3.   *If $m < n$, a system of m linear equations in n unknowns cannot have a unique solution.*

There is another way of looking at the solubility of a system of equations. As before, let **A** be an m by n matrix of rank r. We now view **A** as a collection of m rows [cf. (5.6.3)]. Thus we write

$$A = \begin{pmatrix} a^{(1)} \\ a^{(2)} \\ \vdots \\ a^{(m)} \end{pmatrix}$$

where $a^{(i)} = (a_{i1}, a_{i2}, \ldots, a_{in})$ $(i = 1, 2, \ldots, m)$. A linear relation between the rows of **A**, such as

$$\omega_1 a^{(1)} + \omega_2 a^{(2)} + \ldots + \omega_m a^{(m)} = 0 \tag{5.10.3}$$

can be expressed as a matrix equation

$$u'A = 0, \tag{5.10.4}$$

where

$$u' = (\omega_1, \omega_2, \ldots, \omega_m). \tag{5.10.5}$$

Conversely, any non-zero vector **u** which satisfies (5.10.4), gives rise to a relation between the rows of **A**.

Now suppose that the system (5.10.1) is soluble. Thus there exists a vector **p** such that

$$Ap = h. \tag{5.10.6}$$

If **u** is a solution of (5.10.4) then left-multiplication of (5.10.6) by **u**′ yields

$$u'h = \omega_1 h_1 + \omega_2 h_2 + \ldots + \omega_m h_m = 0.$$

In other words, any relation that holds for the rows of **A** must also hold for the components of the vector **h**. It is an interesting fact, which we quote without proof, that the converse is also true. Hence we enunciate the

Second Criterion of Solubility

The system $\mathbf{Ax} = \mathbf{h}$ is soluble if and only if the vector **h** has the property that $\mathbf{u'h} = \mathbf{0}$ whenever $\mathbf{u'A} = \mathbf{0}$.

EXAMPLE 5.10.1.   We illustrate the use of this criterion by considering the system

$$
\begin{aligned}
x_1 \qquad\quad + 2x_3 + 3x_4 - \ x_5 &= 2 \\
2x_1 + \ x_2 \qquad\quad + 2x_4 \qquad\ &= 1 \\
-4x_1 - 3x_2 + 4x_3 \qquad\quad - 2x_5 &= 1 \\
x_1 + \ x_2 - 2x_3 - \ x_4 + \ x_5 &= -1,
\end{aligned}
\tag{5.10.7}
$$

whose matrix, **A**, is the same as that of the homogeneous system (5.9.6), which is therefore the associated homogeneous system of (5.10.7). The terms on the right of (5.10.7) can be gathered in a column vector **h**, where

$$\mathbf{h'} = (2, 1, 1, -1).$$

There are two independent relations between the rows of **A**, namely

$$
\begin{aligned}
2\mathbf{a}^{(1)} - 3\mathbf{a}^{(2)} - \mathbf{a}^{(3)} \qquad\quad &= \mathbf{0} \\
-\ \mathbf{a}^{(1)} + \ \mathbf{a}^{(2)} \qquad\quad - \mathbf{a}^{(4)} &= \mathbf{0}.
\end{aligned}
\tag{5.10.8}
$$

The coefficients in these relations form the row vectors

$$\mathbf{u_1'} = (2, -3, -1, 0) \quad \text{and} \quad \mathbf{u_2'} = (-1, 1, 0, -1)$$

respectively. On verifying that

$$\mathbf{u_1'h} = \mathbf{u_2'h} = 0$$

we conclude that the system (5.10.7) is soluble.

Since we have already found the general solution [see Theorem 5.9.1] of the associated homogeneous system, it only remains to discover one particular solution of (5.10.7). By virtue of the relations (5.10.8) two of the equations may be discarded. We decide to delete the third and fourth equations and obtain the reduced, but equivalent, system

$$
\begin{aligned}
x_1 \qquad\quad + 2x_3 + 2x_4 - x_5 &= 2 \\
2x_1 + x_2 \qquad\quad + 2x_4 \qquad\ &= 1.
\end{aligned}
$$

For a particular solution it suffices to put $x_3 = x_4 = x_5 = 0$. Then x_1 and x_2 must satisfy

$$
\begin{aligned}
x_1 \qquad\ &= 2 \\
2x_1 + x_2 &= 1,
\end{aligned}
$$

that is $x_1 = 2$, $x_2 = -3$. Hence the vector **p** given by

$$\mathbf{p}' = (2, -3, 0, 0, 0)$$

is a particular solution, and the general solution is obtained by adding to it the vector **y** specified in (5.9.7) and (5.9.8), thus

$$x_1 = 2 \quad - 2t_1 + 3t_3 + \quad t_3$$
$$x_2 = -3 + 4t_1 + 4t_2 - 2t_3$$
$$x_3 = t_1$$
$$x_4 = t_2$$
$$x_5 = t_3.$$

5.11. LINEAR MAPS

The concept of a map (mapping) plays a fundamental role in modern mathematics. Though frequently couched in abstract language the idea is derived from the familar notion of a geographical map: a space X (a suitable region of the earth's surface) is depicted in a space Y (a rectangular piece of paper) in such a way that certain relevant features of X, such as the ratio of distances or angles, are preserved in the transfer from X to Y.

We are here concerned with maps of a vector space X into a vector space Y, where both X and Y are defined over the same field $\mathbb{F}$ of scalars. Such a map is denoted by

$$\alpha: X \rightarrow Y$$

and it may be symbolically described as in Figure 5.11.1. Thus with each vector **x** there is associated a unique vector **y**, called the *image* of **x** under α. This image is usually written $\mathbf{y} = \alpha(\mathbf{x})$ or else $\mathbf{y} = \mathbf{x}\alpha$. For our purposes it is slightly more convenient to use the latter notation. The map is termed *linear* if it preserves vector addition and multiplication by a scalar. This means that we stipulate the following properties of α: (i) if $\mathbf{x}_1$ and $\mathbf{x}_2$ are any two vectors of X, then

$$(\mathbf{x}_1 + \mathbf{x}_2)\alpha = \mathbf{x}_1\alpha + \mathbf{x}_2\alpha, \tag{5.11.1}$$

that is, the image of a sum of two vectors is the sum of the images, and (ii) if λ is any scalar, then

$$(\lambda\mathbf{x})\alpha = \lambda(\mathbf{x}\alpha). \tag{5.11.2}$$

Figure 5.11.1

The rules (i) and (ii) can be expressed in the single formula

$$(\lambda \mathbf{x}_1 + \mu \mathbf{x}_2)\alpha = \lambda(\mathbf{x}_1\alpha) + \mu(\mathbf{x}_2\alpha), \qquad (5.11.3)$$

where $\mathbf{x}_1, \mathbf{x}_2 \in X$ and $\lambda, \mu \in \mathbb{F}$.

On putting λ equal to zero in (5.11.2) we see that every linear map sends the zero vector of X into the zero vector of Y.

For the sake of brevity we shall henceforth adopt the convention that 'map' means 'linear map'.

By definition the map α acts on each vector of X, but we do not in general assume that α is one-to-one. Thus two different vectors of X may have the same image in Y, that is, it may happen that

$$\mathbf{x}_1\alpha = \mathbf{x}_2\alpha, \qquad \mathbf{x}_1 \neq \mathbf{x}_2.$$

Indeed there exists a trivial map, known as the *zero map* and denoted by o, which sends every vector of X into the zero vector of Y, thus

$$\mathbf{x}o = \mathbf{0}, \quad \text{for all } x \in X. \qquad (5.11.4)$$

With every map $\alpha: X \to Y$ there is associated (i) a subspace of X called the *kernel* of α, abbreviated kerα, and (ii) a subspace of Y called the *image* of α, written imα. We shall presently define these subspaces in detail but point out that they may be improper subspaces, consisting of the whole space or reducing to the zero vector only.

(i) kerα: The kernel of α consists of those vectors $\mathbf{u}$ of X which are mapped into the zero vector of Y, thus

$$\text{ker}\alpha = \{\mathbf{u} \in X \mid \mathbf{u}\alpha = \mathbf{0}\}. \qquad (5.11.5)$$

This is indeed a subspace of X (see Definition 5.5.1); for if $\mathbf{u}\alpha = \mathbf{0}$ and $\mathbf{v}\alpha = \mathbf{0}$, then

$$(\lambda\mathbf{u} + \mu\mathbf{v})\alpha = \mathbf{0} \quad (\lambda, \mu \in \mathbb{F}).$$

Evidently, ker$\alpha = X$ if and only if α is the zero map. The other extreme case, when ker$\alpha = \{\mathbf{0}\}$, is more interesting; then $\mathbf{u}\alpha = \mathbf{0}$ implies that $\mathbf{u} = \mathbf{0}$. Such a map is one-to-one [see § 1.4.2], because if $\mathbf{x}_1\alpha = \mathbf{x}_2\alpha$, then $(\mathbf{x}_1 - \mathbf{x}_2)\alpha = \mathbf{0}$ and $\mathbf{x}_1 - \mathbf{x}_2$ lies in the kernel, whence $\mathbf{x}_1 = \mathbf{x}_2$. A one-to-one map is called a *monomorphism* or it is said to be *injective*. Hence we have the rule

A linear map $\alpha: X \to Y$ is injective if and only if ker $\alpha = \{\mathbf{0}\}$.   (5.11.6)

(ii) The collection of those vectors of Y which are images of some vector of X, forms a subspace of Y. For if $\mathbf{y}_1 = \mathbf{x}_1\alpha$ and $\mathbf{y}_2 = \mathbf{x}_2\alpha$ then $\lambda\mathbf{y}_1 + \mu\mathbf{y}_2 = (\lambda\mathbf{x}_1 + \mu\mathbf{x}_2)\alpha$. This subspace is called the image of α, thus

$$\text{im } \alpha = \{\mathbf{y} \in Y \mid \mathbf{y} = \mathbf{x}\alpha, \mathbf{x} \in X\}.$$

If im $\alpha = Y$, then each vector of Y is the image of some vector of X. Such a map is called an *epimorphism* or it is said to be *surjective*. Hence we have the rule

A linear map $\alpha: X \to Y$ is surjective if and only if im $\alpha = Y$.   (5.11.8)

A map which is both injective and surjective is called *bijective* or an *isomor-phism*. Thus if there exists an isomorphism $\alpha: X \to Y$, the vectors of X and Y are placed into a one-to-one correspondence in such a way that all vector opera-tions in X are mirrored in Y in a reversible manner. From an abstract point of view the vector spaces X and Y are distinguished only in their notation but have otherwise the same properties.

We shall now describe the linear map $\alpha: X \to Y$ in greater detail when the vector spaces X and Y are of finite dimensions (over the same field $\mathbb{F}$). Thus suppose that

$$\dim X = m, \qquad \dim Y = n$$

and choose bases

$$X = [\mathbf{e}_1, \mathbf{e}_2, \ldots, \mathbf{e}_m], \qquad Y = [\mathbf{f}_1, \mathbf{f}_2, \ldots, \mathbf{f}_n]. \tag{5.11.9}$$

An arbitrary vector $\mathbf{x}$ of X can then be written in the form

$$\mathbf{x} = x_1\mathbf{e}_1 + x_2\mathbf{e}_2 + \ldots + x_m\mathbf{e}_m,$$

and the linearity of α implies that

$$\mathbf{x}\alpha = x_1(\mathbf{e}_1\alpha) + x_2(\mathbf{e}_2\alpha) + \ldots + x_m(\mathbf{e}_m\alpha).$$

Hence the image of $\mathbf{x}$ under α can be computed as soon as we know the images of the basic vectors, that is $\mathbf{e}_1\alpha, \mathbf{e}_2\alpha, \ldots, \mathbf{e}_m\alpha$. Each $\mathbf{e}_i\alpha$ $(i = 1, 2, \ldots, m)$ is a certain vector in Y and is therefore a linear combination of $\mathbf{f}_1, \mathbf{f}_2, \ldots, \mathbf{f}_n$. Thus the action of α on X is completely determined by equations of the form

$$\mathbf{e}_1\alpha = a_{11}\mathbf{f}_1 + a_{12}\mathbf{f}_2 + \ldots + a_{1n}\mathbf{f}_n$$

$$\mathbf{e}_2\alpha = a_{21}\mathbf{f}_1 + a_{22}\mathbf{f}_2 + \ldots + a_{2n}\mathbf{f}_n$$

$$\ldots\ldots\ldots\ldots\ldots\ldots\ldots\ldots\ldots\ldots\ldots\ldots \tag{5.11.10}$$

$$\mathbf{e}_m\alpha = a_{m1}\mathbf{f}_1 + a_{m2}\mathbf{f}_2 + \ldots + a_{mn}\mathbf{f}_n.$$

More briefly, we may say that, relative to the bases (5.11.9), the linear map is described by the m by n matrix

$$\mathbf{A} = \begin{pmatrix} a_{11} & a_{12} & \cdots & a_{1n} \\ a_{21} & a_{22} & \cdots & a_{2n} \\ \vdots & \vdots & & \vdots \\ a_{m1} & a_{m2} & \cdots & a_{mn} \end{pmatrix}. \tag{5.11.11}$$

All properties of the map α are reflected in properties of the matrix $\mathbf{A}$ and conversely. In particular, suppose that $\mathbf{A}$ is of rank r and denote the rows of $\mathbf{A}$ by

$$\mathbf{a}^{(1)}, \mathbf{a}^{(2)}, \ldots, \mathbf{a}^{(m)}$$

[cf. (5.6.2)]. If there exists a linear relation

$$\lambda_1\mathbf{a}^{(1)} + \lambda_2\mathbf{a}^{(2)} + \ldots + \lambda_m\mathbf{a}^{(m)} = \mathbf{0}$$

between the rows of $\mathbf{A}$, then it follows from (5.11.10) that we have a relation

$$\lambda_1(\mathbf{e}_1\alpha) + \lambda_2(\mathbf{e}_2\alpha) + \ldots + \lambda_m(\mathbf{e}_m\alpha) = \mathbf{0}$$

between the images of $\mathbf{e}_1, \mathbf{e}_2, \ldots, \mathbf{e}_m$, and conversely. Thus the problem of computing imα is equivalent to determining the subspace of Y which is generated by the vectors $\mathbf{e}_1\alpha, \mathbf{e}_2\alpha, \ldots, \mathbf{e}_m\alpha$, that is

$$\text{im}\alpha = \{\mathbf{e}_1\alpha, \mathbf{e}_2\alpha, \ldots, \mathbf{e}_m\alpha\}. \tag{5.11.12}$$

When Y is identified with $\mathbb{R}^n$ [see Example 5.2.2], and α is described by the matrix $\mathbf{A}$, then imα is the subspace generated by $\mathbf{a}^{(1)}, \mathbf{a}^{(2)}, \ldots, \mathbf{a}^{(m)}$. In particular

$$\dim \text{im}\alpha = \text{rank } \mathbf{A}. \tag{5.11.13}$$

Next, we shall describe a method for computing the kernel of α. Suppose that

$$\mathbf{u} = u_1\mathbf{e}_1 + u_2\mathbf{e}_2 + \ldots + u_m\mathbf{e}_m$$

lies in kerα, that is

$$\mathbf{u}\alpha = u_1(\mathbf{e}_1\alpha) + u_2(\mathbf{e}_2\alpha) + \ldots + u_m(\mathbf{e}_m\alpha) = \mathbf{0}.$$

On substituting from (5.11.10) and collecting terms in $\mathbf{f}_1, \mathbf{f}_2, \ldots, \mathbf{f}_n$ we obtain that

$$\sum_{j=1}^{n} (u_1 a_{1j} + u_2 a_{2j} + \ldots + u_m a_{mj})\mathbf{f}_j = \mathbf{0}.$$

Since the vectors $\mathbf{f}_j$ are linearly independent [see remark (2) following Definition 5.4.1], it follows that the coefficients $u_1, u_2, \ldots, u_m$ satisfy the system of equations

$$u_1 a_{11} + u_2 a_{21} + \ldots + u_m a_{m1} = 0$$

$$u_1 a_{12} + u_2 a_{22} + \ldots + u_m a_{m2} = 0$$

$$\cdots\cdots\cdots\cdots\cdots\cdots\cdots\cdots\cdots$$

$$u_1 a_{1n} + u_2 a_{2n} + \ldots + u_m a_{mn} = 0.$$

In the language of maps the following terms are used:

DEFINITION 5.11.1.   If $\alpha: X \to Y$ is a linear map between finite-dimensional vector spaces, then

$$\text{rank}\alpha = \dim \text{im}\alpha \tag{5.11.14}$$

$$\text{nullity}\alpha = \dim \text{ker}\alpha. \tag{5.11.15}$$

We deduce from Theorem 5.9.1 that the nullity of α is equal to $m - r$, and we may record this result in the form:

$$\text{nullity}\alpha + \text{rank}\alpha = \dim X. \tag{5.11.16}$$

We illustrate these concepts with the following example:

EXAMPLE 5.11.1.   The vector spaces X and Y of dimensions 3 and 5 are referred to bases as follows:

$$X = [\mathbf{e}_1, \mathbf{e}_2, \mathbf{e}_3], \qquad Y = [\mathbf{f}_1, \mathbf{f}_2, \mathbf{f}_3, \mathbf{f}_4, \mathbf{f}_5].$$

Consider the linear map $\alpha: X \to Y$ specified by the equations

$$
\begin{aligned}
\mathbf{e}_1 \alpha &= \mathbf{f}_1 + 3\mathbf{f}_2 - \mathbf{f}_3 && + 4\mathbf{f}_5 \\
\mathbf{e}_2 \alpha &= 2\mathbf{f}_1 + \mathbf{f}_2 && + 2\mathbf{f}_4 - \mathbf{f}_5 \\
\mathbf{e}_3 \alpha &= 4\mathbf{f}_1 - 3\mathbf{f}_2 + 2\mathbf{f}_3 + 6\mathbf{f}_4 - 11\mathbf{f}_5.
\end{aligned}
\tag{5.11.17}
$$

In order to determine $\ker \alpha$ we seek scalars u_1, u_2, u_3 such that $(u_1\mathbf{e}_1 + u_2\mathbf{e}_2 + u_3\mathbf{e}_3)\alpha = \mathbf{0}$. To this end multiply the equations (5.11.17) by u_1, u_2, u_3 and add. In the resulting sum we equate the coefficients of $\mathbf{f}_1, \mathbf{f}_2, \ldots, \mathbf{f}_5$ to zero. Thus we obtain the five equations:

$$
\begin{aligned}
R_1 &= u_1 + 2u_2 + 4u_3 = 0 \\
R_2 &= 3u_1 + u_2 - 3u_3 = 0 \\
R_3 &= -u_1 + 2u_3 = 0 \\
R_4 &= + 2u_2 + 6u_3 = 0 \\
R_5 &= 4u_1 - u_2 - 11u_3 = 0.
\end{aligned}
$$

Next, we use R_1 to eliminate u_1 from R_2, R_3, R_4 and R_5 (pivotal condensation, see Example 5.3.1). This leads to the equivalent system

$$
\begin{aligned}
R_1 &= u_1 + 2u_2 + 4u_3 = 0 \\
R_2 - 3R_1 &= - 5u_2 - 15u_3 = 0 \\
R_3 + R_1 &= 2u_2 + 6u_3 = 0 \\
R_4 &= 2u_2 + 6u_3 = 0 \\
R_5 - 4R_1 &= - 9u_2 - 27u_3 = 0.
\end{aligned}
$$

It is now evident that the last three equations are redundant being proportional to the second. The latter may be simplified by dividing throughout by -5. Hence it remains only to solve the two equations

$$u_1 + 2u_2 + 4u_3 = 0$$

$$u_2 + 3u_3 = 0.$$

One of the unknowns, say u_3, is disposable [see Example 5.9.1] and we put $u_3 = t$, where t is arbitrary. Then

$$u_2 = -3u_3 = -3t$$

$$u_1 = -2u_2 - 4u_3 = 6t - 4t = 2t.$$

Hence the most general vector in $\ker\alpha$ is of the form

$$t(2\mathbf{e}_1 - 3\mathbf{e}_2 + \mathbf{e}_3),$$

and it is evident that

$$\dim \ker\alpha = 1.$$

Next we determine $\operatorname{im}\alpha$. In accordance with the rule (5.11.12) we have that

$$\operatorname{im}\alpha = \{\mathbf{e}_1\alpha, \mathbf{e}_2\alpha, \mathbf{e}_3\alpha\}. \tag{5.11.18}$$

But the three generating vectors named in this equation are not linearly independent. For since $2\mathbf{e}_1 - 3\mathbf{e}_2 + \mathbf{e}_3$ lies in the kernel of α we have that

$$2\mathbf{e}_1\alpha - 3\mathbf{e}_2\alpha + \mathbf{e}_3\alpha = \mathbf{0}.$$

Hence one of the vectors in (5.11.18) is redundant. For example we may write

$$\mathbf{e}_3\alpha = -2\mathbf{e}_1\alpha + 3\mathbf{e}_2\alpha$$

and therefore delete the generator $\mathbf{e}_3\alpha$. Evidently the vectors $\mathbf{e}_1\alpha$ and $\mathbf{e}_2\alpha$ are not proportional and are, therefore, linearly independent, as can be seen by inspection of (5.11.17). Thus we can state that

$$\operatorname{im}\alpha = [\mathbf{e}_1\alpha, \mathbf{e}_2\alpha],$$

which shows that

$$\dim \operatorname{im}\alpha = 2.$$

This confirms (5.11.16) in the present example, namely

$$\dim \ker\alpha + \dim \operatorname{im}\alpha = 3.$$

In terms of the basis of Y the vectors of $\operatorname{im}\alpha$ are expressed in the form

$$(\lambda + 2\mu)\mathbf{f}_1 + (3\lambda + \mu)\mathbf{f}_2 - \lambda\mathbf{f}_3 + 2\mu\mathbf{f}_4 + (4\lambda - \mu)\mathbf{f}_5,$$

where λ and μ are arbitrary scalars.

5.12.   CHANGE OF BASIS

We recapitulate the main results of the preceding section: in order to describe a linear map

$$\alpha: X \to Y \tag{5.12.1}$$

in algebraical terms we choose bases for the vector spaces X and Y over the given field of scalars. Thus let

$$\mathscr{E}: \mathbf{e}_1, \mathbf{e}_2, \ldots, \mathbf{e}_m \tag{5.12.2}$$

be a basis for X, and let

$$\mathscr{F}: \mathbf{f}_1, \mathbf{f}_2, \ldots, \mathbf{f}_n \tag{5.12.3}$$

be a basis for Y. Then the action of α is specified by the equations

$$\mathbf{e}_i\alpha = \sum_{j=1}^{n} a_{ij}\mathbf{f}_j \quad (i = 1, 2, \ldots, m). \tag{5.12.4}$$

The information may be presented in tabular form as in Table 5.12.1. The ith row in the table tells us how to express the image of e_i in terms of the basis.

	f_1	f_2	...	f_n
$e_1\alpha$	a_{11}	a_{12}	...	a_{1n}
$e_2\alpha$	a_{21}	a_{22}	...	a_{2n}
$\vdots$	$\vdots$	$\vdots$		$\vdots$
$e_i\alpha$	a_{i1}	a_{i2}	...	a_{in}
$\vdots$	$\vdots$	$\vdots$		$\vdots$
$e_m\alpha$	a_{m1}	a_{m2}		a_{mn}

Table 5.12.1

Granted that particular bases have been chosen for X and Y, any linear map (5.12.1) is virtually synonymous with an m by n matrix $\mathbf{A}$, the entries of which appear in the Table 5.12.1. Conversely, any such matrix can be used to define a linear map. We express this mutual relationship symbolically as

$$\alpha \leftrightarrow \mathbf{A}(\mathscr{E},\mathscr{F}). \tag{5.12.5}$$

The choice of the bases for X and Y is not inherent in the definition of α. Thus without altering the action of α in any way we may replace $\mathscr{E}$ and $\mathscr{F}$ by different bases. Suppose that X is now referred to a basis

$$\mathscr{E}' : e_1', e_2', \ldots, e_m'$$

and Y to a basis

$$\mathscr{F}' : f_1', f_2', \ldots, f_n'.$$

Analogously to (5.12.4) we then have equations of the form

$$e_s'\alpha = \sum_{t=1}^{n} b_{st}f_t' \quad (s = 1, 2, \ldots, m). \tag{5.12.6}$$

The matrix

$$\mathbf{B} = (b_{st})$$

represents the action of α as well as $\mathbf{A}$ does, and the question naturally arises as to the relationship between those two matrices.

We recall from (5.4.4) that the transition from one basis to another, for the same vector space, is described by an invertible matrix. Thus the change of basis for X is given by

$$\mathscr{E}' \to \mathscr{E} : e_s' = \sum_{i=1}^{m} p_{si}e_i \quad (s = 1, 2, \ldots, m), \tag{5.12.7}$$

involving an invertible matrix

$$\mathbf{P} = (p_{si})$$

of order m [see § 6.4]. Similarly the equation

$$\mathscr{F} \to \mathscr{F}': \mathbf{f}_j = \sum_{t=1}^{n} q_{jt}\mathbf{f}'_t \quad (j = 1, 2, \ldots, n) \tag{5.12.8}$$

exhibit the connection between the two bases of Y, where

$$\mathbf{Q} = (q_{jt})$$

is an invertible matrix of order n. It is for formal reasons only that we set up the change of bases for the two spaces in this unsymmetric manner, namely $\mathscr{E}' \to \mathscr{E}$ and $\mathscr{F} \to \mathscr{F}'$. The inverse matrices $\mathbf{P}^{-1}$ and $\mathbf{Q}^{-1}$ specify the transitions $\mathscr{E} \to \mathscr{E}'$ and $\mathscr{F}' \to \mathscr{F}$ respectively. A simple calculation, which we omit, will establish that the relation between $\mathbf{A}$ and $\mathbf{B}$ is given by the equation

$$\mathbf{B} = \mathbf{PAQ}. \tag{5.12.9}$$

Conversely, if two m by n matrices $\mathbf{A}$ and $\mathbf{B}$ satisfy (5.12.9), with suitable invertible matrices $\mathbf{P}$ and $\mathbf{Q}$, then $\mathbf{A}$ and $\mathbf{B}$ represent the same linear map relative to different bases. The idea is made explicit by the following

DEFINITION 5.12.1. The m by n matrices $\mathbf{A}$ and $\mathbf{B}$ are said to be *equivalent* if there exist invertible matrices $\mathbf{P}$ and $\mathbf{Q}$ of orders m and n respectively [see Theorem 6.4.2] such that $\mathbf{B} = \mathbf{PAQ}$.

The set of all m by n matrices has now been split into disjoint equivalence classes [see (1.3.5)]. Two problems arise:

(i) to reduce $\mathbf{A}$ to a particularly simple or 'canonical' form by a suitable choice of $\mathbf{P}$ and $\mathbf{Q}$ [cf. § 7.6];

(ii) to compute the matrices $\mathbf{P}$ and $\mathbf{Q}$ referred to in (i).

We recall that by (5.6.8),

$$r(\mathbf{PAQ}) = r(\mathbf{A}),$$

so that $\mathbf{A}$ has the same rank as its canonical form. The following theorem gives a satisfactory answer to the first problem.

THEOREM 5.12.1. ((P, Q)-*theorem for matrices*). *Let* $\mathbf{A}$ *be an* m *by* n *matrix of rank* r. *Then there exist invertible matrices* $\mathbf{P}$ *and* $\mathbf{Q}$ *of orders* m *by* n *respectively, such that*

$$\mathbf{PAQ} = \begin{pmatrix} \mathbf{I}_r & \mathbf{O}_{r,n-r} \\ \mathbf{O}_{m-r,r} & \mathbf{O}_{m-r,n-r} \end{pmatrix}, \tag{5.12.10}$$

where $\mathbf{I}_r$ *is the unit matrix of order* r *and, generally,* $\mathbf{O}_{kl}$ *denotes the zero matrix with* k *rows and* l *columns* [see (6.2.3), § 6.2(iii), *and, for the definition of a partitioned matrix,* § 6.6].

Remarks: (1) The formula (5.12.10) refers to the general case in which $r < m$ and $r < n$. When $r = m$ the second row of matrices is deleted from the partitioned matrix on the right. Similarly, the second column is omitted when $r = n$.

Of course, when $m = n = r$, the canonical form reduces to $\mathbf{I}_n$, since we may then put $\mathbf{P} = \mathbf{A}^{-1}$, $\mathbf{Q} = \mathbf{I}_n$.

(2) The matrices $\mathbf{P}$ and $\mathbf{Q}$ are not uniquely determined. But they can always be chosen from the field of numbers in which the entries of $\mathbf{A}$ lie. For example if the latter are rational numbers [see § 2.4.1], it may be assumed that the co-efficients of $\mathbf{P}$ and $\mathbf{Q}$ are also rational.

(3) The canonical form of $\mathbf{A}$ is completely determined by the integers m, n and r. Hence we have the following remarkable result.

COROLLARY 5.12.2. *Two m by n matrices are equivalent if and only if they have the same rank.*

Unfortunately, in most cases it is laborious to determine a pair of matrices $\mathbf{P}, \mathbf{Q}$ which reduce $\mathbf{A}$ to its canonical form. One method consists in choosing suitable bases which are adapted to the kernel and image space associated with the linear transformation defined by $\mathbf{A}$ [see A. M. Tropper, p. 36]. We shall here describe a procedure which is more amenable to routine calculations.

Let $\mathbf{A}$ be an m by n matrix of rank r. Then precisely r rows of $\mathbf{A}$ are linearly independent [see § 5.6]. By permuting the rows, if necessary, we can ensure that the first r rows are linearly independent; this means that $\mathbf{A}$ will be replaced by $\mathbf{P}_1\mathbf{A}$, where $\mathbf{P}_1$ is a permutation matrix [see Definition 6.7.1]. The first r rows now form an r by n matrix of rank r, which possesses r linearly independent columns. We bring these columns into the first r positions by subjecting the columns to a suitable permutation, which amounts to a right multiplication by a permutation matrix $\mathbf{Q}_1$. Thus we shall henceforth consider the matrix

$$\mathbf{B} = \mathbf{P}_1\mathbf{A}\mathbf{Q}_1 = \begin{pmatrix} \mathbf{K} & \mathbf{L} \\ \mathbf{M} & \mathbf{N} \end{pmatrix}, \tag{5.12.11}$$

where $\mathbf{K}$ is an r by r invertible matrix [see § 6.4] and $\mathbf{L}, \mathbf{M}, \mathbf{N}$ are of types $r \times (n - r)$, $(m - r) \times r$ and $(m - r) \times (n - r)$ respectively [see § 6.2]. We now define

$$\mathbf{P}_2 = \begin{pmatrix} \mathbf{I} & \mathbf{O} \\ -\mathbf{MK}^{-1} & \mathbf{I} \end{pmatrix}, \quad \mathbf{Q}_2 = \begin{pmatrix} \mathbf{K}^{-1} & -\mathbf{K}^{-1}\mathbf{L} \\ \mathbf{O} & \mathbf{I} \end{pmatrix} \tag{5.12.12}$$

where the symbols $\mathbf{I}$ and $\mathbf{O}$ denote unit and zero matrices of appropriate types; for example, the two $\mathbf{I}$'s in $\mathbf{P}_2$ refer to unit matrices of orders r and $m - r$ respectively. By straightforward matrix algebra we verify that

$$\mathbf{P}_2\mathbf{B}\mathbf{Q}_2 = \begin{pmatrix} \mathbf{I} & \mathbf{O} \\ \mathbf{O} & \mathbf{N} - \mathbf{MK}^{-1}\mathbf{L} \end{pmatrix}. \tag{5.12.13}$$

Now by (5.6.8)

$$\text{rank } \mathbf{P}_2\mathbf{BO}_2 = \text{rank } \mathbf{B} = \text{rank } \mathbf{A} = r.$$

Using (5.6.9) we infer that, automatically,

$$\mathbf{N} - \mathbf{MK}^{-1}\mathbf{L} = \mathbf{O};$$

for if not, the rank of (5.12.13) would be greater than r. Combining (5.12.11) and (5.12.1) and putting $\mathbf{P} = \mathbf{P_2P_1}$, $\mathbf{Q} = \mathbf{Q_1Q_2}$ we obtain the desired reduction, namely

$$\mathbf{PAQ} = \begin{pmatrix} \mathbf{I} & \mathbf{O} \\ \mathbf{O} & \mathbf{O} \end{pmatrix}.$$

EXAMPLE 5.12.1. We use the matrix

$$\mathbf{A} = \left(\begin{array}{cc|ccc} 1 & 0 & 2 & 3 & -1 \\ 2 & 1 & 0 & 2 & 0 \\ \hline -4 & -3 & 4 & 0 & -2 \\ 1 & 1 & -2 & -1 & 1 \end{array} \right) = \begin{pmatrix} \mathbf{K} & \mathbf{L} \\ \mathbf{M} & \mathbf{N} \end{pmatrix}$$

mentioned in Example 5.6.2. Its rank is known to be 2 and, in the partitioning of $\mathbf{A}$, the matrix

$$\mathbf{K} = \begin{pmatrix} 1 & 0 \\ 2 & 1 \end{pmatrix}$$

is already non-singular [see § 6.12]. It is therefore unnecessary to carry out the preliminary permutations of rows and columns, so that $\mathbf{P} = \mathbf{P_2}$ and $\mathbf{Q} = \mathbf{Q_2}$. Now, using the method described in § 6.4(i), we see that

$$\mathbf{K}^{-1} = \begin{pmatrix} 1 & 0 \\ -2 & 1 \end{pmatrix},$$

and the formulae (5.12.12) yield

$$\mathbf{P} = \begin{pmatrix} 1 & 0 & 0 & 0 \\ 0 & 1 & 0 & 0 \\ -2 & 3 & 1 & 0 \\ 1 & -1 & 0 & 1 \end{pmatrix}, \quad \mathbf{Q} = \begin{pmatrix} 1 & 0 & -2 & -3 & 1 \\ -2 & 1 & 4 & 4 & -2 \\ 0 & 0 & 1 & 0 & 0 \\ 0 & 0 & 0 & 1 & 0 \\ 0 & 0 & 0 & 0 & 1 \end{pmatrix}.$$

It may be verified that

$$\mathbf{PAQ} = \begin{pmatrix} \mathbf{I} & \mathbf{O} \\ \mathbf{O} & \mathbf{O} \end{pmatrix},$$

where $\mathbf{I}$ is the unit matrix of order 2.

5.13. LINEAR TRANSFORMATIONS

In the preceding sections we studied linear maps from a vector space X into a vector space Y. An interesting situation arises when $X = Y$. Thus we shall now consider a linear map

$$\alpha: X \to X$$

of X into itself. Such a map is often called a *linear transformation* of X. If X is referred to the basis

$$\mathscr{E}: \mathbf{e}_1, \mathbf{e}_2, \ldots, \mathbf{e}_n$$

then α is specified by the equations

$$\mathbf{e}_1\alpha = a_{11}\mathbf{e}_1 + a_{12}\mathbf{e}_2 + \ldots + a_{1n}\mathbf{e}_n$$
$$\mathbf{e}_2\alpha = a_{21}\mathbf{e}_1 + a_{21}\mathbf{e}_2 + \ldots + a_{2n}\mathbf{e}_n$$

$$\cdots\cdots\cdots\cdots\cdots\cdots\cdots$$

$$\mathbf{e}_n\alpha = a_{n1}\mathbf{e}_1 + a_{n2}\mathbf{e}_2 + \ldots + a_{nn}\mathbf{e}_n.$$

(5.13.1)

In other words, relative to $\mathscr{E}$ the linear transformation is described by a square matrix

$$\mathbf{A} = (a_{ij}).$$

Now suppose we change the basis of X to

$$\mathscr{E}': \mathbf{e}_1', \mathbf{e}_2', \ldots, \mathbf{e}_n'.$$

The transition from $\mathscr{E}$ to $\mathscr{E}'$ is given by a set of equations

$$\mathbf{e}_r = \sum_{s=1}^{n} p_{rs}\mathbf{e}_s' \quad (r = 1, 2, \ldots, n),$$

where the n by n matrix $\mathbf{P} = (p_{rs})$ is invertible [see § 5.4]. Relative to $\mathscr{E}'$ the linear transformation is specified by equations

$$\mathbf{e}_i'\alpha = \sum_{j=1}^{n} b_{ij}\mathbf{e}_j' \quad (i = 1, 2, \ldots, n)$$

involving an n by n matrix

$$\mathbf{B} = (b_{ij}).$$

It can be shown that the relationship between $\mathbf{A}$ and $\mathbf{B}$ is expressed by the matrix equation

$$\mathbf{B} = \mathbf{P}^{-1}\mathbf{A}\mathbf{P}.$$

(5.13.2)

DEFINITION 5.13.1.   The matrices $\mathbf{A}$ and $\mathbf{B}$ are said to be *similar*, if there exists an invertible matrix $\mathbf{P}$ such that $\mathbf{B} = \mathbf{P}^{-1}\mathbf{A}\mathbf{P}$. We can now state that *similar matrices represent the same linear transformation*.

In § 7.6 we shall discuss the problem of choosing the basis for X, that is the matrix $\mathbf{P}$, in such a way that $\mathbf{B}$ in (5.13.2) is reduced to a convenient shape (canonical form).

A linear transformation $\alpha: X \rightarrow X$ is said to be *non-singular* or *invertible* if it is a bijective map [see § 1.4.2] of X on to X. If $\mathbf{A}$ is a non-singular matrix, then so is every matrix similar to $\mathbf{A}$ [see Theorem 6.4.2], and the following proposition is almost self-evident:

PROPOSITION 5.13.1.   *The linear transformation* $\alpha: X \to X$ *is non-singular if and only if it is represented by a non-singular matrix relative to any basis. If* α *is represented by the non-singular matrix* **A**, *then relative to the same basis* α^{-1} *is represented by the matrix* $\mathbf{A}^{-1}$.

   The identity transformation is represented by the unit matrix relative to every basis.

We deduce from (5.6.8) that

$$r(\mathbf{P}^{-1}\mathbf{A}\mathbf{P}) = r(\mathbf{A}),$$

and in § 7.4 it is shown that

$$\text{tr}\,(\mathbf{P}^{-1}\mathbf{A}\mathbf{P}) = \text{tr}\,(\mathbf{A}).$$

Hence we may simply speak about the rank and the trace of a linear transformation α and put

$$r(\alpha) = r(\mathbf{A}), \qquad \text{tr}\,\alpha = \text{tr}\,\mathbf{A}, \tag{5.13.3}$$

since the choice of basis does not affect these numbers. The definitions of kerα and imα given in 5.11(i) and (ii) remain unaltered when $\alpha: X \to X$ is a linear transformation of X into itself. Thus

$$\ker\alpha = \{x \in X \mid \mathbf{x}\alpha = \mathbf{0}\}$$

$$\text{im}\alpha = \{\mathbf{y} \in X \mid \mathbf{y} = \mathbf{z}\alpha \text{ for some } \mathbf{z} \in X\}.$$

W.L.

BIBLIOGRAPHY

Cohn, P. M. (1958). *Linear Equations*, Routledge & Kegan Paul.
Hoffman, K. and Kunze, K. (1961). *Linear Algebra*, Prentice-Hall.
Marcus, M. and Minc, H. (1968). *Elementary Linear Algebra*, Macmillan.
Morris, A. O. (1978). *Linear Algebra*, Van Nostrand Reinhold.
Tropper, A. M. (1969). *Linear Algebra*, Nelson.

CHAPTER 6

Matrices and Determinants

6.1. MOTIVATION

Suppose we have two sets of linear equations [see (5.7.1)]

$$
\begin{aligned}
y_1 &= 3x_1 + 4x_2 \\
y_2 &= -2x_1 + x_2 \\
y_3 &= x_1 - x_2
\end{aligned}
\qquad (6.1.1)
$$

and

$$
\begin{aligned}
z_1 &= 2x_1 + 3x_2 \\
z_2 &= x_1 + x_2 \\
z_3 &= -x_1 + 4x_2,
\end{aligned}
\qquad (6.1.2)
$$

and we wish to express $y_1 + z_1$, $y_2 + z_2$, $y_3 + z_3$ in terms of x_1 and x_2. Thus

$$
\begin{aligned}
y_1 + z_1 &= 5x_1 + 7x_2 \\
y_2 + z_2 &= -x_1 + 2x_2 \\
y_3 + z_3 &= 3x_2.
\end{aligned}
\qquad (6.1.3)
$$

Clearly, the simple calculations we have carried out do not depend on the notations for the variables; they would be the same if x, y, z were replaced by other letters or symbols. All that is involved are the schemes of coefficients or *matrices*, which are the essential features of (6.1.1) and (6.1.2). If we denote the matrices of (6.1.1), (6.1.2) and (6.1.3) by

$$
A = \begin{pmatrix} 3 & 4 \\ -2 & 1 \\ 1 & -1 \end{pmatrix}, \quad
B = \begin{pmatrix} 2 & 3 \\ 1 & 1 \\ -1 & 4 \end{pmatrix}, \quad
C = \begin{pmatrix} 5 & 7 \\ -1 & 2 \\ 0 & 3 \end{pmatrix}
$$

respectively, we observe that C has been obtained by adding corresponding entries of A and B. This suggests writing

$$
C = A + B,
$$

and we call C the *sum* of A and B.

153

Now consider a different problem. Suppose we have the two sets of equations

$$y_1 = 3x_1 + 4x_2$$
$$y_2 = -2x_1 + x_2 \qquad\qquad (6.1.4)$$
$$y_3 = x_1 - x_2,$$

and

$$x_1 = 2u_1 - u_2 + u_3$$
$$x_2 = u_1 + 3u_2 - u_4 \qquad\qquad (6.1.5)$$

and we wish to eliminate x_1 and x_2 so that y_1, y_2, y_3 are expressed directly in terms of u_1, u_2, u_3, u_4.

Thus

$$y_1 = 3(2u_1 - u_2 + u_3) + 4(u_1 + 3u_2 - u_4) = 10u_1 + 9u_2 + 3u_2 - 4u_4$$
$$y_2 = -2(2u_1 - u_2 + u_3) + (u_1 + 3u_2 - u_4) = -3u_1 + 5u_2 - 2u_2 - u_4$$
$$y_3 = 2u_1 - u_2 + u_3 - (u_1 + 3u_2 - u_4) = u_1 - 4u_2 + u_3 + u_4.$$
$$(6.1.6)$$

Let us denote the matrices of (6.1.4), (6.1.5) and (6.1.6) by

$$\mathbf{K} = \begin{pmatrix} 3 & 4 \\ -2 & 1 \\ 1 & -1 \end{pmatrix}, \quad \mathbf{L} = \begin{pmatrix} 2 & -1 & 1 & 0 \\ 1 & 3 & 0 & -1 \end{pmatrix},$$

$$\mathbf{M} = \begin{pmatrix} 10 & 9 & 3 & -4 \\ -3 & 5 & -2 & -1 \\ 1 & -4 & 1 & 1 \end{pmatrix}$$

respectively. How do we describe the formation of $\mathbf{M}$ in relation to $\mathbf{K}$ and $\mathbf{L}$?

For this purpose we require the following concept: an *r-tuple* is a collection of r numbers

$$\mathbf{a} = (a_1, a_2, \ldots, a_r),$$

written in a definite order. If

$$\mathbf{b} = (b_1, b_2, \ldots, b_r)$$

is another *r*-tuple, we define the *inner* (or *scalar*) *product* of $\mathbf{a}$ and $\mathbf{b}$ by

$$\mathbf{a} . \mathbf{b} = a_1 b_1 + a_2 b_2 + \ldots + a_r b_r. \qquad\qquad (6.1.7)$$

Thus out of two *r*-tuples we have constructed a single number, or *scalar*, given by (6.1.7). Notice that the inner product can be formed only if the two sets have the same number of components, r in our case.

In each matrix, the coefficients that are aligned horizontally, are called a *row*, while the entries that stand in the same vertical line, are called a *column*. For example, the rows of **K** are

$$(3, 4), \quad (-2, 1), \quad (1, -1), \tag{6.1.8}$$

and the columns of **L** are

$$\begin{pmatrix} 2 \\ 1 \end{pmatrix}, \quad \begin{pmatrix} -1 \\ 3 \end{pmatrix}, \quad \begin{pmatrix} 1 \\ 0 \end{pmatrix}, \quad \begin{pmatrix} 0 \\ -1 \end{pmatrix}. \tag{6.1.9}$$

Disregarding the horizontal display, we view (6.1.9) as a set of 2-tuples and proceed to form the inner product of each member of (6.1.8) with each member of (6.1.9). In this way we obtain all the entries of **M**. More precisely, the first row of **M** consists of the inner products of (3, 4) with all the members of (6.1.9), for example

$$(3, 4).(2, 1) = \quad 6 + \quad 4 = 10$$
$$(3, 4).(-1, 3) = -3 + 12 = \quad 9,$$

and so on. Taking $(-2, 1)$ instead of $(3, 4)$ yields the second row of **M**, for example

$$(-2, 1).(2, 1) = -4 + 1 = -3,$$
$$(-2, 1).(-1, 3) = \quad 2 + 3 = \quad 5,$$

and the composition of $(1, -1)$ with (6.1.9) gives the third row of **M**. We call **M** the *product* of **K** and **L**, and we write

$$\mathbf{M} = \mathbf{KL}.$$

6.2. MATRIX ALGEBRA

We shall now develop the ideas suggested in the preceding section and describe the laws of matrix algebra in general terms.

By an *m* by *n* (in symbols: $m \times n$) *matrix over a domain* $\mathbb{D}$ we mean a rectangular array of elements of $\mathbb{D}$

$$\mathbf{A} = \begin{pmatrix} a_{11} & a_{12} & \cdots & a_{1n} \\ a_{21} & a_{22} & \cdots & a_{2n} \\ \vdots & \vdots & & \vdots \\ a_{m1} & a_{m2} & \cdots & a_{mn} \end{pmatrix}$$

consisting of *m* rows and *n* columns. Here we have used the double suffix notation; thus a_{12} (read: *a* one-two, not *a* twelve) signifies that a_{12} stands at the intersection of the first row and second column of **A**. Generally a_{ij} occupies the place where the *i*th row and the *j*th column meet. For many purposes it suffices to write more briefly

$$\mathbf{A} = (a_{ij}) \quad (i = 1, 2, \ldots, m; j = 1, 2, \ldots, n),$$

always adhering to the convention that the first suffix refers to the row and the second suffix to the column of **A**. The a_{ij} are called the *elements* or *entries* or *coefficients* of **A**.

In many applications the domain $\mathbb{D}$ is the field of real numbers [see § 2.6.1], in which case we speak of *real matrices*. When $\mathbb{D} = \mathbb{C}$, we are concerned with *complex* matrices [see § 2.7.1]. In some contexts we are dealing with *integral (integer)* matrices ($\mathbb{D} = \mathbb{Z}$) [see § 2.2.1] or with *polynomial matrices*, in which each $a_{ij} = a_{ij}(x)$ is a polynomial in one or several indeterminates [see § 14.1 and § 14.15]. Again, there is an interesting theory about non-negative matrices, where $\mathbb{D} = \mathbb{R}^+$, the set of all real numbers ≥ 0 [see § 7.11].

However, unless the contrary is stated, it will always be assumed that $\mathbb{D}$ is some field $\mathbb{F}$ so that division by elements of $\mathbb{F}$ is possible [see § 2.4.2].

Arthur Cayley (1821–95), who introduced the notion of a matrix, is reported to have been sceptical about the usefulness of his brilliant invention. He compared a matrix to a map that remains neatly folded in our pocket when it is not needed: but the map has to be spread out as soon as we want to obtain any practical information from it.

Nevertheless, experience has shown that a great deal of time and thought can be saved, if we think of matrices as mathematical entities in their own right. We shall define operations for matrices, such as addition and multiplication, which resemble the familiar rules of arithmetic. Although these manipulations are originally expressed in terms of the coefficients, the general laws of matrix algebra do not refer to the individual coefficients, just as the bricklayer, in the pursuit of his craft, need not remember that each brick is composed of numerous molecules.

We shall now state the formal rules of matrix algebra.

(i) *Equality*. Two matrices, **A** and **B**, are equal if and only if they have the same coefficients in corresponding positions. In particular, equal matrices must have the same number of rows and the same number of columns, or, as we shall say more briefly, must be of the same *type* $m \times n$. Thus if $\mathbf{A} = (a_{ij})$ and $\mathbf{B} = (b_{ij})$ are $m \times n$ matrices, then the matrix equation

$$\mathbf{A} = \mathbf{B}$$

is equivalent to the mn numerical equations

$$a_{ij} = b_{ij} \quad (i = 1, 2, \ldots, m; j = 1, 2, \ldots, n).$$

For example,

$$\begin{pmatrix} 2 & 1 \\ 3 & 4 \end{pmatrix} \quad \text{and} \quad \begin{pmatrix} 2 & 1 & 0 \\ 3 & 4 & 0 \end{pmatrix}$$

are distinct matrices, and so are

$$\begin{pmatrix} 2 & 1 \\ 3 & 4 \end{pmatrix} \quad \text{and} \quad \begin{pmatrix} 2 & 3 \\ 2 & 4 \end{pmatrix}.$$

(ii) *Addition*. If $\mathbf{A} = (a_{ij})$ and $\mathbf{B} = (b_{ij})$ are $m \times n$ matrices, the sum is defined by

$$\mathbf{A} + \mathbf{B} = (a_{ij} + b_{ij}).$$

Similarly,

$$\mathbf{A} - \mathbf{B} = (a_{ij} - b_{ij}).$$

Thus matrices can be added or subtracted if and only if they are of the same type.

(iii) A matrix all of whose entries are zero, is called a *zero matrix*, and is denoted by $\mathbf{0}$. More accurately, we should write $\mathbf{0}_{mn}$ to indicate the appropriate numbers of rows and columns. But in most cases these can be inferred from the context, and the abbreviated notation will suffice.

(iv) A 1 by n matrix

$$\mathbf{a} = (a_1 \quad a_2 \quad \dots \quad a_n)$$

is called a *row-vector* of order n, and an n by 1 matrix

$$\mathbf{b} = \begin{pmatrix} b_1 \\ b_2 \\ \vdots \\ b_n \end{pmatrix}$$

is called a *column-vector* of order n. Of course, $\mathbf{a}$ and $\mathbf{b}$ are n-tuples over the underlying field $\mathbb{F}$ and may therefore be regarded as elements of the vector space $\mathbb{F}^n$ [see Example 5.2.2].

(v) *Scalar multiplication*. When we are simultaneously concerned with numbers and other mathematical objects, such as matrices or vectors it is customary to refer to numbers as *scalars* [cf. § 5.1]. If $\mathbf{A} = (a_{ij})$ is an $m \times n$ matrix and k a scalar, the product $k\mathbf{A}$ is defined as the matrix whose coefficients are ka_{ij}, that is

$$k\mathbf{A} = (ka_{ij}) = \begin{pmatrix} ka_{11} & ka_{12} & \dots & ka_{1n} \\ ka_{21} & ka_{22} & \dots & ka_{2n} \\ \vdots & \vdots & \ddots & \vdots \\ ka_{m1} & ka_{m2} & \dots & ka_{mn} \end{pmatrix}.$$

For example,

$$3 \begin{pmatrix} 2 & -1 \\ 4 & 0 \end{pmatrix} = \begin{pmatrix} 6 & -3 \\ 12 & 0 \end{pmatrix}.$$

Of course, $k\mathbf{A}$ may equally well be written $\mathbf{A}k$.

(vi) *Matrix multiplication*. As was indicated in § 6.1 the multiplication of matrices is motivated by the procedure of substituting one set of linear equations into another. Let

$$\mathbf{A} = \begin{pmatrix} a_{11} & a_{12} & \dots & a_{1r} \\ a_{21} & a_{22} & \dots & a_{2r} \\ \vdots & \vdots & \ddots & \vdots \\ a_{m1} & a_{m2} & \dots & a_{mr} \end{pmatrix} \quad \text{and} \quad \mathbf{B} = \begin{pmatrix} b_{11} & b_{12} & \dots & b_{1n} \\ b_{21} & b_{22} & \dots & b_{2n} \\ \vdots & \vdots & \ddots & \vdots \\ b_{r1} & b_{r2} & \dots & b_{rn} \end{pmatrix}$$

be an $m \times r$ and an $r \times n$ matrix respectively. The ith row of $\mathbf{A}$ is the r-tuple

$$\mathbf{a}_i = (a_{i1}, a_{i2}, \ldots, a_{ir}) \quad (i = 1, 2, \ldots, m),$$

and the jth column of $\mathbf{B}$ is the r-tuple

$$\mathbf{b}^{(j)} = (b_{ij}, b_{2j}, \ldots, b_{rj}) \quad (j = 1, 2, \ldots, n),$$

written horizontally. The inner product of $\mathbf{a}_i$ and $\mathbf{b}^{(j)}$, as defined by (6.1.7), is

$$\mathbf{a}_i \cdot \mathbf{b}^{(j)} = a_{i1}b_{1j} + a_{i2}b_{2j} + \ldots + a_{ir}b_{rj} \qquad (6.2.1)$$

$(i = 1, 2, \ldots, m; j = 1, 2, \ldots, n)$. The $m \times n$ matrix whose elements are $\mathbf{a}_i \cdot \mathbf{b}^j$, is called the *product* of $\mathbf{A}$ by $\mathbf{B}$, thus

$$\mathbf{AB} = (\mathbf{a}_i \cdot \mathbf{b}^{(j)}).$$

It is important to observe that the matrix product can be defined only when the number of columns in the first factor is equal to the number of rows in the second factor. Thus the types involved in a matrix product obey the rule:

$$m \times r \quad times \quad r \times n \quad gives \quad m \times n. \qquad (6.2.2)$$

To put it briefly, matrices are multiplied by composing the rows of the first factor with the columns of the second factor, with the understanding that composition means the formation of inner products. Although, at first sight, this definition of multiplication appears to be rather complicated, it is nevertheless the most fruitful way to obtain significant applications. A few illustrative examples will help the reader to become adept in the row by column multiplication of matrices; in each case it should be verified that the rule (6.2.2) is obeyed.

EXAMPLES 6.2.1

1. $\begin{pmatrix} 1 & 2 & 3 \\ 3 & 2 & 1 \end{pmatrix} \begin{pmatrix} -1 & 0 \\ 0 & 1 \\ 1 & 0 \end{pmatrix} = \begin{pmatrix} 2 & 2 \\ -2 & 2 \end{pmatrix}$

2. $\begin{pmatrix} 1 & 0 \\ 2 & 0 \end{pmatrix} \begin{pmatrix} 0 & 0 \\ 1 & 3 \end{pmatrix} = \begin{pmatrix} 0 & 0 \\ 0 & 0 \end{pmatrix}.$

3. $(2 \quad 1 \quad 3) \begin{pmatrix} 4 \\ 1 \\ 2 \end{pmatrix} = 15.$

4. $\begin{pmatrix} a_1 & a_2 \\ b_1 & b_2 \end{pmatrix} \begin{pmatrix} x_1 \\ x_2 \end{pmatrix} = \begin{pmatrix} a_1x_1 + a_2x_2 \\ b_1x_1 + b_2x_2 \end{pmatrix}.$

5. $(a_1 \quad a_2 \quad \ldots \quad a_r) \begin{pmatrix} b_1 \\ b_2 \\ \vdots \\ b_r \end{pmatrix} = a_1b_1 + a_2b_2 + \ldots + a_rb_r.$

6.
$$\begin{pmatrix} b_1 \\ b_2 \\ \vdots \\ b_r \end{pmatrix} (a_1 \quad a_2 \quad \ldots \quad a_r) = \begin{pmatrix} b_1 a_1 & b_1 a_2 & \ldots & b_1 a_r \\ b_2 a_2 & b_2 a_2 & \ldots & b_2 a_r \\ \vdots & \vdots & \ddots & \vdots \\ b_r a_1 & b_r b_2 & \ldots & b_r a_r \end{pmatrix}.$$

We must draw attention to one important aspect in which matrix multiplication differs from ordinary multiplication: even when both **AB** and **BA** are defined, these two products are, in general, distinct. For example, if

$$\mathbf{A} = \begin{pmatrix} 1 & 0 \\ 1 & 1 \end{pmatrix}, \qquad \mathbf{B} = \begin{pmatrix} 1 & 1 \\ 0 & 1 \end{pmatrix},$$

then

$$\mathbf{AB} = \begin{pmatrix} 1 & 1 \\ 1 & 2 \end{pmatrix}, \qquad \mathbf{BA} = \begin{pmatrix} 2 & 1 \\ 1 & 1 \end{pmatrix}.$$

(vii) *Square matrices.* An n by n matrix is called a *square* matrix of *order n* (less frequently *degree n*). Let

$$\mathbf{A} = \begin{pmatrix} a_{11} & a_{12} & \ldots & a_{1n} \\ a_{21} & a_{22} & \ldots & a_{2n} \\ \vdots & \vdots & \ddots & \vdots \\ a_{n1} & a_{n2} & \ldots & a_{nn} \end{pmatrix}$$

be a typical matrix of order n. We call

$$a_{11}, a_{22}, \ldots, a_{nn} \tag{6.2.3}$$

the *(main) diagonal* of **A**. The sum of the diagonal entries is called the *trace* of **A**, abbreviated tr**A**, thus

$$\operatorname{tr} \mathbf{A} = a_{11} + a_{22} + \ldots + a_{nn}. \tag{6.2.4}$$

Any entry that does not lie on the diagonal, that is a_{ij} with $i \neq j$, is said to be an *off-diagonal* entry. The entries

$$a_{12}, a_{23}, \ldots, a_{n-1,n} \tag{6.2.5}$$

are called the (first) *superdiagonal* of **A**, and the entries

$$a_{21}, a_{32}, \ldots, a_{n,n-1} \tag{6.2.6}$$

form the (first) *subdiagonal* of **A**.

For each positive integer n we define the *unit matrix* of order n by

$$\mathbf{I}_n = \begin{pmatrix} 1 & 0 & 0 & \ldots & 0 \\ 0 & 1 & 0 & \ldots & 0 \\ 0 & 0 & 1 & \ldots & 0 \\ & \vdots & & \ddots & \vdots \\ 0 & 0 & 0 & \ldots & 1 \end{pmatrix} = (\delta_{ij}), \tag{6.2.7}$$

where

$$\delta_{ij} = \begin{cases} 0 & \text{if } i \neq j \\ 1 & \text{if } i = j \end{cases}. \tag{6.2.8}$$

The symbol introduced in (6.2.7) is called the *Kronecker delta*. When the order of the unit matrix can be inferred from the context, we often write $\mathbf{I}$ instead of $\mathbf{I}_n$. Multiplying by the unit matrix is analogous to multiplying by unity in ordinary algebra. Thus if $\mathbf{A}$ is an arbitrary m by n matrix,

$$\mathbf{A}\mathbf{I}_n = \mathbf{I}_m\mathbf{A} = \mathbf{A}. \tag{6.2.9}$$

6.3. FORMAL LAWS

The operations of matrix addition and multiplication satisfy certain general laws, which resemble those of ordinary algebra [cf. § 2.3]. For example,

$$\mathbf{A} + \mathbf{B} = \mathbf{B} + \mathbf{A} \qquad (commutative\ law\ of\ addition) \tag{6.3.1}$$

$$(\mathbf{A} + \mathbf{B}) + \mathbf{C} = \mathbf{A} + (\mathbf{B} + \mathbf{C}) \quad (associative\ law\ of\ addition) \tag{6.3.2}$$

$$(\mathbf{AB})\mathbf{C} = \mathbf{A}(\mathbf{BC}) \qquad (associative\ law\ of\ multiplication) \tag{6.3.3}$$

$$\left.\begin{array}{l} \mathbf{A}(\mathbf{B} + \mathbf{C}) = \mathbf{AB} + \mathbf{AC} \\ (\mathbf{A} + \mathbf{B})\mathbf{C} = \mathbf{AC} + \mathbf{BC} \end{array}\right\} \quad (distributive\ laws) \tag{6.3.4}$$

While, in general, $\mathbf{AB} \neq \mathbf{BA}$, it may happen that in exceptional cases

$$\mathbf{AB} = \mathbf{BA}. \tag{6.3.5}$$

We then say that the particular matrices $\mathbf{A}$ and $\mathbf{B}$ *commute*.

When $\mathbf{A}$ is a square matrix, we write $\mathbf{A}^2$ instead of $\mathbf{AA}$, and

$$\mathbf{A}^3 = \mathbf{A}\mathbf{A}^2 = \mathbf{A}^2\mathbf{A} = \mathbf{AAA},$$

and similarly for higher powers of $\mathbf{A}$. The *laws of indices*

$$\mathbf{A}^r\mathbf{A}^s = \mathbf{A}^{r+s}$$

$$(\mathbf{A}^r)^s = \mathbf{A}^{rs}$$

are satisfied for all positive exponents.

By combining powers of $\mathbf{A}$ we can form *polynomials* [see (14.1.1)] in $\mathbf{A}$, thus

$$c_0\mathbf{A}^r + c_1\mathbf{A}^{r-1} + \ldots + c_{r-1}\mathbf{A} + c_r\mathbf{I},$$

where $c_0, c_1, \ldots, c_r$ are arbitrary scalars. Notice that the 'constant term' is a scalar multiple of the unit matrix.

6.4. THE INVERSE MATRIX

When we are concerned with rational, real or complex numbers or, more generally, with any field $\mathbb{F}$, then each non-zero element a has an inverse [see § 2.4.2]. Thus if $a \neq 0$, there exists an element b such that

$$ab = ba = 1. \tag{6.4.1}$$

Unfortunately, in the algebra of matrices the concept of an inverse is less straightforward. As we shall see, not every matrix has an inverse. The obvious analogue of (6.4.1) is the following

DEFINITION 6.4.1.   The matrix **B** is said to be the *inverse* of the matrix **A** if

$$\mathbf{AB} = \mathbf{BA} = \mathbf{I}, \tag{6.4.2}$$

where **I** is the unit matrix of the appropriate order.

We note some immediate consequences:
1. Since **I** is a square matrix, any matrices **A** and **B** which satisfy (6.4.2) are necessarily square.
2. If **A** has an inverse, then this inverse is unique; for suppose that we also have that

$$\mathbf{AC} = \mathbf{CA} = \mathbf{I}.$$

Then we can evaluate the triple product **BAC** in two ways, namely

$$\mathbf{BAC} = \mathbf{B(AC)} = \mathbf{BI} = \mathbf{B}$$

$$= \mathbf{(BA)C} = \mathbf{IC} = \mathbf{C},$$

whence **B** = **C**. The inverse of **A**, if it exists, will be denoted by $\mathbf{A}^{-1}$.
3. Since the conditions (6.4.2) are symmetric in **A** and **B**, we deduce that if **B** is the inverse of **A**, then **A** is the inverse of **B**, that is

$$(\mathbf{A}^{-1})^{-1} = \mathbf{A}. \tag{6.4.3}$$

4. If $\mathbf{A}_1$ and $\mathbf{A}_2$ are *invertible* so is $\mathbf{A}_1\mathbf{A}_2$ and

$$(\mathbf{A}_1\mathbf{A}_2)^{-1} = \mathbf{A}_2^{-1}\mathbf{A}_1^{-1}. \tag{6.4.4}$$

Similarly, if each of the matrices $\mathbf{A}_1, \mathbf{A}_2, \ldots, \mathbf{A}_k$ is invertible, then we have the rule

$$(\mathbf{A}_1\mathbf{A}_2\ldots\mathbf{A}_k)^{-1} = \mathbf{A}_k^{-1}\ldots\mathbf{A}_2^{-1}\mathbf{A}_1^{-1},$$

where the reversing of the factors should be noted.
5. If **x** is a non-zero column vector which annihilates **A** on the right, then **A** is not invertible. For suppose that

$$\mathbf{Ax} = \mathbf{0}, \qquad \mathbf{x} \neq \mathbf{0}.$$

If **A** had an inverse we could infer that $\mathbf{0} = \mathbf{A}^{-1}(\mathbf{Ax}) = \mathbf{Ix} = \mathbf{x}$, which contradicts the hypothesis about **x**.

Similarly, if **A** is annihilated by a non-zero row vector on the left, then **A** cannot possess an inverse. For example, if

$$\mathbf{A} = \begin{pmatrix} 2 & -4 \\ -3 & 6 \end{pmatrix}, \qquad \mathbf{y} = (3 \quad 2),$$

then $\mathbf{yA} = \mathbf{0}$, whence **A** is not invertible.

As a matter of fact, for a square matrix these two modes of annihilation are equivalent, that is we have the

PROPOSITION 6.4.1. *Let* $\mathbf{A}$ *be a square matrix. If there exists a non-zero column vector* $\mathbf{x}$ *such that* $\mathbf{Ax} = \mathbf{0}$, *then there exists a non-zero row vector* $\mathbf{y}$ *such that* $\mathbf{yA} = \mathbf{0}$, *and conversely.*

The following nomenclature is customary:

DEFINITION 6.4.2. A square matrix $\mathbf{A}$ is said to be *singular* if there exists a non-zero column vector $\mathbf{x}$ such that $\mathbf{Ax} = \mathbf{0}$. If no such vector exists, the matrix $\mathbf{A}$ is called *non-singular*. Thus $\mathbf{A}$ is non-singular if and only if $\mathbf{Ax} = \mathbf{0}$ implies that $\mathbf{x} = \mathbf{0}$ [see also Proposition 6.12.1].

In view of Proposition 6.4.1, the distinction between singular and non-singular matrices can equally well be formulated in terms of row-vectors, that is, the square matrix $\mathbf{A}$ is non-singular if and only if $\mathbf{yA} = \mathbf{0}$ implies that $\mathbf{y} = \mathbf{0}$. We have seen that no inverse can exist when the matrix is annihilated by a non-zero vector. It is significant that the converse of this proposition is also true. Thus we have the following

THEOREM 6.4.2. *The matrix* $\mathbf{A}$ *has an inverse if and only if the equation* $\mathbf{Ax} = \mathbf{0}$ *always implies that* $\mathbf{x} = \mathbf{0}$; *in other words: a matrix is invertible if and only if it is non-singular.*

The phenomenon of a matrix that is non-zero but non-invertible is related to the notion of a *zero divisor* [see § 2.13].

The computation of the inverse matrix, if it exists, is one of the most important tasks of linear algebra. We shall mention several methods but only the first of these, known as pivotal condensation, is suitable for 'real life' numerical matrices. The procedure is fully described elsewhere in this work [III § 4.5], and we confine ourselves here to a recapitulation of the main features. The other methods, which are less laborious, apply only in special circumstances and are more valuable in theoretical work.

(i) *Pivotal condensation.* Suppose we wish to invert the n by n matrix $\mathbf{A}$. To this end we set up an n by $2n + 1$ matrix

$$\mathbf{A}_0 = (\mathbf{A} \vdots \mathbf{I} \vdots \mathbf{c}). \tag{6.4.5}$$

The first n columns, which we shall call the *first panel* of $\mathbf{A}_0$ are occupied by the original matrix $\mathbf{A}$, the next n columns, or *second panel*, consist of the unit matrix of order n; the final column, $\mathbf{c}$, which is known as the *check column* contains all the row-sums of the first and second panels. The rows of $\mathbf{A}_0$, that is the rows of both panels and of the check column, are subjected to a series of elementary operations, which are of three kinds:

(α) permutation of the rows [see § 8.1.1];

(β) multiplication of one particular row by a non-zero scalar;

(γ) adding a scalar multiple of one row to another row; thus if the rows of A_0 are denoted by $R_1, R_2, \ldots, R_n$ we may choose a pair of distinct suffixes p, q and replace R_p by $R_p + \lambda R_q$, where λ is an arbitrary scalar.

The object of carrying out these operations is to change A_0 into the matrix

$$B_0 = (I \vdots B \vdots d), \qquad (6.4.6)$$

in which the first panel is the unit matrix of order n. If this transition from A_0 to B_0 can be achieved by a sequence of elementary row operations, then it will turn out that

$$B = A^{-1}.$$

However, the procedure breaks down if, at any stage, the first panel contains a zero row or a zero column. In such a situation the initial matrix A is not invertible.

As an illustration, we shall find the inverse of

$$A = \begin{pmatrix} 0 & 1 & 1 \\ 1 & 0 & 1 \\ 1 & 1 & 0 \end{pmatrix}. \qquad (6.4.7)$$

Despite its simplicity the example suffices to bring out the main features of the method. The Table 6.4.1 shows the details of the calculations. The rows are

R_1	0	1	1	1	0	0	3
R_2	1	0	1	0	1	0	3
R_3	1	1	0	0	0	1	3
$R_4 = R_2$	1	0	1	0	1	0	3
$R_5 = R_1$	0	1	1	1	0	0	3
$R_6 = R_3$	1	1	0	0	0	1	3
$R_7 = R_4$	1	0	1	0	1	0	3
$R_8 = R_5$	0	1	1	1	0	0	3
$R_9 = R_6 - R_4$	0	1	-1	0	-1	1	0
$R_{10} = R_7$	1	0	1	0	1	0	3
$R_{11} = R_8$	0	1	1	1	0	0	3
$R_{12} = R_9 - R_8$	0	0	-2	-1	-1	1	-3
$R_{13} = R_{10} + \frac{1}{2}R_{12}$	1	0	0	$-\frac{1}{2}$	$\frac{1}{2}$	$\frac{1}{2}$	$\frac{3}{2}$
$R_{14} = R_{11} + \frac{1}{2}R_{12}$	0	1	0	$\frac{1}{2}$	$-\frac{1}{2}$	$\frac{1}{2}$	$\frac{3}{2}$
$R_{15} = -\frac{1}{2}R_{12}$	0	0	1	$\frac{1}{2}$	$\frac{1}{2}$	$-\frac{1}{2}$	$\frac{3}{2}$

Table 6.4.1

labelled $R_1, R_2, \ldots, R_{18}$ with an indication as to how each row has been derived from its predecessors.

As already mentioned, the entries in the last column take part in the row operations. At each stage it should be checked that these entries are in fact the appropriate row sums. The last band is now of the form (6.4.6), and we have found that

$$\mathbf{A}^{-1} = \begin{pmatrix} -\frac{1}{2} & \frac{1}{2} & \frac{1}{2} \\ \frac{1}{2} & -\frac{1}{2} & \frac{1}{2} \\ \frac{1}{2} & \frac{1}{2} & -\frac{1}{2} \end{pmatrix}. \tag{6.4.8}$$

(ii) *Polynomial equations.* Let $\mathbf{A}$ be a square matrix of order n. The collection of powers, that is the matrices

$$\mathbf{I}, \mathbf{A}, \mathbf{A}^2, \mathbf{A}^3, \ldots$$

cannot be linearly independent [see Proposition 5.4.1]. Thus there exists a positive integer m and scalars $\alpha_1, \alpha_2, \ldots, \alpha_m$ such that

$$\mathbf{A}^m + \alpha_1 \mathbf{A}^{m-1} + \ldots + \alpha_{m-1}\mathbf{A} + \alpha_m \times \mathbf{I} = \mathbf{0}. \tag{6.4.9}$$

Expressed differently, we may say that $\mathbf{A}$ satisfies a polynomial equation [see (14.1.1)]. In fact, as we shall see, every matrix satisfies its own characteristic equation [see Theorem 7.5.1]. Suppose that we happen to know a polynomial equation (6.4.9) for $\mathbf{A}$ in which $\alpha_m \neq 0$. Then we find that

$$\mathbf{A}^{-1} = (-1/\alpha_m)(\mathbf{A}^{m-1} + \alpha_1\mathbf{A}^{m-2} + \ldots + \alpha_{m-1}\mathbf{I}),$$

that is $\mathbf{A}^{-1}$ is expressed as a polynomial in $\mathbf{A}$. The evaluation of the latter is often simpler than the computation of $\mathbf{A}^{-1}$ by pivotal condensation. Using the same example as before, we write

$$\mathbf{A} = \mathbf{E} - \mathbf{I},$$

where

$$\mathbf{E} = \begin{pmatrix} 1 & 1 & 1 \\ 1 & 1 & 1 \\ 1 & 1 & 1 \end{pmatrix}.$$

Now $\mathbf{E}$ has the property that

$$\mathbf{E}^2 = 3\mathbf{E}.$$

Therefore

$$\mathbf{A}^2 = (\mathbf{E} - \mathbf{I})^2 = \mathbf{E}^2 - 2\mathbf{E} + \mathbf{I} = \mathbf{E} + \mathbf{I} = \mathbf{A} + 2\mathbf{I}.$$

Hence

$$\mathbf{A}^{-1} = \tfrac{1}{2}(\mathbf{A} - \mathbf{I}),$$

which agrees with (6.4.8).

(iii) *The adjugate matrix*. This method uses determinants and is explained more fully in section 6.12 of this chapter.

The formula

$$\mathbf{A}^{-1} = (\det \mathbf{A})^{-1}(\text{adj } \mathbf{A}) = (\det \mathbf{A})^{-1}(A_{ij}) \qquad (6.4.10)$$

has the merit of furnishing an explicit expression for the inverse. But, as a rule, the computation of the adjugate matrix is even more laborious than the method of pivotal condensation. Hence this formula cannot be recommended for practical purposes; nevertheless, the result is of considerable theoretical interest. Clearly (6.4.10) can only be used when $\det \mathbf{A} \neq 0$, and the converse is also true. When $n = 2$, the result is so simple that it can be committed to memory. Thus when

$$\mathbf{A} = \begin{pmatrix} a & b \\ c & d \end{pmatrix}, \qquad ad - bc \neq 0$$

we have that

$$\mathbf{A}^{-1} = (ad - bc)^{-1}\begin{pmatrix} d & -b \\ -c & a \end{pmatrix}. \qquad (6.4.11)$$

In section 6.12 we shall give an example of a three-rowed matrix.

When symmetry is present, the calculations tend to become simpler. Reverting once more to the matrix

$$\mathbf{A} = \begin{pmatrix} 0 & 1 & 1 \\ 1 & 0 & 1 \\ 1 & 1 & 0 \end{pmatrix},$$

we find that the cofactors [see § 6.11] of $\mathbf{A}$ are as follows:

$$A_{11} = A_{22} = A_{33} = \begin{vmatrix} 0 & 1 \\ 1 & 0 \end{vmatrix} = -1,$$

$$A_{12} = A_{21} = -\begin{vmatrix} 1 & 1 \\ 1 & 0 \end{vmatrix} = 1,$$

$$A_{13} = \begin{vmatrix} 1 & 0 \\ 1 & 1 \end{vmatrix} = 1, \qquad A_{31} = \begin{vmatrix} 1 & 1 \\ 0 & 1 \end{vmatrix} = 1,$$

$$A_{23} = A_{32} = -\begin{vmatrix} 0 & 1 \\ 1 & 1 \end{vmatrix} = 1.$$

Since by (6.11.2)

$$\det \mathbf{A} = a_{11}A_{11} + a_{12}A_{12} + a_{13}A_{13} = 2,$$

we obtain that

$$\mathbf{A}^{-1} = \tfrac{1}{2}\begin{pmatrix} -1 & 1 & 1 \\ 1 & -1 & 1 \\ 1 & 1 & -1 \end{pmatrix},$$

as before.

(iv) *Linear equations.* Let

$$\mathbf{x} = \begin{pmatrix} x_1 \\ x_2 \\ \vdots \\ x_n \end{pmatrix}$$

be a column vector of indeterminates and put

$$\mathbf{Ax} = \mathbf{y}. \tag{6.4.12}$$

We may regard this equation as a map which sends the vector $\mathbf{x}$ into the vector $\mathbf{y}$ [see § 5.11]. When $\mathbf{A}$ has an inverse, the map may be inverted [see Proposition 5.13.1], and we can express $\mathbf{x}$ in terms of $\mathbf{y}$ with the aid of the matrix $\mathbf{A}^{-1}$, that is

$$\mathbf{x} = \mathbf{A}^{-1}\mathbf{y}. \tag{6.4.13}$$

The solution of the system (6.4.12) is discussed in section 5.8. But in some cases the equations can be solved from first principles and when this has been done, the coefficients of the components of $\mathbf{y}$ in (6.4.13) will display the matrix $\mathbf{A}^{-1}$ explicitly.

EXAMPLE 6.4.1.   Find the inverse of

$$\mathbf{A} = \begin{pmatrix} 1 & 0 & 0 & 0 \\ 1 & 1 & 0 & 0 \\ 0 & 2 & 1 & 0 \\ 0 & 0 & 3 & 1 \end{pmatrix}.$$

Written out in full the system (6.4.12) becomes

$$\begin{aligned} x_1 &= y_1 \\ x_1 + x_2 &= y_2 \\ 2x_2 + x_3 &= y_3 \\ 3x_3 + x_4 &= y_4. \end{aligned}$$

Solving for x we obtain that

$$\begin{aligned} x_1 &= y_1 \\ x_2 &= y_2 - x_1 = -y_1 + y_2 \\ x_3 &= y_3 - 2x_2 = 2y_1 - 2y_2 + y_3 \\ x_4 &= y_4 - 3x_3 = -6y_1 + 6y_2 - 3y_3 + y_4, \end{aligned}$$

whence we read off that

$$\mathbf{A}^{-1} = \begin{pmatrix} 1 & 0 & 0 & 0 \\ -1 & 1 & 0 & 0 \\ 2 & -2 & 1 & 0 \\ -6 & 6 & -3 & 1 \end{pmatrix}.$$

6.5. TRANSPOSITION

This is a matrix operation which has no analogue in elementary algebra. If $\mathbf{A}$ is an $m \times n$ matrix, the *transpose* of $\mathbf{A}$, denoted by $\mathbf{A}'$, is the $n \times m$ matrix derived from $\mathbf{A}$ by interchanging the roles of rows and columns. For example,

$$\begin{pmatrix} 2 & 3 & 1 \\ 0 & 1 & 4 \end{pmatrix}' = \begin{pmatrix} 2 & 0 \\ 3 & 1 \\ 1 & 4 \end{pmatrix},$$

$$\begin{pmatrix} 1 & 2 \\ -1 & 0 \end{pmatrix}' = \begin{pmatrix} 1 & -1 \\ 2 & 0 \end{pmatrix},$$

$$(a_1 \quad a_2 \quad \dots \quad a_n)' = \begin{pmatrix} a_1 \\ a_2 \\ \vdots \\ a_n \end{pmatrix}.$$

Generally, if

$$\mathbf{A} = \begin{pmatrix} a_{11} & a_{12} & \dots & a_{1n} \\ a_{21} & a_{22} & \dots & a_{2n} \\ \vdots & \vdots & \ddots & \vdots \\ a_{m1} & a_{m2} & \dots & a_{mn} \end{pmatrix}.$$

then

$$\mathbf{A}' = \begin{pmatrix} a_{11} & a_{21} & \dots & a_{m1} \\ a_{12} & a_{22} & \dots & a_{m2} \\ \vdots & \vdots & \ddots & \vdots \\ a_{1n} & a_{2n} & \dots & a_{mn} \end{pmatrix}.$$

There is a simple rule concerning the transpose of a product. It states that

$$(\mathbf{AB})' = \mathbf{B}'\mathbf{A}'. \tag{6.5.1}$$

If $\mathbf{A}$ has an inverse, so also has $\mathbf{A}'$ and

$$(\mathbf{A}')^{-1} = (\mathbf{A}^{-1})'. \tag{6.5.2}$$

Evidently, if

$$\mathbf{a} = (a_1 \quad a_2 \quad \dots \quad a_n)$$

is a row-vector [see § 6.2.iv], then

$$\mathbf{a}' = \begin{pmatrix} a_1 \\ a_2 \\ \vdots \\ a_n \end{pmatrix}$$

is a column-vector, and if

$$\mathbf{b} = \begin{pmatrix} b_1 \\ b_2 \\ \vdots \\ b_n \end{pmatrix}$$

is a column-vector, then

$$\mathbf{b}' = (b_1 \quad b_2 \quad \dots \quad b_n)$$

is a row-vector. We observe that in these circumstances [see Example 6.2.1(5)]

$$\mathbf{ab} = a_1 b_1 + a_2 b_2 + \dots + a_n b_n \tag{6.5.3}$$

is a 1 by 1 matrix, that is a single element of the underlying field. Notice that (6.5.3) is identical with the inner product defined in (6.1.7). If $\mathbf{b}$ had been a row-vector, like $\mathbf{a}$, then the right-hand side of (6.5.3) could be expressed as $\mathbf{ab}'$ or equally as $\mathbf{ba}'$. In particular we have that

$$\mathbf{aa}' = a_1{}^2 + a_2{}^2 + \dots + a_n{}^2. \tag{6.5.4}$$

Similarly,

$$\mathbf{b}'\mathbf{b} = b_1{}^2 + b_2{}^2 + \dots + b_n{}^2.$$

Notice that in matrix algebra the factors in $\mathbf{ab}$, and in $\mathbf{aa}'$ are not separated by a dot.

We mention that some authors denote the transpose of $\mathbf{A}$ by $\mathbf{A}^T$ or by $\mathbf{A}^t$.

6.6. PARTITIONED MATRICES

The matrices we have considered so far have entries in a field $\mathbb{F}$ [see § 6.2]. However, it is not difficult to see that most of the laws of matrix algebra remain valid when the entries of the matrix are themselves matrices of suitable types provided that appropriate safeguards are observed which render addition and multiplication of such elements meaningful. A matrix of matrices is called a *partitioned matrix*. For example, the 3 by 5 matrix

$$\mathbf{M} = \begin{pmatrix} a_1 & b_1 & c_1 & d_1 & e_1 \\ a_2 & b_2 & c_2 & d_2 & e_2 \\ a_3 & b_3 & c_3 & d_3 & e_3 \end{pmatrix}$$

may be written as

$$\mathbf{M} = \begin{pmatrix} \mathbf{A} & \mathbf{B} \\ \mathbf{C} & \mathbf{D} \end{pmatrix},$$

where

$$A = \begin{pmatrix} a_1 & b_1 \\ a_2 & b_2 \end{pmatrix}, \qquad B = \begin{pmatrix} c_1 & d_1 & e_1 \\ c_2 & d_2 & e_2 \end{pmatrix},$$

$$C = (a_3 \quad b_3), \qquad D = (c_3 \quad d_3 \quad e_3).$$

Alternatively, we may use the partitioning

$$M = \begin{pmatrix} P & Q & R \\ X & Y & Z \end{pmatrix},$$

where

$$P = \begin{pmatrix} a_1 \\ a_2 \end{pmatrix}, \qquad Q = \begin{pmatrix} b_1 & c_1 \\ b_2 & c_2 \end{pmatrix}, \qquad R = \begin{pmatrix} d_1 & e_1 \\ d_2 & e_2 \end{pmatrix},$$

$$X = (a_3), \qquad Y = (b_3 \quad c_3), \qquad Z = (d_3 \quad e_3).$$

However, there is an important restriction on the manner in which a matrix may be partitioned: the horizontal and vertical lines must go straight through the whole matrix; thus the pattern

$$M = \left(-- \begin{array}{c|c|c} & & \\ \hline & & \\ \hline & & \end{array} -- \right)$$

is allowed, but

$$M = \left(-- \begin{array}{c|c} & \\ \hline & \end{array} -- \right)$$

is inadmissible. In general, an m by n matrix A may be partitioned as

$$A = \begin{pmatrix} A_{11} & A_{12} & \cdots & A_{1t} \\ A_{21} & A_{22} & \cdots & A_{2t} \\ \vdots & \vdots & \ddots & \vdots \\ A_{s1} & A_{s2} & \cdots & A_{st} \end{pmatrix} = (A_{ij}),$$

where A_{ij} is of type e_i by f_j subject to the conditions

$$\sum_i e_i = m, \qquad \sum_j f_j = n.$$

Let

$$B = (B_{ij})$$

be another partitioned matrix where B_{ij} is of type g_i by h_j. Then summation can be carried out in the form

$$A + B = (A_{ij} + B_{ij}) \tag{6.6.1}$$

provided that

$$e_i = g_i \quad (i = 1, 2, \ldots, s)$$

and

$$f_j = h_j \quad (j = 1, 2, \ldots, t).$$

Similarly, we can write

$$\mathbf{AB} = (\mathbf{C}_{ik}), \tag{6.6.2}$$

provided that $f_j = g_j$, in which case

$$\mathbf{C}_{ik} = \sum_j \mathbf{A}_{ij}\mathbf{B}_{jk}.$$

If taken with a grain of salt, the following rule of thumb adequately summarizes the situation: *partitioned matrices may be added and multiplied analogously to matrices over a field as long as all matrix operations make sense.*

A little more care is needed when we wish to transpose a partitioned matrix, the general formula being

$$\mathbf{A}' = (\mathbf{A}_{ij})' = (\mathbf{A}'_{ji}), \tag{6.6.3}$$

that is, not only is each block transposed but the arrangement of the blocks is also transposed; for example,

$$\begin{pmatrix} \mathbf{P} & \mathbf{Q} & \mathbf{R} \\ \mathbf{X} & \mathbf{Y} & \mathbf{Z} \end{pmatrix}' = \begin{pmatrix} \mathbf{P}' & \mathbf{X}' \\ \mathbf{Q}' & \mathbf{Y}' \\ \mathbf{R}' & \mathbf{Z}' \end{pmatrix}. \tag{6.6.4}$$

The handling of partitioned matrices is simplest when all constituent blocks are square matrices of the same order.

6.7.   MATRICES OF SPECIAL TYPES

In this section we summarize the definitions and main properties of some of the most important classes of matrices. Throughout we shall be concerned with *square* matrices of a fixed order n, and unless otherwise specified, the indices i, j, k, l range from 1 to n.

In some cases reference will be made to results about determinants which will be discussed more fully in subsequent sections of this chapter.

(i) The *unit* or *identity matrix* $\mathbf{I}_n = \mathbf{I} = (\delta_{ij})$, where the symbol

$$\delta_{ij} = \begin{cases} 1, & \text{if } i = j \\ 0, & \text{if } i \neq j \end{cases} \tag{6.7.1}$$

was introduced in (6.2.8). A scalar multiple, $k\mathbf{I}$, of the unit matrix is called a *scalar matrix*.

(ii) *The (natural) matrix units.* This is a set of n^2 matrices $\mathbf{E}_{ij}$. For given values of i and j, the $(i, j\text{th})$th element of $\mathbf{E}_{ij}$ is equal to unity while all other elements are zero. For example, when $n = 4$,

$$\mathbf{E}_{13} = \begin{pmatrix} 0 & 0 & 1 & 0 \\ 0 & 0 & 0 & 0 \\ 0 & 0 & 0 & 0 \\ 0 & 0 & 0 & 0 \end{pmatrix}, \quad \mathbf{E}_{31} = \begin{pmatrix} 0 & 0 & 0 & 0 \\ 0 & 0 & 0 & 0 \\ 1 & 0 & 0 & 0 \\ 0 & 0 & 0 & 0 \end{pmatrix}.$$

We note that generally

(i) $$\mathbf{I} = \mathbf{E}_{11} + \mathbf{E}_{22} + \ldots + \mathbf{E}_{nn}.$$

(ii) If $\mathbf{A} = (a_{ij})$ is an arbitrary n by n matrix, then

$$\mathbf{A} = (a_{ij}) = \sum_{i,j=1}^{n} a_{ij} \mathbf{E}_{ij}.$$

(iii) The matrix units satisfy the relations

$$\mathbf{E}_{ij} \mathbf{E}_{kl} = \delta_{jk} \mathbf{E}_{il}.$$

(iii) *Triangular matrices.* Matrices of the form

$$\mathbf{B} = \begin{pmatrix} b_{11} & b_{12} & b_{13} & \cdots & b_{1n} \\ 0 & b_{22} & b_{23} & \cdots & b_{2n} \\ 0 & 0 & b_{33} & \cdots & b_{3n} \\ \vdots & \vdots & \vdots & \ddots & \vdots \\ 0 & 0 & 0 & \cdots & b_{nn} \end{pmatrix}, \quad \mathbf{C} = \begin{pmatrix} c_{11} & 0 & 0 & \cdots & 0 \\ c_{21} & c_{22} & 0 & \cdots & 0 \\ c_{31} & c_{32} & c_{33} & \cdots & 0 \\ \vdots & \vdots & \vdots & \ddots & \vdots \\ c_{n1} & c_{n2} & c_{n3} & \cdots & c_{nn} \end{pmatrix} \quad (6.7.2)$$

are called *upper* and *lower* triangular respectively.

The sum and the product of two upper (lower) triangular matrices is again an upper (lower) triangular matrix.

(iv) *Diagonal matrices.* We use the notation

$$\mathbf{D} = \operatorname{diag}(a_1, a_2, \ldots, a_n) = \begin{pmatrix} a_1 & 0 & 0 & \cdots & 0 \\ 0 & a_2 & 0 & \cdots & 0 \\ 0 & 0 & a_3 & \cdots & 0 \\ \vdots & \vdots & \vdots & \ddots & \vdots \\ 0 & 0 & 0 & \cdots & a_n \end{pmatrix}. \quad (6.7.3)$$

Note that, if $\mathbf{G} = \operatorname{diag}(b_1, b_2, \ldots, b_n)$ is another diagonal matrix of order n, then $\mathbf{DG} = \operatorname{diag}(a_1 b_1, a_2 b_2, \ldots, a_n b_n)$ and so all diagonal matrices commute with each other.

Extending the terminology to partitioned matrices we may introduce *block-diagonal matrices*

$$\text{diag}\,(\mathbf{A}_1, \mathbf{A}_2, \ldots, \mathbf{A}_k) = \begin{pmatrix} \mathbf{A}_1 & 0 & 0 & \ldots & 0 \\ 0 & \mathbf{A}_2 & 0 & \ldots & 0 \\ \vdots & \vdots & & \ddots & \vdots \\ 0 & 0 & 0 & \ldots & \mathbf{A}_k \end{pmatrix} \qquad (6.7.4)$$

where $\mathbf{A}_1, \mathbf{A}_2, \ldots, \mathbf{A}_k$ are arbitrary square matrices. Block-diagonal matrices, of course, do not in general commute.

(v) *Symmetric matrices* (usually real). The matrix $\mathbf{S} = (s_{ij})$ is said to be *symmetric* if it is equal to its transpose. Thus

$$\mathbf{S}' = \mathbf{S}, \qquad (6.7.5)$$

that is

$$s_{ij} = s_{ji} \quad (i, j = 1, 2, \ldots, n). \qquad (6.7.6)$$

With each symmetric matrix there is associated a quadratic form [see (9.1.4)]

$$Q(\mathbf{x}) = \mathbf{x}'\mathbf{S}\mathbf{x} = \sum_{i,j} s_{ij} x_i x_j, \qquad (6.7.7)$$

where $\mathbf{x}$ is a column vector of indeterminates. We call $\mathbf{S}$ positive (negative) definite, semi-definite or indefinite according as the corresponding quadratic form is positive (negative) definite, semi-definite or indefinite [see § 9.2].

(vi) *Skew-symmetric (anti-symmetric) matrices* (usually real). They are defined by $\mathbf{A}' = -\mathbf{A}$. Thus if $\mathbf{A} = (a_{ij})$ is skew-symmetric, then

$$a_{ij} = -a_{ji}, \qquad (6.7.8)$$

whence $a_{ij} = 0$.

When n is odd, then $\det \mathbf{A} = 0$; when n is even, then $\det \mathbf{A}$ is a perfect square in the coefficients a_{ij} [see § 6.14(iv)].

(vii) *Orthogonal matrices* (usually real). The matrix $\mathbf{R}$ is orthogonal if

$$\mathbf{R}'\mathbf{R} = \mathbf{I}, \qquad (6.7.9)$$

whence

$$\mathbf{R}' = \mathbf{R}^{-1}, \qquad \mathbf{R}\mathbf{R}' = \mathbf{I}.$$

If $\mathbf{x}$ and $\mathbf{y}$ are $n \times 1$ column vectors [see § 6.2(iv)] with the property that

$$\mathbf{y} = \mathbf{R}\mathbf{x},$$

then, by (6.5.1),

$$\mathbf{y}'\mathbf{y} = \mathbf{x}'\mathbf{x},$$

that is

$$\sum_i y_i^2 = \sum_i x_i^2.$$

The rows (columns) of an orthogonal matrix form an orthonormal system of vectors [see § 10.2].

The product of two orthogonal matrices is an orthogonal matrix.

By taking determinants [see (6.10.3)] of (6.7.9) it follows that the determinant of an orthogonal matrix has the value $+1$ or -1; if det $\mathbf{R} = 1$, we say that $\mathbf{R}$ is a *proper orthogonal* matrix.

(viii) *Hermitian matrices* (after the French mathematician Charles Hermite, 1822–1901). We now assume that the entries of the matrix are complex numbers. The matrix $\mathbf{H} = (h_{ij})$ is called *Hermitian* if

$$\overline{\mathbf{H}}' = \mathbf{H}, \tag{6.7.10}$$

where the bar denotes the complex conjugate [see (2.7.10)], that is

$$h_{ij} = \bar{h}_{ji} \quad (i, j = 1, 2, \ldots, n).$$

In particular, $\bar{h}_{ii} = h_{ii}$. Thus the diagonal entries of an Hermitian matrix are necessarily real. It is customary to use the notation

$$\mathbf{A}^* = \overline{\mathbf{A}}', \tag{6.7.11}$$

for the *conjugate transpose* or *adjoint* of an arbitrary, possibly rectangular, matrix. The condition for an Hermitian matrix is then expressed by

$$\mathbf{H}^* = \mathbf{H}. \tag{6.7.12}$$

With every Hermitian matrix there is associated an Hermitian form [see (9.3.2)]

$$\phi(\mathbf{z}) = \mathbf{z}^*\mathbf{H}\mathbf{z} = \sum_{i,j} h_{ij} \bar{z}_i z_j,$$

where $\mathbf{z}$ is a column vector of (complex) indeterminates. It is noteworthy that, for every $\mathbf{z}$, the value of $\phi(\mathbf{z})$ is a real number. Hermitian forms are classified as positive (negative) definite, semi-definite or indefinite [see § 9.3], and this classification is transferred to the matrix $\mathbf{H}$.

(ix) *Skew Hermitian matrices.* The matrix $\mathbf{K} = (k_{ij})$ is said to be skew-Hermitian if $\overline{\mathbf{K}}' = -\mathbf{K}$. Hence $i\mathbf{K}$ is Hermitian.

(x) *Unitary matrices.* A unitary matrix is defined by the equation

$$\overline{\mathbf{U}}'\mathbf{U} = \mathbf{I}, \tag{6.7.13}$$

whence it follows that $\mathbf{U}\overline{\mathbf{U}}' = \mathbf{I}$. The rows of $\mathbf{U}$ form a system of orthonormal vectors in $\mathbb{C}^n$ [see Definition 10.2.1] as also do the columns of $\mathbf{U}$.

Unitary matrices play a fundamental role in the theory of metric vector spaces over the complex field [see § 10.2], notably in relation to the diagonalization of Hermitian forms (Principal Axes Theorem, [see Theorem 7.8.3]).

(xi) *Permutation matrices.* A permutation, π, on the set $S = \{1, 2, \ldots, n\}$ is a map of S onto itself [see § 8.1]. We use the notation

$$\pi = \begin{pmatrix} 1 & 2 & \cdots & n \\ 1\pi & 2\pi & \cdots & n\pi \end{pmatrix} \tag{6.7.14}$$

to indicate that the image of i under π is equal to $i\pi$. Thus $1\pi, 2\pi, \ldots, n\pi$ is a rearrangement of $1, 2, \ldots, n$. For example,

$$\pi = \begin{pmatrix} 1 & 2 & 3 & 4 \\ 2 & 4 & 3 & 1 \end{pmatrix} \tag{6.7.15}$$

is a permutation on four objects.

Let

$$\mathbf{x} = \begin{pmatrix} x_1 \\ x_2 \\ \vdots \\ x_n \end{pmatrix}$$

be a vector whose components are indeterminates. We define the action of π on $\mathbf{x}$ by

$$\mathbf{x}\pi = \begin{pmatrix} x_{1\pi} \\ x_{2\pi} \\ \vdots \\ x_{n\pi} \end{pmatrix}; \tag{6.7.16}$$

in other words, π permutes the indeterminates in accordance with (6.7.14). Now observe that there exists a unique matrix $\mathbf{P}$ such that

$$\mathbf{x}\pi = \mathbf{P}\mathbf{x}. \tag{6.7.17}$$

EXAMPLE 6.7.1.   When π is given by (6.7.15) we have that

$$\mathbf{P} = \begin{pmatrix} 0 & 1 & 0 & 0 \\ 0 & 0 & 0 & 1 \\ 0 & 0 & 1 & 0 \\ 1 & 0 & 0 & 0 \end{pmatrix},$$

because it is readily verified that each side of (6.7.17) is then equal to the vector

$$\begin{pmatrix} x_2 \\ x_4 \\ x_3 \\ x_1 \end{pmatrix}.$$

In the general case the matrix that corresponds to π, is given by

$$\mathbf{P} = (\delta_{i\pi, j}), \tag{6.7.18}$$

where the Kronecker delta δ_{ij} is defined in (6.2.8). This means that in the ith row of $\mathbf{P}$ the entry for which $j = i\pi$ is equal to unity, all other entries being zero.

DEFINITION 6.7.1. The matrix described in (6.7.18) is called the *permutation matrix* corresponding to the permutation π.

A permutation is characterized by the fact that in each row and in each column precisely one entry is equal to unity while all other entries are zero.

All permutation matrices are orthogonal.

If $\mathbf{P}$ and $\mathbf{Q}$ are permutation matrices, so is $\mathbf{PQ}$. The determinant of $\mathbf{P}$ is equal to 1 or -1 according as the corresponding permutation is even or odd [see Definition 8.1.4].

(xii) *Shearing matrices.* Matrices of this type are given by the formula

$$\mathbf{S}_{ij}(\lambda) = \mathbf{I} + \lambda \mathbf{E}_{ij} \quad (i \neq j), \tag{6.7.19}$$

where λ is an arbitrary scalar and $\mathbf{E}_{ij}$ is defined in (ii), of this section. When interpreted as a transformation of a vector space [§ 5.13], the matrix (6.7.19) describes a shearing motion parallel to one of the coordinate axes [see (5.4.3)]. For example, when $n = 3$,

$$\mathbf{S}_{23}(\lambda) = \begin{pmatrix} 1 & 0 & 0 \\ 0 & 1 & \lambda \\ 0 & 0 & 1 \end{pmatrix}.$$

Note that

$$(\mathbf{S}_{ij}(\lambda))^{-1} = \mathbf{S}_{ij}(-\lambda). \tag{6.7.20}$$

(xiii) *Idempotent matrices.* The matrix $\mathbf{E} = (e_{ij})$ is said to be *idempotent* if it satisfies the equation

$$\mathbf{E}^2 = \mathbf{E}. \tag{6.7.21}$$

The zero matrix and the unit matrix are trivial examples of idempotent matrices. In all other cases an idempotent matrix is of rank r [see (5.6.4)], where

$$0 < r < n.$$

Using the trace of $\mathbf{E}$, defined by

$$\text{tr } \mathbf{E} = e_{11} + e_{22} + \ldots + e_{nn},$$

[see (6.2.4)], we note that the relation

$$\text{tr } \mathbf{E} = \text{rank } \mathbf{E}, \tag{6.7.22}$$

is valid for all idempotent matrices.

(xiv) *Nilpotent matrices.* A matrix $\mathbf{N}$ is said to be *nilpotent* if there exists a positive integer k such that

$$\mathbf{N}^k = \mathbf{0}. \tag{6.7.23}$$

If (6.7.23) is satisfied, but if $\mathbf{N}^{k-1} \neq \mathbf{0}$, then $\mathbf{N}$ is said to be *nilpotent of order k*.

An important type of nilpotent matrix is furnished by

$$\mathbf{K} = \begin{pmatrix} 0 & a_{12} & a_{13} & \cdots & a_{1n} \\ 0 & 0 & a_{23} & \cdots & a_{2n} \\ \vdots & \vdots & & \ddots & \vdots \\ 0 & 0 & 0 & \cdots & a_{n-1,n} \\ 0 & 0 & 0 & \cdots & 0 \end{pmatrix}, \tag{6.7.24}$$

that is, an upper triangular matrix, in which all diagonal entries are zero. Such a matrix is nilpotent of order n provided that each entry in the 'superdiagonal'

$$a_{12}, a_{23}, \ldots, a_{n-1,n}$$

is non-zero.

If $\mathbf{N}$ is nilpotent of order k, so also is its transpose $\mathbf{N}'$. Hence a lower triangular matrix with zero entries in the diagonal is nilpotent.

A special kind of nilpotent matrix is given by

$$\mathbf{J}_n = \begin{pmatrix} 0 & 1 & 0 & \cdots & 0 \\ 0 & 0 & 1 & \cdots & 0 \\ \vdots & \vdots & \vdots & \ddots & \vdots \\ 0 & 0 & 0 & \cdots & 1 \\ 0 & 0 & 0 & \cdots & 0 \end{pmatrix}. \tag{6.7.25}$$

If $k, l, m, \ldots$ are integers [see § 2.2.1] such that

$$k \ge l \ge m \ge \ldots,$$

then the 'block-diagonal' matrix

$$\mathbf{N} = \text{diag}(\mathbf{J}_k, \mathbf{J}_l, \mathbf{J}_m, \ldots)$$

is nilpotent of order k.

(xv) *Elementary matrices.* Consider the matrix equation

$$\mathbf{B} = \mathbf{TA}, \tag{6.7.26}$$

in which $\mathbf{A}$ and $\mathbf{B}$ have m and n rows respectively, the number of columns in either matrix being irrelevant for our purpose.

The matrix $\mathbf{T} = (t_{ij})$ is then necessarily of type m by n. If we write

$$\mathbf{A} = \begin{pmatrix} \mathbf{a}_1' \\ \mathbf{a}_2' \\ \vdots \\ \mathbf{a}_m' \end{pmatrix}, \qquad \mathbf{B} = \begin{pmatrix} \mathbf{b}_1' \\ \mathbf{b}_2' \\ \vdots \\ \mathbf{b}_n' \end{pmatrix},$$

where $\mathbf{a}_i'$ and $\mathbf{b}_i'$ are row vectors (and therefore $\mathbf{a}_i$ and $\mathbf{b}_i$ are column vectors), then (6.7.26) can be viewed as relations between the rows of $\mathbf{A}$ and $\mathbf{B}$, namely

$$\mathbf{b}_i' = \sum_{j=1}^{n} t_{ij} \mathbf{a}_j' \quad (i = 1, 2, \ldots, m).$$

In the process of pivotal condensation the rows of a given matrix $\mathbf{A}$ are subjected to a series of operations which are summarized in section 6.4(i). We point out here that each operation is equivalent to pre-multiplication by a suitable matrix factor chosen as follows:

(α) in order to carry out the permutation π on the rows of $\mathbf{A}$ replace $\mathbf{A}$ by $\mathbf{PA}$ where $\mathbf{P}$ is the permutation matrix that corresponds to π [see Definition 6.7.1].

(β) multiplying the ith row of $\mathbf{A}$ by the scalar k is equivalent to pre-multiplying $\mathbf{A}$ by the diagonal matrix

$$\mathbf{D}_i(k) = \text{diag}\,(1,\ldots, 1, k, 1,\ldots, 1),$$

in which k occupies the ith position on the diagonal.

(γ) using the shearing matrix $\mathbf{S}_{ij}(\lambda)$ [see (xii)] we find that $\mathbf{S}_{ij}(\lambda)\mathbf{A}$ can be obtained from $\mathbf{A}$ by changing $\mathbf{a}_i$ into $\mathbf{a}_i + \lambda\mathbf{a}_j$ $(i \neq j)$.

If we wish to operate on the columns rather than on the rows, we are interested in the relationship between $\mathbf{A}'$ and $\mathbf{B}'$. Since by (6.5.1), equation (6.7.26) is equivalent to

$$\mathbf{B}' = \mathbf{A}'\mathbf{T}',$$

we deduce that column operators analogous to (α), (β), (γ) amount to post-multiplication by $\mathbf{P}'$, $\mathbf{D}_i(k)$ and $\mathbf{S}_{ji}(\lambda)$ $(=(\mathbf{S}_{ij}(\lambda))')$ respectively. In particular, if the rows and columns of $\mathbf{A}$ are subjected to the same permutation, then $\mathbf{A}$ is replaced by $\mathbf{PAP}'$, where $\mathbf{P}$ is the appropriate permutation matrix.

The matrices $\mathbf{P}$, $\mathbf{D}_i(k)$ and $\mathbf{S}_{ij}(\lambda)$ are sometimes referred to as *elementary matrices*.

(xvi) *Periodic matrices*. The matrix $\mathbf{A}$ is said to be *periodic of period k*, where k is a non-negative integer, if

$$\mathbf{A}^k = \mathbf{I}, \qquad \mathbf{A}^{k-1} \neq \mathbf{I}. \tag{6.7.27}$$

Periodic matrices are always invertible; indeed,

$$\mathbf{A}^{-1} = \mathbf{A}^{k-1}.$$

6.8. MOTIVATION FOR DETERMINANTS

A system of n linear equations in n unknowns can be written in the form [see (5.7.2)]

$$a_{11}x_1 + a_{12}x_2 + \ldots + a_{1n}x_n = h_1$$
$$a_{21}x_1 + a_{22}x_1 + \ldots + a_{2n}x_n = h_2$$
$$\vdots \qquad \vdots \qquad\qquad \vdots \qquad \vdots \tag{6.8.1}$$
$$a_{n1}x_1 + a_{n2}x_2 + \ldots + a_{nn}x_n = h_n.$$

In matrix notation [see (5.7.4)] this is condensed into

$$\mathbf{Ax} = \mathbf{h}, \tag{6.8.2}$$

where

$$\mathbf{A} = (a_{ij})$$

is the matrix of coefficients. The column vectors

$$\mathbf{x} = \begin{pmatrix} x_1 \\ x_2 \\ \vdots \\ x_n \end{pmatrix}, \qquad \mathbf{h} = \begin{pmatrix} h_1 \\ h_2 \\ \vdots \\ h_n \end{pmatrix}$$

comprise the n unknowns and the n right-hand members of (6.8.2) respectively. When $\mathbf{A}$ and $\mathbf{h}$ are given, the system has a unique solution provided that $\mathbf{A}$ is a regular matrix [see Theorem 5.8.1]. This solution may be written as

$$\mathbf{x} = \mathbf{A}^{-1}\mathbf{h} \tag{6.8.3}$$

in the symbolism of matrix algebra.

We have learned how $\mathbf{x}$ can be computed by the method of pivotal condensation [see Example 5.8.1] in a concrete situation. Whilst this procedure is appropriate for numerical calculations, it would be of theoretical interest to obtain a formula which expresses the unknowns $x_1, x_2, \ldots, x_n$ explicitly in terms of the given quantities a_{ij} $(i, j = 1, 2, \ldots, n)$ and $h_1, h_2, \ldots, h_n$. We illustrate the idea by displaying the solution of the problem when only two or three unknowns are involved.

Thus, when $n = 2$, we are concerned with the system

$$a_{11}x_1 + a_{12}x_2 = h_1$$
$$a_{21}x_2 + a_{22}x_2 = h_2. \tag{6.8.4}$$

Provided that

$$D = a_{11}a_{22} - a_{12}a_{21} \neq 0, \tag{6.8.5}$$

the (unique) solution can be written as

$$x_1 = \frac{h_1 a_{22} - h_2 a_{12}}{a_{11}a_{22} - a_{12}a_{21}}, \qquad x_2 = \frac{a_{11}h_2 - a_{21}h_1}{a_{11}a_{22} - a_{12}a_{21}}.$$

The condition (6.8.5) is crucial for the theory of linear equations: The system (6.8.4) is regular if and only if (6.8.5) is satisfied [see § 5.8.II]. The expression D is called the *determinant* of the two by two matrix

$$\mathbf{A} = \begin{pmatrix} a_{11} & a_{12} \\ a_{21} & a_{22} \end{pmatrix}. \tag{6.8.6}$$

We say that D is computed by 'cross-multiplying' the 4 entries of $\mathbf{A}$.

Several notations for the determinant are in common use. We mention in particular

$$D = \det \mathbf{A} = \begin{vmatrix} a_{11} & a_{12} \\ a_{21} & a_{22} \end{vmatrix} = \det |a_{ij}| \tag{6.8.7}$$

Notice that the solution (6.8.3) can now be expressed as quotients of determinants, namely

$$x_1 = \frac{\begin{vmatrix} h_1 & a_{12} \\ h_2 & a_{22} \end{vmatrix}}{\begin{vmatrix} a_{11} & a_{12} \\ a_{21} & a_{22} \end{vmatrix}}, \qquad x_2 = \frac{\begin{vmatrix} a_{11} & h_1 \\ a_{21} & h_2 \end{vmatrix}}{\begin{vmatrix} a_{11} & a_{12} \\ a_{21} & a_{22} \end{vmatrix}}.$$

This is a special case of Cramer's Rule [see (6.12.2)].

EXAMPLE 6.8.1.   The solution of

$$3x_1 - 4x_2 = 2$$

$$x_1 + 2x_2 = 1$$

is

$$x_1 = \frac{\begin{vmatrix} 2 & -4 \\ 1 & 2 \end{vmatrix}}{\begin{vmatrix} 3 & -4 \\ 1 & 2 \end{vmatrix}} = \tfrac{8}{10} = \tfrac{4}{5},$$

$$x_2 = \frac{\begin{vmatrix} 3 & 2 \\ 1 & 1 \end{vmatrix}}{\begin{vmatrix} 3 & -4 \\ 1 & 2 \end{vmatrix}} = \tfrac{1}{10}.$$

When $n = 3$, the calculations are still not too tedious, though they would rapidly become so if the number of unknowns were to increase further.

Thus in order that the system

$$a_{11}x_1 + a_{12}x_2 + a_{13}x_3 = h_1$$

$$a_{21}x_1 + a_{22}x_2 + a_{23}x_3 = h_2$$

$$a_{31}x_1 + a_{32}x_2 + a_{33}x_3 = h_3$$

may have a unique solution it is necessary and sufficient that $D \neq 0$, where

$$D = a_{11}a_{22}a_{33} - a_{11}a_{23}a_{32} - a_{12}a_{21}a_{33} + a_{12}a_{23}a_{31} + a_{13}a_{21}a_{32} - a_{13}a_{22}a_{31}.$$
$$(6.8.8)$$

We call D the determinant of the matrix

$$\mathbf{A} = \begin{pmatrix} a_{11} & a_{12} & a_{13} \\ a_{21} & a_{22} & a_{23} \\ a_{31} & a_{32} & a_{33} \end{pmatrix},$$

and we write

$$D = \det \mathbf{A} = \begin{vmatrix} a_{11} & a_{12} & a_{13} \\ a_{21} & a_{22} & a_{23} \\ a_{31} & a_{32} & a_{33} \end{vmatrix}.$$

6.9. FORMAL DEFINITION OF A DETERMINANT

At first sight the expression given in (6.8.8) looks bewildering, and it is hard to see how it can be generalized. We observe that the right-hand side of (6.8.8) consists of six terms. Disregarding the minus signs for a moment we note that each term is of the form

$$a_{1i}a_{2j}a_{3k},$$

where i, j, k is a permutation of 1, 2, 3. We know [from Example 8.1.4] that there are six such permutations namely

$$\pi_1 = \begin{pmatrix} 1 & 2 & 3 \\ 1 & 2 & 3 \end{pmatrix}, \qquad \pi_2 = \begin{pmatrix} 1 & 2 & 3 \\ 1 & 3 & 2 \end{pmatrix}, \qquad \pi_3 = \begin{pmatrix} 1 & 2 & 3 \\ 2 & 1 & 3 \end{pmatrix}$$

$$\pi_4 = \begin{pmatrix} 1 & 2 & 3 \\ 2 & 3 & 1 \end{pmatrix}, \qquad \pi_5 = \begin{pmatrix} 1 & 2 & 3 \\ 3 & 1 & 2 \end{pmatrix}, \qquad \pi_6 = \begin{pmatrix} 1 & 2 & 3 \\ 3 & 2 & 1 \end{pmatrix}. \tag{6.9.1}$$

It is now plain that the six permutations are in one-to-one correspondence with the terms of D. Thus the permutation

$$\pi = \begin{pmatrix} 1 & 2 & 3 \\ i & j & k \end{pmatrix}$$

is associated with the product $a_{1i}a_{2j}a_{3k}$ and conversely. It only remains to settle the question of sign. We recall [see § 8.1.3] that permutations are classified as even or odd. For example, among the six permutations listed in (6.9.1) π_1, π_4 and π_5 are even, while π_2, π_3 and π_6 are odd. The mystery of the formula (6.8.8) is now revealed if we attach a plus or a minus term to a product according as the corresponding permutation is even or odd.

The foregoing definition can easily be extended to determinants of order n. Let

$$\mathbf{A} = \begin{pmatrix} a_{11} & a_{12} & \cdots & a_{1n} \\ a_{21} & a_{22} & \cdots & a_{2n} \\ \vdots & \vdots & \ddots & \vdots \\ a_{n1} & a_{n2} & \cdots & a_{nn} \end{pmatrix} \tag{6.9.2}$$

be an arbitrary n by n matrix. A permutation of degree n may be written

$$\pi = \begin{pmatrix} 1 & 2 & \cdots & n \\ 1\pi & 2\pi & \cdots & n\pi \end{pmatrix},$$

where $1\pi, 2\pi, \ldots, n\pi$ are the symbols $1, 2, \ldots, n$ in some order. We put [see Definition 8.1.4]

$$\varepsilon(\pi) = \begin{cases} 1 & \text{if } \pi \text{ is even} \\ -1 & \text{if } \pi \text{ is odd.} \end{cases}$$

We are now ready to enunciate the formal definition of any determinant:

Let $\mathbf{A} = (a_{ij})$ be an n by n matrix. Then the determinant of $\mathbf{A}$ is given by

$$\det \mathbf{A} = \sum_{\pi} \varepsilon(\pi) a_{1,1\pi} a_{2,2\pi} \cdots a_{n,n\pi} \qquad (6.9.3)$$

where the summation ranges over all $n!$ permutations of the symbols $1, 2, \ldots, n$ [see (3.7.1)].

A more explicit notation is

$$\det \mathbf{A} = \begin{vmatrix} a_{11} & a_{12} & \cdots & a_{1n} \\ a_{21} & a_{22} & \cdots & a_{2n} \\ \vdots & \vdots & \ddots & \vdots \\ a_{n1} & a_{n2} & \cdots & a_{nn} \end{vmatrix}. \qquad (6.9.4)$$

Remarks

(i) Careful distinction should be made between a matrix and its determinant. The notations used in (6.9.2) and (6.9.4) must not be confused. The determinant is a single number associated with a square matrix. It may well happen that distinct matrices have the same determinant.

(ii) The formula (6.9.3) involves $n!$ terms. Even for moderate values of n this is an uncomfortably large number. Except in very special circumstances the formal definition is unsuitable for the numerical evaluation of a determinant.

(iii) No attempt will be made to define the determinant of a non-square matrix. When $n = 3$, but for no other value of n, there exists a simple rule: We call a_{11}, a_{22}, a_{33} the *positive* diagonal (or *main* diagonal [see (6.2.3)]) of

$$\mathbf{A} = \begin{pmatrix} a_{11} & a_{12} & a_{13} \\ a_{21} & a_{22} & a_{23} \\ a_{31} & a_{32} & a_{33} \end{pmatrix}.$$

Then the three positive terms in (6.8.8) correspond to the positive diagonal and to the vertices of the two triangles whose bases are parallel to the positive diagonal opposite to the positions of a_{13} and a_{31} respectively; thus the positive terms are described in the picture

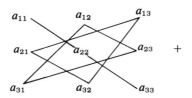

Similarly, the *negative diagonal* a_{13}, a_{22}, a_{31} together with the analogous two triangles correspond to the negative terms of (6.8.8), as shown below

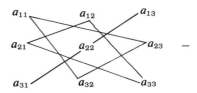

6.10.   GENERAL LAW FOR DETERMINANTS

In order to facilitate the notation it is convenient to think of an n by n matrix as a sequence of n column vectors. Thus we write

$$\mathbf{A} = (\mathbf{a}_1, \mathbf{a}_2, \ldots, \mathbf{a}_j, \ldots, \mathbf{a}_n),$$

where

$$\mathbf{a}_j = \begin{pmatrix} a_{1j} \\ a_{2j} \\ \vdots \\ a_{nj} \end{pmatrix} \quad (j = 1, 2, \ldots, n),$$

and we denote the determinant of $\mathbf{A}$ by

$$D = D(\mathbf{a}_1, \ldots, \mathbf{a}_n). \tag{6.10.1}$$

The following general laws can be derived from the definition given in (6.9.3).

I. If one column is multiplied by a fixed scalar λ, then the value of the determinant is multiplied by λ, thus

$$D(\mathbf{a}_1, \ldots, \lambda\mathbf{a}_j, \ldots, \mathbf{a}_n) = \lambda D(\mathbf{a}_1, \ldots, \mathbf{a}_j, \ldots, \mathbf{a}_n).$$

Hence a determinant is always equal to zero if it contains a zero column.

II. The determinant is an 'additive function' of any particular column, that is

$$D(\mathbf{a}_1, \ldots, \mathbf{b}_j + \mathbf{c}_j, \ldots, \mathbf{a}_n) = D(\mathbf{a}_1, \ldots, \mathbf{b}_j, \ldots, \mathbf{a}_n)$$
$$+ D(\mathbf{a}_1, \ldots, \mathbf{c}_j, \ldots, \mathbf{a}_n).$$

It should be observed that only one column, the jth column in our notation, is written as the sum of two columns. In fact we put $\mathbf{a}_j = \mathbf{b}_j + \mathbf{c}_j$, but the other columns remained unchanged. The result would be more complicated if each column were expressed as the sum of two columns. In general,

$$\det (\mathbf{A} + \mathbf{B}) \neq \det \mathbf{A} + \det \mathbf{B}. \tag{6.10.2}$$

III. If two columns are interchanged, then the value of the determinant is multiplied by -1, that is

$$D(\mathbf{a}_1, \ldots, \mathbf{a}_i, \ldots, \mathbf{a}_j, \ldots, \mathbf{a}_n) = -D(\mathbf{a}_1, \ldots, \mathbf{a}_j, \ldots, \mathbf{a}_i, \ldots, \mathbf{a}_n).$$

More generally, if the columns undergo a permutation ρ, then the determinant is multiplied by $\varepsilon(\rho)$. We recall that $\varepsilon(\rho) = -1$, when ρ is a transposition [see Proposition 8.1.4(ii)]. A striking consequence of this property is the fact that a determinant is zero if it contains two identical columns or, more generally, if it contains two proportional columns. For example,

$$\begin{vmatrix} a_1 & b_1 & 2a_1 & c_1 \\ a_2 & b_2 & 2a_2 & c_2 \\ a_3 & b_3 & 2a_3 & c_3 \\ a_4 & b_4 & 2a_4 & c_4 \end{vmatrix} = 0.$$

IV. The value of a determinant is unchanged if one column is modified by adding to it an arbitrary multiple of another column, thus

$$D(\mathbf{a}_1, \ldots, \mathbf{a}_i, \ldots, \mathbf{a}_j, \ldots, \mathbf{a}_n) = D(\mathbf{a}_1, \ldots, \mathbf{a}_i + \lambda \mathbf{a}_j, \ldots, \mathbf{a}_j, \ldots, \mathbf{a}_n).$$

This rule gives a great deal of scope for simplifying the entries in a determinant without changing the value of the determinant. The method of pivotal condensation [see § 6.4(i)] is based on this property. For example, we have that

$$\begin{vmatrix} 101 & 102 & 1 \\ 102 & 104 & 2 \\ 103 & 106 & 3 \end{vmatrix} = \begin{vmatrix} 101 & 1 & 1 \\ 102 & 2 & 2 \\ 103 & 3 & 3 \end{vmatrix} = 0,$$

because we may subtract the first column from the second column without changing the value of the determinant, leaving the first and third columns unaltered. The new determinant has two identical columns and therefore vanishes.

V. A matrix and its transpose have the same determinant, that is

$$\det \mathbf{A} = \det \mathbf{A}'. \tag{6.10.3}$$

For example,

$$\begin{vmatrix} a_1 & b_1 & c_1 \\ a_2 & b_2 & c_2 \\ a_3 & b_3 & c_3 \end{vmatrix} = \begin{vmatrix} a_1 & a_2 & a_3 \\ b_1 & b_2 & b_3 \\ c_1 & c_2 & c_3 \end{vmatrix}.$$

As a consequence of this rule all the properties of a determinant which we have stated in I, II, III and IV in terms of columns hold equally for rows. For example, a determinant is zero if two of its rows are proportional. Again, a determinant is multiplied by -1 when two of its rows are interchanged.

Finally, there is an interesting rule for the determinant of matrix product:

VI. The multiplication theorem:

$$\det (\mathbf{AB}) = (\det \mathbf{A})(\det \mathbf{B}), \tag{6.10.4}$$

where $\mathbf{A}$ and $\mathbf{B}$ are arbitrary n by n matrices.

6.11. COFACTORS

In this section we present several ways in which a determinant of order n can be expressed as a sum of determinants of lower order. For brevity, we write

$$D = \det \mathbf{A},$$

where

$$\mathbf{A} = \begin{pmatrix} a_{11} & \cdots & a_{1j} & \cdots & a_{1n} \\ \vdots & & \vdots & & \vdots \\ a_{i1} & \cdots & a_{ij} & \cdots & a_{in} \\ \vdots & & \vdots & & \vdots \\ a_{n1} & \cdots & a_{nj} & \cdots & a_{nn} \end{pmatrix}.$$

If we delete the ith row and the jth column from $\mathbf{A}$, we obtain a matrix of $n - 1$ rows and columns. Its determinant will be denoted by

$$M_{ij} = \begin{vmatrix} a_{11} & & & a_{1n} \\ \text{---} & \text{---} & \text{---} & \text{---} \\ a_{n1} & & & a_{nn} \end{vmatrix} \text{---} i,$$

$$j$$

where the dashed lines indicate the row and the column that have been omitted. We call M_{ij} an $(n - 1)$-*rowed minor* of D. As i and j can take any one of the values $1, 2, \ldots, n$, we obtain altogether n^2 such minors. Finally, we put

$$A_{ij} = (-1)^{i+j} M_{ij} \tag{6.11.1}$$

$(i, j = 1, 2, \ldots, n)$, and we call A_{ij} the *cofactor* of a_{ij} in D. Thus the cofactor of a_{ij} is the $(n - 1)$-rowed determinant obtained by crossing out the row and the column which intersect in a_{ij} and attaching the *position sign* $(-1)^{i+j}$ to the resulting determinant.

By way of illustration we list the nine cofactors of the three-rowed determinant

$$\begin{vmatrix} a_{11} & a_{12} & a_{13} \\ a_{21} & a_{22} & a_{23} \\ a_{31} & a_{32} & a_{33} \end{vmatrix},$$

thus

$$A_{11} = a_{22}a_{33} - a_{23}a_{32}, \qquad A_{12} = -(a_{21}a_{33} - a_{23}a_{31}),$$

$$A_{13} = a_{21}a_{32} - a_{22}a_{31}, \qquad A_{21} = -(a_{12}a_{33} - a_{13}a_{32}),$$

$$A_{22} = a_{11}a_{33} - a_{13}a_{31}, \qquad A_{23} = -(a_{11}a_{32} - a_{12}a_{31}),$$

$$A_{31} = a_{12}a_{23} - a_{13}a_{22}, \qquad A_{32} = -(a_{11}a_{23} - a_{13}a_{21}),$$

$$A_{33} = a_{11}a_{22} - a_{12}a_{21}.$$

Generally, the position signs may be memorized from the chess board scheme

$$\begin{bmatrix} + & - & + & \cdots \\ - & + & - & \cdots \\ + & - & + & \cdots \\ \cdots\cdots & & + \end{bmatrix}.$$

In particular, it should be noted that the positive sign appears in all diagonal positions.

We are now ready to describe the expansion formulae mentioned at the beginning of this section. The first row of $\mathbf{A}$ may be regarded as a row vector

$$\mathbf{a}^{(1)} = (a_{11}, a_{12}, \ldots, a_{1n}).$$

The corresponding cofactors may likewise be arranged as a row vector thus

$$\mathbf{A}^{(1)} = (A_{11}, A_{12}, \ldots, A_{1n}).$$

Then it turns out that the inner product [see (6.1.7)] of these vectors is equal to the determinant, thus

$$D = a_{11}A_{11} + a_{12}A_{12} + \ldots + a_{1n}A_{1n}. \tag{6.11.2}$$

Referring to (6.11.2) we say that D has been expanded with respect to the first row. Analogous formulae exist for any one of the n rows. Thus we have an expansion of D with respect to the ith row, namely

$$D = a_{i1}A_{i1} + a_{i2}A_{i2} + \ldots + a_{in}A_{in} \tag{6.11.3}$$

$(1 \le i \le n)$. Similarly, there exists an expansion with respect to any one of the n columns, thus

$$D = a_{1j}A_{1j} + a_{2j}A_{2j} + \ldots + a_{nj}A_{nj} \tag{6.11.4}$$

$(1 \le j \le n)$.

EXAMPLE. 6.11.1. Expand

$$D = \begin{vmatrix} 1 & 0 & 2 \\ 3 & 1 & -1 \\ 4 & 0 & 3 \end{vmatrix} \tag{6.11.5}$$

in turn with respect to the (i) first row, (ii) first column and (iii) second column. We find that

(i) $D = 1. \begin{vmatrix} 1 & -1 \\ 0 & 3 \end{vmatrix} - 0. \begin{vmatrix} 3 & -1 \\ 4 & 3 \end{vmatrix} + 2 \begin{vmatrix} 3 & 1 \\ 4 & 0 \end{vmatrix} = 3 - 0 + 2(-4) = -5,$

(ii) $D = 1. \begin{vmatrix} 1 & -1 \\ 0 & 3 \end{vmatrix} - 3. \begin{vmatrix} 0 & 2 \\ 0 & 3 \end{vmatrix} + 4 \begin{vmatrix} 0 & 2 \\ 1 & -1 \end{vmatrix} = 3 - 0 + 4(-2) = -5,$

(iii) $D = -0. \begin{vmatrix} 3 & -1 \\ 4 & 3 \end{vmatrix} + 1. \begin{vmatrix} 1 & 2 \\ 4 & 3 \end{vmatrix} - 0. \begin{vmatrix} 1 & 2 \\ 3 & -1 \end{vmatrix} = 0 - 5 - 0 = -5.$

Clearly, in this example the last mode of expansion is the most economical one, as only one minor has to be evaluated. Generally, it is advantageous to expand with respect to a row or column that contains a large number of zero entries.

Further relations between the elements of a determinant and its cofactors are obtained by forming the inner product of a row (column) with the cofactors of a different row (column). In all these cases the result is zero. Thus, we have that

$$a_{i1}A_{k1} + a_{i2}A_{k2} + \ldots + a_{in}A_{kn} = 0 \qquad (6.11.6)$$

if $i \neq k$, and

$$a_{1j}A_{1k} + a_{2j}A_{2k} + \ldots + a_{nj}A_{nk} = 0, \qquad (6.11.7)$$

if $j \neq k$. For example, if we form the inner product of the first row with the co-factors of the third row in the determinant (6.11.5), we find that

$$1. \begin{vmatrix} 0 & 2 \\ 1 & -1 \end{vmatrix} - 0. \begin{vmatrix} 1 & 2 \\ 3 & -1 \end{vmatrix} + 2 \begin{vmatrix} 1 & 0 \\ 3 & 1 \end{vmatrix} = -2 - 0 + 2 = 0.$$

In order to describe the various relations more concisely, it is convenient to introduce the matrix

$$\text{adj } \mathbf{A} = \begin{pmatrix} A_{11} & A_{21} & \cdots & A_{n1} \\ A_{12} & A_{22} & \cdots & A_{n2} \\ \vdots & \vdots & \ddots & \vdots \\ A_{1n} & A_{2n} & \cdots & A_{nn} \end{pmatrix}, \qquad (6.11.8)$$

which is called the *adjugate* of A (some authors call this matrix the adjoint of A; but this term is used by us in a different sense [see (6.7.11)]. The equations (6.11.3), (6.11.4), (6.11.6) and (6.11.7) can now be condensed into the matrix equations

$$\left. \begin{array}{l} A(\text{adj } \mathbf{A}) = D.\mathbf{I} \\ (\text{adj } \mathbf{A})A = D.\mathbf{I} \end{array} \right\}, \qquad (6.11.9)$$

where the product of the matrices $\mathbf{A}$ and adj $\mathbf{A}$ is formed in the usual way with due regard of the matrix factors. The right-hand side of (6.11.9) is a scalar multiple of the unit matrix, thus

$$D.\mathbf{I} = \begin{pmatrix} D & 0 & \cdots & 0 \\ 0 & D & \cdots & 0 \\ \vdots & \vdots & \ddots & \vdots \\ 0 & 0 & \cdots & D \end{pmatrix}.$$

It is noteworthy that $\mathbf{A}$ and adj $\mathbf{A}$ always commute.

EXAMPLE 6.11.2. When

$$\mathbf{A} = \begin{pmatrix} 1 & 0 & 2 \\ 3 & 1 & -1 \\ 4 & 0 & 3 \end{pmatrix},$$

it is found that

$$\text{adj } \mathbf{A} = \begin{pmatrix} 3 & 0 & -2 \\ -13 & -5 & 7 \\ -4 & 0 & 1 \end{pmatrix}, \tag{6.11.10}$$

and it can be verified that

$$\begin{pmatrix} 1 & 0 & 2 \\ 3 & 1 & -1 \\ 4 & 0 & 3 \end{pmatrix} \begin{pmatrix} 3 & 0 & -2 \\ -13 & -5 & 7 \\ -4 & 0 & 1 \end{pmatrix}$$
$$= \begin{pmatrix} 3 & 0 & -2 \\ -13 & -5 & 7 \\ -4 & 0 & 1 \end{pmatrix} \begin{pmatrix} 1 & 0 & 2 \\ 3 & 1 & -1 \\ 4 & 0 & 3 \end{pmatrix} = \begin{pmatrix} -5 & 0 & 0 \\ 0 & -5 & 0 \\ 0 & 0 & -5 \end{pmatrix}.$$

6.12. CRAMER'S RULE

Recall from Definition 6.4.2 that a square matrix $\mathbf{A}$ is *singular* if there exists a non-zero column vector $\mathbf{x}$ such that $\mathbf{Ax} = \mathbf{0}$. If $\mathbf{A} = (a_{ij})\,(i, j = 1, 2, \ldots, n)$ and $\mathbf{x} = (x_1, x_2, \ldots, x_n)'$ then this is equivalent to the n linear equations

$$a_{11}x_1 + a_{12}x_2 + \ldots + a_{1n}x_n = 0$$
$$a_{21}x_1 + a_{22}x_2 + \ldots + a_{2n}x_n = 0$$
$$\vdots$$
$$a_{n1}x_1 + a_{n2}x_2 + \ldots + a_{nn}x_n = 0$$

and the process of pivotal condensation [cf. Example 5.8.1] eventually leads to the relations

$$x_i[\det \mathbf{A}] = 0 \quad (i = 1, 2, \ldots, n).$$

Since not all x_i are zero we must therefore have det $\mathbf{A} = 0$. For example, when $n = 2$ we have

$$a_{11}x_1 + a_{12}x_2 = 0$$

$$a_{21}x_1 + a_{22}x_2 = 0$$

and the scheme for pivotal condensation is

	x_1	x_2		
R_1	a_{11}	a_{12}	0	$a_{11} + a_{12}$
R_2	a_{21}	a_{22}	0	$a_{21} + a_{22}$
$R_3 = a_{21}R_1$	$a_{21}a_{11}$	$a_{21}a_{12}$	0	$a_{21}(a_{11} + a_{12})$
$R_4 = a_{11}R_2$	$a_{11}a_{21}$	$a_{11}a_{22}$	0	$a_{11}(a_{21} + a_{22})$

We deduce from $R_4 - R_3$ that

$$x_2(a_{11}a_{22} - a_{21}a_{12}) = 0$$

and a similar computation leads to the equation

$$x_1(a_{11}a_{22} - a_{21}a_{12}) = 0,$$

whence

$$a_{11}a_{22} - a_{21}a_{12} = 0$$

since $\mathbf{x}$ is non-zero.

Thus a definition which is equivalent to Definition 6.4.2 is

DEFINITION 6.12.1.   A square matrix $\mathbf{A}$ is *non-singular* if and only if det $\mathbf{A} \neq 0$. The equation (6.11.9) makes it evident that the inverse of a non-singular matrix is given by

$$\mathbf{A}^{-1} = \frac{1}{\det \mathbf{A}} (\text{adj } \mathbf{A}). \tag{6.12.1}$$

Although this formula is of considerable interest, it is hardly ever useful for numerical computation [see § 6.4(iii)]. We have shown in section 6.4(i) how the inverse, if it exists, can be found by pivotal condensation. It was also remarked in Theorem 6.4.2, that the concept of a regular or invertible matrix coincides with that of a non-singular matrix. Thus *a matrix possesses an inverse if and only if its determinant is non-zero*. We take the matrix $\mathbf{A}$ of Example 6.11.2 as an illustration.

EXAMPLE 6.12.1.   Substituting (6.11.10) in (6.12.1) we find that

$$\mathbf{A}^{-1} = \left(-\tfrac{1}{5}\right) \begin{pmatrix} 3 & 0 & -2 \\ -13 & -5 & 7 \\ -4 & 0 & 1 \end{pmatrix}.$$

The formula for the inverse leads to the explicit solution of a square system of linear equations, to which we have alluded in section 6.8. Thus if A is an n by n matrix of non-zero determinant, the system

$$Ax = h$$

has the unique solution

$$x = (\det A)^{-1} (\text{adj } A)h.$$

Using the notation (6.10.1) we obtain the following expression for the ith component of x.

$$x_i = \frac{D(a_1, \ldots, a_{i-1}, h, a_{i+1}, \ldots, a_n)}{D(a_1, \ldots, a_{i-1}, a_i, a_{i+1}, \ldots, a_n)} \tag{6.12.2}$$

$(i = 1, 2, \ldots, n)$. This is known as *Cramer's Rule* and is one of the earliest discoveries (1750) in theory of determinants. One of the advantages of this method is the fact that each unknown can be computed independently of the others. For example, if we are given the system of equations

$$x_1 \qquad\quad + 2x_3 = 2$$
$$3x_1 + x_2 - \quad x_3 = 1$$
$$4x_1 \qquad\quad + 3x_3 = 0$$

we find that

$$x_2 = \frac{\begin{vmatrix} 1 & 2 & 2 \\ 3 & 1 & -1 \\ 4 & 0 & 3 \end{vmatrix}}{\begin{vmatrix} 1 & 0 & 2 \\ 3 & 1 & -1 \\ 4 & 0 & 3 \end{vmatrix}} = \tfrac{31}{5}.$$

But on the whole Cramer's rule is only of theoretical interest, as it would involve excessive calculations in most cases. As we have already stressed in section 5.8, systems of linear equations should be solved by pivotal condensation except in special circumstances.

6.13. MINORS

The notion of a minor, introduced in section 6.11 can be generalized. For this purpose it is convenient to denote the entries of a matrix by $a(i, j)$. Let

$$A = \begin{vmatrix} a(1, 1) & \cdots & a(1, n) \\ \vdots & & \vdots \\ a(m, 1) & \cdots & a(m, n) \end{vmatrix}$$

be an m by n matrix. Choose two integers s and t such that

$$1 \le s \le m \quad \text{and} \quad 1 \le t \le n$$

and select s rows specified by the suffixes

$$k_1 < k_2 < \ldots < k_s \qquad (6.13.1)$$

and t columns specified by the suffixes

$$l_1 < l_2 < \ldots < l_t. \qquad (6.13.2)$$

The matrix whose entries stand in the intersections of the selected rows and columns is given by

$$\mathbf{S}\begin{pmatrix} k_1 & \cdots & k_s \\ l_1 & \cdots & l_t \end{pmatrix} = \begin{pmatrix} a(k_1, l_1) & \cdots & a(k_1, l_t) \\ \vdots & & \vdots \\ a(k_s, l_1) & \cdots & a(k_s, l_t) \end{pmatrix}. \qquad (6.13.3)$$

Any matrix cut out of $\mathbf{A}$ in this way is called a *submatrix* of $\mathbf{A}$. There are

$$\binom{m}{s}\binom{n}{t}$$

submatrices of type s by t that can be extracted from $\mathbf{A}$. Now suppose that $s = t$. Then we can form the determinant of (6.13.3) and write

$$M\begin{pmatrix} k_1 & \cdots & k_s \\ l_1 & \cdots & l_s \end{pmatrix} = \det \mathbf{S}\begin{pmatrix} k_1 & \cdots & k_s \\ l_1 & \cdots & l_s \end{pmatrix}. \qquad (6.13.4)$$

Any determinant defined in this way is called a *minor of A*.

The following special cases should be mentioned: when the sets of suffixes for the selected rows and columns are the same we obtain *principal submatrices* and *principal minors* of $\mathbf{A}$, typified by

$$\mathbf{S}\begin{pmatrix} k_1, \ldots, k_s \\ k_1, \ldots, k_s \end{pmatrix} \quad \text{and} \quad M\begin{pmatrix} k_1, \ldots, k_s \\ k_1, \ldots, k_s \end{pmatrix} \qquad (6.13.5)$$

respectively. Their diagonals are part of the diagonal of $\mathbf{A}$. For example, when $m \geq 4$ and $n \geq 4$, then

$$\begin{vmatrix} a(2, 2) & a(2, 4) \\ a(4, 2) & a(4, 4) \end{vmatrix}$$

is a minor of order 2. The submatrices or minors cut out of the left-hand upper corner of $\mathbf{A}$ are called the *leading submatrices* and *leading minors* respectively, thus

$$\mathbf{S}\begin{pmatrix} 1, 2, \ldots, s \\ 1, 2, \ldots, s \end{pmatrix} \quad \text{and} \quad M\begin{pmatrix} 1, 2, \ldots, s \\ 1, 2, \ldots, s \end{pmatrix} \qquad (6.13.6)$$

are the leading submatrix and the leading minor of order s. For example,

$$\begin{vmatrix} a(1, 1) & a(1, 2) & a(1, 3) \\ a(2, 1) & a(2, 2) & a(2, 3) \\ a(3, 1) & a(3, 2) & a(3, 3) \end{vmatrix}$$

is the leading minor of order 3.

There is an interesting connection between the minors and the rank of $\mathbf{A}$ [see § 5.6]:

PROPOSITION 6.13.1.   *The matrix $\mathbf{A}$ is of rank r if and only if it possesses at least one non-zero minor of order r, while all minors of order $r + 1$ vanish.*

EXAMPLE 6.13.1.   The matrix

$$\begin{pmatrix} 0 & 1 & 0 & 0 \\ 0 & 0 & 0 & 1 \\ 0 & 0 & 0 & 0 \end{pmatrix}$$

is of rank 2, because it possesses a non-zero minor of order 2, namely

$$M\begin{pmatrix} 1 & 2 \\ 2 & 4 \end{pmatrix}$$

while all minors of order 3 are zero.

6.14. SOME SPECIAL DETERMINANTS

(i) For a diagonal matrix

$$\mathbf{A} = \text{diag}\,(a_1, a_2, \ldots, a_n) = \begin{pmatrix} a_1 & 0 & \ldots & 0 \\ 0 & a_2 & \ldots & 0 \\ & & \ddots & \\ 0 & 0 & \ldots & a_n \end{pmatrix}$$

we have that

$$\det \mathbf{A} = a_1 a_2 \ldots a_n. \tag{6.14.1}$$

(ii) More generally, the determinant of a triangular matrix

$$\mathbf{T} = \begin{pmatrix} a_{11} & a_{12} & a_{13} & \ldots & a_{1n} \\ 0 & a_{22} & a_{23} & \ldots & a_{2n} \\ 0 & 0 & a_{33} & \ldots & a_{3n} \\ 0 & 0 & 0 & \ldots & a_{nn} \end{pmatrix}$$

is given by

$$\det \mathbf{T} = a_{11} a_{22} \ldots a_{nn} \tag{6.14.2}$$

(iii) Let $x_1, x_2, \ldots, x_n$ be indeterminates. Then

$$V = \begin{vmatrix} x_1^{n-1} & x_2^{n-1} & \ldots & x_n^{n-1} \\ x_1^{n-2} & x_2^{n-2} & \ldots & x_n^{n-2} \\ x_1 & x_2 & \ldots & x_n \\ 1 & 1 & & 1 \end{vmatrix}$$

is called the *Vandermonde determinant* of $x_1, x_2, \ldots, x_n$. It can be shown that

$$V = \prod_{i<j} (x_i - x_j) \tag{6.14.3}$$

where the product involves all $\frac{1}{2}n(n - 1)$ differences between the x's. Hence $V = 0$ if and only if at least two x's coincide.

(iv) The determinant of a skew-symmetric matrix, that is

$$S_n = \begin{vmatrix} 0 & a_{12} & a_{13} & \cdots & a_{1n} \\ -a_{12} & 0 & a_{23} & \cdots & a_{2n} \\ -a_{13} & -a_{23} & 0 & \cdots & a_{3n} \\ & & & & \\ -a_{1n} & -a_{2n} & -a_{3n} & \cdots & 0 \end{vmatrix},$$

is equal to zero when n is odd, but is equal to a perfect square in the entries a_{ij} when n is even. For example,

$$S_2 = \begin{vmatrix} 0 & a_{12} \\ -a_{12} & 0 \end{vmatrix} = a_{12}{}^2,$$

$$S_3 = \begin{vmatrix} 0 & a_{12} & a_{13} \\ -a_{12} & 0 & a_{23} \\ -a_{13} & -a_{23} & 0 \end{vmatrix} = 0,$$

$$S_4 = \begin{vmatrix} 0 & a_{12} & a_{13} & a_{14} \\ -a_{12} & 0 & a_{23} & a_{24} \\ -a_{13} & -a_{23} & 0 & a_{34} \\ -a_{14} & -a_{24} & -a_{34} & 0 \end{vmatrix} = (a_{12}a_{34} - a_{13}a_{24} + a_{14}a_{23})^2.$$

(v) In some cases the determinant of a partitioned matrix can be found by evaluating determinants of smaller orders. Let

$$\mathbf{A} = \begin{pmatrix} \mathbf{A}_{11} & \mathbf{A}_{12} & \cdots & \mathbf{A}_{1k} \\ 0 & \mathbf{A}_{22} & \cdots & \mathbf{A}_{2k} \\ \vdots & \vdots & \ddots & \vdots \\ 0 & 0 & \cdots & \mathbf{A}_{kk} \end{pmatrix}, \tag{6.14.4}$$

where $\mathbf{A}_{11}, \mathbf{A}_{22}, \ldots, \mathbf{A}_{kk}$ are square matrices, not necessarily of the same order. Then

$$\det \mathbf{A} = (\det \mathbf{A}_{11})(\det \mathbf{A}_{22})\ldots(\det \mathbf{A}_{kk}). \tag{6.14.5}$$

A similar formula holds when all blocks above the diagonal are zero while those below the diagonal are arbitrary.

The situation is more complicated when there are non-zero blocks both below and above the diagonal. Consider the simplest case of a matrix

$$\mathbf{V} = \begin{pmatrix} \mathbf{V}_1 & \mathbf{C}_1 \\ \mathbf{C}_2 & \mathbf{V}_2 \end{pmatrix} \tag{6.14.6}$$

of order $2n$, where $\mathbf{V}_1, \mathbf{V}_2, \mathbf{C}_1, \mathbf{C}_2$ are matrices of order n. We make the additional assumption that

$$\det \mathbf{V}_1 \neq 0. \tag{6.14.7}$$

Then it can be shown that

$$\det \mathbf{V} = \det \mathbf{V}_1 (\det (\mathbf{V}_2 - \mathbf{C}_2 \mathbf{V}_1^{-1} \mathbf{C}_1)$$
$$= \det (\mathbf{V}_1 \mathbf{V}_2 - \mathbf{V}_1 \mathbf{C}_2 \mathbf{V}_1^{-1} \mathbf{C}_1). \tag{6.14.8}$$

Incidentally, if $\mathbf{V}_1$ commutes either with $\mathbf{C}_1$ or with $\mathbf{C}_2$ we obtain the curious formula

$$\begin{vmatrix} \mathbf{V}_1 & \mathbf{C}_1 \\ \mathbf{C}_2 & \mathbf{V}_2 \end{vmatrix} = \det (\mathbf{V}_1 \mathbf{V}_2 - \mathbf{C}_2 \mathbf{C}_1), \tag{6.14.9}$$

which resembles the elementary result (6.8.5) for a determinant of order 2.

6.15.   THE KRONECKER PRODUCT

The matrix product which was defined in section 6.2 is by far the most important way of multiplying matrices. But other laws of composition are possible and useful. One of these is due to L. Kronecker (1823–91).

DEFINITION 6.15.1.   Let $\mathbf{A} = (a_{ij})$ be an m by n matrix and $\mathbf{B} = (b_{kl})$ a p by q matrix. Then the *Kronecker product* or *tensor product* or *direct product* of $\mathbf{A}$ and $\mathbf{B}$ is defined by

$$\mathbf{A} \times \mathbf{B} = \begin{pmatrix} a_{11}\mathbf{B} & a_{12}\mathbf{B} & \cdots & a_{1n}\mathbf{B} \\ a_{21}\mathbf{B} & a_{22}\mathbf{B} & \cdots & a_{2n}\mathbf{B} \\ \vdots & \vdots & \ddots & \vdots \\ a_{m1}\mathbf{B} & a_{m2}\mathbf{B} & \cdots & a_{mn}\mathbf{B} \end{pmatrix}. \tag{6.15.1}$$

Thus $\mathbf{A} \times \mathbf{B}$ has mp rows and nq columns. In general, $\mathbf{A} \times \mathbf{B} \neq \mathbf{B} \times \mathbf{A}$, but there exists a permutation matrix $\mathbf{J}$ [see Definition 6.7.1], which depends only on m, n, p and q, but not on $\mathbf{A}$ and $\mathbf{B}$ such that

$$\mathbf{J}'(\mathbf{A} \times \mathbf{B})\mathbf{J} = \mathbf{B} \times \mathbf{A}.$$

The most significant property of the Kronecker product is given in the following theorem:

THEOREM 6.15.1.   *Let $\mathbf{A}, \mathbf{B}, \mathbf{C}$ and $\mathbf{D}$ be matrices of types $m \times n$, $p \times q$, $n \times s$ and $q \times t$ respectively. Then*

$$(\mathbf{A} \times \mathbf{B})(\mathbf{C} \times \mathbf{D}) = \mathbf{AC} \times \mathbf{BD}. \tag{6.15.2}$$

Finally we mention a property that applies to square matrices only:

PROPOSITION 6.15.2.   *When A and B are square matrices of orders n and p respectively we have that*

$$\text{tr } (\mathbf{A} \times \mathbf{B}) = (\text{tr } \mathbf{A})(\text{tr } \mathbf{B}) \qquad (6.15.3)$$

and

$$\det (\mathbf{A} \times \mathbf{B}) = (\det \mathbf{A})^p (\det \mathbf{B})^n. \qquad (6.15.4)$$

W.L.

BIBLIOGRAPHY

Cohn, P. M. (1958). *Linear Equations*, Routledge & Kegan Paul.
Hoffman, K. and Kunze, K. (1961). *Linear Algebra*, Prentice-Hall.
Marcus, M. and Minc, H. (1968). *Elementary Linear Algebra*, Macmillan.
Morris, A. O. (1978). *Linear Algebra*, Van Nostrand Reinhold.
Tropper, A. M. (1969). *Linear Algebra*, Nelson.

CHAPTER 7

Eigenvalues

7.1. GEOMETRICAL MOTIVATION

We consider the space $\mathbb{F}^n$ which consists of all column vectors of order n with components in $\mathbb{F}$ [see Example 5.2.2]. In most applications $\mathbb{F}$ is either the field of real numbers or the field of complex numbers [see § 2.6.2 and § 2.7.1].

Let $\mathbf{A}$ be a fixed n by n matrix with coefficients in $\mathbb{F}$ [see § 6.2]. Now $\mathbf{A}$ induces a linear map of $\mathbb{F}$ into itself [see § 5.13] in which a vector $\mathbf{x}$ has an image $\mathbf{y}$ given by

$$\mathbf{y} = \mathbf{A}\mathbf{x}. \tag{7.1.1}$$

Although it would be natural to ask whether this map possesses invariant points, that is vectors coinciding with their images, it is in fact more significant to study *invariant lines* through the origin (Figure 7.1.1): a line through 0 is specified by a non-zero vector $\mathbf{u}$ and any point on this line has a position vector $k\mathbf{u}$, where k is an arbitrary scalar $(-\infty < k < \infty)$. The line is said to be invariant under the action of $\mathbf{A}$ if every point on it has an image that again lies on the line. This will happen if and only if

$$\mathbf{A}\mathbf{u} = \alpha\mathbf{u}, \tag{7.1.2}$$

where α is a suitable scalar associated with $\mathbf{A}$, because (7.1.2) implies that

$$A(k\mathbf{u}) = \alpha(k\mathbf{u}) \quad (-\infty < k < \infty).$$

We shall allow α to take complex values, even when A is real, since we are concerned with finding any non-zero (possibly complex) vector $\mathbf{u}$ which satisfies (7.1.2). Moreover, we accept that α may be zero provided that there exists a non-zero vector $\mathbf{u}$ such that $\mathbf{A}\mathbf{u} = \mathbf{0}$. The notion of an invariant line is reformulated as follows:

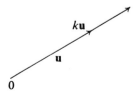

Figure 7.1.1

DEFINITION 7.1.1.   The complex number α is said to be an *eigenvalue* of **A** if there exists a non-zero vector **u** such that $\mathbf{Au} = \alpha\mathbf{u}$. Then **u** is called a column-*eigenvector* belonging to α.

There is no need to prefer column vectors to row vectors [see § 6.5]. Thus we have the analogous

DEFINITION 7.1.1'.   The complex number α is said to be an eigenvalue of **A** if there exists a non-zero row-vector $\mathbf{v}'$ such that $\mathbf{v}'\mathbf{A} = \alpha\mathbf{v}'$. Then $\mathbf{v}'$ is called a row-eigenvector of **A** belonging to α.

It will follow from Proposition 7.2.1 that the same set of eigenvalues will emerge from either definition.

Remark on nomenclature: 'Eigen' is a German word for 'proper' (in the sense of property), so that 'eigenvalue of **A**' means a value that belongs to **A**.

The ugly hybrid word was introduced into the scientific literature by quantum physicists who were perhaps unaware that English algebraists had coined the term '*latent root*' half-a-century before the advent of quantum theory. Other authors, mainly American, use 'characteristic root'. But, regrettably, it seems that 'eigenvalue' is now firmly established in the mathematical literature.

We recapitulate some of the points made in this introductory section:
1. Eigenvectors are, by definition, non-zero; but zero eigenvalues are admitted.
2. Eigenvectors are not uniquely determined; for if **u** belongs to α, in the sense of (7.1.2), so does $k\mathbf{u}$ provided that $k \neq 0$.
3. Eigenvalues and eigenvectors are defined for square matrices only.

7.2. SIMPLE ALGEBRAIC PROPERTIES

The computation of eigenvalues for a given matrix **A** is an important but often difficult problem. The numerical aspect will be treated elsewhere [see III, § 4.6].

We present here some theoretical results which will elucidate the concept and, in some cases, simplify the tasks of finding the eigenvalue. Let

$$p(t) = a_0 t^m + a_1 t^{m-1} + \ldots + a_{m-1} t + a_m$$

be a polynomial in an indeterminate t [see (14.1.1)]. For a square matrix **A** of order n [see § 6.2(vii)] define

$$p(\mathbf{A}) = a_0 \mathbf{A}^m + a_1 \mathbf{A}^{m-1} + \ldots + a_{m-1} \mathbf{A} + a_m \mathbf{I}.$$

PROPOSITION 7.2.1.   (i) *If* **u** *is a non-zero vector such that* $\mathbf{Au} = \alpha\mathbf{u}$, *then* $p(\mathbf{A}) = p(\alpha)\mathbf{u}$. *Hence* $p(\alpha)$ *is an eigenvalue of* $p(\mathbf{A})$ *with the same eigenvector that belongs to the eigenvalue* α *of* **A**.

(ii) *If* **A** *is invertible* [*see* § 6.4], *then* $\alpha \neq 0$ *and* $\mathbf{A}^{-1}\mathbf{u} = \alpha^{-1}\mathbf{u}$, *which implies that* α^{-1} *is an eigenvalue of* $\mathbf{A}^{-1}$.

We mention two trivial cases: all the eigenvalues of $\mathbf{0}$ [see § 6.2(iii)] are zero and all the eigenvalues of $\mathbf{I}$ [see (6.2.7)] are unity; in either case every non-zero vector is an eigenvector.

Some useful information can be obtained when we happen to know a polynomial equation

$$f(\mathbf{A}) = \mathbf{0}$$

satisfied by $\mathbf{A}$:

PROPOSITION 7.2.2. *If $f(\mathbf{A}) = \mathbf{0}$, then every eigenvalue α of $\mathbf{A}$ satisfies the equation $f(\alpha) = 0$, that is the eigenvalues of $\mathbf{A}$ occur among the roots of the equation $f(t) = 0$.*

EXAMPLE 7.2.1. The matrix E is said to be idempotent if $\mathbf{E}^2 = \mathbf{E}$. Hence every eigenvalue α of E must satisfy the equation $\alpha^2 = \alpha$, so that the only possible eigenvalues are zero and unity.

It must be stressed that the converse of Proposition 7.2.2 does not hold: from the fact that $f(\mathbf{A}) = \mathbf{0}$ we cannot conclude that each root of $f(t) = 0$ is an eigenvalue of $\mathbf{A}$. We shall return to this topic in section 7.5.

If α is an eigenvalue of $\mathbf{A}$ and $\mathbf{u}$ a corresponding column eigenvector, then the defining equation (7.1.2) can be written as

$$(\alpha\mathbf{I} - \mathbf{A})\mathbf{u} = \mathbf{0}.$$

This vector equation is equivalent to the statement that the system of homogeneous equations [see § 5.9].

$$(\alpha\mathbf{I} - \mathbf{A})\mathbf{x} = \mathbf{0} \tag{7.2.1}$$

has a non-trivial solution $\mathbf{x} = \mathbf{u}$, where $\mathbf{x}$ is a column vector of n unknowns. By the theory of linear equations [see Theorem 5.8.1] a non-trivial solution exists if and only if the matrix of coefficients, that is $\alpha\mathbf{I} - \mathbf{A}$, is singular (non-invertible), or equivalently [see § 6.12]

$$\det(\alpha\mathbf{I} - \mathbf{A}) = 0. \tag{7.2.2}$$

We have therefore reached the following fundamental conclusion.

THEOREM 7.2.3. *The eigenvalues of $\mathbf{A}$ are the roots of the equation*

$$\det(t\mathbf{I} - \mathbf{A}) = 0, \tag{7.2.3}$$

where t is an indeterminate.

Since the determinant of a matrix is equal to the determinant of its transpose [see (6.10.3)] we note

PROPOSITION 7.2.4. *The matrices $\mathbf{A}$ and $\mathbf{A}'$ have the same eigenvalues (though not necessarily the same eigenvectors).*

7.3. THE CHARACTERISTIC POLYNOMIAL

On expanding the determinant [see (6.9.3)]

$$\det(t\mathbf{I} - \mathbf{A}) = \begin{vmatrix} t - a_{11} & -a_{12} & -a_{13} & \cdots & -a_{1n} \\ -a_{21} & t - a_{22} & -a_{23} & \cdots & -a_{2n} \\ -a_{31} & -a_{32} & t - a_{33} & \cdots & -a_{3n} \\ \vdots & \vdots & \vdots & \ddots & \vdots \\ -a_{n1} & -a_{n2} & -a_{n3} & \cdots & t - a_{nn} \end{vmatrix} \qquad (7.3.1)$$

and collecting terms with equal powers of t we can write [see (14.1.1)]

$$\det(t\mathbf{I} - \mathbf{A}) = t^n - c_1 t^{n-1} + c_2 t^{n-2} - \ldots + (-1)^n c_n. \qquad (7.3.2)$$

The coefficients $c_1, c_2, \ldots, c_n$ are somewhat complicated expressions involving the n^2 coefficients a_{ij}, which we shall discuss presently. The alternating signs in (7.3.2) are introduced for convenience and are a matter of definition. It is customary to call

$$h_{\mathbf{A}}(t) = \det(t\mathbf{I} - \mathbf{A})$$

the *characteristic polynomial* of $\mathbf{A}$. When no confusion can arise, the suffix $\mathbf{A}$ may be omitted.

By virtue of the Fundamental Theorem of Algebra [see Theorem 14.5.2] it is (theoretically) possible to express $h(t)$ as a product of linear factors, thus

$$h(t) = (t - \alpha_1)(t - \alpha_2) \ldots (t - \alpha_n) \qquad (7.3.3)$$

displaying the roots of the equation $h(t) = 0$, which, in general, are complex numbers. In our case they are the eigenvalues of $\mathbf{A}$. It should be noted that the factors in (7.3.3) need not be distinct so that $\mathbf{A}$ may have *multiple* eigenvalues. Also, it follows from (7.3.3) that a real matrix may have complex eigenvalues, as was indicated in section 7.1. The following result is often used.

PROPOSITION 7.3.1.   *If α is a complex eigenvalue of a real matrix, then so is $\bar{\alpha}$* [see (2.7.10)] *moreover α and $\bar{\alpha}$ occur with the same multiplicity.*

In order to obtain expressions for the coefficients in (7.3.2) we require the notion of a *principal minor* of $\mathbf{A}$: let k be a fixed integer satisfying $1 \leq k \leq n$, and choose k integers $i_1, i_2, \ldots, i_k$ such that

$$1 \leq i_1 < i_2 < \ldots < i_k \leq n. \qquad (7.3.4)$$

The k-rowed determinant

$$M(i_1, i_2, \ldots, i_k) = \begin{vmatrix} a_{i_1 i_1} & a_{i_1 i_2} & \cdots & a_{i_1 i_k} \\ a_{i_2 i_1} & a_{i_2 i_2} & \cdots & a_{i_2 i_k} \\ \vdots & \vdots & \ddots & \vdots \\ a_{i_k i_1} & a_{i_k i_2} & \cdots & a_{i_k i_k} \end{vmatrix} \qquad (7.3.5)$$

is called the principal minor corresponding to the set of indices (7.3.4) [cf. (6.13.5)]. Notice that the diagonal elements of (7.3.5) are part of the diagonal of det **A**. There are $\binom{n}{k}$ principal minors of order k [see (3.8.1)]. We now state the following result:

PROPOSITION 7.3.2. *If the characteristic polynomial of* **A** *is written in the form* (7.3.2), *then*

$$c_k = \text{sum of all principal minors of order } k.$$

In particular

$$c_1 = a_{11} + a_{22} + \ldots + a_{nn}, \qquad (7.3.6)$$

$$c_2 = \begin{vmatrix} a_{11} & a_{12} \\ a_{21} & a_{22} \end{vmatrix} + \ldots + \begin{vmatrix} a_{n-1,n-1} & a_{n-1,n} \\ a_{n,n-1} & a_{nn} \end{vmatrix} \qquad (7.3.7)$$

comprising $\frac{1}{2}n(n-1)$ terms. The formulae for $c_3, c_4, \ldots, c_{n-1}$ are less useful, but it is important to know that

$$c_n = \det \mathbf{A}, \qquad (7.3.8)$$

which can be seen directly from (7.3.2) by putting $t = 0$.

The coefficient c_1 plays an important part in the more advanced branches of algebra and therefore deserves a name of its own [cf. (6.2.4)]:

DEFINITION 7.3.1. The trace of the n by n matrix $\mathbf{A} = (a_{ij})$ is defined as

$$\text{tr } \mathbf{A} = a_{11} + a_{22} + \ldots + a_{nn}.$$

By expanding the right-hand side of (7.3.3) and collecting powers of t we see that the coefficients $c_1, c_2, \ldots, c_n$ in (7.3.2) are the elementary symmetric functions [cf. (14.6.2)] of the eigenvalues, thus

$$c_1 = \text{tr } \mathbf{A} = \alpha_1 + \alpha_2 + \ldots + \alpha_n \qquad (7.3.9)$$

$$c_2 = \alpha_1\alpha_2 + \alpha_1\alpha_3 + \ldots + \alpha_{n-1}\alpha_n$$

$$\cdots$$

$$c_k = \alpha_1\alpha_2 \ldots \alpha_k + \ldots + \alpha_{n-k+1}\alpha_{n-k+2} \cdots \alpha_n$$

(the sum of the $\binom{n}{k}$ products of k factors)

$$\cdots$$

$$c_n = \det \mathbf{A} = \alpha_1\alpha_2 \ldots \alpha_n. \qquad (7.3.10)$$

The equations (7.3.9) and (7.3.10) are especially noteworthy. In particular, we deduce from (7.3.10) and Definition 6.12.1:

PROPOSITION 7.3.3.   *The matrix $\mathbf{A}$ is non-singular if and only if all its eigenvalues are non-zero.*

EXAMPLE 7.3.1.   For the upper triangular matrix

$$\mathbf{T} = \begin{pmatrix} t_{11} & t_{12} & \cdots & t_{1n} \\ 0 & t_{22} & \cdots & t_{2n} \\ \vdots & \vdots & \ddots & \vdots \\ 0 & 0 & \cdots & t_{nn} \end{pmatrix}$$

we have that

$$\det(t\mathbf{I} - \mathbf{T}) = (t - t_{11})(t - t_{22})\ldots(t - t_{nn}),$$

and an analogous result holds for the lower triangular matrix $\mathbf{T}'$. Hence the eigenvalues of $\mathbf{T}$ and of $\mathbf{T}'$ are the diagonal elements

$$t_{11}, t_{22}, \ldots, t_{nn},$$

which need not be distinct. We observe that the off-diagonal elements have no influence on the eigenvalues. Incidentally, we infer that there are infinitely many matrices (over an infinite field) all having the same set of eigenvalues, and that there are matrices with a prescribed set of eigenvalues.

As a special case we note that the eigenvalues of the diagonal matrix [see (6.7.3)]

$$D = \text{diag}\,(\alpha_1, \alpha_2, \ldots, \alpha_n)$$

are equal to $\alpha_1, \alpha_2, \ldots, \alpha_n$.

EXAMPLE 7.3.2.   Let

$$\mathbf{A} = \begin{pmatrix} 1 & 3 & 1 \\ 1 & -1 & -1 \\ 1 & 2 & 0 \end{pmatrix}. \tag{7.3.11}$$

Its characteristic polynomial is given by

$$h(t) = \begin{vmatrix} t-1 & -3 & -1 \\ -1 & t+1 & 1 \\ -1 & -2 & t \end{vmatrix}$$

$$= t^3 + \left\{ \begin{vmatrix} 1 & 3 \\ 1 & -1 \end{vmatrix} + \begin{vmatrix} 1 & 1 \\ 1 & 0 \end{vmatrix} + \begin{vmatrix} -1 & -1 \\ 2 & 0 \end{vmatrix} \right\} t - \begin{vmatrix} 1 & 3 & 1 \\ 1 & -1 & -1 \\ 1 & 2 & 0 \end{vmatrix}$$

$$= t^3 + \{-4 - 1 + 2\}t - \{-3 + 2 + 1 + 2\}$$

$$= t^3 - 3t - 2 = (t+1)^2(t-2).$$

Hence the eigenvalues of A are 2 (once) and -1 (twice). Note that $\operatorname{tr} A = 0$.

The explicit formula for c_k given in Proposition 7.3.2 is particularly useful when the matrix is of low rank [see (5.6.4)].

EXAMPLE 7.3.3. Let

$$a = (a_1, a_2, \ldots, a_n)' \quad \text{and} \quad b = (b_1, b_2, \ldots, b_n)'$$

be two non-zero column-vectors and put [see § 6.2(vi)]

$$A = ab' = \begin{pmatrix} a_1b_1 & a_1b_2 & \cdots & a_1b_n \\ a_2b_1 & a_2b_2 & \cdots & a_2b_n \\ \vdots & \vdots & \ddots & \vdots \\ a_nb_1 & a_nb_2 & \cdots & a_nb_n \end{pmatrix}.$$

Then

$$c_1 = \operatorname{tr} A = \sum_{i=1}^{n} a_ib_i = a'b = b'a,$$

but all minors of order greater than unity are equal to zero. Hence $c_2 = c_3 = \ldots = c_n = 0$ and

$$h(t) = t^n - c_1 t^{n-1}. \tag{7.3.12}$$

Therefore, when $c_1 \neq 0$, the eigenvalues are equal to c_1 (once) and zero ($n - 1$ times). When $c_1 = 0$,

$$h(t) = t^n,$$

and all eigenvalues are zero. In this case we have that

$$A \neq 0, \qquad A^2 = ab'ab' = 0,$$

that is A is nilpotent of order two [see § 6.7(xiv)]. The formula (7.3.12) can be used to evaluate the determinant

$$\Delta = \begin{vmatrix} a & b & b & \cdots & b \\ b & a & b & \cdots & b \\ \vdots & \vdots & \vdots & \ddots & \vdots \\ b & b & b & \cdots & a \end{vmatrix}. \tag{7.3.13}$$

To this end write the diagonal entries as

$$t + b,$$

where $t = a - b$ and regard (7.3.13) as the characteristic polynomial of the matrix

$$\begin{pmatrix} -b & -b & \cdots & -b \\ -b & -b & \cdots & -b \\ \vdots & \vdots & \ddots & \vdots \\ -b & -b & \cdots & -b \end{pmatrix}.$$

This matrix is of rank unity, and its trace is $-nb$. Hence by (7.3.12),

$$\Delta = (a - b)^n + nb(a - b)^{n-1}.$$

Example 7.3.1 is capable of a significant generalization. We say that $\mathbf{A}$ appears in *reduced form* if

$$\mathbf{A} = \begin{pmatrix} \mathbf{B} & \mathbf{C} \\ \mathbf{0} & \mathbf{D} \end{pmatrix}, \tag{7.3.14}$$

where $\mathbf{B}$ and $\mathbf{D}$ are square matrices, not necessarily of the same order, $\mathbf{0}$ is a block of zeros and $\mathbf{C}$ is a matrix of the appropriate size. More precisely, let $\mathbf{B}$ be a square matrix of order r, $\mathbf{D}$ a square matrix of order s, $\mathbf{0}$ a zero matrix of order s by r and $\mathbf{C}$ a matrix order r by s, where $n = r + s$. Since [by (6.14.5)]

$$\det (t\mathbf{I}_n - \mathbf{A}) = \det (t\mathbf{I}_r - \mathbf{B}) \det (t\mathbf{I}_s - \mathbf{D}),$$

we deduce that the eigenvalues of $\mathbf{A}$ are those of $\mathbf{B}$ and $\mathbf{D}$ taken together. More generally,

$$\mathbf{A} = \begin{pmatrix} \mathbf{A}_{11} & \mathbf{A}_{12} & \cdots & \mathbf{A}_{1n} \\ \mathbf{0}_{21} & \mathbf{A}_{22} & \cdots & \mathbf{A}_{2n} \\ \vdots & \vdots & \ddots & \vdots \\ \mathbf{0}_{n1} & \mathbf{0}_{n2} & \cdots & \mathbf{A}_{nn} \end{pmatrix},$$

be a partitioned matrix in which $\mathbf{A}_{11}, \mathbf{A}_{22}, \ldots, \mathbf{A}_{nn}$ are square matrices, not necessarily of the same order, while $\mathbf{A}_{ij}$ $(i < j)$ are arbitrary and $\mathbf{0}_{ij}$ $(i > j)$ are zero matrices of the appropriate dimensions. Then the eigenvalues of $\mathbf{A}$ are those of $\mathbf{A}_{11}, \mathbf{A}_{22}, \ldots, \mathbf{A}_{nn}$ taken together.

EXAMPLE 7.3.4.   Let

$$\mathbf{A} = \begin{pmatrix} 1 & 2 & a & b \\ 1 & 0 & c & d \\ 0 & 0 & -1 & 2 \\ 0 & 0 & -1 & 1 \end{pmatrix}.$$

Then

$$\det (t\mathbf{I} - \mathbf{A}) = \begin{vmatrix} t-1 & -2 \\ -1 & t \end{vmatrix} \cdot \begin{vmatrix} t+1 & -2 \\ 1 & t-1 \end{vmatrix}$$

$$= (t^2 - t - 2)(t^2 + 1)$$

$$= (t - 2)(t + 1)(t - i)(t + i).$$

Hence the eigenvalues are $2, -1, i, -i$.

As we have seen in (7.3.2), the characteristic polynomial of an n by n matrix $\mathbf{A}$ is a monic polynomial of degree n.

Conversely, if

$$f(t) = b_0 + b_1 t + b_2 t^2 + \ldots + b_{n-1} t^{n-1} + t^n \tag{7.3.15}$$

is a preassigned monic polynomial of degree n, there always exists a matrix $\mathbf{B}$ such that

$$\det (t\mathbf{I} - \mathbf{B}) = f(t).$$

One such matrix is given by

$$\mathbf{B} = \begin{pmatrix} 0 & 1 & 0 & \cdots & 0 & 0 \\ 0 & 0 & 1 & \cdots & 0 & 0 \\ & \vdots & & \ddots & & \vdots \\ 0 & 0 & 0 & & 0 & 1 \\ -b_0 & -b_1 & -b_2 & & -b_{n-2} & -b_{n-1} \end{pmatrix}. \tag{7.3.16}$$

DEFINITION 7.3.2.   The matrix $\mathbf{B}$ presented in (7.3.16) is called the *companion matrix* of the polynomial (7.3.15).

7.4. SIMILAR MATRICES

We recall that the matrices $\mathbf{A}$ and $\mathbf{B}$ are said to be similar, if there exists a non-singular matrix $\mathbf{P}$ such that

$$\mathbf{B} = \mathbf{P}^{-1}\mathbf{A}\mathbf{P},$$

[see Definition 5.13.1]. The significance of this concept stems from the fact that similar matrices describe the same linear map

$$\theta : V \rightarrow V \tag{7.4.1}$$

of a vector space V into itself, albeit relative to different bases. The transition between the two bases of V is expressed by the matrix $\mathbf{P}$ [see § 5.13].

Now it is easy to see that similar matrices have the same characteristic polynomial. For

$$\mathbf{P}^{-1}(t\mathbf{I} - \mathbf{A})\mathbf{P} = t\mathbf{P}^{-1}\mathbf{I}\mathbf{P} - \mathbf{P}^{-1}\mathbf{A}\mathbf{P}.$$

So

$$\mathbf{P}^{-1}(t\mathbf{I} - \mathbf{A})\mathbf{P} = t\mathbf{I} - \mathbf{B},$$

whence on taking determinants [see (6.10.4)]

$$(\det \mathbf{P}^{-1}) \det (t\mathbf{I} - \mathbf{A}) \det \mathbf{P} = \det (t\mathbf{I} - \mathbf{B}).$$

Since $(\det \mathbf{P}^{-1})(\det \mathbf{P}) = \det (\mathbf{P}^{-1}\mathbf{P}) = 1$ we deduce that

$$h_{\mathbf{A}}(t) = h_{\mathbf{B}}(t), \tag{7.4.2}$$

as asserted. We formulate the result as follows:

THEOREM 7.4.1.   *Similar matrices have the same eigenvalues, multiplicities being taken into account.*

As a corollary we note that

$$\operatorname{tr} (\mathbf{P}^{-1}\mathbf{A}\mathbf{P}) = \operatorname{tr} \mathbf{A},$$

either side being equal to the sum of the eigenvalues, and

$$\det (\mathbf{P}^{-1}\mathbf{A}\mathbf{P}) = \det \mathbf{A},$$

since the determinant equals the product of the eigenvalues.

In view of Theorem 7.4.1 we can state that the eigenvalues are a property of the linear map θ, since they do not depend on the basis of V.

Unfortunately, the converse of Theorem 7.4.1 is not true: two matrices which have the same eigenvalues need not be similar. For example, the matrices

$$\mathbf{J} = \begin{pmatrix} 1 & 1 \\ 0 & 1 \end{pmatrix} \quad \text{and} \quad \mathbf{I} = \begin{pmatrix} 1 & 0 \\ 0 & 1 \end{pmatrix}$$

have the same eigenvalues, namely unity counted twice. But for every non-singular $\mathbf{P}$ we have that

$$\mathbf{P}^{-1}\mathbf{I}\mathbf{P} = \mathbf{I} \neq \mathbf{J}.$$

If $\mathbf{A}$ has eigenvalues $\alpha_1, \alpha_2, \ldots, \alpha_n$, then $\mathbf{A}$ has the same eigenvalues as the diagonal matrix [see (6.7.3)]

$$\mathbf{D} = \operatorname{diag}(\alpha_1, \alpha_2, \ldots, \alpha_n).$$

We are interested in situations where $\mathbf{A}$ and $\mathbf{D}$ are, after all, similar.

DEFINITION 7.4.1.   The matrix $\mathbf{A}$ is said to be *diagonalizable* if there exists a non-singular matrix $\mathbf{P}$ [see Definition 6.4.2] such that

$$\mathbf{P}^{-1}\mathbf{A}\mathbf{P} = \operatorname{diag}(\alpha_1, \alpha_2, \ldots, \alpha_n),$$

where $\alpha_1, \alpha_2, \ldots, \alpha_n$ are then necessarily the eigenvalues of A. We note that the matrix $\mathbf{J}$, mentioned above is not diagonalizable.

An important class of diagonalizable matrix is described in the following.

THEOREM 7.4.2.   *Suppose that* $\mathbf{A}$ *has distinct eigenvalues* $\alpha_1, \alpha_2, \ldots, \alpha_n$. *For each* α_i *choose an eigenvector* $\mathbf{p}_i$, *thus*

$$\mathbf{A}\mathbf{p}_i = \alpha_i\mathbf{p}_i \quad (i = 1, 2, \ldots, n). \tag{7.4.3}$$

Then the matrix $\mathbf{P} = (\mathbf{p}_1, \mathbf{p}_2, \ldots, \mathbf{p}_n)$ *is non-singular and*

$$\mathbf{P}^{-1}\mathbf{A}\mathbf{P} = \operatorname{diag}(\alpha_1, \alpha_2, \ldots, \alpha_n). \tag{7.4.4}$$

We illustrate this theorem by an example which underlines the unpalatable fact that even for a simple matrix the process of diagonalization may involve rather heavy computations.

EXAMPLE 7.4.1.   Let

$$\mathbf{A} = \begin{pmatrix} 1 & 2 & 0 \\ -1 & 0 & -1 \\ 0 & 2 & -1 \end{pmatrix}.$$

We begin by finding the eigenvalues of **A**. Since

$$\det(t\mathbf{I} - \mathbf{A}) = \begin{vmatrix} t-1 & -2 & 0 \\ 1 & t & 1 \\ 0 & -2 & t+1 \end{vmatrix} = t(t^2 + 3),$$

it follows that the eigenvalues of **A** are given by

$$\alpha_1 = 0, \qquad \alpha_2 = i\sqrt{3}, \qquad \alpha_3 = -i\sqrt{3}. \tag{7.4.5}$$

Next, we set up the vector equation (7.4.3) for each eigenvalue in turn:
(1) $\alpha_1 = 0$: Let $\mathbf{p}_1 = (x_1, y_1, z_1)'$, where $\mathbf{p}_1$ is a column vector. Then

$$\begin{aligned} x_1 + 2y_1 \quad\quad &= 0 \\ -x_1 \quad\quad - z_1 &= 0 \\ 2y_1 - z_1 &= 0. \end{aligned} \tag{7.4.6}$$

Observe that the third equation is the sum of the first two equations. Thus the system of equations is redundant [see Example 5.9.1]. This will be the case for each of the systems (7.4.3) and, indeed guarantees the existence of (non-zero) eigenvectors. The equations (7.4.6) are solved by

$$\mathbf{p}_1 = (2, -1, -2)', \tag{7.4.7}$$

or by any vector that is a scalar multiple of $\mathbf{p}_1$ [see Theorem 5.9.1].
(2) $\alpha_2 = i\sqrt{3}$: We put $\mathbf{p}_2' = (x_2, y_2, z_2)'$. The equations (7.4.3) now become

$$\begin{aligned} E_1 &\equiv (i\sqrt{3} - 1)x_2 \quad - 2y_2 \quad\quad\quad\quad = 0 \\ E_2 &\equiv \quad\quad x_2 + i\quad 3y_2 + z_2 \quad\quad = 0 \\ E_3 &\equiv \quad\quad\quad\quad - 2y_2 + (i\sqrt{3} + 1)z_2 = 0. \end{aligned}$$

As before, the equations are not independent [see § 5.8]; indeed we have that

$$\tfrac{1}{4}(1 + i\sqrt{3})E_1 + E_2 - \tfrac{1}{4}(1 - i\sqrt{3})E_3 = 0.$$

A non-zero solution is furnished by

$$\mathbf{p}_2 = (1 + i\sqrt{3}, -2, -1 + i\sqrt{3})'. \tag{7.4.8}$$

(3) $\alpha_3 = -i\sqrt{3}$: We denote the corresponding eigenvector by $\mathbf{p}_3 = (x_3, y_3, z_3)'$. There is no need to repeat the calculations in detail. Since the coefficients of $\det(t\mathbf{I} - \mathbf{A})$ are real, every pair of a complex eigenvalue and eigenvector must be matched by a corresponding conjugate complex pair [see § 2.7.1]. Thus we may put

$$\mathbf{p}_3 = (1 - i\sqrt{3}, -2, -1 - i\sqrt{3})'. \tag{7.4.9}$$

Finally, we construct the transforming matrix **P** by juxtaposing the three column-eigenvectors displayed in (7.4.7), (7.4.8) and (7.4.9), thus

$$\mathbf{P} = \begin{pmatrix} 2 & 1 + i\sqrt{3} & 1 - i\sqrt{3} \\ -1 & -2 & -2 \\ -2 & -1 + i\sqrt{3} & -1 - i\sqrt{3} \end{pmatrix}.$$

By simple matrix algebra it may be verified that

$$\mathbf{AP} = \mathbf{P} \operatorname{diag}(0, i\sqrt{3}, -i\sqrt{3}),$$

which is equivalent to (7.4.4).

Another class of diagonalizable matrices is described in the following.

DEFINITION 7.4.2.   A square matrix $\mathbf{A}$ is said to be *normal* if it commutes with its Hermitian conjugate $\mathbf{A}^*$ $(=\overline{\mathbf{A}}')$ [see (6.7.11)], that is if

$$\mathbf{AA}^* = \mathbf{A}^*\mathbf{A}.$$

Normal matrices include the Hermitian [see (6.7.10)], real symmetric [see (6.7.5)], unitary [see (6.7.13)] and real orthogonal matrices [see (6.7.9)]. We quote

PROPOSITION 7.4.3.   *All normal matrices are diagonalizable.*

This result will be significantly strengthened in section 7.7 (Theorem 7.7.2).

7.5.   MATRIX EQUATIONS

The set $M_n(\mathbb{F})$ of all n by n matrices over $\mathbb{F}$ forms a vector space, in which composition is taken to be vector addition [see Definition 5.2.1]. The dimension of $M_n(\mathbb{F})$ is equal to n^2 [see Definition 5.4.1]; for in this connection it is immaterial that the n^2 elements a_{ij} are arranged in a square rather than in a single row or column. According to a general result [see Proposition 5.4.1] about finite-dimensional vector spaces, every collection comprising more than n^2 matrices is linearly dependent. In particular, the matrices

$$\mathbf{I}, \mathbf{A}, \mathbf{A}^2, \ldots, \mathbf{A}^{n^2}$$

are linearly dependent, that is [see Definition 5.3.1] there exist scalars

$$a_0, a_1, a_2, \ldots, a_{n^2},$$

not all zero such that

$$a_0\mathbf{I} + a_1\mathbf{A} + a_2\mathbf{A}^2 + \ldots + a_{n^2}\mathbf{A}^{n^2} = \mathbf{0}. \tag{7.5.1}$$

The left-hand side of (7.5.1) is a polynomial in $\mathbf{A}$ of degree not exceeding n^2. In other words, every square matrix satisfies some polynomial equation. Among all the polynomial equations for a particular matrix $\mathbf{A}$ there is bound to exist one of minimal degree, say m, and without loss of generality we may assume that the coefficient of $\mathbf{A}^m$ in this equation is equal to unity, thus

$$\mu(\mathbf{A}) = \mathbf{A}^m + b_1\mathbf{A}^{m-1} + \ldots + b_{m-1}\mathbf{A} + b_m\mathbf{I} = \mathbf{0}. \tag{7.5.2}$$

Introducing an indeterminate t we call

$$\mu(t) = t^m + b_1 t^{m-1} + \ldots + b_{m-1}t + b_m \tag{7.5.3}$$

the *minimum polynomial* of **A**. It is uniquely determined by **A**, being characterized as the equation of least degree for **A** with leading coefficient unity. Two
further properties of the minimum polynomial are noted in the following.

PROPOSITION 7.5.1.   *Let $\mu(t)$ be the minimal polynomial of* **A**.
   (i) *If $f(t)$ is any polynomial such that $f(\mathbf{A}) = \mathbf{0}$, then $f(t) = \mu(t)g(t)$, where
      $g(t)$ is a suitable polynomial, that is f is divisible by μ.*
   (ii) *The roots of $\mu(t) = 0$ are precisely the eigenvalues of* **A** *when multiplicities
      are disregarded.*

EXAMPLE 7.5.1.   If $\mathbf{E}^2 = \mathbf{E}, \mathbf{E} \neq \mathbf{0}, \mathbf{E} \neq \mathbf{I}$, then the minimal polynomial of
E is $\mu(t) = t^2 - t$.

   The foregoing remarks do not help us to determine the minimum polynomial.
A theoretical answer to this question will be given later [see Theorem 7.6.3]. All
we can say at present is that $m \leq n^2$. But there is a celebrated result which,
among other things, makes it evident that $m \leq n$.

THEOREM 7.5.1.   (*Cayley-Hamilton*). *Every square matrix satisfies its own
characteristic equation, that is if $h(t) = \det(t\mathbf{I} - \mathbf{A})$, then*

$$h(\mathbf{A}) = \mathbf{0}.$$

EXAMPLE 7.5.2.   As we saw in Example 7.3.2 the characteristic polynomial
of the matrix

$$\mathbf{A} = \begin{pmatrix} 1 & 3 & 1 \\ 1 & -1 & -1 \\ 1 & 2 & 0 \end{pmatrix}$$

is $t^3 - 3t - 2$. Accordingly, we have that

$$\mathbf{A}^3 - 3\mathbf{A} - 2\mathbf{I} = \mathbf{0}.$$

EXAMPLE 7.5.3.   When

$$\mathbf{N} = \begin{pmatrix} 0 & 1 & 0 \\ 0 & 0 & 0 \\ 0 & 0 & 0 \end{pmatrix},$$

we find that the characteristic polynomial is given by

$$h_{\mathbf{N}}(t) = \begin{vmatrix} t & -1 & 0 \\ 0 & t & 0 \\ 0 & 0 & t \end{vmatrix} = t^3.$$

   By virtue of the Cayley-Hamilton theorem we have that $\mathbf{N}^3 = \mathbf{0}$. However, it
is easily seen that **N** already satisfies $\mathbf{N}^2 = \mathbf{0}$. In fact

$$\mu(t) = t^2$$

is the minimal polynomial of **N**.

7.6.   THE JORDAN CANONICAL FORM

We revert to the question arising from section 7.4 as to when two n by n matrices $\mathbf{A}$ and $\mathbf{B}$ are similar, which we express in symbols as

$$\mathbf{A} \sim \mathbf{B}. \qquad (7.6.1)$$

Of course, we could simply use the definition 5.13.1 and say that (7.6.1) holds if and only if there exists a non-singular matrix $\mathbf{P}$ satisfying

$$\mathbf{AP} = \mathbf{PB}.$$

But this is hardly a feasible procedure, and a different approach is required.

Starting with a given matrix $\mathbf{A}$ we envisage the set of all matrices similar to $\mathbf{A}$. This set is called the *equivalence class* of $\mathbf{A}$ and is denoted by $(\mathbf{A})$. Our aim is to select a particularly simple and unique member of $(\mathbf{A})$, which will be termed the *canonical form* of $\mathbf{A}$ (under similarity). We already know from Theorem 7.4.1 that all members of $(\mathbf{A})$ have the same eigenvalues and, incidentally, also the same rank [see (5.6.8)].

First, assume that the eigenvalues of $\mathbf{A}$ are distinct and that they are arranged in a definite order

$$\alpha_1, \alpha_2, \ldots, \alpha_n. \qquad (7.6.2)$$

Then by Theorem 7.4.2 each member of $(\mathbf{A})$ is similar to

$$\mathbf{D} = \text{diag}\,(\alpha_1, \alpha_2, \ldots, \alpha_n),$$

and this diagonal matrix will be designated as the canonical form of $\mathbf{A}$.

The case of distinct eigenvalues having been disposed of we shall henceforth assume that $\mathbf{A}$ has multiple eigenvalues. The situation is then much more complicated as diagonalization may no longer be possible. It will turn out that a convenient canonical form has the shape

$$\mathbf{C} = \text{diag}\,(\mathbf{C}_1, \mathbf{C}_2, \ldots, \mathbf{C}_s) = \begin{pmatrix} \mathbf{C}_1 & 0 & \cdots & 0 \\ 0 & \mathbf{C}_2 & \cdots & 0 \\ \vdots & \vdots & \ddots & \vdots \\ 0 & 0 & \cdots & \mathbf{C}_s \end{pmatrix}, \qquad (7.6.3)$$

in which square matrices, not necessarily of the same order, appear as diagonal blocks, all other blocks being zero.

For ease of notation we shall write (7.6.3) as

$$\mathbf{C} = \mathbf{C}_1 + \mathbf{C}_2 + \ldots + \mathbf{C}_s$$

and say that $\mathbf{C}$ is the *direct sum* of $\mathbf{C}_1, \mathbf{C}_2, \ldots, \mathbf{C}_s$. If $\mathbf{C}$ is of order n and $\mathbf{C}_1, \mathbf{C}_2, \ldots, \mathbf{C}_s$ are of orders $k_1, k_2, \ldots, k_s$ respectively, then

$$n = k_1 + k_2 + \ldots + k_s.$$

For a complex number α and a positive integer k we define the *Jordan block* of order k as

$$\mathbf{J}(\alpha; k) = \begin{pmatrix} \alpha & 1 & 0 & \cdots & 0 \\ 0 & \alpha & 1 & \cdots & 0 \\ 0 & 0 & & \cdots & \alpha \end{pmatrix}, \tag{7.6.4}$$

having α along the diagonal and unity immediately above it. By Example 7.3.1 each eigenvalue of $\mathbf{J}(\alpha; k)$ is equal to α and

$$\det \mathbf{J}(\alpha; k) = \alpha^k.$$

More generally, we consider the *composite Jordan matrix*

$$\mathbf{J}_s = \mathbf{J}(\alpha; k_1, k_2, \ldots, k_s)$$
$$= \mathbf{J}(\alpha; k_1) + \mathbf{J}(\alpha; k_2) + \ldots + \mathbf{J}(\alpha; k_s), \tag{7.6.5}$$

which is the direct sum of Jordan blocks.

For the sake of definiteness [see Theorem 7.9.1] we shall always suppose that

$$k_1 \geq k_2 \geq \ldots \geq k_s.$$

The most important property of Jordan matrices is the fact that they are never similar to each other unless they are actually identical, that is

$$\mathbf{J}(\alpha; k_1, k_2, \ldots, k_s) \sim \mathbf{J}(\beta; l_1, l_2, \ldots, l_t)$$

if and only if

$$\alpha = \beta, \quad s = t, \quad k_i = l_i \ (i = 1, 2, \ldots, s).$$

This is one of the deepest results in Linear Algebra and we must accept it here without proof. We begin our discussion by considering matrices all of whose eigenvalues are equal. Thus suppose that the characteristic polynomial of $\mathbf{K}$ is given by

$$\det (t\mathbf{I} - \mathbf{K}) = (t - \alpha)^n. \tag{7.6.6}$$

THEOREM 7.6.1.   *Let* $\mathbf{K}$ *be an n by n matrix whose sole eigenvalue is equal to α. Then there exists a unique set of positive integers $k_1, k_2, \ldots, k_s$ such that*

$$k_1 \geq k_2 \geq \ldots \geq k_s,$$
$$k_1 + k_2 + \ldots + k_s = n,$$
$$\mathbf{K} \sim \mathbf{J}(\alpha; k_1, k_2, \ldots, k_s). \tag{7.6.7}$$

The right-hand side of (7.6.7) is the canonical form of $\mathbf{K}$. The polynomials

$$(t - \alpha)^{k_1}, (t - \alpha)^{k_2}, \ldots, (t - \alpha)^{k_s}$$

are called the *elementary divisors* of K corresponding to α. We note that their product is equal to the characteristic polynomial of $\mathbf{K}$.

If $\mathbf{L}$ is another matrix with the property that

$$\det (t\mathbf{I} - \mathbf{L}) = (t - \alpha)^n,$$

then $\mathbf{K}$ and $\mathbf{L}$ are similar if and only if they have the same elementary divisors (corresponding to α).

So far we have only asserted the existence of the canonical form (7.6.7), but we have not indicated how it can be determined. In most cases no importance is attached to the transforming matrix $\mathbf{P}$ in the equation

$$\mathbf{P}^{-1}\mathbf{K}\mathbf{P} = \mathbf{J}(\alpha; k_1, k_2, \ldots, k_s).$$

It is, however, of considerable interest to know the elementary divisors, or, equivalently, the exponents

$$k_1, k_2, \ldots, k_s,$$

which are called the *Segre characteristic* of K corresponding to α. (It is still assumed that (7.6.6) holds.)

We shall here describe a method which is based on the supposition that α is known and that it is possible to compute the ranks of $\mathbf{K} - \alpha\mathbf{I}$ and its successive powers. As in (5.6.4) we shall denote the rank of $\mathbf{X}$ by $r(\mathbf{X})$.

The reason for examining the rank becomes apparent when we study a single Jordan block $\mathbf{J}_1 = \mathbf{J}(\alpha; k)$. We note that

$$\mathbf{J}_1 - \alpha\mathbf{I} = \begin{pmatrix} 0 & 1 & 0 & \ldots & 0 \\ 0 & 0 & 1 & \ldots & 0 \\ & & & & \\ 0 & 0 & 0 & \ldots & 1 \\ 0 & 0 & 0 & \ldots & 0 \end{pmatrix}$$

is a nilpotent matrix of order k [see (6.7.23)]. Its rank and the rank of its powers are easily determined. In fact, we have that

$$r((\mathbf{J}_1 - \alpha\mathbf{I})^h) = k - h \quad (h = 0, 1, \ldots, k),$$

where we have adopted the convention that $\mathbf{X}^0 = \mathbf{I}$ for every square matrix $\mathbf{X}$.

Next, we consider the matrix

$$\mathbf{J}_s = \mathbf{J}(\alpha; k_1, k_2, \ldots, k_s),$$

defined in (7.6.5), with a view to finding its rank and the rank of its powers. By a repeated use of the rules (which may be easily verified from (7.6.3))

$$(\mathbf{A} \dotplus \mathbf{B})^h = \mathbf{A}^h \dotplus \mathbf{B}^h$$

and

$$r(\mathbf{A} \dotplus \mathbf{B}) = r(\mathbf{A}) + r(\mathbf{B})$$

we can express the rank of

$$(\mathbf{J}_s - \alpha\mathbf{I})^h \quad (h = 0, 1, 2, \ldots)$$

in terms of the Segre characteristic of $\mathbf{J}_s$.

The following typical example explains the procedure.

EXAMPLE 7.6.1.   Let

$$\mathbf{J}_6 = \mathbf{J}(\alpha; 5) + \mathbf{J}(\alpha; 5) + \mathbf{J}(\alpha; 3) + \mathbf{J}(\alpha; 3) + \mathbf{J}(\alpha; 3) + \mathbf{J}(\alpha; 1).$$

Then

$$r_0 = r((\mathbf{J}_7 - \alpha\mathbf{I})^0) = 5 + 5 + 3 + 3 + 3 + 1 = 20 = n$$

$$r_1 = r((\mathbf{J}_7 - \alpha\mathbf{I})) = 4 + 4 + 2 + 2 + 2 \qquad = 14$$

$$r_2 = r((\mathbf{J}_7 - \alpha\mathbf{I})^2) = 3 + 3 + 1 + 1 + 1 \qquad = 9$$

$$r_3 = r((\mathbf{J}_7 - \alpha\mathbf{I})^3) = 2 + 2 \qquad\qquad\quad = 4$$

$$r_4 = r((\mathbf{J}_7 - \alpha\mathbf{I})^4) = 1 + 1 \qquad\qquad\quad = 2$$

$$r_5 = r((\mathbf{J}_7 - \alpha\mathbf{I})^5) = \qquad\qquad\qquad\quad = 0.$$

Our problem is to calculate the Segre characteristic by using the above integers $r_0, r_1, \ldots$. To this end we suppose that $\mathbf{K}$ satisfies (7.6.6), whence by Theorem (7.6.1),

$$\mathbf{K} \sim \mathbf{J}_s = \mathbf{J}(\alpha; k_1, k_2, \ldots, k_s), \qquad (7.6.8)$$

where the Segre characteristic is as yet unknown.

It follows from (7.6.8) that

$$\mathbf{K} - \alpha\mathbf{I} \sim \mathbf{J}_s - \alpha\mathbf{I}.$$

Since similar matrices have the same rank [see (5.6.8)], it follows that

$$r((\mathbf{K} - \alpha\mathbf{I}))^h = r((\mathbf{J}_s - \alpha\mathbf{I})^h) \quad (h = 0, 1, 2, \ldots).$$

Now assume that the integers

$$r_n = r((\mathbf{K} - \alpha\mathbf{I})^h) \quad (h = 0, 1, 2, \ldots)$$

have been computed. It will be observed that $n = r_0 > r_1 > r_2 > r_3 > r_4 > r_5 = 0$. A rather less obvious property is that

$$r_0 - r_1 \geq r_1 - r_2 \geq r_2 - r_3 \geq \ldots \geq 0.$$

We introduce the notation

$$\Delta_h = r_h - r_{n+1} \quad (h = 0, 1, \ldots)$$

and call $\Delta_0, \Delta_1, \ldots$ the sequence of *rank differences*. It will be noted that the rank differences have been found directly from the matrix $\mathbf{K} - \alpha\mathbf{I}$ without a knowledge of the Segre characteristic.

For sufficiently great values of h both r_n and Δ_h become zero, and we shall confine ourselves to positive values of Δ_h. It should be observed that

$$\Delta_0 + \Delta_1 + \Delta_2 + \ldots = r_0 = n.$$

At this stage it is appropriate to interpose a digression on *partitions* [see also §3.9]: Let n be a fixed positive integer. We say that the integers $n_1, n_2, \ldots, n_t$ form a partition of n if

$$n_1 \geq n_2 \geq \ldots \geq n_t > 0$$

and

$$n = n_1 + n_2 + \ldots + n_t. \tag{7.6.9}$$

The number of parts, that is t, is arbitrary. It is convenient to represent a partition by a diagram of nodes arranged in rows whose initial members are vertically aligned: the first row consists of n_1 nodes, the second of n_2 nodes, the third of n_3 nodes, and so on.

EXAMPLE 7.6.2.   The partition of 12 given by the equation

$$12 = 5 + 3 + 2 + 2$$

corresponds to the diagram

.   .   .   .   .

.   .   .

.   .

.   .

With each partition of n we associate the *conjugate partition* obtained by reading the diagram of nodes by columns rather than by rows. Thus in the above example the conjugate partition is

$$12 = 4 + 4 + 2 + 1 + 1.$$

We are now ready to formulate our main result:

THEOREM 7.6.2.   *Let* $\mathbf{K}$ *be an n by n matrix having* α *as its sole eigenvalue. Put*

$$r_h = r((\mathbf{K} - \alpha \mathbf{I})^h),$$

$$\Delta_h = r_h - r_{h+1},$$

$(h = 0, 1, 2, \ldots)$. *Then*

$$\Delta_0 \geq \Delta_1 \geq \Delta_2 \geq \ldots,$$

and there exists an integer t such that

$$\Delta_t > 0, \qquad \Delta_{t+1} = 0,$$

and

$$n = \Delta_0 + \Delta_1 + \ldots + \Delta_t \tag{7.6.10}$$

is a partition of n. The Segre characteristic of **K** *with respect to* α *is the partition*

$$n = k_1 + k_2 + \ldots + k_s$$

which is conjugate to (7.6.10). *Thus*

$$\mathbf{K} \sim \mathbf{J}(\alpha; k_1, k_2, \ldots, k_s).$$

Moreover, $t = k_1$ *and* $s = \Delta_0$.

We shall now remove the restriction that the matrix has only one eigenvalue. We consider the most general case of an n by n matrix **A** which has distinct eigenvalues $\alpha, \beta, \ldots, \omega$ occurring with multiplicities $p, q, \ldots, z$ respectively, thus

$$\det(t\mathbf{I} - \mathbf{A}) = (t - \alpha)^p (t - \beta)^q \ldots (t - \omega)^z.$$

Then the canonical form for **A** under similarity is the direct sum [see (7.6.3)] of canonical forms of the type (7.6.7), one corresponding to each eigenvalue, that is

$$\mathbf{A} \sim \mathbf{J}(\alpha; k_1, k_2, \ldots, k_u)$$
$$+ \mathbf{J}(\beta; l_1, l_2, \ldots, l_v)$$
$$\ldots$$
$$+ \mathbf{J}(\omega; m_1, m_2, \ldots, m_w). \qquad (7.6.11)$$

We now have a Segre characteristic for each eigenvalue, and this characteristic is a partition of the multiplicity of the eigenvalue. Thus

$$p = k_1 + k_2 + \ldots + k_u$$
$$q = l_1 + l_2 + \ldots + l_v$$
$$\ldots$$
$$z = m_1 + m_2 + \ldots + m_w.$$

Also, it is evident that

$$n = p + q + \ldots + z. \qquad (7.6.12)$$

Fortunately, the procedure outlined in Theorem 7.6.2 suffices to determine the Segre characteristic for each eigenvalue. For example, in order to find the Segre characteristic for β we put

$$r_h(\beta) = r((\beta\mathbf{I} - \mathbf{A})^h),$$
$$\Delta_h(\beta) = r_h(\beta) - r_{h+1}(\beta) \quad (h = 0, 1, 2, \ldots).$$

Then we have a partition

$$q = \Delta_0(\beta) + \Delta_1(\beta) + \ldots$$

and its conjugate partition, say

$$q = l_1 + l_2 + \ldots$$

furnishes the Segre characteristic for β.

EXAMPLE 7.6.3.   As an illustration we consider the matrix

$$\mathbf{A} = \begin{pmatrix} 1 & 3 & 1 \\ 1 & -1 & -1 \\ 1 & 2 & 0 \end{pmatrix}$$

which we discussed in Example 7.3.2. Its eigenvalues are

$$\alpha = -1 \text{ (twice)}, \qquad \beta = 2.$$

In order to compute the Segre characteristic for -1 we have to find the numbers

$$r_h = r((-\mathbf{I} - \mathbf{A})^h) = r((\mathbf{I} + \mathbf{A}))^h \quad (h = 0, 1, \ldots).$$

Now

$$\mathbf{I} + \mathbf{A} = \begin{pmatrix} 2 & 3 & 1 \\ 1 & 0 & -1 \\ 1 & 2 & 1 \end{pmatrix},$$

and by pivotal condensation or otherwise it is easy to show that this matrix is of rank two [cf. Example 5.6.2], thus

$$r_0 = 3, \qquad r_1 = 2.$$

Next,

$$(\mathbf{I} + \mathbf{A})^2 = \begin{pmatrix} 8 & 8 & 0 \\ 1 & 1 & 0 \\ 5 & 5 & 0 \end{pmatrix},$$

which is clearly of rank unity, so that

$$r_2 = 1,$$

and it is not hard to see that all subsequent powers are likewise of rank unity. Hence

$$\Delta_0 = 3 - 2 = 1, \qquad \Delta_1 = 2 - 1 = 1,$$

and we have the partition

$$2 = 1 + 1$$

for the multiplicity of the eigenvalue -1; this partition has the diagram

$$\begin{matrix} \cdot \\ \cdot \end{matrix},$$

and the conjugate partition is given by the equation

$$2 = 2,$$

whence there is a single Jordan block $\mathbf{J}(-1; 2)$.

Since the eigenvalue 2 occurs with multiplicity unity, it is obvious that it can give rise only to the one-dimensional block $J(2; 1)$.

We conclude that the canonical form of A, under similarity, is given by

$$A \sim J(-1; 2) \dotplus J(2; 1) = \begin{pmatrix} -1 & 1 & 0 \\ 0 & -1 & 0 \\ 0 & 0 & 2 \end{pmatrix}.$$

Amongst the more theoretical applications we mention the following result, where we use the notation introduced in (7.6.4) and (7.6.8):

THEOREM 7.6.3. *Let $\alpha, \beta, \ldots, \omega$ be the distinct eigenvalues of A and suppose that the largest Jordan blocks that correspond to the various eigenvalues are of dimensions $k_1, l_1, \ldots, m_1$. Then the minimal polynomial satisfied by A is*

$$(t - \alpha)^{k_1}(t - \beta)^{l_1} \ldots (t - \omega)^{m_1}.$$

7.7. UNITARY (ORTHOGONAL) EQUIVALENCE

We recall from Definition 5.13.1 that the complex matrices A and B are similar if there exists a non-singular matrix P such that

$$B = P^{-1}AP.$$

In fact, P describes the transition from one basis of $\mathbb{C}^n$ to another basis and any non-singular matrix P is admissible [see § 5.13].

When the space is endowed with an inner product

$$(\mathbf{x}, \mathbf{y}) = \mathbf{x}^*\mathbf{y}$$

where $\mathbf{x}^* = \bar{\mathbf{x}}'$ is the transpose of the conjugate complex [see (6.7.11)], that is when we are dealing with the complex Euclidean space $\mathbb{E}^n$ [see § 10.1] then only unitary matrices may be used for changing the basis [see § 10.2]. Since by definition a unitary matrix satisfies

$$\mathbf{U}^{-1} = \mathbf{U}^*, \tag{7.7.1}$$

the notion of equivalence has to be modified as follows:

DEFINITION 7.7.1. The matrix B is said to be *unitarily equivalent* to A if there exists a unitary matrix U such that

$$B = U^*AU. \tag{7.7.2}$$

By virtue of (7.7.1) it is evident that if two matrices are unitarily equivalent, then they are similar and, therefore, have the same eigenvalues.

We do not endeavour to solve the problem of a canonical form under unitary equivalence for an arbitrary complex matrix. But we mention a partial answer which has important consequences.

THEOREM 7.7.1. *(I. Schur's triangle theorem). Let* **A** *be an n by n complex matrix. Then there exists a unitary matrix* **U** *such that*

$$\mathbf{U^*AU} = \begin{pmatrix} \alpha_1 & c_{12} & c_{13} & \cdots & c_{1n} \\ 0 & \alpha_2 & c_{23} & \cdots & c_{2n} \\ 0 & 0 & \alpha_3 & \cdots & c_{3n} \\ \vdots & \vdots & \vdots & \ddots & \vdots \\ 0 & 0 & 0 & \cdots & \alpha_n \end{pmatrix}, \qquad (7.7.3)$$

is an upper triangular matrix, in which the diagonal elements are the eigenvalues of **A**.

It is not claimed that the coefficients c_{ij} are, in general, unique. But it can be shown that, for a special class of matrices, all these coefficients are necessarily zero.

When **N** is normal [Definition 7.4.2], so is **U*NU**, where **U** is an arbitrary unitary matrix; in other words, normality is preserved under unitary transformation. Now a simple calculation shows that an upper triangular matrix cannot be normal unless all off-diagonal coefficients vanish. Hence Theorem 7.7.1 yields the important corollary.

THEOREM 7.7.2. *All normal matrices can be diagonalized by a unitary transformation, that is if* **N** *is normal, there exists a unitary matrix* **U** *such that*

$$\mathbf{U^*NU} = \text{diag}\,(\nu_1, \nu_2, \ldots, \nu_n),$$

where $\nu_1, \nu_2, \ldots, \nu_n$ *are the eigenvalues of* **N**.

Over the complex field the commonest types of normal matrices are the Hermitian matrices (**H*** = **H**) and the unitary matrices (**U*** = **U**$^{-1}$) [see § 6.7(viii) and (x)].

When a matrix is both normal and real and when its eigenvalues are also real, then the transformation to diagonal form may be accomplished by a real orthogonal matrix **R**; (**R'** = **R**$^{-1}$) [see § 6.7(vii)]

$$\mathbf{R'NR} = \text{diag}\,(\nu_1, \nu_2, \ldots, \nu_n).$$

DEFINITION 7.7.2. When **N** and **R** are real, we shall say that

$$\mathbf{N} \quad \text{and} \quad \mathbf{R'NR}$$

are *orthogonally equivalent*.

7.8. THE PRINCIPAL AXES THEOREM

The eigenvalues and eigenvectors of a Hermitian matrix [see § 6.7(viii)] possess the following fundamental properties.

THEOREM 7.8.1.   *Let* **H** *be an n by n Hermitian matrix. Then*
   (i) *The eigenvalues of* **H** *are real;*
   (ii) *eigenvectors that belong to distinct eigenvalues are orthogonal, that is if* **p** *and* **q** *are non-zero vectors such that*

$$\mathbf{Hp} = \alpha\mathbf{p}, \qquad \mathbf{Hq} = \beta\mathbf{q}, \qquad \alpha \neq \beta, \tag{7.8.1}$$

then [*see* (6.7.11)]

$$\mathbf{p}^*\mathbf{q} = 0. \tag{7.8.2}$$

The proof of (i) is given in Proposition 10.3.1.

We have seen in Theorem 7.7.2 that every normal matrix is unitarily equivalent to a diagonal matrix. But this diagonal matrix is not unique except in the trivial case when all eigenvalues are equal. For example, let **A** be a 2 by 2 normal matrix with distinct eigenvalues α, β. Then there exists a unitary matrix **U** such that

$$\mathbf{U}^*\mathbf{AU} = \mathrm{diag}\,(\alpha, \beta).$$

If we put

$$\mathbf{T} = \begin{pmatrix} 0 & 1 \\ 1 & 0 \end{pmatrix},$$

it is easy to check that

$$(\mathbf{UT})^*\mathbf{A}(\mathbf{UT}) = \mathrm{diag}\,(\beta, \alpha).$$

Moreover, $\mathbf{V} = \mathbf{UT}$ is unitary, and we have demonstrated that **A** is unitarily equivalent to two distinct diagonal forms.

Generally, if

$$\mathbf{A} \sim \mathrm{diag}\,(\alpha_1, \alpha_2, \ldots, \alpha_n) \tag{7.8.3}$$

it can be shown that also

$$\mathbf{A} \sim \mathrm{diag}\,(\alpha_{k_1}, \alpha_{k_2}, \ldots, \alpha_{k_n}),$$

where $\alpha_{k_1}, \alpha_{k_2}, \ldots, \alpha_{k_n}$ are the eigenvalues of **A** in a preassigned order.

When **H** is Hermitian, the eigenvalues are real and can be arranged in decreasing order of magnitude. With this convention the canonical form (7.8.3) is rendered unique.

In applications we are mainly concerned with real matrices. Theorem 7.7.2 then takes the following form:

THEOREM 7.8.2.   (*Principal Axes Theorem*): *Let* **A** *be a real symmetric matrix* [*see* (6.7.5)] *with eigenvalues*

$$\alpha_1 \geq \alpha_2 \geq \ldots \geq \alpha_n. \tag{7.8.4}$$

Then there exists a real orthogonal **R** *such that*

$$\mathbf{R}'\mathbf{AR} = \mathrm{diag}\,(\alpha_1, \alpha_2, \ldots, \alpha_n),$$

which is the canonical form of **A**.

The geometric-sounding name will be explained later in this section.

Next, we shall describe a method for finding the transforming matrix $\mathbf{R}$. To this end we assume that the (harder) problem of determining the eigenvalues has already been solved. Two cases have to be distinguished:

(i) *The eigenvalues* (7.8.4) *are distinct:* For each α_i $(i = 1, 2, \ldots, n)$ the system of equations

$$(\mathbf{A} - \alpha_i\mathbf{I})\mathbf{x} = \mathbf{0} \tag{7.8.5}$$

has a non-trivial real solution, say $\mathbf{x} = \mathbf{u}_i$. Then $\mathbf{u}_i'\mathbf{u}_i$ is a non-zero scalar and the vector

$$\mathbf{p}_i = (\mathbf{u}_i'\mathbf{u}_i)^{-1/2}\mathbf{u}_i \tag{7.8.6}$$

also satisfies (7.8.5). We say that $\mathbf{p}_i$ is a *normalized* eigenvector belonging to α_i. By virtue of (7.8.2) and (7.8.6) we have that

$$\mathbf{p}_i'\mathbf{p}_j = \delta_{ij} = \begin{cases} 1, & \text{when } i = j \\ 0, & \text{when } i \neq j' \end{cases}$$

[cf. (6.7.1) for the definition of δ_{ij}] which is equivalent to the statement that the matrix

$$\mathbf{R} = [\mathbf{p}_1, \mathbf{p}_2, \ldots, \mathbf{p}_n]$$

is orthogonal. It is easy to verify that

$$\mathbf{R}'\mathbf{A}\mathbf{R} = \text{diag}\,(\alpha_1, \alpha_2, \ldots, \alpha_n).$$

We may summarize the procedure as follows: When the eigenvalues are distinct, the transforming matrix is constructed by taking as its columns the n normalized eigenvectors.

EXAMPLE 7.8.1.   Let

$$\mathbf{A} = \begin{pmatrix} 1 & 1 & -2 \\ 1 & 2 & -1 \\ -2 & -1 & 1 \end{pmatrix}.$$

We find that

$$\det(t\mathbf{I} - \mathbf{A}) = \begin{vmatrix} t-1 & -1 & 2 \\ -1 & t-2 & 1 \\ 2 & 1 & t-1 \end{vmatrix} = (t^2 - 1)(t - 4).$$

Hence the eigenvalues of $\mathbf{A}$ are

$$\alpha_1 = 4, \qquad \alpha_2 = 1, \qquad \alpha_3 = -1.$$

The normalized eigenvectors are computed as follows:

$\alpha_1 = 4$:
$$3x_1 - x_2 + 2x_3 = 0$$
$$-x_1 + 2x_2 + x_3 = 0$$
$$2x_1 + x_2 + 3x_3 = 0,$$

$$\mathbf{x}' = (1, 1, -1), \qquad \mathbf{p}'_1 = \frac{1}{\sqrt{3}}(1, 1, -1).$$

$\alpha_2 = 1$:
$$-x_2 + 2x_3 = 0$$
$$-x_1 - x_2 + x_3 = 0$$
$$2x_1 + x_2 = 0,$$

$$\mathbf{x}' = (-1, 2, 1), \qquad \mathbf{p}'_2 = \frac{1}{\sqrt{6}}(-1, 2, 1).$$

$\alpha_3 = -1$:
$$2x_1 - x_2 + 2x_3 = 0$$
$$-x_1 - 3x_2 + x_3 = 0$$
$$2x_1 + x_2 - 2x_3 = 0,$$

$$\mathbf{x}' = (1, 0, 1), \qquad \mathbf{p}'_3 = \frac{1}{\sqrt{2}}(1, 0, 1).$$

It may be verified that

$$\mathbf{R} = (\mathbf{p}_1, \mathbf{p}_2, \mathbf{p}_3) = \frac{1}{\sqrt{6}} \begin{pmatrix} \sqrt{2} & -1 & \sqrt{3} \\ \sqrt{2} & 2 & 0 \\ -\sqrt{2} & 1 & \sqrt{3} \end{pmatrix}.$$

is an orthogonal matrix and that

$$\mathbf{R}'\mathbf{A}\mathbf{R} = \text{diag}(4, 1, -1).$$

(ii) *The matrix possesses multiple eigenvalues.* Suppose that α is an m-fold eigenvalue of $\mathbf{A}$. Then it can be proved that $\alpha\mathbf{I} - \mathbf{A}$ is of rank $n - m$ so that the system of equations

$$(\alpha I - A)\mathbf{x} = 0 \qquad (7.8.7)$$

has m linearly independent solutions, say $\mathbf{u}_1, \mathbf{u}_2, \ldots, \mathbf{u}_m$ [see Theorem 5.9.1]. By an application of the Gram-Schmidt process [see Theorem 10.2.1] these vectors may be replaced by a set of vectors of the form

$$\mathbf{v}_1 = \lambda_1 \mathbf{u}_1$$
$$\mathbf{v}_2 = \lambda_2(\mathbf{u}_2 + \mu_{21}\mathbf{u}_1)$$
$$\mathbf{v}_3 = \lambda_3(\mathbf{u}_3 + \mu_{32}\mathbf{u}_2 + \mu_{31}\mathbf{u}_1)$$
$$\mathbf{v}_m = \lambda_m(\mathbf{u}_m + \mu_{m,m-1}\mathbf{u}_{m-1} + \ldots + \mu_{m,1}\mathbf{u}_1),$$

which are solutions of (7.8.7) and, in addition, satisfy the relations

$$\mathbf{v}_i'\mathbf{v}_j = \delta_{ij} \quad (i, j = 1, 2, \ldots, m),$$

where δ_{ij} is defined in (6.7.1).

Thus we have the result that corresponding to an m-fold eigenvector of $\mathbf{A}$, there exists a set of m *orthonormal* ($=$ mutually orthogonal and normalized [see Definition 10.2.1]), eigenvectors. If this procedure is carried out for each eigenvalue in turn, we obtain a complete set of n orthonormal vectors. The matrix $\mathbf{R}$ that has these vectors as columns, is the transforming matrix required.

EXAMPLE 7.8.2.   Let

$$\mathbf{A} = \begin{pmatrix} 3 & 1 & 1 \\ 1 & 3 & 1 \\ 1 & 1 & 3 \end{pmatrix}.$$

We find that

$$\det (t\mathbf{I} - \mathbf{A}) = (t - 2)^2(t - 5).$$

Hence the eigenvalues of $\mathbf{A}$ are

$$\alpha_1 = 5, \qquad \alpha_2 = \alpha_3 = 2.$$

A set of orthonormal eigenvectors is determined as follows. $\alpha_1 = 5$: The equations (7.8.7) become

$$2x_1 - x_2 - x_3 = 0$$
$$-x_1 + 2x_2 - x_3 = 0$$
$$-x_1 - x_2 + 2x_3 = 0$$

and have the normalized solution

$$\mathbf{p}_1' = \frac{1}{\sqrt{3}}(1, 1, 1).$$

$\alpha_2 = \alpha_3 = 2$: As expected, the system of equations is of rank $3 - 2 = 1$ and effectively reduces to the single equation

$$-x_1 - x_2 - x_3 = 0.$$

We may start with any pair of linearly independent solutions, for example

$$\mathbf{u}' = (1, -1, 0), \qquad \mathbf{v}' = (1, 0, -1).$$

In accordance with the Gram-Schmidt process an orthogonal pair of solutions may be obtained by putting

$$\mathbf{w} = \mathbf{v} + k\mathbf{u}$$

and choosing the scalar k in such a way that

$$0 = \mathbf{u}'\mathbf{w} = \mathbf{u}'\mathbf{v} + k\mathbf{u}'\mathbf{u},$$

that is

$$0 = 1 + 2k,$$

$$k = -\tfrac{1}{2}.$$

Thus

$$\mathbf{w}' = (\tfrac{1}{2}, \tfrac{1}{2}, -1).$$

Normalizing $\mathbf{u}'$ and $\mathbf{w}'$ we obtain

$$\mathbf{p}_2' = \frac{1}{\sqrt{2}}(1, -1, 0), \qquad \mathbf{p}_3' = \frac{1}{\sqrt{6}}(1, 1, -2),$$

which yields the transforming matrix

$$\mathbf{R} = (\mathbf{p}_1, \mathbf{p}_2, \mathbf{p}_3) = \frac{1}{\sqrt{6}}\begin{pmatrix} \sqrt{2} & \sqrt{3} & 1 \\ \sqrt{2} & -\sqrt{3} & 1 \\ \sqrt{2} & 0 & -2 \end{pmatrix}.$$

The study of a real symmetric matrix $\mathbf{A}$ can be viewed in a different light if we consider the quadratic form associated with $\mathbf{A}$ [see (6.7.7)]. Thus let

$$q(\mathbf{x}) = \mathbf{x}'\mathbf{A}\mathbf{x}$$

$$= a_{11}x_1^2 + \ldots + a_{nn}x_n^2 + 2a_{12}x_1x_2 + \ldots + 2a_{n-1,n}x_{n-1}x_n, \quad (7.8.8)$$

where

$$\mathbf{A} = (a_{ij}), \qquad a_{ij} = a_{ji}, \quad \text{and} \quad \mathbf{x}' = (x_1, x_2, \ldots, x_n).$$

In the present context we admit only orthogonal transformations. Thus we may put

$$\mathbf{x} = \mathbf{R}\mathbf{y}, \tag{7.8.9}$$

where $\mathbf{R}$ is a real orthogonal matrix [see (6.7.9)]. Hence

$$\mathbf{x}'\mathbf{x} = \mathbf{y}'\mathbf{y}, \tag{7.8.10}$$

that is

$$x_1^2 + x_2^2 + \ldots + x_n^2 = y_1^2 + y_2^2 + \ldots + y_n^2,$$

or, more briefly,

$$|\mathbf{x}| = |\mathbf{y}|,$$

where the 'length' of a vector $\mathbf{x}$ is defined by

$$|\mathbf{x}| = (\mathbf{x}'\mathbf{x})^{1/2} = (x_1^2 + \ldots + x_n^2)^{1/2}.$$

The equation

$$q(\mathbf{x}) = 1 \tag{7.8.11}$$

represents a surface in n-dimensional space. It is known as a *central quadric* [see V, § 2.12]. It is analogous to central conics (ellipse, hyperbola) in a plane. By Theorem 7.8.2, we can choose **R** in such a way that

$$\mathbf{R'AR} = \mathbf{D} = \text{diag}\,(\alpha_1, \alpha_2, \ldots, \alpha_n).$$

On substituting (7.8.9) in (7.8.8) we find that

$$q(\mathbf{x}) = \mathbf{y'R'ARy} = \mathbf{y'Dy}$$

$$= \alpha_1 y_1^2 + \alpha_2 y_2^2 + \ldots + \alpha_n y_n^2.$$

It is customary to say that $q(\mathbf{x})$ has been transformed into a 'sum of squares', ignoring the fact that coefficients are attached to the squares. The equation (7.8.11) now becomes

$$\alpha_1 y_1^2 + \alpha_2 y_2^2 + \ldots + \alpha_n y_n^2 = 1.$$

When the equation of the surface has been reduced to this form the coordinate axes coincide with the principal axes of the surface [see V, § 2.4]. This explains the name of Theorem 7.8.2.

Analogous results can be established over the complex field. They are summarized as follows:

THEOREM 7.8.3. (*Complex Principal Axes Theorem*)
  (i) *Let* **H** *be an n by n Hermitian matrix with eigenvalues*

$$\alpha_1 \geq \alpha_2 \geq \ldots \geq \alpha_n.$$

Then there exists a unitary matrix **U** *such that*

$$\mathbf{U^*HU} = \text{diag}\,(\alpha_1, \alpha_2, \ldots, \alpha_n).$$

  (ii) *Every Hermitian form* [*see* § 6.7(viii)]

$$h = \sum_{i,j=1}^{n} h_{ij}\bar{z}_i z_j = \mathbf{z^*Hz}$$

can be transformed into a sum of squares

$$h = \sum_{i=1}^{n} \alpha_i |w_i|^2,$$

where $\mathbf{z} = \mathbf{Uw}$, **U** *being a unitary matrix.*

7.9. EXTREMAL PROPERTIES

In this section we are concerned with the *set of values* which a real quadratic form [see (9.1.4)]

$$q(\mathbf{x}) = \mathbf{x'Ax} \tag{7.9.1}$$

attains when the vector **x** ranges over all possible real vectors. Transferring the

Definition 7.7.2 from matrices to the corresponding forms we say that the forms

$$q(\mathbf{x}) = \mathbf{x}'\mathbf{A}\mathbf{x} \quad \text{and} \quad s(\mathbf{y}) = \mathbf{y}'(\mathbf{R}'\mathbf{A}\mathbf{R})\mathbf{y}$$

are equivalent, where $\mathbf{R}$ is an arbitrary real orthogonal matrix [see (6.7.9)]. It is easy to see that equivalent forms have the same set of values. Indeed, suppose that λ is attained by q, that is there exists a vector $\mathbf{u}_1$ such that

$$q(\mathbf{u}_1) = \mathbf{u}_1'\mathbf{A}\mathbf{u}_1 = \lambda.$$

If we now put $\mathbf{v}_1 = \mathbf{R}^{-1}\mathbf{u}_1$, then

$$s(\mathbf{v}_1) = \mathbf{v}_1'\mathbf{R}'\mathbf{A}\mathbf{R}\mathbf{v}_1 = \mathbf{u}_1'\mathbf{A}\mathbf{u}_1 = \lambda,$$

which shows that λ is attained by s. Conversely, every value that is attained by s is also attained by q.

Evidently, the most convenient way to study the set of values is to replace 7.9.1 by its canonical form. Thus we choose $\mathbf{R}$ in such a way that $q(\mathbf{x})$ is equivalent to

$$s(\mathbf{y}) = \alpha_1 y_1^2 + \alpha_2 y_2^2 + \ldots + \alpha_n y_n^2, \tag{7.9.2}$$

where $\alpha_1, \alpha_2, \ldots, \alpha_n$ are the eigenvalues of $\mathbf{A}$, subject to the convention

$$\alpha_1 \geq \alpha_2 \geq \ldots \geq \alpha_n, \tag{7.9.3}$$

and where $y_1, y_2, \ldots, y_n$ are independent variables. A knowledge of the eigenvalues enables us to decide whether a quadratic form is positive (negative) definite or indefinite [see Theorem 9.2.2]. Again, for this purpose we may use the canonical form (7.9.2). The result is summarized in

THEOREM 7.9.1. *The quadratic form* $\mathbf{x}'\mathbf{A}\mathbf{x}$ *is*
  (i) *positive (negative) definite if all eigenvalues of* $\mathbf{A}$ *are positive (negative);*
  (ii) *positive (negative) semi-definite if at least one eigenvalue is zero while the other eigenvalues are positive (negative);*
  (iii) *indefinite if there are both positive and negative eigenvalues.*

Remark
  In concrete cases it is usually preferable to use the methods of section 9.2 to examine the definiteness of a quadratic form; for the problem can be solved without a prior determination of the eigenvalues.
  On combining (7.9.2) and (7.9.3) we obtain that

$$\alpha_n(y_1^2 + y_2^2 + \ldots + y_n^2) \leq s(\mathbf{y}) \leq \alpha_1(y_1^2 + y_2^2 + \ldots + y_n^2),$$

or, more succinctly,

$$\alpha_n \mathbf{y}'\mathbf{y} \leq s(\mathbf{y}) \leq \alpha_1 \mathbf{y}'\mathbf{y}.$$

We recall that $s(\mathbf{y})$ takes the same values as $\mathbf{x}'\mathbf{A}\mathbf{x}$. Also since $\mathbf{x} = \mathbf{R}\mathbf{y}$, where $\mathbf{R}$ is an orthogonal matrix, we have that

$$\mathbf{x}'\mathbf{x} = \mathbf{y}'\mathbf{y}.$$

Hence we can express our result in terms of $\mathbf{x}$ by stating that every non-zero vector $\mathbf{x}$ satisfies the inequalities

$$\alpha_n \le \frac{\mathbf{x}'\mathbf{Ax}}{\mathbf{x}'\mathbf{x}} \le \alpha_1.$$

The quotient in the middle remains unaltered if we replace $\mathbf{x}$ by $k\mathbf{x}$ where k is an arbitrary non-zero scalar. In particular, we can choose k such that $k\mathbf{x}$ is a unit vector [see Definition 10.2.1]; alternatively, we may assume from the outset that $\mathbf{x}$ is a unit vector. The results are collected in the following

THEOREM 7.9.2.   *Let $\mathbf{A}$ be a real symmetric matrix. Then*

$$\max_{|\mathbf{x}|>0} \frac{\mathbf{x}'\mathbf{Ax}}{\mathbf{x}'\mathbf{x}} = \alpha_1, \qquad \min_{|\mathbf{x}|>0} \frac{\mathbf{x}'\mathbf{Ax}}{\mathbf{x}'\mathbf{x}} = \alpha_n,$$

where α_1 and α_n are the greatest and the least eigenvalue of $\mathbf{A}$ respectively. Alternatively

$$\max_{|\mathbf{x}|=1} \mathbf{x}'\mathbf{Ax} = \alpha_1, \qquad \min_{|\mathbf{x}|=1} \mathbf{x}'\mathbf{Ax} = \alpha_n.$$

While this theorem characterizes the greatest and the least eigenvalue as solutions of extremal problems, the question may be raised whether similar properties hold for the intermediate eigenvalues. Although the answer is rather more involved, it is still worth recording here.

THEOREM 7.9.3.   *Let $\mathbf{A}$ be a real symmetric matrix [see (6.7.5)] with eigenvalues*

$$\alpha_1 \ge \alpha_2 \ge \ldots \ge \alpha_n.$$

Then α_k ($k > 1$) is the least value which the maximum of $\mathbf{x}'\mathbf{Ax}$ attains when $\mathbf{x}$ is a unit vector ($|\mathbf{x}| = 1$) subject to $k - 1$ arbitrary linear constraints

$$\sum_{j=1}^{n} c_{ij}x_j = 0 \quad (i = 1, 2, \ldots, k - 1). \tag{7.9.4}$$

Paraphrasing the statement we let $\mathbf{C} = (c_{ij})$ be the coefficient matrix of the constraints [see (5.7.3)]. For a fixed $\mathbf{C}$, denote by $\mu(\mathbf{C})$ the maximum of $\mathbf{x}'\mathbf{Ax}$ when $\mathbf{x}$ ranges over all unit vectors subject to (7.9.4); then vary $\mathbf{C}$ so that $\mu(\mathbf{C})$ becomes as small as possible; the number so obtained is equal to α_k.

It is unlikely that this theorem will be used to compute α_k, but some interesting general deductions can be made.

If the last row and the last column of $\mathbf{A}$ are deleted we obtain the $(n - 1)$ by $(n - 1)$ real symmetric matrix

$$\mathbf{A}_n = \begin{pmatrix} a_{11} & a_{12} & \cdots & a_{1,n-1} \\ a_{21} & a_{22} & \cdots & a_{2,n-1} \\ \vdots & \vdots & \ddots & \vdots \\ a_{n-1,1} & a_{n-1,2} & \cdots & a_{n-1,n-1} \end{pmatrix} (a_{ij} = a_{ji})$$

which may be called the *n*th *section* of $\mathbf{A}$. Similarly, if the *k*th row and the *k*th column are deleted we obtain the *k*th section of $\mathbf{A}$.

The eigenvalues of a section interlace those of $\mathbf{A}$ in the following sense.

THEOREM 7.9.4. *Let $\mathbf{A}$ be an n by n real symmetric matrix with eigenvalues*

$$\alpha_1 \geq \alpha_2 \geq \ldots \geq \alpha_n.$$

For a fixed value of k ($1 \leq k \leq n$), denote the eigenvalues of $\mathbf{A}_k$ by

$$\beta_1 \geq \beta_2 \geq \ldots \geq \beta_{n-1}.$$

Then

$$\alpha_1 \geq \beta_1 \geq \alpha_2 \geq \beta_2 \geq \ldots \geq \alpha_{n-1} \geq \beta_{n-1} \geq \alpha_n.$$

7.10. SPECTRAL DECOMPOSITION

In this section we assume that $\mathbf{A}$ is a diagonalizable matrix with eigenvalues $\alpha_1, \alpha_2, \ldots, \alpha_n$ which need not be either real or distinct [see Definition 7.4.1]. Thus there exists a non-singular matrix $\mathbf{P}$ such that

$$\mathbf{P}^{-1}\mathbf{A}\mathbf{P} = \text{diag}(\alpha_1, \alpha_2, \ldots, \alpha_n) = D. \tag{7.10.1}$$

Resolving $\mathbf{P}$ into columns, we write

$$\mathbf{P} = (\mathbf{p}_1, \mathbf{p}_2, \ldots, \mathbf{p}_n).$$

Similarly, we resolve $\mathbf{P}^{-1}$ into rows, that is

$$\mathbf{P}^{-1} = \begin{pmatrix} \mathbf{q}_1' \\ \mathbf{q}_2' \\ \vdots \\ \mathbf{q}_n' \end{pmatrix},$$

where $\mathbf{q}_1, \mathbf{q}_2, \ldots, \mathbf{q}_n$ are the columns of $(\mathbf{P}^{-1})'$. Since $\mathbf{P}^{-1}\mathbf{P} = \mathbf{I}$, we have that

$$\mathbf{q}_i'\mathbf{p}_j = \delta_{ij} \quad (i, j = 1, 2, \ldots, n),$$

where the symbol δ_{ij} is defined in (6.7.1). For each value of *i* define the *n* by *n* matrix

$$\mathbf{E}_i = \mathbf{p}_i\mathbf{q}_i' \quad (i = 1, 2, \ldots, n). \tag{7.10.2}$$

These *n* matrices satisfy the relations

$$\mathbf{E}_i\mathbf{E}_j = \delta_{ij}\mathbf{E}_i \quad (i, j = 1, 2, \ldots, n), \tag{7.10.3}$$

that is $\mathbf{E}_i^2 = \mathbf{E}_i$, and $\mathbf{E}_i\mathbf{E}_j = \mathbf{E}_j\mathbf{E}_i = 0$ if $i \neq j$; for this reason, (7.10.2) is called a set of mutually orthogonal *idempotents*, a matrix $\mathbf{J}$ being called idempotent if $\mathbf{J}^2 = \mathbf{J}$.

The matrix equation (7.10.1) is equivalent to

$$\mathbf{A} = \mathbf{P}\mathbf{D}\mathbf{P}^{-1},$$

that is,

$$\mathbf{A} = \alpha_1 \mathbf{E}_1 + \alpha_2 \mathbf{E}_2 + \ldots + \alpha_n \mathbf{E}_n. \tag{7.10.4}$$

This formula is called the *spectral decomposition of* $\mathbf{A}$. The term 'spectrum' is synonymous with 'set of eigenvalues' and has been suggested by physical applications. The spectral decomposition can be computed provided that we know the eigenvalues of $\mathbf{A}$ and the transforming matrix $\mathbf{P}$ together with its inverse.

When $\mathbf{A}$ is real and symmetric, $\mathbf{P}$ may be taken to be real orthogonal so that $(\mathbf{P}^{-1})' = \mathbf{P}$. Hence in this case $\mathbf{q}_i = \mathbf{p}_i$ $(i = 1, 2, \ldots, n)$ and the idempotents

$$\mathbf{E}_i = \mathbf{p}_i \mathbf{p}_i'$$

are themselves real symmetric matrices.

The spectral decomposition is a useful tool for evaluating functions of a matrix. This is based on the observation that, by virtue of (7.10.3),

$$\mathbf{A}^2 = \alpha_1{}^2 \mathbf{E}_1 + \alpha_2{}^2 \mathbf{E}_2 + \ldots + \alpha_n{}^2 \mathbf{E}_n.$$

By induction it follows that

$$\mathbf{A}^k = \alpha_1{}^k \mathbf{E}_1 + \alpha_2{}^k \mathbf{E}_2 + \ldots + \alpha_n{}^k \mathbf{E}_n.$$

This equation may be multiplied throughout by a scalar c_k and then summed over k. Thus if

$$p(t) = c_0 + c_1 t + c_2 t^2 + \ldots + c_m t^m$$

is an arbitrary polynomial in t [see (14.1.1)] we obtain that

$$p(\mathbf{A}) = p(\alpha_1)\mathbf{E}_1 + p(\alpha_2)\mathbf{E}_2 + \ldots + p(\alpha_n)\mathbf{E}_n. \tag{7.10.5}$$

Equations analogous to (7.10.5) also hold for some matrix functions other than polynomials. For example, the matrix

$$\mathbf{B} = \alpha_1{}^{1/2} \mathbf{E}_1 + \alpha_2{}^{1/2} \mathbf{E}_2 + \ldots + \alpha_n{}^{1/2} \mathbf{E}_n \tag{7.10.6}$$

satisfies the equation $\mathbf{B}^2 = \mathbf{A}$. We may therefore call $\mathbf{B}$ a square root of $\mathbf{A}$ and write

$$\mathbf{B} = \sqrt{\mathbf{A}} \quad \text{or} \quad \mathbf{B} = \mathbf{A}^{1/2}.$$

Of course, if $\mathbf{A} \neq 0$, the matrix $\mathbf{B}$ is not uniquely determined since the sign of each $\alpha_i{}^{1/2}$ is at our disposal. But we note that when $\mathbf{A}$ is positive definite (all eigenvalues positive by Theorem 7.9.1), then we can choose $\mathbf{B}$ to be positive definite. Hence a positive definite matrix possesses at least one positive definite square root.

More generally, let λ be an arbitrary real number. Then we define

$$\mathbf{A}^\lambda = \alpha_1{}^\lambda \mathbf{E}_1 + \alpha_2{}^\lambda \mathbf{E}_2 + \ldots + \alpha_n{}^\lambda \mathbf{E}_n,$$

assuming that no eigenvalue is zero when $\lambda < 0$.

Again, let

$$f(z) = c_0 + c_1 z + c_2 z^2 + \ldots + c_m z^m + \ldots$$

be a power series which converges when $|z| < r$ [see IV, § 1.10]. Then we may define

$$f(\mathbf{A}) = c_0\mathbf{I} + c_1\mathbf{A} + c_2\mathbf{A}^2 + \ldots + c_m\mathbf{A}^m + \ldots$$

provided that the series on the right-hand side converges to a finite matrix-limit. When A is diagonalizable, it is easily seen that convergence is assured if each eigenvalue α satisfies $|\alpha| < r$. Indeed we then have that

$$f(\mathbf{A}) = \sum_{i=1}^{n} f(\alpha_i)\mathbf{E}_i.$$

The most interesting case is that of the exponential function. The series

$$\exp \mathbf{A} = \sum_{m=0}^{\infty} \mathbf{A}^m/m! \tag{7.10.7}$$

converges for each $\mathbf{A}$ [see IV, Example 1.10.1]. If $\mathbf{A}$ is diagonalizable,

$$\exp \mathbf{A} = \sum_{i=1}^{n} (\exp \alpha_i)\mathbf{E}_i.$$

The formula (7.10.7) holds whether $\mathbf{A}$ is diagonalizable or not.

In the study of linear differential equations it is important to know whether

$$\lim_{t \to \infty} \exp (t\mathbf{A})$$

exists, where t is a scalar variable. The following result is relevant in this connection:

PROPOSITION 7.10.1.   *If all eigenvalues of* $\mathbf{A}$ *have negative real parts we call* $\mathbf{A}$ stable, *and*

$$\lim \exp (t\mathbf{A}) = \mathbf{0} \quad \text{as} \quad t \to \infty.$$

More information about this topic can be found in S. Barnett and C. Storey (1970).

7.11.   NON-NEGATIVE MATRICES

We say that the matrix $\mathbf{M} = (m_{ij})$ is *positive*, and we write $\mathbf{M} > \mathbf{0}$ if $m_{ij} > 0$ for all i and j. Similarly, $\mathbf{M}$ is said to be *non-negative* and we write $\mathbf{M} \geq \mathbf{0}$ if $m_{ij} \geq 0$ for all i and j.

At the beginning of this century interesting properties of non-negative matrices were discovered by O. Perron and G. Frobenius. Their results were extended and proved in a more elegant fashion by H. Wielandt (1950), whose account we follow here. We require a preliminary concept:

DEFINITION 7.11.1.   A square matrix $\mathbf{A}$ is said to be *decomposable* if it is of the form

$$\begin{pmatrix} \mathbf{B} & \mathbf{C} \\ \mathbf{0} & \mathbf{D} \end{pmatrix}, \tag{7.11.1}$$

where **B** and **D** are square matrices, not necessarily of the same order, or else if there exists a permutation matrix π [see Definition 6.7.1] such that $\pi'\mathbf{A}\pi$ is of the form (7.11.1). If no such permutation matrix exists, then **A** is called *indecomposable*.

Remark

This concept must not be confused with that of reducibility. The latter [see (7.3.14)] means that the shape (7.11.1) may be attained by a transformation $\mathbf{A} \rightarrow \mathbf{P}^{-1}\mathbf{AP}$, where **B** is an arbitrary invertible matrix.

It is clear that if **A** is decomposable, its eigenvalues are those of **B** and **D** taken together. For a study of eigenvalues it is therefore sufficient to restrict attention to indecomposable matrices.

EXAMPLE 7.11.1.   The matrix

$$\mathbf{A} = \begin{pmatrix} 1 & 1 & 2 \\ 0 & 1 & 0 \\ 2 & 3 & 1 \end{pmatrix}$$

is decomposable because if

$$\pi = \begin{pmatrix} 0 & 1 & 0 \\ 0 & 0 & 1 \\ 1 & 0 & 0 \end{pmatrix},$$

then

$$\pi'\mathbf{A}\pi = \begin{pmatrix} 1 & 2 & 3 \\ 2 & 1 & 1 \\ 0 & 0 & 1 \end{pmatrix}.$$

The main results are collected in the following theorem.

THEOREM 7.11.1.   (*Perron-Frobenius-Wielandt*). *Let* **A** *be an n by n non-negative indecomposable matrix. Then*

(i) **A** *possesses a simple positive eigenvalue* ρ.

(ii) *Every eigenvalue* α *of* **A** *satisfies* $|\alpha| \leq \rho$.

(iii) *The eigenvector* **u** *corresponding to* ρ *may be chosen in such a way that* **u** > 0.

(iv) ρ *is the only eigenvalue which is associated with a non-negative eigenvector.*

Suppose now that **A** *has precisely* k *eigenvalues of maximal modulus* [*see* (2.7.13)] *say* $\rho\ (=\rho_0),\ \rho_1, \ldots, \rho_{k-1}$. *Then*

(v) $\rho_h = \rho \exp(2\pi i h/k)\ (h = 0, 1, \ldots, k-1)$ [cf. (2.7.32)].

(vi) *When the* n *eigenvalues of* **A** *are plotted in the complex plane* [*see* § 2.7], *the configuration admits of a rotation about the origin through an angle of* $2\pi/k$ *but not through a smaller angle* [*cf. Figure* 2.7.11].

(vii) *There exists a permutation matrix [see Definition 6.7.1] (which may be the identity) such that*

$$\pi' A \pi = \begin{pmatrix} 0 & A_{12} & 0 & \dots & 0 \\ 0 & 0 & A_{23} & \dots & 0 \\ & & & & \\ 0 & 0 & 0 & \dots & A_{k-1,k} \\ A_{ki} & 0 & 0 & \dots & 0 \end{pmatrix}$$

is partitioned into blocks in such a way that the diagonal consists of square zero blocks, not necessarily of the same order.

The quantity ρ can be characterized by an interesting maximum-minimum principle, namely,

$$\rho = \max_{x>0} \min_{j} \frac{\sum_j a_{ij} x_j}{x_i},$$

where the quotient on the right is interpreted as $+\infty$ if $x_i = 0$.

The most important case is that in which $k = 1$. Such a matrix is called *primitive*. Thus a primitive non-negative matrix has precisely one eigenvalue of maximal modulus. This eigenvalue is positive and possesses a positive eigenvector.

The following criterion helps to decide whether a non-negative matrix is primitive.

PROPOSITION 7.11.2. *If $A \geq 0$, then A is primitive if and only if there exists a positive integer p such that*

$$A^p > 0.$$

EXAMPLE 7.11.2. Let

$$A = \begin{pmatrix} \frac{1}{2} & \frac{1}{3} & 0 \\ 0 & \frac{1}{3} & \frac{1}{2} \\ \frac{1}{2} & \frac{1}{3} & \frac{1}{2} \end{pmatrix}. \tag{7.11.2}$$

This is an instance of a *stochastic matrix* [see II, 19], that is a non-negative matrix in which the column-sums or, alternatively, the row-sums are equal to unity.

In the present case, it is obvious by inspection that $A^2 > 0$. Hence A is primitive by Proposition 7.11.2.

Let

$$e' = (1, 1, 1).$$

Then the fact that all column-sums of A are equal to unity may be expressed by the equation

$$e'A = e',$$

or, equivalently, by

$$\mathbf{A'e = e}.$$

Thus **e** is an eigenvector of **A'** with eigenvalue unity. Now **A** and **A'** have the same eigenvalues [see Proposition 7.2.4] though not necessarily the same eigenvectors. In particular, unity is an eigenvalue of **A**, and there exists a non-zero vector **u** such that

$$\mathbf{(I - A)u = 0}.$$

Solving this system of linear equations [see § 5.9] we find that a possible choice for **u** (or **u'**) is given by

$$\mathbf{u'} = (2, 3, 4).$$

Since this vector is positive we infer from Theorem 7.11.2(iv) that $\rho = 1$ is the maximal eigenvalue. As a matter of fact, the other two eigenvalues are $(1 \pm i\sqrt{2})/6$ and have modulus $(\sqrt{3})/6 \; (<1)$.

7.12.   SIMULTANEOUS DIAGONALIZATION

We shall briefly consider the following question: given a set of matrices

$$S = \{\mathbf{A, B, C}, \ldots\}, \tag{7.12.1}$$

[see Definition 6.4.2] under what conditions does there exist a non-singular matrix **P** such that each of the matrices

$$\mathbf{P^{-1}AP}, \qquad \mathbf{P^{-1}BP}, \qquad \mathbf{P^{-1}CP}, \ldots$$

is in diagonal form. Of course, we must suppose that all members of S are individually diagonalizable [see Definition 7.4.1]. Also, it should be noted that if the pair **A, B** commutes, so does the pair $\mathbf{P^{-1}AP}$, $\mathbf{P^{-1}BP}$, and conversely. Accordingly [see § 6.7(iv)] we must further postulate that the members of S commute in pairs. It turns out that these conditions are sufficient. We formulate the result for real symmetric matrices [see (6.7.5)]:

THEOREM 7.12.1.   *If the real symmetric matrices*

$$\mathbf{A, B, C}, \ldots$$

commute in pairs, then there exists a real orthogonal matrix **R** [*see* (6.7.9)] *such that each of the matrices*

$$\mathbf{R'AR}, \qquad \mathbf{R'BR}, \qquad \mathbf{R'CR}, \ldots$$

is in diagonal form.

A rather different problem of simultaneous diagonalization is concerned with two real symmetric matrices one of which is positive definite [see § 6.7(v)]. No assumption is made about commutativity, but in the following theorem the transforming matrix **Q** need no longer be orthogonal.

THEOREM 7.12.2. *Let* **A** *and* **B** *be real symmetric matrices and assume that* **A** *is positive definite. Then there exists a non-singular matrix* **Q** *such that*

$$Q'AQ = I, \qquad Q'BQ = D, \tag{7.12.2}$$

where **D** *is a diagonal matrix. (As in § 6.7(v), the matrix* **A** *is said to be positive definite if and only if the quadratic form* $x'Ax$ *is positive definite [see § 9.2].)*

The following technique may be employed for carrying out the reduction. Set up the *generalized eigenequation*

$$\det(tA - B) = 0. \tag{7.12.3}$$

It can be shown that the roots of this equation are real. For the moment let us assume that they are also distinct, say $\alpha_1 > \alpha_2 > \ldots > \alpha_n$.

For each α_i, there exists a non-zero vector u_i such that

$$\alpha_i A u_i = B u_i \quad (i = 1, 2, \ldots, n). \tag{7.12.4}$$

It will be found that, when $i \neq j$,

$$u_j' A u_i = 0.$$

Again, $u_i' A u_i > 0$ because **A** is positive-definite [see Definition 9.2.1]. Thus we may normalize the solution (differently from previous practice) by putting

$$q_i = (u_i' A u_i)^{-1/2} u_i.$$

Then

$$q_i' A q_j = \delta_{ij} \tag{7.12.5}$$

[see (6.7.1)] and

$$Q = (q_1, q_2, \ldots, q_n)$$

is the matrix which accomplishes the transformation (7.12.2). In fact,

$$D = \text{diag}(\alpha_1, \alpha_2, \ldots, \alpha_n).$$

When α is an m-fold root of (7.12.3), the solution of (7.12.4) includes m linearly independent vectors [cf. Theorem 5.9.1]. These may be chosen so as to satisfy (7.12.5).

It is often advantageous to replace (7.12.4) by the equivalent equations

$$(\alpha_i I - A^{-1}B)u_i = 0 \quad (i = 1, 2, \ldots, n). \tag{7.12.6}$$

The problem is now reduced to finding the eigenvalues and corresponding eigenvectors of the matrix

$$C = A^{-1}B,$$

which, however, might no longer be symmetric.

The procedure is similar to that explained in section 7.8, except that the eigenvectors are normalized by the conditions

$$q_i' A q_i = 1 \quad (i = 1, 2, \ldots, n). \tag{7.12.7}$$

EXAMPLE 7.12.1.   Let

$$A = \begin{pmatrix} 2 & 1 & 1 \\ 1 & 2 & 1 \\ 1 & 1 & 2 \end{pmatrix}, \qquad B = \begin{pmatrix} 0 & 1 & 0 \\ 1 & 0 & 0 \\ 0 & 0 & 0 \end{pmatrix}.$$

Then

$$C = A^{-1}B = \tfrac{1}{4} \begin{pmatrix} -1 & 3 & 0 \\ 3 & -1 & 0 \\ -1 & -1 & 0 \end{pmatrix}.$$

The eigenvalues of C are the roots of the equation

$$\det (tI - C) = t[(t + \tfrac{1}{4})^2 - (\tfrac{3}{4})^2] = 0,$$

that is

$$\alpha_1 = \tfrac{1}{2}, \qquad \alpha_2 = 0, \qquad \alpha_3 = -1.$$

For each α_i we solve (7.12.6) and denote by q_i the solution that also satisfies (7.12.7). It is found that

$$q'_1 = \tfrac{1}{2}(1, 1, -1)$$

$$q'_2 = \frac{1}{\sqrt{2}}(0, 0, 1)$$

$$q'_3 = \frac{1}{\sqrt{2}}(1, -1, 0).$$

Hence the transforming matrix, that is $Q = (q_1, q_2, q_3)$, becomes

$$Q = \tfrac{1}{2} \begin{pmatrix} 1 & 0 & \sqrt{2} \\ 1 & 0 & -\sqrt{2} \\ -1 & \sqrt{2} & 0 \end{pmatrix}.$$

Indeed it is easy to verify that

$$Q'AQ = I, \qquad Q'BQ = \text{diag}\,(\tfrac{1}{2}, 0, -1).$$

7.13.   INEQUALITIES FOR EIGENVALUES

Since by Theorem 7.2.3 the eigenvalues of the matrix $A = (a_{ij})$ are the roots of a polynomial of degree n, their precise determination will, as a rule, pose a difficult numerical problem. It is therefore of interest to establish simple inequalities which throw some light on the location of the eigenvalues in the complex plane. A good account of this topic can be found in L. Mirsky's text (1955), pp. 210–12; pp. 309–11, and we shall adopt his notation here.

It is convenient to introduce the numbers

$$R_k = |a_{k1}| + |a_{k2}| + \ldots + |a_{kk}| + \ldots + |a_{kn}| \qquad (7.13.1)$$

[see (2.7.13)] and

$$\rho_k = R_k - |a_{kk}| \tag{7.13.2}$$

($k = 1, 2, \ldots, n$), which are the kth absolute row-sum and the kth off-diagonal absolute row-sum respectively.

The following special type of matrix is relevant to our discussion:

DEFINITION 7.13.1. The n by n matrix $\mathbf{A} = (a_{ij})$ is said to be *diagonally dominant* if

$$|a_{kk}| > \rho_k \tag{7.13.3}$$

($k = 1, 2, \ldots, n$), that is if in each row the diagonal element dominates the sum of all the off-diagonal elements in absolute value.

The next proposition, which is easy to establish, expresses a basic fact.

PROPOSITION 7.13.1. *The determinant of a diagonally dominant matrix is never zero.*

Now if α is an eigenvalue of $\mathbf{A}$, then

$$\det (\alpha \mathbf{I} - \mathbf{A}) = 0. \tag{7.13.4}$$

It is clear that α cannot be so large that the matrix $\alpha \mathbf{I} - \mathbf{A}$ becomes diagonally dominant, as this would be incompatible with (7.13.4).

PROPOSITION 7.13.2. *If α is an eigenvalue of $\mathbf{A} = (a_{ij})$, then there exists at least one integer k, where $1 \leq k \leq n$, such that*

$$|\alpha - a_{kk}| \leq \rho_k.$$

Since $\mathbf{A}$ and $\mathbf{A}'$ have the same eigenvalues [see Proposition 7.2.4], analogous statements are valid about the columns of $\mathbf{A}$. Thus let

$$C_k = |a_{1k}| + \ldots + |a_{kk}| + \ldots + |a_{nk}|, \tag{7.13.5}$$

$$\gamma_k = C_k - |a_{kk}|$$

($k = 1, 2, \ldots, n$). Then we have

PROPOSITION 7.13.3. *If α is an eigenvalue of $\mathbf{A} = (a_{ij})$, there exists at least one integer k, where $1 < k < n$, such that*

$$|\alpha - a_{kk}| \leq \gamma_k.$$

The assertions of Propositions 7.13.2 and 7.13.3 are not strong unless the off-diagonal elements are small. For example, we can state that none of the eigenvalues of

$$\begin{pmatrix} c & \varepsilon & 0 & 0 \\ \varepsilon & c & \varepsilon & 0 \\ 0 & \varepsilon & c & \varepsilon \\ 0 & 0 & \varepsilon & c \end{pmatrix}$$

differs from c by more than $2|\varepsilon|$.

An even weaker inequality is contained in

PROPOSITION 7.13.4. *Let*

$$R = \max_{1 \le k \le n} R_k, \qquad C = \max_{1 \le k \le n} C_k,$$

where R_k and C_k are defined in (7.13.1) and (7.13.5). Then every eigenvalue of **A** *satisfies*

$$|\alpha| \le R \quad \text{and} \quad |\alpha| \le C.$$

Applying this result to the columns of (7.11.2), we find that $|\alpha| \le 1$, which confirms one part of the Perron-Frobenius-Wielandt theorem of section 7.11.

An inequality of a different type is obtained by applying the theory of Hermitian forms [see § 9.3]. When **A** is an arbitrary (complex) matrix of order m by n, then

$$h = \mathbf{z}^*\mathbf{A}^*\mathbf{A}\mathbf{z}$$

is a Hermitian form in the n indeterminates $z_1, z_2, \ldots, z_n$ [see (9.3.2)], where

$$\mathbf{z}' = (z_1, z_2, \ldots, z_n)$$

and

$$\mathbf{z}^* = (\bar{z}_1, \bar{z}_2, \ldots, \bar{z}_n)$$

is the complex conjugate of $\mathbf{z}'$ [see (2.7.10), and cf. (6.7.11)].

As we have indicated in Theorem 7.8.3, the theory of Hermitian forms is analogous to, and indeed a generalization of, the theory of real quadratic forms. In particular, if μ is the greatest eigenvalue of $\mathbf{A}^*\mathbf{A}$, then it follows

$$\mathbf{z}^*\mathbf{A}^*\mathbf{A}\mathbf{z} \le \mu\mathbf{z}^*\mathbf{z},$$

for every vector $\mathbf{z}$.

Now suppose that α is an eigenvalue of **A** and that $\mathbf{u}$ ($\ne \mathbf{0}$) is an eigenvector belonging to α. Then

$$\mathbf{A}\mathbf{u} = \alpha\mathbf{u}.$$

Taking the conjugate transpose we obtain that

$$\mathbf{u}^*\mathbf{A}^* = \bar{\alpha}\mathbf{u}^*,$$

whence

$$\mathbf{u^*A^*Au} = |\alpha|^2\mathbf{u^*u} \le \mu\mathbf{u^*u}.$$

Thus we have

PROPOSITION 7.13.5. *If α is an eigenvalue of $\mathbf{A}$ and if μ is the greatest eigenvalue of $\mathbf{A^*A}$, then*

$$|\alpha|^2 \le \mu.$$

W.L.

BIBLIOGRAPHY

Barnett, S. and Storey, S. (1970) *Matrix Methods in Stability Theory*, Nelson.

Cohn, P. M. (1958). *Linear Equations*, Routledge & Kegan Paul.

Hoffman, K. and Kunze, K. (1961). *Linear Algebra*, Prentice-Hall.

Marcus, M. and Minc, H. (1968), *Elementary Linear Algebra*, Macmillan.

Mirsky, L. (1955). *An Introduction to Linear Algebra*, Oxford University Press.

Morris, A. O. (1978). *Linear Algebra*, Van Nostrand Reinhold.

Tropper, A. M. (1969). *Linear Algebra*, Nelson.

Wielandt, H. (1950). Unzerlegbare, nicht negative Matrizen (*Mathematische Zeitschrift*, **52**, 643–648).

CHAPTER 8

Groups

INTRODUCTION

Some of the most significant mathematical concepts, in particular that of a group, arose from considerations of the set of *operators* of a given kind on a certain space. Operators (such as, for example, continuous functions whose domain is the vector space $\mathbb{R}^n$ [see IV, § 5.3]) may be regarded as entities in their own right in so far as they can be combined together to produce another entity of the same type. That is, if f and g are operators on the space V then so is fg, where the product is the composition of f and g, suitably defined (see, for example, the 'composition of functions' in (1.4.3)). Furthermore, this process of composition obeys certain general laws; for example, that of *associativity* which states that

$$f(gh) = (fg)h = fgh$$

[see, for example, (1.4.4)]. On the other hand a law that is not, in general, obeyed by composition of operators is that of *commutativity*. That is, for two operators f and g we have, in general, that

$$fg \neq gf.$$

Failure of the commutative law for operators may be illustrated by the two operators 'picking up the telephone receiver' and 'dialling the number'... it is clear that the end result is dependent on the order in which the operators are performed!

There are, in fact, four basic laws which are satisfied by the composition of operators in many important cases, and it was these four axioms (to be expounded in the second section of this chapter) which led to the development of the abstract concept of a group. In the first section of this chapter we illustrate these ideas with the set of all permutations of degree n, which are operators on a finite set of n elements.

8.1.   PERMUTATIONS

8.1.1.   The Algebra of Permutations

Let Ω be a set consisting of n distinct objects, which we often denote either by digits or by letters, say

$$\Omega = \{1, 2, \ldots, n\}.$$

Then a permutation on Ω is a mapping of Ω onto itself [see §§ 1.4.2]. More precisely:

DEFINITION 8.1.1.   A *permutation* π of degree n is an operation on the set $\{1, 2, \ldots, n\}$ which sends the digit i to the digit $i\pi$ in such a way that $i\pi \neq j\pi$ when $i \neq j$. We write

$$\pi = \begin{pmatrix} 1 & 2 & \ldots & n \\ 1\pi & 2\pi & \ldots & n\pi \end{pmatrix}, \tag{8.1.1}$$

where the numbers $1\pi, 2\pi, \ldots, n\pi$ are an arrangement of the numbers $1, 2, \ldots, n$.

Since there are $n!$ distinct arrangements of n objects [see (3.7.1)] there are $n!$ distinct permutation operations on the set $\{1, 2, \ldots, n\}$. When writing a permutation in the form (8.1.1) the order in which the numbers on the top row are written is irrelevant. For example, the permutation of degree 5 given by

$$\pi = \begin{pmatrix} 1 & 2 & 3 & 4 & 5 \\ 3 & 5 & 4 & 1 & 2 \end{pmatrix}$$

in which $1\pi = 3$, $2\pi = 5$, etc., may equally well be written in the form

$$\pi = \begin{pmatrix} 4 & 2 & 3 & 1 & 5 \\ 1 & 5 & 4 & 3 & 2 \end{pmatrix}.$$

There are other methods of writing permutations: one may simply regard a permutation as an arrangement, as in § 3.7, or one can employ the concise notation of Proposition 8.1.1. But the form (8.1.1) has been introduced to illustrate that a permutation is really a mapping, which can therefore be combined with another such permutation to form a product which is again a permutation of the same degree.

DEFINITION 8.1.2.   The *product* $\pi\rho$ of two permutations π and ρ of degree n is defined by applying first π and then ρ, so that $\pi\rho$ sends the digit i to the digit $i\pi\rho$. If

$$\rho = \begin{pmatrix} 1 & 2 & \ldots & n \\ 1\rho & 2\rho & \ldots & n\rho \end{pmatrix}$$

one prepares it for multiplication by π by writing it in the form

$$\rho = \begin{pmatrix} 1\pi & 2\pi & \ldots & n\pi \\ (1\pi)\rho & (2\pi)\rho & \ldots & (n\pi)\rho \end{pmatrix}$$

so that the first row of ρ is equal to the second row of π. Thus

$$\pi\rho = \begin{pmatrix} 1 & 2 & \dots & n \\ 1\pi & 2\pi & \dots & n\pi \end{pmatrix}\begin{pmatrix} 1\pi & 2\pi & \dots & n\pi \\ (1\pi)\rho & (2\pi)\rho & \dots & (n\pi)\rho \end{pmatrix}$$

$$= \begin{pmatrix} 1 & 2 & \dots & n \\ 1\pi\rho & 2\pi\rho & \dots & n\pi\rho \end{pmatrix}.$$

EXAMPLE 8.1.1. If

$$\pi = \begin{pmatrix} 1 & 2 & 3 & 4 \\ 3 & 1 & 2 & 4 \end{pmatrix}$$

and

$$\rho = \begin{pmatrix} 1 & 2 & 3 & 4 \\ 1 & 3 & 4 & 2 \end{pmatrix}$$

we write ρ in the form

$$\rho = \begin{pmatrix} 3 & 1 & 2 & 4 \\ 4 & 1 & 3 & 2 \end{pmatrix}$$

and then

$$\pi\rho = \begin{pmatrix} 1 & 2 & 3 & 4 \\ 3 & 1 & 4 & 2 \end{pmatrix}\begin{pmatrix} 3 & 1 & 4 & 2 \\ 4 & 1 & 3 & 2 \end{pmatrix} = \begin{pmatrix} 1 & 2 & 3 & 4 \\ 4 & 1 & 3 & 2 \end{pmatrix}.$$

Note that, in general, permutations are not commutative. That

$$\pi\rho \neq \rho\pi$$

for some permutations of the same degree may be illustrated by the permutations in the above example, since

$$\rho\pi = \begin{pmatrix} 1 & 2 & 3 & 4 \\ 1 & 3 & 4 & 2 \end{pmatrix}\begin{pmatrix} 1 & 3 & 4 & 2 \\ 3 & 2 & 4 & 1 \end{pmatrix} = \begin{pmatrix} 1 & 2 & 3 & 4 \\ 3 & 2 & 4 & 1 \end{pmatrix}$$

but

$$\pi\rho = \begin{pmatrix} 1 & 2 & 3 & 4 \\ 4 & 1 & 3 & 2 \end{pmatrix}.$$

However, it has already been observed that mappings obey the associative law of composition [see (1.4.4)] and so, in particular, any three permutations π, ρ and σ of degree n satisfy

$$(\pi\rho)\sigma = \pi(\rho\sigma) \qquad\qquad (8.1.2)$$

and such a product is written $\pi\rho\sigma$. Analogous results hold for the product of an arbitrary number of permutations of the same degree.

EXAMPLE 8.1.2

$$\left\{\begin{pmatrix} 1 & 2 & 3 & 4 \\ 2 & 3 & 4 & 1 \end{pmatrix}\begin{pmatrix} 1 & 2 & 3 & 4 \\ 3 & 4 & 1 & 2 \end{pmatrix}\right\}\begin{pmatrix} 1 & 2 & 3 & 4 \\ 2 & 3 & 1 & 4 \end{pmatrix}$$

$$= \begin{pmatrix} 1 & 2 & 3 & 4 \\ 4 & 1 & 2 & 3 \end{pmatrix}\begin{pmatrix} 1 & 2 & 3 & 4 \\ 2 & 3 & 1 & 4 \end{pmatrix} = \begin{pmatrix} 1 & 2 & 3 & 4 \\ 4 & 2 & 3 & 1 \end{pmatrix}$$

and

$$\begin{pmatrix} 1 & 2 & 3 & 4 \\ 2 & 3 & 4 & 1 \end{pmatrix}\left\{\begin{pmatrix} 1 & 2 & 3 & 4 \\ 3 & 4 & 1 & 2 \end{pmatrix}\begin{pmatrix} 1 & 2 & 3 & 4 \\ 2 & 3 & 1 & 4 \end{pmatrix}\right\}$$

$$= \begin{pmatrix} 1 & 2 & 3 & 4 \\ 2 & 3 & 4 & 1 \end{pmatrix}\begin{pmatrix} 1 & 2 & 3 & 4 \\ 1 & 4 & 2 & 3 \end{pmatrix} = \begin{pmatrix} 1 & 2 & 3 & 4 \\ 4 & 2 & 3 & 1 \end{pmatrix}.$$

We define the permutation π^2 (π 'squared') to be the product of π with itself and by property (8.1.2) we can meaningfully interpret the permutation π^r as the product of π with itself r times. The powers of a permutation π obey the following law of indices:

$$\pi^r \pi^s = \pi^{r+s}. \tag{8.1.3}$$

We shall now provide a more concise method of representing a permutation, and for this purpose we need the following definition:

DEFINITION 8.1.3.   A *cycle of length m* is a permutation of degree n of the form

$$\begin{pmatrix} A_1 & A_2 & A_3 & \ldots & A_{m-1} & A_m & A_{m+1} & \ldots & A_n \\ A_2 & A_3 & A_4 & \ldots & A_m & A_1 & A_{m+1} & \ldots & A_n \end{pmatrix}$$

where each A_i denotes any one of the digits $1, 2, \ldots, n$.

Thus a cycle of length m sends the digit A_i to the digit A_{i+1} for $i = 1, 2, \ldots, m - 1$, sends A_m to A_1 and leaves the digits A_j fixed for $j = m + 1, \ldots, n$. A cycle of length unity means that a digit is fixed, and a cycle of length two is called a *transposition*. Cycles of length m are usually written in the conventional form

$$(A_1, A_2, \ldots, A_m);$$

but, since cycles of length unity are not featured in such a representation, a little confusion may arise if the degree of the cycle is disregarded. Thus, for example,

$$(1 \quad 2 \quad 3 \quad 4) = \begin{pmatrix} 1 & 2 & 3 & 4 \\ 2 & 3 & 4 & 1 \end{pmatrix}$$

as a permutation of degree 4, but as a permutation of degree 5,

$$(1 \quad 2 \quad 3 \quad 4) = \begin{pmatrix} 1 & 2 & 3 & 4 & 5 \\ 2 & 3 & 4 & 1 & 5 \end{pmatrix}.$$

Two cycles $(A_1, A_2, \ldots, A_m)$ and $(B_1, B_2, \ldots, B_r)$ are *disjoint* if they have no element in common, so that $A_i \neq B_j$ for $i = 1, \ldots, m$ and $j = 1, \ldots, r$. We can now give a proposition which provides a convenient method of representing permutations.

PROPOSITION 8.1.1. *Any permutation π may be written as a product of disjoint cycles.*

This can be done as follows: starting with any digit, say 1, find its image under π, 1π. Then find the image $1\pi^2$ of 1π and continue the process until the digit 1 is reached as an image, say $1\pi^s = 1$. Incidently, the set of digits

$$1, 1\pi, 1\pi^2, \ldots, 1\pi^{s-1} \tag{8.1.4}$$

is called the *orbit* of 1 under π. Now start with another digit which is not in the orbit of 1 and continue until all digits have been exhausted. The distinct orbits which this process produces then constitute the disjoint cycles in the representation of π.

EXAMPLE 8.1.3

$$\begin{pmatrix} 1 & 2 & 3 & 4 & 5 & 6 & 7 \\ 6 & 4 & 7 & 5 & 2 & 1 & 3 \end{pmatrix}$$

$$= \begin{pmatrix} 1 & 6 & 2 & 4 & 5 & 3 & 7 \\ 6 & 1 & 4 & 5 & 2 & 7 & 3 \end{pmatrix}$$

$$= \begin{pmatrix} 1 & 6 & 2 & 3 & 4 & 5 & 7 \\ 6 & 1 & 2 & 3 & 4 & 5 & 7 \end{pmatrix} \begin{pmatrix} 2 & 4 & 5 & 1 & 3 & 6 & 7 \\ 4 & 5 & 2 & 1 & 3 & 6 & 7 \end{pmatrix} \begin{pmatrix} 3 & 7 & 1 & 2 & 4 & 5 & 6 \\ 7 & 3 & 1 & 2 & 4 & 5 & 6 \end{pmatrix}$$

$$= (1 \quad 6)(2 \quad 4 \quad 5)(3 \quad 7).$$

It has already been mentioned that permutations do not, in general, commute. However, disjoint cycles *do* commute and hence so do permutations whose representations in disjoint cycles have no element in common. For example, the permutations of degree ten

$$\pi = \begin{pmatrix} 1 & 2 & 3 & 4 & 5 & 6 & 7 & 8 & 9 & 10 \\ 4 & 2 & 1 & 3 & 5 & 8 & 7 & 6 & 9 & 10 \end{pmatrix}$$

$$= (1 \quad 4 \quad 3)(6 \quad 8)$$

and

$$\rho = \begin{pmatrix} 1 & 2 & 3 & 4 & 5 & 6 & 7 & 8 & 9 & 10 \\ 1 & 10 & 3 & 4 & 9 & 6 & 7 & 8 & 5 & 2 \end{pmatrix}$$

$$= (2 \quad 10)(5 \quad 9)$$

do commute.

PROPOSITION 8.1.2.   *A cycle of length m may be written as a product of m − 1 transpositions which, in general, are not disjoint. Neither will the factors in this product be unique, but in particular*

$$(A_1 A_2 \dots A_m) = (A_1 A_2)(A_1 A_3) \dots (A_1 A_m). \qquad (8.1.5)$$

PROPOSITION 8.1.3.   *A cycle σ of length m and degree n has the property*

$$\sigma^m = \iota$$

where

$$\iota = \begin{pmatrix} 1 & 2 & \dots & n \\ 1 & 2 & \dots & n \end{pmatrix}$$

is the identity permutation *of degree n.*

8.1.2. Properties of Permutations

We can now use the foregoing discussion to abstract certain properties that the operators 'permutations of degree n' have in their own right; that is, properties which are independent of the objects upon which the permutations act. These are as follows:

   (i) The set of all permutations of degree n is *closed* under the binary operation 'product of permutations'. That is to say that [see § 1.5] the product of two permutations of degree n is again a permutation of degree n.

  (ii) We have observed that

$$\pi(\rho\sigma) = (\pi\rho)\sigma = \pi\rho\sigma,$$

that is, composition of permutations obeys the associative law (Example 8.1.2 illustrates this fact).

 (iii) There exists a permutation ι, called the *identity permutation of degree n,* which leaves every digit fixed. That is

$$\iota = \begin{pmatrix} 1 & 2 & \dots & n \\ 1 & 2 & \dots & n \end{pmatrix}. \qquad (8.1.6)$$

 (iv) To every permutation π of degree n, given by

$$\pi = \begin{pmatrix} 1 & 2 & \dots & n \\ 1\pi & 2\pi & \dots & n\pi \end{pmatrix}$$

there exists a unique permutation π^{-1} of degree n, which may be written in the form

$$\pi^{-1} = \begin{pmatrix} 1\pi & 2\pi & \dots & n\pi \\ 1 & 2 & \dots & n \end{pmatrix}, \qquad (8.1.7)$$

having the property

$$\pi\pi^{-1} = \pi^{-1}\pi = \iota$$

where ι denotes the identity permutation of degree n. The permutation π^{-1} is called the *inverse* of π.

It is because of these four properties that the set of all permutations of degree n is said to form a *group* under the binary operation 'product of permutations' [see Definition 8.2.1] which we call the *full symmetric group of degree n* and usually denote by S_n. From the remark following Definition 8.1.1 we know that the *order* of this group [see Definition 8.2.3] is $n!$

EXAMPLE 8.1.4. There are six $(= 3!)$ different permutations of degree 3 which constitute the group S_3. These are

$$\iota = \begin{pmatrix} 1 & 2 & 3 \\ 1 & 2 & 3 \end{pmatrix}, \quad \begin{pmatrix} 1 & 2 & 3 \\ 2 & 3 & 1 \end{pmatrix}, \quad \begin{pmatrix} 1 & 2 & 3 \\ 3 & 1 & 2 \end{pmatrix},$$

$$\begin{pmatrix} 1 & 2 & 3 \\ 2 & 1 & 3 \end{pmatrix}, \quad \begin{pmatrix} 1 & 2 & 3 \\ 3 & 2 & 1 \end{pmatrix} \quad \text{and} \quad \begin{pmatrix} 1 & 2 & 3 \\ 1 & 3 & 2 \end{pmatrix}$$

or, in cyclic form,

$$\iota, \quad (1\ 2\ 3), \quad (1\ 3\ 2), \quad (1\ 2), \quad (1\ 3) \quad \text{and} \quad (2\ 3).$$

It is easy to verify that

$$(1\ 2\ 3)^2 = (1\ 3\ 2)$$

and that

$$(1\ 2\ 3)^3 = (1\ 2)^2 = (1\ 3)^\iota = (2\ 3)^2 = \iota.$$

8.1.3. Even and Odd Permutations

To introduce the concept of *parity* of a permutation we consider the action of a permutation of degree n on the expression

$$\Delta = (x_1 - x_2)(x_1 - x_3)\ldots(x_1 - x_n)$$
$$\times (x_2 - x_3)\ldots(x_2 - x_n)$$
$$\times (x_{n-1} - x_n)$$
$$= \prod_{1 \le i < j \le n} (x_i - x_j) \tag{8.1.8}$$

called the *difference product* of the indeterminates $x_1, \ldots, x_n$. Let π be a permutation of the n objects $x_1, \ldots, x_n$, so that

$$x_i \pi = x_{i\pi}$$

and we may write

$$\pi = \begin{pmatrix} x_1 & x_2 & \ldots & x_n \\ x_{1\pi} & x_{2\pi} & \ldots & x_{n\pi} \end{pmatrix}.$$

Denote by $\Delta\pi$ the difference product Δ of $x_1, \ldots, x_n$ taken after the indeterminates have been permuted by π so that

$$\pi : \Delta \mapsto \Delta\pi$$

where

$$\Delta\pi = (x_{1\pi} - x_{2\pi})(x_{1\pi} - x_{3\pi})\dots(x_{1\pi} - x_{n\pi})$$
$$\times (x_{2\pi} - x_{3\pi})\dots(x_{2\pi} - x_{n\pi})$$
$$\vdots$$
$$\times (x_{(n-1)\pi} - x_{n\pi}).$$

Then it is easy to verify that

$$\Delta\pi = \pm\Delta$$

and so we may write

$$\Delta\pi = \varepsilon(\pi)\Delta$$

where the function

$$\varepsilon(\pi) = \pm 1 \qquad\qquad\qquad (8.1.9)$$

is called the *alternating character* of π.

DEFINITION 8.1.4. The permutation π is *even* if $\varepsilon(\pi) = 1$ and *odd* if $\varepsilon(\pi) = -1$.

We now give two important properties of the function ε which will enable the parity of a permutation to be easily calculated. The first of these properties is a direct result of the definition of product of permutations and the second follows easily from Proposition 8.1.2:

PROPOSITION 8.1.4
(i) The function ε defined in (8.1.9) is a multiplicative function, that is

$$\varepsilon(\pi\rho) = \varepsilon(\pi)\varepsilon(\rho)$$

for any two permutations π and ρ of degree n.
(ii) The alternating character of a cycle of length m takes the value $(-1)^{m-1}$.

The parity of a permutation π can now be calculated as follows: decompose π into disjoint cycles, as

$$\pi = \sigma_1\sigma_2\dots\sigma_r$$

say. Then by property (i) above

$$\varepsilon(\pi) = \varepsilon(\sigma_1)\varepsilon(\sigma_2)\dots\varepsilon(\sigma_r). \qquad\qquad (8.1.10)$$

Now, if the cycle σ_i has length l_i, we have

$$\varepsilon(\sigma_i) = (-1)^{l_i-1} \qquad\qquad\qquad (8.1.11)$$

by property (ii), and together (8.1.10) and (8.1.11) will give the parity of π.

EXAMPLE 8.1.5.   Let

$$\pi = \begin{pmatrix} 1 & 2 & 3 & 4 & 5 & 6 & 7 & 8 & 9 & 10 \\ 3 & 1 & 2 & 7 & 8 & 10 & 5 & 4 & 6 & 9 \end{pmatrix}$$
$$= (1 \quad 3 \quad 2)(4 \quad 7 \quad 5 \quad 8)(6 \quad 10 \quad 9).$$

Then

$$\varepsilon(\pi) = (-1)^2(-1)^3(-1)^2 = (-1)^7 = -1,$$

so π is an odd permutation.

8.2.   GROUPS

8.2.1.   The Group Concept

In this section we use the properties of permutations set out in section 8.1.2 to formulate the abstract concept of a group. For this purpose we need the idea of a *binary operation* on a set [see § 1.5] which associates a *product* '*ab*' with each pair (a, b) of elements in the set. In symbols, the binary operation may be written

$$(a, b) \mapsto ab. \tag{8.2.1}$$

For example, in § 8.1.2 the set was the set of all permutations of degree n and the product was defined by product of permutations.

DEFINITION 8.2.1.   A *group* is a non-empty set of elements upon which a binary operation (8.2.1) has been defined which satisfies the following four axioms:

G1: The product of every pair of elements of G is again an element of G. In symbols,

$$ab \in G \quad \forall a, b \in G$$

and we say that G is *closed* under the binary operation (8.2.1).

G2: The operation (8.2.1) is *associative* on G, that is

$$(ab)c = a(bc) = abc \quad \forall a, b, c \in G.$$

G3: There exists a unique *identity* element e in G which satisfies

$$ae = ea = a \quad \forall a \in G.$$

G4: Every element $a \in G$ has a unique *inverse* $a^{-1} \in G$ which satisfies

$$aa^{-1} = a^{-1}a = e.$$

The notation ab which we used in Definition 8.2.1 for the product under the operation (8.2.1) is a conventional notation; we shall use this throughout the rest of this chapter with the exception of certain specific examples. For instance, in example 8.2.1 the binary operation is the usual operation of addition and the 'product' under this operation is written $a + b$. In the Example 8.2.1 and in general the additive notation for a group operation requires that the identity element be denoted by 0 and the inverse of an element a by $-a$.

The following are immediate consequences of the Definition 8.2.1:

PROPOSITION 8.2.1

(i) *Every group satisfies the left cancellation law*

$$ab = ac \Rightarrow b = c \tag{8.2.2}$$

for all elements a, b, c in the group, and likewise for cancellation on the right.

(ii) *Every linear equation of the form*

$$ax = b \qquad a, b \in G$$

has the unique solution

$$x = a^{-1}b$$

in the group. Similarly, if $ya = b$ then $y = ba^{-1}$ is the unique solution in the group. Note that $a^{-1}b$ is, in general, different from ba^{-1}, and so the notation b/a is not allowed.

(iii) *The inverse of the product ab of two group elements a and b is given by*

$$(ab)^{-1} = b^{-1}a^{-1}. \tag{8.2.3}$$

(iv) *The only solution of the equation*

$$x^2 = x$$

is

$$x = e.$$

That is, the identity element e is the only idempotent *element in the group.*

(v) *All powers, positive or negative, of any element $a \in G$ commute, that is*

$$a^k a^l = a^l a^k \quad l, k \in \mathbb{Z} \tag{8.2.4}$$

where $a^0 = e$, the identity element of G.

(vi) *The inverse of a power of an element $a \in G$ is equal to the power of the inverse a^{-1} of a. That is*

$$(a^m)^{-1} = (a^{-1})^m \tag{8.2.5}$$

and we denote this element by a^{-m}.

EXAMPLE 8.2.1.   The set $\mathbb{Z}$ of integers [see § 2.2.1] forms a group under the binary operation of addition defined by

$$(a, b) \mapsto a + b \qquad a, b \in \mathbb{Z}.$$

The integer 0 forms the identity element of this group and every non-zero integer n has the unique inverse $-n$. However, $\mathbb{Z}$ does not form a group under 'multiplication', defined by

$$(a, b) \mapsto ab \qquad a, b \in \mathbb{Z},$$

since, even though 1 acts as an identity element on this set, the number $1/n$ is not an integer when $n \in \mathbb{Z}$ and $n \neq \pm 1$.

EXAMPLE 8.2.2.   The set $\mathbb{C}^*$ of all non-zero complex numbers [see § 2.7.1] forms a group under the binary operation of multiplication of complex numbers defined by [see (2.7.5)]

$$(\alpha + i\beta, \gamma + i\delta) \mapsto (\alpha\gamma - \beta\delta) + i(\beta\gamma + \alpha\delta) \quad \alpha, \beta, \gamma, \delta \in \mathbb{R}.$$

For the (complex) number 1 acts as an identity, and the element $\alpha + i\beta \in \mathbb{C}^*$ has the unique inverse

$$(\alpha + i\beta)^{-1} = \frac{\alpha - i\beta}{\alpha^2 + \beta^2} = \frac{\alpha}{\alpha^2 + \beta^2} + i\left(\frac{-\beta}{\alpha^2 + \beta^2}\right)$$

in $\mathbb{C}^*$.

EXAMPLE 8.2.3.   The set of all $m \times n$ matrices over a field $\mathbb{F}$ [see § 6.2] forms a group under 'addition of matrices' [see § 6.2(ii)], in which the zero $m \times n$ matrix $0_{m,n}$ [see § 6.2(iii)] acts as an identity and the inverse of the matrix A is the $m \times n$ matrix $-A$.

EXAMPLE 8.2.4.   The set of all square matrices of degree n with entries in a field $\mathbb{F}$ [see § 6.2(vii)] which are also non-singular [see Definition 6.4.2] form a group under the binary operation of matrix multiplication [see § 6.2(vi)]. The unit matrix I_n of degree n is the identity and the inverse of the square matrix A is the matrix A^{-1} [see § 6.4]. We call this group the *general linear group of degree n* over $\mathbb{F}$ and denote it $GL_n(\mathbb{F})$.

EXAMPLE 8.2.5.   The set of all symmetry operations on a given object in $\mathbb{R}^3$ forms a group under 'composition of functions'. These groups are discussed in detail in Volume V, Chapter 11.

There is one particular class of groups called the *Abelian* groups for which an elaborate theory has been developed, but we do not enter into any details of this theory in this chapter. However, the 'Basis theorem for Abelian groups' [see Theorem 8.2.13] is mentioned, since many properties which are particular to Abelian groups follow from this fundamental theorem.

DEFINITION 8.2.2.   A group G is said to be *Abelian* if every pair of elements in G commute. Abelian groups are usually written additively, so that in symbols, G is Abelian if

$$a + b = b + a \quad \forall a, b \in G. \tag{8.2.6}$$

Accordingly, a non-Abelian group is one for which (8.2.6) does not hold for *every* pair of elements, although in every group there are always some elements which commute. For example, every element will commute with its inverse, by definition, and the identity element always commutes with every element of the group.

In this chapter all groups are assumed to be non-Abelian unless a statement is made to the contrary.

DEFINITION 8.2.3.   The number of distinct elements in a group G is called the *order* of G, and G is said to be *finite* or *infinite* according as this number is finite or infinite. If a finite group has order n, we write

$$|G| = n.$$

EXAMPLE 8.2.6.   The set of all integers $\mathbb{Z}$, the set of all non-zero complex numbers $\mathbb{C}^*$ and the set of all $m \times n$ matrices over an (infinite) field $\mathbb{F}$ all from infinite Abelian groups under the binary operations defined in Examples 8.2.1–8.2.3. The group $GL_n(\mathbb{F})$ of Example 8.2.4 is also infinite, if $\mathbb{F}$ is infinite, but it is non-Abelian since multiplication of matrices is non-commutative. Finally, the symmetry groups of Example 8.2.5 are usually finite and non-Abelian.

EXAMPLE 8.2.7.   The complete set of residues

$$\{0, 1, \ldots, n - 1\}$$

modulo n where n is a natural number [see § 2.1], forms an Abelian group of order n under the operation $+_n$ of ordinary addition followed by reduction modulo n [see I, § 4.3.2]. The element 0 is the identity and the inverse of an element a in this group is that number b such that $0 \le b \le n - 1$ and $b \equiv -a \pmod{n}$.

Similarly, the set of all column vectors of the form [see § 6.5]

$$\mathbf{a} = (a_1, a_2, \ldots, a_r)'$$

where each a_i is a residue modulo n forms an Abelian group of order n^r under the operation $\oplus_n$ consisting of the addition of two vectors followed by reduction of each component modulo n. Thus, if

$$\mathbf{b} = (b_1, b_2, \ldots, b_r)',$$

then

$$\mathbf{a} \oplus_n \mathbf{b} = (a_1 +_n b_1, a_2 +_n b_2, \ldots, a_r +_n b_r)'.$$

The zero vector

$$\mathbf{0} = (0, 0, \ldots, 0)'$$

is the identity element of this group and the vector $\mathbf{a}$ has a unique inverse in this group given by the vector

$$\mathbf{c} = (c_1, c_2, \ldots, c_r)'$$

where each c_i is a residue modulo n and

$$c_i \equiv -a_i \pmod{n}.$$

8.2.2. Group Tables

A convenient method of representing a finite group, which illustrates how a binary operation imposes the structure of a group upon the underlying set, is provided by the scheme of a *group table*, illustrated as follows

	g_1	g_2	$\cdots$	g_l	$\cdots$	g_n
g_1						
g_2						
$\vdots$						
g_k			$\cdots\cdots\cdots\cdots\cdots$	$g_k g_l$		
$\vdots$						
g_n						

Table 8.2.1

The first row and first column of such a table consist of the n elements of the finite group which are written in the same order, usually with the unit in the first place. Now we form an $n \times n$ array by taking all possible products

$$g_i g_j \qquad (i, j = 1, \ldots, n)$$

of group elements. From the closure axiom of Definition 8.2.1, we see that every element of this array is again an element of the group. Moreover, it follows from the cancellation law (8.2.2) that the elements

$$g_i g_1, g_i g_2, \ldots, g_i g_n$$

in any one row are all distinct. That is to say, each element occurs only once in every row and, equally, in every column, of the array. Consequently, we can regard each row (or column) of the array as a permutation of the n objects $g_1, g_2, \ldots, g_n$ which shows that a group table is a latin square [see V, § 9.5].

Note that a group table is symmetric about its main diagonal (top left to bottom right) if and only if the group is Abelian.

EXAMPLE 8.2.8. The group S_3 of all permutations of degree 3 [see Example 8.1.4] has the group table shown in Table 8.2.2.

	ι	(1 2 3)	(1 3 2)	(1 2)	(1 3)	(2 3)
ι	ι	(1 2 3)	(1 3 2)	(1 2)	(1 3)	(2 3)
(1 2 3)	(1 2 3)	(1 3 2)	ι	(2 3)	(1 2)	(1 3)
(1 3 2)	(1 3 2)	ι	(1 2 3)	(1 3)	(2 3)	(1 2)
(1 2)	(1 2)	(1 3)	(2 3)	ι	(1 2 3)	(1 3 2)
(1 3)	(1 3)	(2 3)	(1 2)	(1 3 2)	ι	(1 2 3)
(2 3)	(2 3)	(1 2)	(1 3)	(1 2 3)	(1 3 2)	ι

Table 8.2.2

8.2.3.   Generators and Relations

Another convenient way of representing groups is provided by their set of 'generators' and the 'relations' which these generators satisfy. Before discussing the general case we describe the generators and relations of the symmetric groups S_3 in a simple example.

EXAMPLE 8.2.9.   The six elements

$$\iota, \quad (1 \ \ 2 \ \ 3), \quad (1 \ \ 3 \ \ 2), \quad (1 \ \ 2), \quad (1 \ \ 3) \ \text{ and } \ (2 \ \ 3)$$

of S_3 satisfy the following equations (exhibited in Table 8.2.2 but which could also be verified directly):

$$(1 \ \ 2 \ \ 3)^2 = (1 \ \ 3 \ \ 2)$$
$$(1 \ \ 2)(1 \ \ 2 \ \ 3) = (1 \ \ 3)$$
$$(1 \ \ 2)(1 \ \ 3 \ \ 2) = (2 \ \ 3)$$

and

$$(1 \ \ 2 \ \ 3)^3 = (1 \ \ 2)^2 = (1 \ \ 3)^2 = \iota. \tag{8.2.7}$$

Therefore, if we denote by a the permutation $(1 \ \ 2 \ \ 3)$ and by b the permutation $(1 \ \ 2)$, the elements of S_3 may be written in the form

$$\iota, \quad a, \quad a^2, \quad b, \quad ba \ \text{ and } \ ba^2. \tag{8.2.8}$$

Moreover, there exists no expression for a in terms of b or vice versa, since if there were, the entire set of elements of S_3 could be expressed in terms of powers of one of these elements, a say, which by (8.2.6) would contradict the fact that S_3 is non-Abelian.

We have shown that the whole of S_3 can be generated by taking powers and products of two of its elements a and b and, furthermore, it follows from (8.2.7) that the elements (8.2.8) are all possible powers and products of a and b. We say that S_3 is 'generated' by a and b, with defining 'relations' (8.2.7). (Incidentally, S_3 may also be generated by $(1 \ \ 3 \ \ 2)$ and $(1 \ \ 3)$, with defining relations $(1 \ \ 3 \ \ 2)^3 = (1 \ \ 3)^2 = (2 \ \ 3)^2 = 2$, since

$$(1 \ \ 3 \ \ 2)^2 = (1 \ \ 2 \ \ 3)$$

and

$$(1 \ \ 3)(1 \ \ 3 \ \ 2) = (1 \ \ 2)$$

so that both a and b, and therefore the whole of S_3, can also be expressed in terms of these two elements.) Thus we write S_3 in terms of its generators and relations, as

$$S_3 = \{a, b | a^3 = b^2 = (ab)^2 = e\}.$$

In the general case, consider a subset

$$S = \{a_1, a_2, a_3, \ldots\}$$

of elements of an arbitrary group G. If G is infinite, S may be either finite or infinite. We say that G is *generated* by S if every element of G can be expressed as a product of a finite number of elements, and inverses of elements, of S, possibly with repetition of the factors. Formally:

DEFINITION 8.2.4.   The subset

$$S = \{a_1, a_2, a_3, \ldots\}$$

of elements of a group G is said to constitute a set of *generators* of G if

$$g = a_{p_1}{}^{\alpha_1} a_{p_2}{}^{\alpha_2} a_{p_3}{}^{\alpha_3} \ldots a_{p_k}{}^{\alpha_k} \quad \forall g \in G$$

where each a_{p_i} is any one of the elements $a_1, a_2, a_3, \ldots$ of S, and $\alpha_i = \pm 1$ for $i = 1, 2, \ldots, k$. We write

$$G = \text{gp}\,\{a_1, a_2, a_3, \ldots\}.$$

If, furthermore, no element of S can be expressed in terms of the other elements of S, we call S an *irredundant* set of generators of G.

DEFINITION 8.2.5.   A group G is said to be *finitely generated* if it is generated by a finite set.

Clearly a finite group is finitely generated, but the converse does not always hold: in fact a finitely generated group has finite order only when the elements of an irredundant set of generators S satisfy certain conditions, called the *relations* of the group. Supposing that

$$S = \{a_1, a_2, \ldots, a_m\}$$

such a relation on the elements of S might typically be

$$a_1 a_2 a_m{}^{-1} a_3{}^{-1} = e,$$

where e denotes the identity element of G. If we use the convention

$$R(a_1, a_2, \ldots, a_m) = e$$

to denote the complete set of relations of a finitely generated group G, we can describe the group as

$$G = \{a_1, a_2, \ldots, a_m \,|\, R(a_1, a_2, \ldots, a_m) = e\}.$$

EXAMPLE 8.2.10 (Cyclic groups).   *Cyclic* groups are those groups which are generated by a single element, say r. If the group is to be finite this generator must satisfy a relation of the form

$$r^n = e,$$

where e is the identity element and $n \in \mathbb{N}$. The group corresponding to this relation is called the *cyclic group of order* n; written

$$C_n = \{r \,|\, r^n = e\},$$

and its elements may be listed as

$$e, r, r^2, \ldots, r^{n-1}.$$

Note that we need include no elements of the form r^{-m} in the above list, since the relation

$$r^n = e$$

implies that $r^{-1} = r^{n-1}$. In fact this is true of any finite group. For if $a \ (\neq e)$ is an element of a finite group, the elements

$$e, a, a^2, \ldots$$

cannot all be distinct so that there must be a relation of the form

$$a^l = a^k, \quad l > k$$

or

$$a^n = e \quad \text{where} \quad n = l - k.$$

This implies that the inverse a^{-1} of a, whence by (8.2.5) inverses of all elements in a finite group, can be expressed in terms of positive powers of its generators.

The group table of C_n is given in Table 8.2.3.

	e	r	r^2	r^3	$\ldots$	r^{n-1}
e	e	r	r^2	r^3	$\ldots$	r^{n-1}
r	r	r^2	r^3	r^4	$\ldots$	e
r^2	r^2	r^3	r^4	r^5	$\ldots$	r
r^3	r^3	r^4	r^5	r^6	$\ldots$	r^2
$\vdots$	$\vdots$	$\vdots$	$\vdots$	$\vdots$		$\vdots$
r^{n-1}	r^{n-1}	e	r	r^2	$\ldots$	r^{n-2}

Table 8.2.3

The *infinite cyclic group*, C_∞ is generated by a single element r which satisfies no relations. Accordingly, the elements of C_∞ are all those elements of the form

$$r^\alpha, \quad \alpha \in \mathbb{Z},$$

that is, the elements of C_∞ are

$$\{\ldots r^{-2}, r^{-1}, e, r, r^2, r^3, \ldots\}$$

and we write

$$C_\infty = \text{gp}\ \{r\}.$$

EXAMPLE 8.2.11 (Dihedral groups).   There is far more than one possibility for a finite group generated by two elements. The *dihedral group* D_n is that group generated by two elements, say a and b, which satisfy the relations $a^n = b^2 = (ab)^2 = e$

$$D_n = \{a, b | a^n = b^2 = (ab)^2 = e\}.$$

Note that $|D_n| = 2n$. For example, the dihedral group D_3 of order 6 has the group table shown in Table 8.2.4.

	e	a	a^2	b	ba	ba^2
e	e	a	a^2	b	ba	ba^2
a	a	a^2	e	ba^2	b	ba
a^2	a^2	e	a	ba	ba^2	b
b	b	ba	ba^2	e	a	a^2
ba	ba	ba^2	b	a^2	e	a
ba^2	ba^2	b	ba	a	a^2	e

Table 8.2.4

In Volume V, Chapter 11, it is shown how the dihedral group D_n arises as the symmetry group of a regular n-sided polygon [see V, § 1.1.4]. We illustrate this method by obtaining the group D_6 of a regular hexagon as follows.

Label the vertices of a regular hexagon by the letters A, B, C, D, E and F and 'represent' the original position of the hexagon by the digits $1, 2, \ldots, 6$, as shown in Figure 8.2.1.

A symmetry operation on the hexagon is an operation which moves the hexagon but leaves it in a position which is indistinguishable from the original position (labels apart). Thus a symmetry operation interchanges certain vertices of the hexagon, and therefore we can represent it by a permutation of the digits $1, 2, \ldots, 6$ [see Definition 8.1.1], which corresponds to the new positions of the vertices. For example, the operation r of rotation, about an axis passing through the centre 0 of the hexagon, through an angle of $\pi/3$ in an anti-clockwise direction, moves vertex A from position 1 to position 2, vertex B from 2 to 3, C from 3 to 4, and so on, as illustrated in Figure 8.2.2.

Thus r may be represented by the permutation

$$r = \begin{pmatrix} 1 & 2 & 3 & 4 & 5 & 6 \\ 2 & 3 & 4 & 5 & 6 & 1 \end{pmatrix} = (1 \quad 2 \quad 3 \quad 4 \quad 5 \quad 6).$$

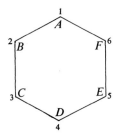

Figure 8.2.1.

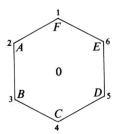

Figure 8.2.2.

Hence

$$r^2 = \begin{pmatrix} 1 & 2 & 3 & 4 & 5 & 6 \\ 3 & 4 & 5 & 6 & 1 & 2 \end{pmatrix} = (1 \quad 3 \quad 5)(2 \quad 4 \quad 6)$$

and

$$r^3 = (1 \quad 4)(2 \quad 5)(3 \quad 6)$$
$$r^4 = (1 \quad 5 \quad 3)(2 \quad 6 \quad 4)$$
$$r^5 = (1 \quad 6 \quad 5 \quad 4 \quad 3 \quad 2)$$

and

$$r^6 = \iota,$$

the identity permutation which leaves every vertex fixed. The operation r represents one of the generators of D_6 in this example. The other generator is provided by a symmetry operation of a different kind, whose effect on the hexagon is illustrated in Figure 8.2.3, the shade being used to indicate that the 'opposite' face of the lamina is now turned upward. This operation, m, may therefore be regarded as a 'flip' about an axis passing through the vertices A and D.

Thus we may write

$$m = \begin{pmatrix} 1 & 2 & 3 & 4 & 5 & 6 \\ 1 & 6 & 5 & 4 & 3 & 2 \end{pmatrix} = (2 \quad 6)(3 \quad 5)$$

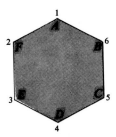

Figure 8.2.3.

whence

$$m^2 = \iota.$$

It can be shown that all other symmetry operations on this figure may be expressed as a product of a power of m with a power of r, so that m and r together generate the whole group. It is also easy to verify that

$$mr = r^5 m$$

$$mr^2 = r^4 m$$

$$\vdots$$

$$mr^5 = rm.$$

We can use the relations

$$r^6 = m^2 = \iota$$

to replace the above five equations by the relations

$$(mr^k)^2 = \iota \quad \text{for } k = 1, 2, \ldots, 5,$$

and it is easy to see that these relations are equivalent to having the single relation

$$(mr)^2 = \iota.$$

For, if we assume that $(mr)^2 = \iota$ we may use the relations

$$m^2 = \iota$$

to infer that

$$rm = mr^{-1}$$

and so

$$r^k m = r^{k-1}(rm) = r^{k-1}mr^{-1} = \ldots = mr^{-k} \quad \forall k \in \mathbb{Z}$$

so that

$$(mr^k)^2 = m(r^k m)r^k = m^2 r^{-k} r^k = \iota.$$

Thus a complete set of relations for r and m is given by

$$r^6 = m^2 = (mr)^2 = \iota$$

and therefore we can write D_6 in the form

$$D_6 = \{r, m | r^6 = m^2 = (rm)^2 = \iota\}.$$

8.2.4. Subgroups and Cosets

DEFINITION 8.2.6.   Let G be a group and G' a subset of G [see § 1.2.3]. If G' also forms a group under the binary operation of G it is called a *subgroup* of G and written

$$G' \leq G.$$

If G' is a proper subset of G consisting of more than one element, then G' is called a *proper subgroup* of G and written

$$G' < G.$$

EXAMPLE 8.2.12.   The dihedral group

$$D_4 = \{a, b \mid a^4 = b^2 = (ab)^2 = e\}$$

of order eight has the proper subgroup

$$\{e, a, a^2, a^3\}$$

which is a cyclic group of order four, and proper subgroups

$$\{e, b\}, \qquad \{e, ab\}, \quad \text{and} \quad \{e, a^2 b\}$$

which are cyclic of order 2. The improper subgroups of D_4 are, of course, the trivial group $\{e\}$ and D_4 itself.

EXAMPLE 8.2.13.   The group S_3 of permutations of degree three has four proper subgroups: one which is cyclic of order 3, given by

$$\{\iota, (1 \quad 2 \quad 3), (1 \quad 3 \quad 2)\},$$

and three that are cyclic of order 2;

$$\{\iota, (1 \quad 2)\}, \quad \{\iota, (1 \quad 3)\} \quad \text{and} \quad \{\iota, (2 \quad 3)\}.$$

The concept of a subgroup of a group is associated with that of a *coset* in the following way:

DEFINITION 8.2.7.   Suppose that $G' = \{a_1, a_2, a_3, \ldots\}$ is a subgroup of a (not necessarily finite) group G. Let a be an arbitrary element of G. Then the sets

$$aG' = \{aa_1, aa_2, aa_3, \ldots\}$$

and

$$G'a = \{a_1 a, a_2 a, a_3 a, \ldots\}$$

are called the *left-coset* and *right-coset* of G' in G, belonging to the element $a \in G$. Note that the coset of a subgroup is a set and not necessarily a subgroup itself.

Later, in section 8.2.6, we shall be considering the case when, for a given element $a \in G$ and a given subgroup G' of G, the left-coset belonging to a is the same as the right-coset. But for the present we look into the case when, for a given subgroup $G' < G$ and for two given elements a and b of G, the two left-cosets aG' and bG' which they define are equal. Now, it is a straightforward verification that the statement

$$\text{'}a \sim b \quad \text{if} \quad aG' = bG\text{'',}$$

defines an equivalence relation [see § 1.3.3] on G. Since any equivalence relation on a set divides this set into disjoint equivalent classes [see (1.3.5)] we have that two left-cosets are either disjoint or equal, that is either

$$aG' \cap bG' = \emptyset$$

or

$$aG' = bG'.$$

A similar argument shows that G may also be decomposed into disjoint equivalence classes of right-cosets. Clearly the group G may be expressed as a union of cosets, one taken from each equivalence class. This is expressed formally in the following:

PROPOSITION 8.2.2.   *Let G' be a subgroup of the group G. Then there exists a set $\{b_1, b_2, \ldots\}$ and a set $\{c_1, c_2, \ldots\}$ both of elements of G, having the properties*

$$G = b_1 G' \cup b_2 G' \cup \ldots \tag{8.2.8}$$

and

$$G = G' c_1 \cup G' c_2 \cup \ldots \tag{8.2.9}$$

where the left-cosets in (8.2.8) are mutually disjoint, as are the right-cosets in (8.2.9).

In both (8.2.8) and (8.2.9) one of the cosets will have to be G' itself and so, if we assume that

$$b_i G' = G' c_j = G' \quad \text{for some } i, j,$$

we must take $b_i, c_j \in G'$. In fact it is usual to take

$$b_1 = c_1 = e$$

where e is the identity element of G.

DEFINITION 8.2.8.   We call the sets $\{b_1, b_2, \ldots\}$ and $\{c_1, c_2, \ldots\}$ a *left-transversal* and a *right-transversal* of G' in G, respectively. Accordingly the decompositions (8.2.8) and (8.2.9) are called *left-coset decompositions* and *right-coset decompositions* of G' in G.

Now, if there are a finite number of disjoint left-cosets in (8.2.8) then there are the same number of disjoint right-cosets in (8.2.9) and conversely [see Ledermann (1973), p. 34]. If this number is k say, we say that the subgroup G' has *finite index* in G and k is called the *index* of G' in G, written

$$[G:G'] = k. \tag{8.2.10}$$

In the case that G is a finite group, all subgroups must obviously have finite index. If $|G| = n$ and $|G'| = m$ then every coset of G' has m elements and so

$$n = km$$

and consequently we have

THEOREM 8.2.3 (*Lagrange's theorem*).   *The order of every subgroup of a finite group divides the order of the group.*

EXAMPLE 8.2.14 (Subgroups of a cyclic group)
(i) All subgroups of C_n, the cyclic group of order n given by

$$C_n = \{r \,|\, r^n = e\}$$

are cyclic, of order dividing n, and
(ii) If $m|n$ then there exists exactly one subgroup of C_n of order m, namely that generated by r^k, where $k = n/m$. Thus if $s = r^k$ the sole subgroup of C_n of order m consists of the elements

$$e, s, s^2, \ldots, s^{m-1}.$$

8.2.5.   Properties of Group Elements

The first property we consider is a consequence of Theorem 8.2.3 (Lagrange's theorem). First we need the definition:

DEFINITION 8.2.9.   Let G be a finite group and let $g \in G$. The *period* (or *order*) of the element g, denoted $o(g)$, is the least positive power of g which is equal to the unit e of G, that is

$$g^{o(g)} = e \qquad\qquad (8.2.11)$$

and $o(g)$ is the least positive power of g for which (8.2.11) holds. In fact it can be shown that

$$o(g)|x \quad \text{if } g^x = e.$$

Also, since the element g generates a cyclic subgroup of G of order $o(g)$, Lagrange's theorem implies that

$$o(g) \,\big|\, |G| \quad \forall g \in G.$$

The second property that we consider in this section defines an equivalence relation on the elements of a group [see § 1.3.3].

DEFINITION 8.2.10.   The elements a and b of a group G are said to be *conjugate in G* if there exists some element $g \in G$ with the property

$$g^{-1}ag = b. \qquad\qquad (8.2.12)$$

That 'is conjugate to' is an equivalence relation is straightforward to verify. As with any equivalence relation on a set, (8.2.12) breaks G up into disjoint classes [see (1.3.5)], which we call the *conjugacy classes* of G, so that the conjugacy class (a) of the element $a \in G$ is the set of all those elements of G into which a may be transformed, under the *conjugacy transform* given by (8.2.12). Thus G may be written as a union of all its distinct conjugacy classes. In fact, it can be shown that the number of elements in every conjugacy class of a finite group divides the order of the group.

EXAMPLE 8.2.14.   The dihedral group D_4 of order eight has elements

$$e, a, a^2, a^3, b, ba, ba^2, ba^3,$$

and it is not difficult to verify that

$$(e) = \{e\}$$
$$(a) = \{a, a^3\} = (a^3)$$
$$(a^2) = \{a^2\}$$
$$(b) = \{b, ba^2\} = (ba^2)$$

and

$$(ba) = \{ba, ba^3\} = (ba^3).$$

Note that two elements which are conjugate in a group G do not necessarily remain conjugate in a subgroup of G.

PROPOSITION 8.2.4.   *Two permutations of degree n are conjugate in S_n [see § 8.1.1] if and only if they have the same cycle pattern.*

As an illustration see Example 8.2.19.
   Finally we give

DEFINITION 8.2.11.   The *centre* $Z(G)$ of a group G is the set of all elements which commute with each element of G:

$$Z(G) = \{a \in G \mid ab = ba, \forall b \in G\}.$$

Clearly $Z(G) \leq G$ and $Z(G) = G$ if and only if G is Abelian. The elements of $Z(G)$ are characterized by the fact that they are clearly invariant under the conjugacy transform (8.2.12). Thus each element of $Z(G)$ forms a conjugacy class on its own and for this reason elements in the centre of a group are sometimes called *self-conjugate* elements of the group.

EXAMPLE 8.2.15.   It follows from Example 8.2.14 that

$$Z(D_4) = \{e, a^2\}.$$

8.2.6.   Normal Subgroup and Factor Groups

In this section we consider the case where, for a given subgroup $G' < G$ and any element $a \in G$, the left- and right-cosets of G' in G belonging to a are equal, in which case they are referred to simply as *cosets* of G' and G and written in the same way as either left- or right-cosets. Correspondingly, the left- and right-transversals in Definition 8.2.8 may also be taken as equal and shall be referred to as the *transversals* of G' and G. In such a case the subgroup G' is called a *normal subgroup* of G. We formulate this definition more precisely in

DEFINITION 8.2.12.   The subgroup G' of G is said to be *normal in G*, and written

$$G' \triangleleft G,$$

if $aG' = G'a$ for every element $a \in G$.

EXAMPLE 8.2.16.   All subgroups of an Abelian group (and in particular of a cyclic group) are normal.

EXAMPLE 8.2.17.   If $[G:G'] = 2$ then $G' \triangleleft G$.

EXAMPLE 8.2.18.   In every group G the trivial group $\{e\}$ and the group G itself are (improper) normal subgroups. If, however, a group has no other normal subgroups (that is, if it has no proper normal subgroups) then it is called *simple*.

   Notice that 'is a normal subgroup of' is not a transitive relation, in contrast to the relation 'is a subgroup of'. That is,

$$A \triangleleft B \triangleleft C \quad \text{does not imply that} \quad A \triangleleft C$$

but

$$A \leq B \leq C \quad \text{does imply that} \quad A \leq C.$$

We give two other ways of looking at normal subgroups.

DEFINITION 8.2.13.   Two subgroups A and B of a group G are said to be *conjugate* if there exists some element $g \in G$ such that the sets $g^{-1}Ag$ and B are equal:

$$g^{-1}Ag = B. \tag{8.2.13}$$

   Thus normal subgroups are transformed into themselves under a conjugacy transform such as (8.2.13). For this reason normal subgroups are sometimes called the *self-conjugate* subgroups of a group. Thus it is not difficult to see that a normal subgroup consists of *complete* conjugacy classes of the group. More precisely:

PROPOSITION 8.2.5.   *$G' \triangleleft G$ if and only if G' is a subgroup which is the union of complete conjugacy classes in G.*

EXAMPLE 8.2.19.   The full symmetric group S_4 [see § 8.1.1] may be written as the union of distinct conjugacy classes:

$$S_4 = (\iota) \cup ((1\ \ 2)) \cup ((1\ \ 2\ \ 3)) \cup ((1\ \ 2\ \ 3\ \ 4)) \cup ((1\ \ 2)(3\ \ 4))$$

where

$$(\iota) = \iota$$

$$((1 \quad 2)) = \{(1 \quad 2), (1 \quad 3), (1 \quad 4), (2 \quad 3), (2 \quad 4), (3 \quad 4)\}$$

$$((1 \quad 2 \quad 3)) = \{(1 \quad 2 \quad 3), (1 \quad 3 \quad 2), (1 \quad 2 \quad 4), (1 \quad 4 \quad 2),$$

$$(1 \quad 3 \quad 4), (1 \quad 4 \quad 3), (2 \quad 3 \quad 4), (2 \quad 4 \quad 3)\}$$

$$((1 \quad 2 \quad 3 \quad 4)) = \{(1 \quad 2 \quad 3 \quad 4), (1 \quad 3 \quad 2 \quad 4), (1 \quad 4 \quad 2 \quad 3),$$

$$(1 \quad 4 \quad 3 \quad 2), (1 \quad 2 \quad 4 \quad 3), (1 \quad 3 \quad 4 \quad 2)\}$$

and

$$((1 \quad 2)(3 \quad 4)) = \{(1 \quad 2)(3 \quad 4), (1 \quad 3)(2 \quad 4), (14)(23)\}.$$

We look for a union of complete classes, including the class (ι), so that the total number of elements divides 24 (by Theorem 8.2.2). In this way it may be verified that S_4 has two normal subgroups:

$$A_4 = (\iota) \cup ((1 \quad 2 \quad 3)) \cup ((1 \quad 2)(3 \quad 4))$$

and

$$V_4 = (\iota) \cup ((1 \quad 2)(3 \quad 4))$$

which is isomorphic to the Klein four group [see Example 8.2.24].

DEFINITION 8.2.14.   Let $G' \leq G$. The *normalizer*, $N(G')$, of G' in G, is defined as the set of those elements in G whose left-cosets and right-cosets of G' and G are equal. That is

$$N(G') = \{a \in G \mid aG' = G'a\}.$$

Clearly $G' \lhd N(G') \leq G$ and $N(G') = G$ if and only if $G' \lhd G$. Thus $N(G')$ measures the 'extent to which G' is a normal subgroup of G'.

EXAMPLE 8.2.20.   It follows from Definition 8.2.1 that $N(Z(G)) = G$, whence

$$Z(G) \lhd G.$$

An important property of normal subgroups is that the complete set of their cosets forms a group under a binary operation to be defined presently. But first, if A and B are any subsets of a group we define the product AB as the set of all elements of the form ab, where $a \in A$, $b \in B$. It is not hard to show that

$$(AB)C = A(BC).$$

When A is a subgroup, the closure axiom of Definition 8.2.1 implies that

$$A^2 = A \tag{8.2.14}$$

Suppose now that $G' \triangleleft G$ so that $aG' = G'a$. This enables us to define a binary operation on the set of all cosets of G' in G by

$$(aG')(bG') = ab(G')^2 = abG'. \tag{8.2.15}$$

Indeed, the above rule is 'well defined': for if

$$aG' = a_1 G' \quad \text{and if} \quad bG' = b_1 G'$$

then

$$abG' = a_1 b_1 G'$$

so that the product is independent of the choice of representative for each coset. Also, the operation (8.2.15) is associative on the set of all cosets of G' in G since

$$aG'(bG'cG') = aG'(bcG') = abcG'$$

and

$$(aG'bG')cG' = (abG')cG' = abcG'.$$

Thirdly,

$$(G')^2 = G'$$

since the coset G' is a group and so by (iv), Proposition 8.2.1, G' acts as an identity on the set of all its cosets. Finally,

$$aG'a^{-1}G' = G'$$

where a^{-1} denotes the inverse of the element a in G, so that the coset aG' has a unique inverse coset given by $a^{-1}G$. We have therefore shown the following:

THEOREM 8.2.6.   *Let G be a group, $N \triangleleft G$ and $[G:N] = r$. If $n_1, \ldots, n_r$ is a transversal of N in G then the set*

$$\{n_1 N, \ldots, n_r N\}$$

forms a group, of order r, under the binary operation defined in (8.2.15). This group is called the quotient *or* factor group *of N in G and denoted G/N.*

8.2.7.   Homomorphisms between Groups

In this section we take a brief look at the various kinds of *structure-preserving maps* that can exist between two groups. Such maps are defined as follows:

DEFINITION 8.2.15.   Let A and B be two groups. The map

$$\theta : A \to B$$

is said to be a *homomorphism* of A into B if

$$\theta(a_1 a_2) = \theta(a_1)\theta(a_2) \quad \forall a_1, a_2 \in A \tag{8.2.16}$$

where the products on each side of (8.2.16) are defined by the respective group operations [see § 1.4.1]. If we put $a_1 = a_2 = e_A$, the unit element of A, in the above equation we have that

$$\theta(e_A) = \theta(e_A)\theta(e_A)$$

so that $\theta(e_A) = e_n$, the unit element B, for every group homomorphism $A \rightarrow B$. The group B is said to be *homomorphic* to A whenever there exists a homo- morphism $A \rightarrow B$.

DEFINITION 8.2.16. The *kernel* of a group homomorphism $\theta : A \rightarrow B$, written ker θ, consists of all those elements of A which are mapped to the unit of B under θ. Thus

$$\text{ker } \theta = \{a \in A \mid \theta(a) = e_B\}$$

and it is easy to verify that

$$\text{ker } \theta \leq A.$$

(In fact we can show that ker $\theta \lhd A$: see Theorem 8.2.8 below.)

DEFINITION 8.2.17. The *image group* of a group homomorphism $\theta: A \rightarrow B$ written im θ, consists of all elements of B which are images of elements of A under θ. Thus

$$\text{im } \theta = \{b \in B \mid b = \theta(a), a \in A\}$$

and again, it is easy to show that

$$\text{im } \theta \leq B.$$

(But im θ is not generally a normal subgroup of B.)

PROPOSITION 8.2.6. *The group homomorphism* $\theta: A \rightarrow B$ *is injective if and only if* ker $\theta = \{e_A\}$ *and surjective if and only if* im $\theta = B$ [see § 1.4.2].

DEFINITION 8.2.18. A homomorphism is called a *monomorphism, epi- morphism* or *isomorphism* according as whether it is injective, surjective or bijective respectively [see § 1.4.2].

DEFINITION 8.2.19. Two groups A and B are called *isomorphic* if there exists an isomorphism between them, and we write

$$A \simeq B.$$

Thus isomorphic groups have the same structure and consequently they will have multiplication tables which are identical, although their elements and/or binary operations may well be of completely different natures.

EXAMPLE 8.2.21

$$D_3 \simeq S_3$$

as can be seen by comparing multiplication tables [see Tables 8.2.2 and 8.2.4]. The isomorphism is defined by the mapping

$$\theta: D_3 \rightarrow S_3$$

where

$$\theta(e) = \iota, \qquad \theta(a) = (1 \quad 2 \quad 3) \quad \text{and} \quad \theta(b) = (1 \quad 2).$$

Note that it suffices to specify the images of the generators of D_3.

The following theorem is important since it provides the existence of an isomorphism between the image group of an arbitrary homomorphism and a certain factor group of its domain.

THEOREM 8.2.8 (*the first isomorphism theorem*).   *Let* $\theta: A \rightarrow B$ *be a group homomorphism. Then*

$$\text{(I)} \quad \ker \theta \lhd A$$

and

$$\text{(II)} \quad A/\ker \theta \simeq \text{im } \theta.$$

Proof.   See Ledermann (1973), Theorem 8, p. 68.

In fact we can say more; for it can be shown that, conversely, every normal subgroup of a group G occurs as the kernel of some homomorphism from G:

THEOREM 8.2.9.   *Let* $G' \lhd G$. *Then there exists an epimorphism*

$$\theta: G \rightarrow G'/G$$

having the property that

$$\ker \theta = G'.$$

Such an epimorphism is defined by the mapping

$$\theta(a) = aG' \quad a \in G. \tag{8.2.17}$$

DEFINITION 8.2.20.   The homomorphism defined in (8.2.17) is called the *projection* or *natural map* of G onto G/G'.

For the remainder of this section we look at a specific type of isomorphism, of particular importance in studying the structure of a given group.

DEFINITION 8.2.21.   An *automorphism* of a group G is an isomorphism

$$\alpha: G \rightarrow G$$

of G onto itself. Thus an automorphism is a permutation of the elements of G which preserves the structure of G:

$$(g_1 g_2)\alpha = (g_1\alpha)(g_2\alpha) \quad g_1, g_2 \in G.$$

The notation $g\alpha$ is used to denote the effect of an automorphism α on the element g, instead of the conventional $\alpha(g)$, since this conforms with the notation used for permutations in § 8.1.1.

PROPOSITION 8.2.10. *The set of all automorphisms of a group G forms another group, called the* automorphism group *of G, under the binary operation 'composition of functions' [see (1.4.3)]. The product $\alpha\beta$ of two automorphisms α and β of G is defined by describing its effect on an element $g \in G$:*

$$g(\alpha\beta) = (g\alpha)\beta. \tag{8.2.18}$$

Proof. Axiom $G1$ follows directly from the Definition 8.2.21. The associativity axiom $G2$ follows since composition of functions is always an associative operation [see § 1.4.2]. The *identity automorphism* $\iota: G \to G$ may be regarded as that automorphism which leaves every element of G fixed, and the last axiom, $G4$ holds since all isomorphisms are bijective mappings and, therefore, have unique inverses. In particular, an automorphism α has the inverse α^{-1} in the automorphism group.

8.2.8. Permutation Groups and Cayleys Theorem

Recall from section 8.1.2 that the set of all permutations of degree n forms a group, S_n, of order $n!$, under the operation 'product of permutations' defined in Definition 8.1.2. We call S_n the *full symmetric group of degree n*, and any subgroup of S_n is called a *permutation group of degree n*. It can be shown that there is a unique subgroup of S_n of index 2 which is therefore, by Example 8.2.17, a normal subgroup. We have:

PROPOSITION 8.2.11. *The $n!/2$ even permutations [see Definition 8.1.4] of degree n form a normal subgroup of S_n, denoted A_n, and called the* alternating group of degree n.

For instance, see Example 8.2.19, where A_4 is displayed.

Cayley's method of representing a group of order n by an $n \times n$ multiplication table demonstrates how each element is mapped on to a permutation of degree n: one simply associates with each element $g \in G$ of the group its corresponding row, regarding it as a permutation of the n objects which constitute the group. If we denote this isomorphism by θ, the permutation corresponding to g shall be denoted $g\theta$.

EXAMPLE 8.2.22. If we denote the six elements

$$e, a, a^2, b, ba, ba^2$$

of the dihedral group D_3 by the digits

$$1, 2, 3, 4, 5, 6$$

in that order, then from Table 8.2.4 we see that

$$a\theta = (1 \quad 2 \quad 3)(4 \quad 6 \quad 5).$$

Above we have described a monomorphism $\theta\colon G \twoheadrightarrow S_n$ which leads us to the following important theorem due to Arthur Cayley:

THEOREM 8.2.12 (*Cayley's theorem*).   *Every finite group of order n is isomorphic to some permutation group of degree n.*

In fact we can go further than this. For, if an arbitrary group G has a subgroup H of index k in G, then each element $g \in G$ corresponds to a permutation of the k right-cosets of H in G in the following manner:

$$g \to \begin{pmatrix} Ha_1 & Ha_2 & \ldots & Ha_k \\ Ha_1 g & Ha_2 g & \ldots & Ha_k g \end{pmatrix}, \tag{8.2.19}$$

and it is not difficult to verify that this correspondence defines a homomorphism of the group G into the symmetric group of degree k. Note that the homomorphism which defines the isomorphism in Cayley's theorem is a monomorphism from G into S_n, whilst the homomorphism (8.2.19) above need not be injective. Formally, we have

THEOREM 8.2.13.   *If a group G has a subgroup of index k in G, there exists a homomorphism*

$$G \to S_k$$

defined by (8.2.19).

EXAMPLE 8.2.23.   By Theorem 8.2.12 (Cayley's theorem) there exists a monomorphism

$$D_3 \twoheadrightarrow S_6$$

from the dihedral group of order 6 into the symmetric group of degree 6. Theorem 8.2.13 tells us that, in addition to this monomorphism, there exist two homomorphisms

$$D_3 \twoheadrightarrow S_3$$
$$D_3 \twoheadrightarrow C_2$$

which are defined by the permutations of degree three and degree two which act on the cosets of the subgroups $\{e, b\}$ and $\{e, a, a^2\}$ of D_3 respectively. It can be shown that the first of these homomorphisms is in fact an isomorphism [see Example 8.2.21].

8.2.9.   Direct Products

Given two groups A and B we can form a new group, $A \times B$, known as the direct product of A and B, which consists of all pairs

$$(a, b)$$

where $a \in A$ and $b \in B$. The binary operation for this direct product group is defined by

$$(a_1, b_1)(a_2, b_2) = (a_1 a_2, b_1 b_2), \qquad (8.2.20)$$

where $a_1 a_2$ is the product of a_1 and a_2 in A and $b_1 b_2$ is the product of b_1 and b_2 in B. That (8.2.20) does make the direct product into a group is easy to see once we have observed that

$$(e_A, e_B)$$

acts as the *identity element* (where e_A and e_B are the identity elements of A and B respectively), and that the element (a, b) of the direct product group has the *inverse*

$$(a^{-1}, b^{-1})$$

in this group (where a^{-1} denotes the inverse of a in A, and b^{-1} the inverse of b in B).

Analogously we define the direct product of a finite set of groups as follows:

DEFINITION 8.2.22.   Let $G_1, \ldots, G_s$ be a set of groups. The *external direct product*

$$G_1 \times G_2 \times \ldots \times G_s$$

of these groups is the group of all s-tuples

$$(g_1, g_2, \ldots, g_s)$$

where $g_i \in G_i$, with group operation

$$(g_1, g_2, \ldots, g_s)(h_1, h_2, \ldots, h_s) = (g_1 h_1, g_2 h_2, \ldots, g_s h_s).$$

EXAMPLE 8.2.24.   The direct product of two cyclic groups of order 2

$$C_2 \simeq \{e_1, a_1\} \quad \text{and} \quad C_2 \simeq \{e_2, a_2\}$$

is the group

$$C_2 \times C_2 \simeq \{(e, e_2), (e_1, a_2), (a_1, e_2), (a_1, a_2)\}$$

which is clearly isomorphic to the group

$$V_4 = \{e, a, b, ab\}$$

under the map

$$(e_1, e_2) \to e, \qquad (e_1, a_2) \to a$$
$$(a_1, e_2) \to b, \qquad (a_1, a_2) \to ab.$$

The group V_4 is called the *Klein four-group* (see also Example 8.2.19).

The above example illustrates the fact that, for finite groups A and B,

$$|A|\,|B| = |A \times B|, \qquad (8.2.21)$$

and the analogous result holds for the direct product of any finite set of finite groups.

A special case of the direct product arises when a group G may be expressed as a direct product of two of its subgroups, H and K. In this case we speak of the *internal direct product* and it arises when

$$G \cong H \times K.$$

Thus an isomorphism $\theta: G \to H \times K$ can be established when the following three conditions are satisfied:

   (i) $g = hk$ holds for every $g \in G$ and some $h \in H$, $k \in K$ so that $G = HK$;
   (ii) $H \cap K = \{e\}$, the identity element of G;
   (iii) H and K commute elementwise, that is $hk = kh$ holds for every $h \in H$ and every $k \in K$.

In these circumstances the elements h, k in (i) are uniquely determined by g and we may define

$$g\theta = (h, k).$$

Notice that H and K must, in fact, be *normal* subgroups of G.

An important application of the internal direct product is to finitely generated Abelian groups, since it can be shown that every finitely generated Abelian group may be expressed as an internal direct product of some of its cyclic subgroups (which are necessarily normal). More precisely:

THEOREM 8.2.14 (*Basis theorem for finitely generated Abelian groups*).   *Every finitely generated Abelian group A is isomorphic to a direct product*

$$G_1 \times G_2 \times \ldots \times G_n \times H_1 \times H_2 \times \ldots \times H_s \qquad (8.2.22)$$

where $G_1, G_2, \ldots, G_n$ are finite cyclic and $H_1, H_2, \ldots, H_s$ are infinite cyclic subgroups of A. Furthermore, it can be shown that the order of each finite cyclic subgroup G_i of A may be taken so that either

   (i) $|G_i|$ *is a power of a prime*
or so that
   (ii) $|G_i| \mid |G_{i+1}|$ *for every $i = 1, 2, \ldots, n - 1$.*

Note.   Either the factors G_i for $i = 1, \ldots, n$, or H_j for $j = 1, \ldots, s$ may be absent from (8.2.22).
Proof.   See Ledermann (1973), p. 90, § 27.

It is beyond the scope of this chapter to give details of a practical technique of effecting a decomposition of the form (8.2.22), but this is given in Ledermann (1973), p. 99, § 29.

8.3.   REPRESENTATIONS OF GROUPS

8.3.1.   Introduction

Before introducing the subject of this section we must distinguish between the terms 'abstract' and 'concrete' for describing a group: *concrete* groups consist of elements of a certain specified kind, such as permutation groups or groups of transformations, but on the other hand, in an *abstract* group there is no particular interpretation attached to its elements, so that the nature of the group elements is arbitrary and only the structure of the group is specified. For example, all the groups mentioned in Examples 8.2.1 to 8.2.5 are concrete but the cyclic group of Example 8.2.10 is abstract.

As we have already seen in § 8.2.2 and § 8.2.3 the structure of an arbitrary group is not determined by the nature of its elements: indeed, through the concept of isomorphism introduced in § 8.2.6, we see that two groups having the same structure may have elements of completely different natures. We are thus led to the following consideration: given an arbitrary abstract group G does there exist a class of concrete groups which will always contain one group which is homomorphic to G? In other words, can the elements of G always be regarded as elements of a certain specified kind? In the case when G is finite an affirmative answer is provided by Cayley's theorem (Theorem 8.2.12), and Theorem 8.2.13 shows that this is also true when an arbitrary group G has a subgroup of finite index. The general idea behind these theorems is to construct a homomorphism

$$G \to S_m$$

called a *permutation representation*, from an abstract group G into the full symmetric group S_m, and hence show that G is homomorphic to a permutation group of degree m. In what follows we shall be considering *matrix representations* of an abstract group which are likewise provided by homomorphisms but this time into a group of matrices. In general we have:

DEFINITION 8.3.1. A representation of an abstract group G is a homomorphism

$$G \to \Gamma$$

of G into the concrete group Γ.

8.3.2. Matrix Representations of Groups

In Example 8.2.4 we showed that the set of all $m \times m$ non-singular matrices over a field $\mathbb{F}$ forms a group, $GL_m(\mathbb{F})$, called the general linear group of degree m over $\mathbb{F}$. In what follows we shall confine our interest exclusively to the case where $\mathbb{F}$ is a subfield [see § 2.4.2] of the complex numbers $\mathbb{C}$ [see § 2.7.1], in which case

$$GL_m(\mathbb{F}) \leq GL_m(\mathbb{C})$$

and $GL_m(\mathbb{F})$ is called a *matrix group of degree m*.

DEFINITION 8.3.2. A *matrix representation of degree (or dimension) m over* $\mathbb{F}$ of an abstract group G is a homomorphism

$$\theta: G \to GL_m(\mathbb{F}).$$

Thus θ assigns an $m \times m$ matrix, say $\mathbf{A}(g)$, to every element $g \in G$. With slight abuse of notation we refer to the 'matrix representation $\mathbf{A}(g)$', with the understanding that, for a particular element $g \in G$, $\mathbf{A}(g)$ is an $m \times m$ non-singular matrix over $\mathbb{F}$. Since θ is a homomorphism it exhibits the following properties [see Definition 8.2.15]:

(i) The unit e of G is mapped to the unit of $GL_m(\mathbb{F})$ under θ, that is

$$\mathbf{A}(e) = \mathbf{I}_m, \tag{8.3.1}$$

where $\mathbf{I}_m$ is the unit matrix of degree m [see (6.2.7)].

(ii) The inverse g^{-1} of an element $g \in G$ is mapped under θ into the inverse of the matrix which represents g, so

$$\mathbf{A}(g^{-1}) = \mathbf{A}(g)^{-1} \quad \forall g \in G \tag{8.3.2}$$

and thirdly

(iii) A representation is a multiplicative function, that is

$$\mathbf{A}(g_1 g_2) = \mathbf{A}(g_1)\mathbf{A}(g_2) \quad \forall g_1, g_2 \in G \tag{8.3.3}$$

where the product on the right is defined by multiplication of matrices [see § 6.2(vi)].

From these properties of representation it follows that one can obtain the entire group of representative matrices by specifying only those matrices which represent the generators of the group. For example, if G is generated by two elements a and b and if another element $c \in G$ is given by

$$c = aba^{-1}b$$

then, by (8.3.2) and (8.3.3), we have

$$\mathbf{A}(c) = \mathbf{A}(a)\mathbf{A}(b)\mathbf{A}(a)^{-1}\mathbf{A}(b).$$

EXAMPLE 8.3.1.   A matrix representation of C_n, the cyclic group of order n [see Example 8.2.10], may be completely specified by assigning a matrix $\mathbf{A}(r) \in GL_m(\mathbb{C})$ to the generator r of C_n with the property

$$\mathbf{A}(r)^n = \mathbf{I}_m,$$

that is, $\mathbf{A}(r)$ is any complex $m \times m$ matrix which is periodic of order dividing n [see (6.7.27)]. Thus if

$$\theta(r) = \mathbf{A}(r)$$

then by (8.3.3) we have

$$\theta(r^2) = \mathbf{A}(r)^2, \ldots, \theta(r^{n-1}) = \mathbf{A}(r)^{n-1} \quad \text{and} \quad \theta(e) = \mathbf{I}_m,$$

as required.

EXAMPLE 8.3.2.   We obtain a representation of degree 2 of the group

$$D_3 = \{a, b | a^3 = b^2 = (ab)^2 = e\}$$

[see Example 8.2.11] by assigning a 2×2 matrix to each of a and b as follows:

$$\mathbf{A}(a) = \begin{pmatrix} 0 & 1 \\ -1 & -1 \end{pmatrix} \quad \text{and} \quad \mathbf{A}(b) = \begin{pmatrix} 0 & 1 \\ 1 & 0 \end{pmatrix}.$$

For then

$$\mathbf{A}(ab) = \begin{pmatrix} 1 & 0 \\ -1 & -1 \end{pmatrix}$$

is the product of $\mathbf{A}(a)$ and $\mathbf{A}(b)$, and it can be verified that

$$(\mathbf{A}(a))^3 = (\mathbf{A}(b))^2 = (\mathbf{A}(ab))^2 = \mathbf{I}_3$$

as required.

EXAMPLE 8.3.3.   A representation of degree 4 of the group

$$V_4 = \{a, b \mid a^2 = b^2 = (ab)^2 = e\}$$

[see Example 8.2.19] may be furnished by assigning

$$\mathbf{A}(a) = \begin{pmatrix} 1 & 0 & 0 & 0 \\ -1 & -1 & 0 & 0 \\ 1 & 2 & 1 & 0 \\ -1 & -3 & -3 & -1 \end{pmatrix} \quad \text{and} \quad \mathbf{A}(b) = \begin{pmatrix} -1 & 0 & 0 & 0 \\ 0 & -1 & 0 & 0 \\ 0 & 0 & -1 & 0 \\ 0 & 0 & 0 & -1 \end{pmatrix}.$$

Then

$$\mathbf{A}(ab) = \begin{pmatrix} -1 & 0 & 0 & 0 \\ 1 & 1 & 0 & 0 \\ -1 & -2 & -1 & 0 \\ 1 & 3 & 3 & 1 \end{pmatrix}$$

and

$$(\mathbf{A}(a))^2 = (\mathbf{A}(b))^2 = (\mathbf{A}(ab))^2 = \mathbf{I}_4$$

as required.

In general, matrix representations are neither epimorphisms nor monomorphisms [see Definition 8.2.18]. Suppose, however, that the representation θ is a monomorphism. Then by Proposition 8.2.6, ker $\theta = \mathbf{I}_m$ so that the identity matrix $\mathbf{I}_m$ represents solely the unit element of the group. Thus we have

DEFINITION 8.3.3.   A *faithful representation* of degree m of an abstract group G is a monomorphism

$$\theta: G \to GL_m(\mathbb{F}),$$

so that $\theta(g) = \mathbf{I}_m$ if, and only if, g is the identity element of G.

EXAMPLE 8.3.4.   All representations in Examples 8.3.1–8.3.3 are faithful.

As an example of a representation which is not faithful consider the three dimensional representation of

$$C_6 = \{r \mid r^6 = e\}$$

given by

$$A(r) = \begin{pmatrix} 1 & 0 & 0 \\ 0 & \omega & 0 \\ 0 & 0 & \omega^2 \end{pmatrix}$$

where

$$\omega = -\tfrac{1}{2} + i\frac{\sqrt{3}}{2}.$$

Then $A(r^3) = I_3$ and $A(r)$ is therefore not faithful. (However, $A(r)$ is faithful if restricted to the subgroup C_3, generated by r^2.)

DEFINITION 8.3.4.   We call a matrix representation θ of degree unity of a group G a *linear representation* of G. Thus θ is a homomorphism of G into $\mathbb{C}^*$ the group of non-zero complex numbers described in Example 8.2.2. The linear representation

$$\theta: G \to \mathbb{C}^*$$

defined by

$$\theta(g) = 1 \quad \forall g \in G \tag{8.3.4}$$

is called the *trivial representation* of G.

DEFINITION 8.3.5.   Given an arbitrary group G, two representations

$$\theta_A: G \to GL_m(\mathbb{F})$$

and

$$\theta_B: G \to GL_m(\mathbb{F})$$

where

$$\theta_A(g) = A(g) \quad \text{and} \quad \theta_B(g) = B(g) \quad \text{for } g \in G$$

are said to be *equivalent*, and written

$$A(g) \sim B(g) \qquad (\text{or } \theta_A \sim \theta_B)$$

if there exists a matrix $T \in GL_m(\mathbb{F})$ such that

$$A(g) = T^{-1}B(g)T \quad \forall g \in G. \tag{8.3.5}$$

Thus each matrix in $A(g)$ is similar to the corresponding matrix in $B(g)$ and, moreover, the similarity transform (8.3.5) is independent of g. In the same way that we need only specify the matrices which represent the generators of G, we need only check the equivalence (8.3.5) for the generators of G. For if

$$\mathbf{A}(a) = \mathbf{T}^{-1}\mathbf{B}(a)\mathbf{T} \quad \text{and} \quad \mathbf{A}(b) = \mathbf{T}^{-1}\mathbf{B}(b)\mathbf{T}$$

then

$$\mathbf{A}(ab) = \mathbf{A}(a)\mathbf{A}(b) = \mathbf{T}^{-1}\mathbf{B}(a)\mathbf{T}\mathbf{T}^{-1}\mathbf{B}(b)\mathbf{T}$$

$$= \mathbf{T}^{-1}\mathbf{B}(a)\mathbf{B}(b)\mathbf{T} = \mathbf{T}^{-1}\mathbf{B}(ab)\mathbf{T}.$$

EXAMPLE 8.3.5.   Let $\mathbf{A}(g)$ be the representation of D_3 given in Example 8.3.2. Then $\mathbf{A}(g)$ is equivalent to the representation $\mathbf{B}(g)$ of D_3 given by

$$\mathbf{B}(a) = \begin{pmatrix} -1 & 1 \\ -1 & 0 \end{pmatrix} \quad \text{and} \quad \mathbf{B}(b) = \begin{pmatrix} 0 & -1 \\ -1 & 0 \end{pmatrix}.$$

For (8.3.5) holds when $g = a$ and $g = b$ and the matrix $\mathbf{T}$ is given by

$$\begin{pmatrix} 0 & -1 \\ 1 & 0 \end{pmatrix}.$$

8.3.3.   Reducibility of Matrix Representations

From now on we assume that the field $\mathbb{F}$ is $\mathbb{C}$ itself.

DEFINITION 8.3.6.   A representation

$$\theta: G \to GL_m(\mathbb{C})$$

is said to be *reducible* if there exists an $m \times m$ non-singular matrix $\mathbf{T}$ such that the equation [see § 6.6]

$$\mathbf{T}^{-1}\mathbf{A}(g)\mathbf{T} = \begin{pmatrix} \mathbf{B}(g) & \mathbf{0} \\ \mathbf{D}(g) & \mathbf{C}(g) \end{pmatrix} \tag{8.3.5}$$

holds for every matrix $\mathbf{A}(g)$ in θ. Otherwise θ is said to be *irreducible*.

The assumption that $\mathbb{F} = \mathbb{C}$ was made because reducibility depends upon the matrix group in which θ is allowed to take values. That is, if the field $\mathbb{K}$ is such that [see § 2.4.2]

$$\mathbb{F} < \mathbb{K} \leq \mathbb{C}$$

so that [see Definition 8.2.6]

$$GL_m(\mathbb{F}) < GL_m(\mathbb{K}) \leq GL_m(\mathbb{C})$$

then it may happen that an irreducible representation

$$\theta: G \to GL_m(\mathbb{F})$$

becomes reducible if it is regarded as a representation

$$\theta: G \to GL_m(\mathbb{K}).$$

Since the matrix $\mathbf{A}(g)$ in (8.3.6) defines a homomorphism we have

$$\begin{pmatrix} \mathbf{B}(g) & \mathbf{0} \\ \mathbf{D}(g) & \mathbf{C}(g) \end{pmatrix} \begin{pmatrix} \mathbf{B}(h) & \mathbf{0} \\ \mathbf{D}(h) & \mathbf{C}(h) \end{pmatrix} = \begin{pmatrix} \mathbf{B}(gh) & \mathbf{0} \\ \mathbf{D}(gh) & \mathbf{C}(gh) \end{pmatrix} \quad \forall g, h \in G.$$

Thus

$$\mathbf{B}(g)\mathbf{B}(h) = \mathbf{B}(gh) \quad \text{and} \quad \mathbf{C}(g)\mathbf{C}(h) = \mathbf{C}(gh) \quad \forall g, h \in G$$

so that $\mathbf{B}(g)$ and $\mathbf{C}(g)$ themselves, define homomorphisms of G into matrix groups of degrees which are determined by the partitioning of the matrix in (8.3.6). That is, if the matrices $\mathbf{B}(g)$ are $r \times r$ so that the matrices $\mathbf{C}(g)$ have degree $m - r$, $\mathbf{B}(g)$ and $\mathbf{C}(g)$ define representations

$$\phi: G \to GL_r(\mathbb{C})$$

and

$$\psi: G \to GL_{m-r}(\mathbb{C})$$

respectively. This explains the use of the term 'reducible'. We can in fact go further than (8.3.5) to show that, in certain circumstances, equation (8.3.5) implies that [see (6.7.4)]

$$\mathbf{A}(g) \sim \text{diag}\,(\mathbf{B}(g), \mathbf{C}(g)),$$

that is, there exists another matrix $\mathbf{S}$ in $GL_m(\mathbb{C})$ such that

$$\mathbf{S}^{-1}\mathbf{A}(g)\mathbf{S} = \begin{pmatrix} \mathbf{B}(g) & \mathbf{0} \\ \mathbf{0} & \mathbf{C}(g) \end{pmatrix} \quad \forall g \in G.$$

More formally we have

THEOREM 8.3.1 (*Maschke's theorem*).   *Let $\theta: G \to GL_m(\mathbb{C})$ be a representation of the group G given by*

$$\theta(g) = \mathbf{A}(g), \quad g \in G.$$

Then if G is finite and if θ is reducible,

$$\mathbf{A}(g) \sim \begin{pmatrix} \mathbf{B}(g) & \mathbf{0} \\ \mathbf{0} & \mathbf{C}(g) \end{pmatrix} \quad \forall g \in G$$

where $\mathbf{B}(g)$ and $\mathbf{C}(g)$ themselves define representations of G.

Proof.   See Ledermann (1977), p. 21, Theorem 1.2.

If either of the representations $\mathbf{B}(g)$ and $\mathbf{C}(g)$ in the theorem above are reducible we can repeat the process of reduction until all diagonal blocks are irreducible representations, which leads us to the following definition:

DEFINITION 8.3.7.   The representation

$$\theta: G \rightarrow GL_m(\mathbb{C})$$

is *completely reducible* if

$$A(g) \sim \text{diag} (A_1(g), A_2(g), \ldots, A_r(g)) \quad (g \in G)$$

holds, where $A_i(g)$ are irreducible representations of G for $i = 1, 2, \ldots, r$.

   With this definition, Theorem 8.3.1 tells us that all reducible representations of a finite group are completely reducible. Thus the problem of finding all possible representations of a finite group reduces to that of determining all the irreducible ones. Notice that linear representations are always irreducible. With this in mind we quote the following theorem:

THEOREM 8.3.2.   *The number of inequivalent irreducible representations of a finite group (over $\mathbb{C}$) is equal to the number of conjugacy classes [see § 8.2.5] in G.*

Proof.   See Ledermann (1977), p. 49, Theorem 2.3.

EXAMPLE 8.3.6.   The dihedral group

$$D_4 = \{a, b \mid a^4 = b^2 = (ab)^2 = e\}$$

has five conjugacy classes, as shown in Example 8.2.14, denoted

$$c_1 = \{e\}; \quad c_2 = \{a, a^2\}; \quad c_3 = \{a^2\}; \quad c_4 = \{b, a^2b\} \quad \text{and} \quad c_5 = \{ab, a^3b\}.$$

Thus there are five inequivalent irreducible representations of D_4. Let us first determine all linear representations. Suppose therefore that

$$\theta: D_4 \rightarrow \mathbb{C}^*$$

is such that

$$\theta(a) = \alpha \quad \text{and} \quad \theta(b) = \beta$$

where α and β are non-zero complex numbers which satisfy the relations

$$\alpha^4 = \beta^2 = (\alpha\beta)^2 = 1.$$

Since α and β commute, these relations imply

$$\alpha^2 = \beta^2 = 1$$

and there are four distinct linear representations of D_4, given by

$$\theta_i: D_4 \rightarrow \mathbb{C}^*$$

where

$$\theta_1(a) = \theta_1(b) = 1 \quad \text{(the trivial representation)}$$
$$\theta_2(a) = 1 \quad \text{and} \quad \theta_2(b) = -1$$
$$\theta_3(a) = -1 \quad \text{and} \quad \theta_3(b) = 1$$

and

$$\theta_4(a) = -1 \quad \text{and} \quad \theta_4(b) = -1.$$

Of course distinct linear representations are inequivalent. For, since $\mathbb{C}^*$ is Abelian, all linear representations commute and therefore are similar (in the sense of (8.3.5)) only to themselves.

The fifth irreducible representation of D_4 is provided by

$$\theta_5: D_4 \to GL_2(\mathbb{C})$$

where $\theta_5(g) = \mathbf{A}(g)$, say, and

$$\mathbf{A}(a) = \begin{pmatrix} 0 & -1 \\ 1 & 0 \end{pmatrix} \quad \text{and} \quad \mathbf{A}(b) = \begin{pmatrix} 0 & 1 \\ 1 & 0 \end{pmatrix}.$$

It is easy to verify that θ_5 is indeed a representation of D_4 (i.e. that $(\mathbf{A}(a))^4 = (\mathbf{A}(b))^2 = (\mathbf{A}(ab))^2 = \mathbf{I}_2$). To show that θ_5 is irreducible we assume otherwise. Suppose there exists a matrix $\mathbf{T}$ in $GL_2(\mathbb{C})$ such that

$$\mathbf{T}^{-1}\mathbf{A}(a)\mathbf{T} = \begin{pmatrix} \alpha & 0 \\ 0 & \beta \end{pmatrix}$$

and

$$\mathbf{T}^{-1}\mathbf{A}(b)\mathbf{T} = \begin{pmatrix} \gamma & 0 \\ 0 & \delta \end{pmatrix}$$

where α, β, γ and δ are complex numbers. Then $\mathbf{T}^{-1}\mathbf{A}(a)\mathbf{T}$ and $\mathbf{T}^{-1}\mathbf{A}(b)\mathbf{T}$, being diagonal matrices, would commute [see § 6.7(iv)] and therefore $\mathbf{A}(a)$ and $\mathbf{A}(b)$ would also commute, which is false. Thus θ_5 must be irreducible.

8.3.4.   Linear Representations of Finite Abelian Groups

If G is a finite Abelian group then each element of G lies in a conjugacy class of its own [see § 8.2.5] and therefore the number of conjugacy classes in G is equal to the order, $|G|$, of G. Thus by Theorem 8.3.2, a finite Abelian group has $|G|$ distinct irreducible representations. In this section we determine all linear representations of G, show that there are $|G|$ of them and therefore that these are the only irreducible representations of G over $\mathbb{C}$. We need the following elementary lemma

LEMMA 8.3.3.   *There are n distinct linear representations, $\theta_1, \theta_2, \ldots, \theta_n$, say, of the cyclic group of order n, C_n, given by*

$$\theta_i(r) = \xi_i$$

where r generates C_n and $\xi_1, \xi_2, \ldots, \xi_n$ are the n nth roots of unity [see (2.7.31)].

EXAMPLE 8.3.7.   There are three distinct linear representations of C_3 given by

$$\theta_1(r) = 1 \quad \text{the trivial representation}$$
$$\theta_2(r) = \omega$$

and

$$\theta_3(r) = \omega^2$$

where

$$\omega = -\tfrac{1}{2} + i\frac{\sqrt{3}}{2} \quad \text{and} \quad \omega^2 = -\tfrac{1}{2} - i\frac{\sqrt{3}}{2} \quad \text{and} \quad \omega^3 = 1.$$

Now, it follows from the basis theorem for finitely generated Abelian groups (Theorem 8.2.14) that a finite Abelian group G say of order n may be decomposed into a direct product of cyclic subgroups $G_i \le G$:

$$G = G_1 \times G_2 \times \ldots \times G_s$$

and every element $x \in G$ is of the form

$$x = (x_1, x_2, \ldots, x_s) \quad x_i \in G_i \quad \text{for } i = 1, 2, \ldots, s.$$

We obtain a linear representation of G from the linear representations of the G_i as follows. Let

$$\lambda_i : G_i \to \mathbb{C}^* \quad \text{for } i = 1, 2, \ldots, s$$

be linear representations of $G_1, G_2, \ldots, G_s$. Define a mapping

$$\lambda : G \to \mathbb{C}^*$$

by

$$\lambda(x) = \lambda_1(x_1)\lambda_2(x_2)\ldots\lambda_s(x_s) \tag{8.3.6}$$

where $x = (x_1, x_2, \ldots, x_s)$ and each $x_i \in G_i$. Then λ is in fact a linear representation of G. For if

$$y = (y_1, y_2, \ldots, y_s), \quad y_i \in G,$$

is another element of G, then

$$\lambda(x)\lambda(y) = \lambda_1(x_1)\lambda_1(y_1)\ldots\lambda_s(x_s)\lambda_s(y_s)$$
$$= \lambda_1(x_1 y_1)\ldots\lambda_s(x_s y_s)$$

since each λ_i is a representation of G_i, and so

$$\lambda(x)\lambda(y) = \lambda(xy)$$

as required.

Moreover, there are precisely n such linear representations of G. For if $|G_i| = n_i$ there are, by Lemma 8.3.3, exactly n_i distinct linear representations of G_i. Thus there are $\prod_{i=1}^{s} n_i$ distinct linear representations (8.3.6) of G, and by (8.2.2) this number is equal to the order of G. We summarize these results in the following:

THEOREM 8.3.4. *Let G be a finite Abelian group of order n and suppose that G has direct product decomposition*

$$G = G_1 \times G_2 \times \ldots \times G_s$$

where G_i denotes a cyclic subgroup of G of order n_i. Then there are n distinct linear representations of G, given by

$$\lambda(x) = \lambda_1(x_1)\lambda_2(x_2) \ldots \lambda_s(x_s)$$

where the element $x \in G$ has the form

$$x = (x_1, x_2, \ldots, x_s), \quad x_i \in G_i,$$

and where λ_i is one of the n_i distinct linear representations of G_i. These are the only irreducible representations of G.

8.3.5.   Group Characters

DEFINITION 8.3.8.   Let G be a finite group and let

$$\theta : G \to GL_m(\mathbb{C})$$

be a representation of G given by

$$\theta(g) = \mathbf{A}(g) \quad \forall g \in G.$$

Associated with θ there is a function

$$\chi_\theta : G \to \mathbb{C}$$

defined by

$$\chi_\theta(g) = \text{trace } \mathbf{A}(g)$$

[see (6.2.4) for definition of the trace of a matrix] and we call the function χ_θ the *character* of the representation θ.

The following properties may be found useful when working with group characters:

(i) $\chi_\theta(e) = \text{degree } \theta$

where e is the identity element of G, and for any $g \in G$,

$$|\chi_\theta(g)| \leq \text{degree } \theta$$

where $|\ |$ denotes the absolute value of a complex number [see (2.7.13)].

(ii) χ_θ is not, generally, a multiplicative function but it is commutative. That is

$$\chi_\theta(gh) \neq \chi_\theta(g)\chi_\theta(h)$$

but

$$\chi_\theta(gh) = \chi_\theta(hg) \quad \forall g, h \in G. \tag{8.3.7}$$

(iii) $\chi_\theta(g^{-1}) = \overline{\chi_\theta(g)} \quad \forall g \in G$ (8.3.8)

where the bar denotes the complex conjugate [see (2.7.10)].

(iv) $\chi_\theta(g) = $ sum of the eigenvalues of $\mathbf{A}(g)$ [see (7.3.9)]. And, finally

(v) If $\mathbf{A}(g)$ is a periodic matrix of order n, that is if $\mathbf{A}(g)^n = \mathbf{I}_m$ then the eigenvalues of $\mathbf{A}(g)$ will all be found amongst the nth roots of unity. [For, by

Proposition 7.2.1(i), if α is an eigenvalue of the periodic matrix $\mathbf{A}(g)$, then α^n is an eigenvalue of $\mathbf{I}_m$ and so $\alpha^n = 1$.]

PROPOSITION 8.3.5

(i) *A character assumes the same value throughout a conjugacy class* [see § 8.2.5] *of G. That is to say, that if a and b are conjugate elements in G, then* $\chi_\theta(a) = \chi_\theta(b)$: *in symbols*

$$a = g^{-1}bg \Rightarrow \chi_\theta(a) = \chi_\theta(b)$$

which is (8.3.6) again, and

(ii) *If θ and ψ are equivalent representations then their character functions are the same. Thus, if $\theta \sim \psi$ then* $\chi_\theta(g) = \chi_\psi(g) \; \forall g \in G$.

It is more a profound fact that the converse of this statement holds in the situation we are concerned with.

THEOREM 8.3.6.   *Let θ and ψ be representations of a finite group over the complex field $\mathbb{C}$. Then*

$$\theta \sim \psi$$

if and only if

$$\chi_\theta(g) = \chi_\psi(g)$$

for all elements g of G.

EXAMPLE 8.3.8.   Denote by θ_A the representation of D_3 given in Example 8.3.2 and by θ_B the representation

$$D_3 \rightarrow GL_2(\mathbb{C})$$

given by

$$\theta_B(g) = \mathbf{B}(g) \quad \forall g \in G$$

and

$$\mathbf{B}(a) = \begin{pmatrix} 0 & -1 \\ 1 & -1 \end{pmatrix} \quad \text{and} \quad \mathbf{B}(b) = \begin{pmatrix} 0 & -1 \\ -1 & 0 \end{pmatrix}.$$

We wish to show that $\theta_A \sim \theta_B$ and so, by Theorem 8.3.6 above, we must show that

$$\chi_{\theta_A}(g) = \chi_{\theta_B}(g)$$

where g ranges through the representatives of all conjugacy classes of D_3. There are three conjugacy classes of D_3:

$$c_1 = \{e\}, \qquad c_2 = \{a, a^2\}, \quad \text{and} \quad c_3 = \{b, ab, a^2b\}$$

so, choosing as class representatives the elements e, a and b, we must show that

$$\chi_{\theta_A}(g) = \chi_{\theta_B}(g) \quad \text{for } g = e, \ g = a \text{ and } g = b.$$

It now follows simply by computing traces that

$$\chi_{\theta_A}(e) = \chi_{\theta_B}(e) = 2,$$

the degree of the representations; moreover

$$\chi_{\theta_A}(a) = \chi_{\theta_B}(a) = -1,$$

and finally

$$\chi_{\theta_A}(b) = \chi_{\theta_B}(b) = 0.$$

Therefore, $\theta_A \sim \theta_B$.

DEFINITION 8.3.9.   Characters associated with irreducible representations are called *simple*, and those of reducible representations are *compound*.

Theorem 8.3.2 states that the number of simple characters of a finite group G is equal to the number of conjugacy classes in G. We customarily denote the conjugacy classes by

$$c_1, c_2, \ldots, c_k,$$

and we assume that these classes contain

$$h_1, h_2, \ldots, h_k$$

members respectively. Thus we have the class formula

$$|G| = h_1 + h_2 + \ldots + h_k.$$

The simple characters will be denoted by

$$\chi^{(1)}, \chi^{(2)}, \ldots, \chi^{(k)}.$$

Since every character is constant throughout a conjugacy class, we shall put

$$\chi_j^{(i)} = \text{value of } \chi^{(i)} \text{ on the class } c_j.$$

It therefore seems natural to display the simple characters of a finite group G in a square array in which $\chi_j^{(i)}$ is the (i, j)th element. Such a scheme is called a *character table*, illustrated as shown in Table 8.3.1.

	c_1	c_2	$\ldots$	c_j	$\ldots$	c_k
$\chi^{(1)}$				$\vdots$		
$\chi^{(2)}$				$\vdots$		
$\vdots$				$\vdots$		
$\chi^{(i)}$	$\cdots$	$\cdots$	$\cdots$	$\chi_j^{(i)}$		
$\vdots$						
$\chi^{(k)}$						

Table 8.3.1

EXAMPLE 8.3.9.   We listed the five conjugacy classes of the dihedral group D_4 in Example 8.3.5, and we also found the five irreducible representations

$\theta_1, \ldots, \theta_5$ of D_4. If we denote by $\chi^{(i)}$ the simple character associated with the representation θ_i, we have shown that D_4 has the character table as shown in Table 8.3.2.

	c_1	c_2	c_3	c_4	c_5
$\chi^{(1)}$	1	1	1	1	1
$\chi^{(2)}$	1	1	1	-1	-1
$\chi^{(3)}$	1	-1	1	1	-1
$\chi^{(4)}$	1	-1	1	-1	1
$\chi^{(5)}$	2	0	-2	0	0

Table 8.3.2

8.3.6. Orthogonality Properties of Group Characters

In this section we give a brief description of a theory which has greatly simplified the problem of finding the simple characters of a finite group.

We begin by defining an *inner product* function on any two complex valued group functions as follows: Let Φ and Ψ be any two complex valued functions whose domain is a finite group of order n, G say [see Definition 1.4.1]. The inner product $\langle \Phi, \Psi \rangle$ of these two functions is defined by

$$\langle \Phi, \Psi \rangle = \frac{1}{n} \sum_{g \in G} \Phi(g)\Psi(g^{-1}). \tag{8.3.9}$$

Since g^{-1} runs through G as g does, we have

$$\langle \Phi, \Psi \rangle = \langle \Psi, \Phi \rangle.$$

DEFINITION 8.3.10. The functions Φ and Ψ on a finite group are said to be *orthogonal* if

$$\langle \Phi, \Psi \rangle = 0.$$

Consider now the inner product of two characters ϕ and ψ. Let ϕ_α be the value of ϕ on the class c_α. Since $\phi(g^{-1})$ is the conjugate complex of $\phi(g)$, we can write $\psi(g^{-1}) = \bar{\psi}_\alpha$ if $g \in c_\alpha$ [see (8.3.8)]. Hence (8.3.9) becomes

$$\langle \phi, \psi \rangle = |G|^{-1} \sum_{\alpha=1}^{k} h_\alpha \phi_\alpha \bar{\psi}_\alpha. \tag{8.3.10}$$

As we shall see, the inner product provides a convenient method for determining whether a given character is simple or compound, and whether or not two simple characters correspond to equivalent representations of G. More precisely we have

THEOREM 8.3.7 (*Orthogonality relations of the first kind*). *Let* $\chi^{(1)}, \chi^{(2)}, \ldots, \chi^{(k)}$
be the complete set of simple characters of the finite group G. Then

$$\langle \chi^{(i)}, \chi^{(j)} \rangle = \delta_{ij} \tag{8.3.12}$$

that is

$$|G|^{-1} \sum_{\alpha=1}^{k} h_\alpha \chi_\alpha^{(i)} \bar{\chi}_\alpha^{(j)} = \delta_{ij} \tag{8.3.13}$$

where δ_{ij} *denotes the Kronecker delta* [*see* (6.7.1)].

There is another kind of orthogonality relation, which in contrast to the first involves all the simple characters simultaneously. For this purpose it is convenient to use the notation

$$x \sim y$$

to indicate that the elements x and y belong to the same conjugacy class of G.

THEOREM 8.3.8 (*Orthogonality relations of the second kind*). *Let G be a finite group with simple characters* $\chi^{(1)}, \chi^{(2)}, \ldots, \chi^{(k)}$. *Then*

$$\sum_{i=1}^{k} \chi^{(i)}(a)\chi^{(i)}(b) = \begin{cases} \dfrac{|G|}{h_\alpha} & \text{if } a \sim b^{-1} \\ 0 & \text{if } a \nsim b^{-1} \end{cases} \tag{8.3.14}$$

where h_α *denotes the number of elements in the conjugacy class of a; alternatively,*

$$\sum_{i=1}^{k} \chi_\alpha^{(i)} \bar{\chi}_\beta^{(i)} = \delta_{\alpha\beta} \frac{|G|}{h_\alpha}$$

where $\chi_\alpha^{(i)}$ *denotes the values of the character* $\chi^{(i)}$ *for the class* c_α.

Let

$$\mathbf{F}^{(1)}, \mathbf{F}^{(2)}, \ldots, \mathbf{F}^{(k)} \tag{8.3.15}$$

be a complete set of irreducible representations of G; of course the representations (8.3.15) are determined only up to equivalence. But their traces will yield the complete set of simple characters $\chi^{(i)}$ ($i = 1, 2, \ldots, k$). The degree of $\mathbf{F}^{(i)}$ is given by

$$f^{(i)} = \chi^{(i)}(e).$$

On putting $a = b = e$ in (8.3.14) and noting that $h_e = 1$, we obtain the important result that

$$\sum_{i=1}^{k} (f^{(i)})^2 = |G|. \tag{8.3.16}$$

The degrees $f^{(i)}$ also satisfy a remarkable arithmetical condition, which, however, is more difficult to establish, namely that

$$\text{each } f^{(i)} \text{ divides } |G|. \tag{8.3.17}$$

We now return to the study of an arbitrary representation $\mathbf{A}(g)$ of G with character ϕ. As we have seen in Definition 8.3.7 it may be assumed that

$$\mathbf{A}(g) \sim \text{diag } (\mathbf{F}^{(p)}(g), \mathbf{F}^{(q)}(g), \ldots, \mathbf{F}^{(t)}(g)) \qquad (8.3.18)$$

where each of the diagonal blocks on the right of (8.3.18) is one of the irreducible representations listed in (8.3.15), possibly with repetitions. The order in which the blocks are arranged is immaterial, but the multiplicity of each irreducible constituent is important. Let us assume that $\mathbf{F}^{(1)}$ occurs e_1 times, $\mathbf{F}^{(2)}$ occurs e_2 times, ..., $\mathbf{F}^{(k)}$ occurs e_k times. Then on taking traces on both sides of (8.3.18) we obtain

$$\phi(g) = \sum_{i=1}^{k} e_i \chi^{(i)}(g). \qquad (8.3.19)$$

Thus every character $\phi(g)$ is given by an expression (8.3.19) with non-negative integral coefficients e_i.

Now it is one of the most striking features of character theory that we can compute the coefficients e_i, solely from a knowledge of ϕ and the character table of G, that is, it is unnecessary to carry out the decomposition of $\mathbf{A}(g)$ into irreducible constituents, which is likely to be a laborious task. For if we take the inner product of both sides with a fixed irreducible character $\chi^{(j)}$ and use the orthogonality relations of the first kind, we find that $\langle \phi, \chi^{(j)} \rangle = e_j$. We record these results in the following

THEOREM 8.3.9. *Every character of the finite group G is of the form (8.3.19), where $e_i \geq 0$ ($i = 1, 2, \ldots, k$) and*

$$e_i = \langle \phi, \chi^{(i)} \rangle. \qquad (8.3.20)$$

Furthermore,

$$\langle \phi, \phi \rangle = \sum_{i=1}^{k} e_i^2, \qquad (8.3.21)$$

and ϕ is a simple character, if and only if

$$\langle \phi, \phi \rangle = 1.$$

Because of the similarity with the theory of periodic functions Theorem 8.3.9 is sometimes referred to as the *Fourier Analysis of characters*. Indeed, (8.3.20) is analogous to the formula for the Fourier coefficients and (8.3.21) corresponds to Parseval's theorem [see IV, Chapter 20].

An illustration of this technique will be given in the next section.

8.3.7. Matrix Representations of Permutation Groups

Associated with every permutation group P of degree n there is a representation of degree n given by assigning to each element $\pi \in P$ the permutation matrix which represents π [see Definition 6.7.1]. We call this representation the *natural representation* of P.

EXAMPLE 8.3.10.   The conjugacy classes of the alternating group A_4 [see Definition 8.2.10] are given by

$$c_1 = \iota, \quad \text{the identity permutation of degree four,}$$
$$c_2 = \{(1 \quad 2)(3 \quad 4), (1 \quad 3)(2 \quad 4), (1 \quad 4)(2 \quad 3)\}$$
$$c_3 = \{(1 \quad 2 \quad 3), (1 \quad 3 \quad 4), (1 \quad 4 \quad 2), (2 \quad 4 \quad 3)\}$$

and

$$c_4 = \{(1 \quad 3 \quad 2), (1 \quad 4 \quad 3), (1 \quad 2 \quad 4), (2 \quad 3 \quad 4)\},$$

where $A_4 = c_1 \cup c_2 \cup c_3 \cup c_4$. Now, if we denote by a the permutation (1   2   3) and by b the permutation (1   2)(3   4), A_4 can also be expressed as follows

$$A_4 = \{a, b | a^3 = b^2 = (ab)^3 = e\}$$

in terms of its generators and relations. Therefore, the natural representation of A_4 is completely specified by

$$\mathbf{P}(a) = \begin{pmatrix} 0 & 0 & 1 & 0 \\ 1 & 0 & 0 & 0 \\ 0 & 1 & 0 & 0 \\ 0 & 0 & 0 & 1 \end{pmatrix} \quad \text{and} \quad \mathbf{P}(b) = \begin{pmatrix} 0 & 1 & 0 & 0 \\ 1 & 0 & 0 & 0 \\ 0 & 0 & 0 & 1 \\ 0 & 0 & 1 & 0 \end{pmatrix}$$

where $\mathbf{P}(a)$ and $\mathbf{P}(b)$ denote the permutation matrices associated with a and b respectively. It is easy to verify that

$$\mathbf{P}(a)^3 = \mathbf{P}(b)^2 = \mathbf{P}(ab)^3 = \mathbf{I}_4.$$

The character corresponding to the natural representation of $\mathbf{P}$, called the *natural character* of $\mathbf{P}$ and denoted by v, is easily calculated since evidently

$$v(\pi) = \text{number of objects left fixed by } \pi. \tag{8.3.22}$$

EXAMPLE 8.3.11.   The natural character of A_4 takes values

$$4, 0, 1 \quad \text{and} \quad 1$$

corresponding to the conjugacy classes c_1, c_2, c_3 and c_4 respectively. Now, since these classes contain 1, 3, 4 and 4 elements respectively, we find by (8.3.10) that

$$(v, v) = \tfrac{1}{12}\{4^2 + 3.0^2 + 4.1^2 + 4.1^2\} = 2,$$

and so by Theorem 8.3.9, the natural representation of A_4 is reducible. Referring to (8.3.21) we deduce that in the Fourier analysis

$$v = e_1 \chi^{(1)} + e_2 \chi^{(2)} + e_3 \chi^{(3)} + e_4 \chi^{(4)}$$

exactly two terms occur with coefficients unity while all other terms are zero. Inner products are best evaluated with the aid of (8.3.10). Thus if $\chi^{(1)}$ is the trivial character, we find that

$$\langle \nu, \chi^{(1)} \rangle = 1.$$

Hence $e_1 = 1$, and we may write

$$\nu = \chi^{(1)} + \chi^{(2)}, \tag{8.3.23}$$

where $\chi^{(2)}$ is an irreducible character other than $\chi^{(1)}$. Solving (8.2.23) for $\chi^{(2)}$ we find that $\chi^{(2)}$ takes the values 3, -1, 0 0 for the conjugacy classes c_1, c_2, c_3, c_4. We can now draw up a provisional character table for A_4 as follows:

	c_1	c_2	c_3	c_4
h	1	3	4	4
$\chi^{(1)}$	1	1	1	1
$\chi^{(2)}$	3	-1	0	0
$\chi^{(3)}$	1	x	α	$\bar{\alpha}$
$\chi^{(4)}$	1	y	β	$\bar{\beta}$

Table 8.3.3

For the formula $12 = f_1^2 + f_2^2 + f_3^2 + f_4^2$ [see (8.3.16)] makes it plain that $f_3 = f_4 = 1$ and we have used the fact that the entries for the columns headed c_3 and c_4 must be complex conjugate because the elements in c_3 and c_4 are inverse to one another [see (8.3.8)]. The relation $\langle \chi^{(2)}, \chi^{(3)} \rangle = 0$ tells us that $0 = 3 \times 1 + 3x(-1)$, whence

$$x = 1.$$

Similarly,

$$y = 1.$$

Next, after a short calculation we find that the equations

$$\langle \chi^{(1)}, \chi^{(3)} \rangle = 0, \qquad \langle \chi^{(3)}, \chi^{(3)} \rangle = 1$$

are equivalent to

$$\alpha + \bar{\alpha} = -1, \qquad \alpha\bar{\alpha} = 1.$$

By interchanging $\chi^{(3)}$ and $\chi^{(4)}$ we can establish that

$$\beta + \bar{\beta} = -1, \qquad \beta\bar{\beta} = 1.$$

Hence we infer [see § 14.5] that α and β are the two roots of the quadratic

$$t^2 + t + 1 = 0.$$

Thus we may choose our notation so that

$$\alpha = -\tfrac{1}{2} + \frac{i\sqrt{3}}{2} = \omega, \qquad \beta = -\tfrac{1}{2} - \frac{i\sqrt{3}}{2} = \omega^2.$$

The final form of the character table for the group A_4 is given below:

	c_1	c_2	c_3	c_4
h	1	3	4	4
$\chi^{(1)}$	1	1	1	1
$\chi^{(2)}$	3	-1	0	0
$\chi^{(3)}$	1	1	ω	ω^2
$\chi^{(4)}$	1	1	ω^2	ω

Table 8.3.4

8.4. OTHER MATHEMATICAL STRUCTURES

Although groups are much the most important of those mathematical structures which consist of a set with one binary operation [for structures with more than one binary operation, see § 2.3 and § 2.4.2] there are others worth mentioning. These are defined as follows.

DEFINITION 8.4.1.   A *groupoid* is a non-empty set upon which a binary operation has been defined. Thus by definition [see § 1.5] a groupoid is a set, closed under a binary operation.

DEFINITION 8.4.2.   A non-empty set with an associative binary operation [see (1.5.2)] is called a *semigroup* and if this set also contains an element which acts as an identity under this operation, the semigroup is called a *monoid*.

EXAMPLE 8.4.1.   The integers [see § 2.2.1] form a monoid under the operation of multiplication, with identity element 1.

EXAMPLE 8.4.2.   The natural numbers [§ 2.1] form a monoid under addition, with identity 0; they also form a monoid under multiplication, with identity 1.

EXAMPLE 8.4.3.   The set of all even integers forms a semigroup under addition, and also under multiplication.

REFERENCES

Ledermann, W. (1972). *Introduction to Group Theory*, Longman.

Ledermann, W. (1977). *Introduction to Group characters*, Cambridge University Press.

CHAPTER 9

Bilinear and Quadratic Forms

9.1. CLASSIFICATION OF QUADRATIC FORMS UNDER CONGRUENCE

We denote the Euclidean plane, endowed with a fixed Cartesian coordinate system [see V, § 1.2.1], by $\mathbb{R}^2$, and Euclidean 3-dimensional space by $\mathbb{R}^3$ [see V, § 2.4]. Thus a point of $\mathbb{R}^2$ is given by an ordered pair of real numbers (x, y) and a point of $\mathbb{R}^3$ by an ordered triplet (x, y, z). Our point of view in this chapter will be that $\mathbb{R}^2$ and $\mathbb{R}^3$ are *sets* whose *elements* are pairs of real numbers (x, y) and triplets of real numbers (x, y, z), respectively. We are then led to a natural generalization to *n-dimensional space* $\mathbb{R}^n$, a set whose elements are ordered *n*-tuples of real numbers $(x_1, x_2, \ldots, x_n)$. (We may, of course, take $n = 1$; this is the real line [see Figure 2.6.1] and it is customary to write $\mathbb{R}$ instead of $\mathbb{R}^1$.)

Certainly $\mathbb{R}^n$ is more than just a set. It has a great deal of structure on it which enables us to do geometry on *n*-dimensional space with the aid of algebra. When our point of view is geometrical we should keep in mind that it is possible to change coordinate systems in $\mathbb{R}^n$ [see § 5.4]; however, we will adopt the view here that the origin of coordinates is fixed.

Let us consider $\mathbb{R}^n$ as a real vector space [see Example 5.2.2]. Then $\mathbb{R}^n$ has the additional structure of an *inner product*, defined by

$$(\mathbf{x}, \mathbf{y}) = x_1 y_1 + x_2 y_2 + \ldots + x_n y_n \qquad (9.1.1)$$

[cf. (6.5.3)] where

$$\mathbf{x} = (x_1, x_2, \ldots, x_n)', \qquad \mathbf{y} = (y_1, y_2, \ldots, y_n)'.$$

We may refer to (9.1.1) as the *standard* inner product on $\mathbb{R}^n$, so that the symbol $\mathbb{R}^n$ refers to the vector space with this inner product. Thus if $n = 3$ and $\mathbf{x} = (1, -2, 4)'$, $\mathbf{y} = (-2, -3, 2)'$, then $(\mathbf{x}, \mathbf{y}) = -2 + 6 + 8 = 12$. Notice that (9.1.1) is associated with the *metric* on $\mathbb{R}^n$ in the following way. First we define a *norm* on $\mathbb{R}^n$ by

$$\|\mathbf{x}\| = \sqrt{(\mathbf{x}, \mathbf{x})}, \qquad (9.1.2)$$

and then the metric ρ on $\mathbb{R}^n$ is given by

$$\rho(\mathbf{x}, \mathbf{y}) = \|\mathbf{x} - \mathbf{y}\|. \qquad (9.1.3)$$

287

This is the usual Euclidean metric (familiar in the cases $n = 2, 3$ [see V, (2.4.2)]), which gives the distance $\rho(\mathbf{x}, \mathbf{y})$ between $\mathbf{x}$ and $\mathbf{y}$ as

$$\rho(\mathbf{x}, \mathbf{y}) = \sqrt{(x_1 - y_1)^2 + (x_2 - y_2)^2 + \ldots + (x_n - y_n)^2}.$$

Now let $\mathbf{A}$ be a real symmetric $(n \times n)$-matrix [see (6.7.5)]. We define the associated real *quadratic form* Q as the function $Q: \mathbb{R}^n \to \mathbb{R}$ [see IV, § 5.3], given by

$$Q(\mathbf{x}) = (\mathbf{A}\mathbf{x}, \mathbf{x}), \qquad \mathbf{x} \in \mathbb{R}^n. \tag{9.1.4}$$

Thus, if $\mathbf{A} = (a_{ij})$, then we have explicitly

$$Q(\mathbf{x}) = \sum_{i,j=1}^n a_{ij} x_i x_j = \sum_{i=1}^n a_{ii} x_i^2 + 2 \sum_{i<j} a_{ij} x_i x_j. \tag{9.1.5}$$

Conversely, to obtain the symmetric matrix $\mathbf{A}$ from the quadratic form Q one enters as the diagonal element a_{ii} the coefficient of x_i^2 in Q and as the element a_{ij}, $i < j$ (which must be the same as the element a_{ji}), one-half of the coefficient of $x_i x_j$ in Q. For example, with $n = 2$, a typical form is $a_{11} x_1^2 + 2a_{12} x_1 x_2 + a_{22} x_2^2$, the associated matrix being

$$\begin{pmatrix} a_{11} & a_{12} \\ a_{21} & a_{22} \end{pmatrix}$$

with $a_{12} = a_{21}$.

We may also associate with the real symmetric matrix $\mathbf{A}$ the *bilinear form* B, defined as the function $B: \mathbb{R}^n \times \mathbb{R}^n \to \mathbb{R}$, given by

$$B(\mathbf{x}, \mathbf{y}) = (\mathbf{A}\mathbf{x}, \mathbf{y}).$$

PROPOSITION 9.1.1. *The inner product* (9.1.1) *satisfies the identity* $(\mathbf{T}\mathbf{x}, \mathbf{y}) = (\mathbf{x}, \mathbf{T}'\mathbf{y})$, *for any* $\mathbf{x}, \mathbf{y} \in \mathbb{R}^n$ *and* $(n \times n)$-*matrix* $\mathbf{T}$, *where* $\mathbf{T}'$ *is the transpose of* $\mathbf{T}$ [*see* § 6.5].

Proof. The inner product is given by $(\mathbf{x}, \mathbf{y}) = \mathbf{x}'\mathbf{y}$. Thus

$$(\mathbf{T}\mathbf{x}, \mathbf{y}) = (\mathbf{T}\mathbf{x})'\mathbf{y} = \mathbf{x}'\mathbf{T}'\mathbf{y} = (\mathbf{x}, \mathbf{T}'\mathbf{y}).$$

We use Proposition 9.1.1 to motivate the introduction of the *congruence* relationship between real symmetric $(n \times n)$-matrices. Let $\mathbf{T}$ be a non-singular $(n \times n)$-matrix [see Definition 6.4.2] and let $\mathbf{y} = \mathbf{T}\mathbf{x}$, $\mathbf{x} \in \mathbb{R}^n$. Then

$$(\mathbf{A}\mathbf{y}, \mathbf{y}) = (\mathbf{A}\mathbf{T}\mathbf{x}, \mathbf{T}\mathbf{x}) = (\mathbf{T}'\mathbf{A}\mathbf{T}\mathbf{x}, \mathbf{x}).$$

This means that the matrices $\mathbf{A}$ and $\mathbf{T}'\mathbf{A}\mathbf{T}$ essentially determine the same quadratic form so that it is natural to consider the equivalence class [see (1.3.5)] of real symmetric $(n \times n)$-matrices $\mathbf{A}$ under the *congruence* relation

$$\mathbf{A} \sim \mathbf{T}'\mathbf{A}\mathbf{T}, \quad \mathbf{T} \text{ non-singular } (n \times n)\text{-matrix}. \tag{9.1.6}$$

We would, of course, be led to the same equivalence relation of congruence if we put the emphasis on bilinear forms rather than quadratic forms. We may, if we wish, speak of congruent quadratic forms, or congruent bilinear forms.

We may classify real symmetric $(n \times n)$-matrices up to congruence as follows. First we show that every such matrix $\mathbf{A}$ is congruent to a matrix of the form

$$\begin{pmatrix} \mathbf{I}_r & 0 & 0 \\ 0 & -\mathbf{I}_s & 0 \\ 0 & 0 & 0 \end{pmatrix} \qquad (9.1.7)$$

where I_m denotes the unit matrix of order m [see § 6.6 and (6.2.7)]. It may then be shown that the integers r, s are uniquely determined by $\mathbf{A}$ and are invariant under congruence. Thus the congruence classes are in one-one correspondence with ordered pairs of non-negative integers (r, s) with $r + s \leq n$ [see § 1.2.6]. Actually it is more customary to speak of the pair $(r + s, r - s)$ as completely specifying the congruence class; then $r + s$ is, of course, the *rank* of A [see (5.6.4)] and $r - s$ is, by definition, the *signature* of A.

Let us write $\mathbf{I}_n(r, s)$ for the matrix (9.1.7). Obvious modifications have to be made when $r = 0$ or $s = 0$ or $r + s = n$; in these cases the first or the second or the third diagonal block in (9.1.7) is missing, together with the corresponding row and column. For example,

$$\mathbf{I}_n(0, s) = \begin{pmatrix} -\mathbf{I}_s & 0 \\ 0 & 0 \end{pmatrix}, \qquad \text{when } s < n,$$

$$\mathbf{I}(n, 0) = \mathbf{I}_n,$$

$$\mathbf{I}(n - 2, 2) = \begin{pmatrix} \mathbf{I}_{n-2} & 0 \\ 0 & -\mathbf{I}_2 \end{pmatrix}, \qquad \text{when } n > 2, \text{ etc.}$$

Then our main theorem may be stated as follows; it is due to the English mathematician J. J. Sylvester.

THEOREM 9.1.2.   *Let $\mathbf{A}$ be a real symmetric $(n \times n)$-matrix. Then there exists a unique pair of non-negative integers r, s, with $r + s \leq n$, such that $\mathbf{A} \sim \mathbf{I}_n(r, s)$, where $\mathbf{I}_n(r, s)$ is the matrix (9.1.7).*

So long as no confusion is to be feared we will write $\mathbf{I}(r, s)$ for $\mathbf{I}_n(r, s)$. We will be content to show how to find the matrix $\mathbf{I}(r, s)$ such that $\mathbf{A} \sim \mathbf{I}(r, s)$. The interested reader may consult a standard text to find the proof that r, s are uniquely determined by $\mathbf{A}$. We know that the eigenvalues of $\mathbf{A}$ are real [see Theorem 7.8.1(i)]. Let them be $\lambda_1, \lambda_2, \ldots, \lambda_n$, where each eigenvalue is taken with the appropriate multiplicity. Moreover, suppose the eigenvalues ordered so that $\lambda_1, \lambda_2, \ldots, \lambda_r$ are positive, $\lambda_{r+1}, \lambda_{r+2}, \ldots, \lambda_{r+s}$ are negative, and the rest are zero. Finally, let us choose eigenvectors $\mathbf{v}_1, \mathbf{v}_2, \ldots, \mathbf{v}_n$, corresponding to the eigenvalues $\lambda_1, \lambda_2, \ldots, \lambda_n$, respectively, such that $\mathbf{v}_1, \mathbf{v}_2, \ldots, \mathbf{v}_n$ form an orthonormal basis for $\mathbb{R}^n$ [see Definition 10.2.2]—we know that this is possible [see Theorem 7.8.1(ii)].

Now set $\mathbf{H} = (\mathbf{v}_1, \mathbf{v}_2, \ldots, \mathbf{v}_n)$, so that $\mathbf{H}$ is a (real) orthogonal $(n \times n)$-matrix [see § 6.7(vii)]. Then

$$(\mathbf{H'AH})_{ij} = (\mathbf{v}_i'\mathbf{A}\mathbf{v}_j) = (\lambda_j\mathbf{v}_i'\mathbf{v}_j) = \begin{cases} \lambda_i, & i = j, \\ 0, & i \neq j. \end{cases}$$

Next, let **S** be the real diagonal matrix

$$\mathbf{S} = \operatorname{diag}(\lambda_1^{-1/2}, \ldots, \lambda_r^{-1/2}, (-\lambda_{r+1})^{-1/2}, \ldots, (-\lambda_{r+s})^{-1/2}, 1, \ldots, 1),$$

where we use the notation

$$\operatorname{diag}(a_1, a_2, \ldots, a_n) = \begin{pmatrix} a_1 & & & \mathbf{0} \\ & a_2 & & \\ & & \ddots & \\ \mathbf{0} & & & a_n \end{pmatrix}$$

to specify a diagonal matrix [see (6.7.3)].

Then, if **T** = **HS**, it is plain that **T** is non-singular [see Definition 6.12.1], and

$$\mathbf{T}'\mathbf{A}\mathbf{T} = \mathbf{I}(r, s),$$

as required.

By studying the form of the matrix **H′AH** in this argument, we obtain the following important result.

PROPOSITION 9.1.3.   *Let* **A** *be a real symmetric* (n × n)*-matrix with* **A** ∼ **I**(r, s). *Then we may choose orthonormal coordinate axes in* $\mathbb{R}^n$ [*see* Definition 10.2.2] *such that the associated quadratic form Q is given by*

$$Q(\mathbf{x}) = \lambda_1 x_1^2 + \ldots + \lambda_r x_r^2 + \lambda_{r+1} x_{r+1}^2 + \ldots + \lambda_{r+s} x_{r+s}^2, \qquad (9.1.8)$$

where $\lambda_1, \ldots, \lambda_{r+s}$ *are the non-zero eigenvalues of* **A**, *counted with their multiplicities. Moreover, we may take* $\lambda_1, \ldots, \lambda_r$ *positive,* $\lambda_{r+1}, \ldots, \lambda_{r+s}$ *negative.*

We say that by this choice of coordinate axes in $\mathbb{R}^n$ we have *diagonalized* the quadratic form Q (and the matrix **A**). We also say that we have *reduced* the quadratic form to diagonal form.

Of course, if we are prepared to accept *any* basis for the vector space $\mathbb{R}^n$, then Theorem 9.1.2 shows that we may express the quadratic form by

$$Q(\mathbf{x}) = x_1^2 + \ldots + x_r^2 - x_{r+1}^2 - \ldots - x_{r+s}^2 \qquad (9.1.9)$$

where $x_1, x_2, \ldots, x_n$ are the coefficients of **x** expressed as a linear combination of the basis elements. Let us say that the quadratic form $Q(\mathbf{x})$ of (9.1.9) is *of type* (r, s). Notice that r, s can be any non-negative integers subject only to the condition $r + s \leq n$.

We may illustrate the possibilities implicit in (9.1.9) in the simple (but important!) case n = 2. Ignoring the zero quadratic form $Q(\mathbf{x}) \equiv 0$, which is, of course, of type (0, 0), we have the following possibilities for the type of the quadratic form $Q(\mathbf{x})$, namely (2, 0), (1, 1), (0, 2), (1, 0), (0, 1). We will give examples of each of these five possibilities. Since we will only be concerned with the case $n = 2$, we will use the symbols x, y for the components of a vector, that is, we shall put

$$\mathbf{x} = (x, y)'.$$

EXAMPLE 9.1.1. Consider the quadratic form $Q \equiv 34x^2 - 24xy + 41y^2$. The associated matrix is

$$\mathbf{A} = \begin{pmatrix} 34 & -12 \\ -12 & 41 \end{pmatrix}.$$

To find the eigenvalues of $\mathbf{A}$ we must solve the equation

$$\begin{vmatrix} 34 - \lambda & -12 \\ -12 & 41 - \lambda \end{vmatrix} = 0,$$

[see (7.2.3)], or

$$\lambda^2 - 75\lambda + 1250 = 0,$$

yielding $\lambda = 25$ or 50 [see (2.7.9)]. An eigenvector $\begin{pmatrix} a \\ b \end{pmatrix}$ corresponding to $\lambda = 25$ must satisfy

$$\begin{pmatrix} 9 & -12 \\ -12 & 16 \end{pmatrix} \begin{pmatrix} a \\ b \end{pmatrix} = \begin{pmatrix} 0 \\ 0 \end{pmatrix};$$

thus $3a = 4b$. Thus a *unit* vector [see Definition 10.2.1] corresponding to $\lambda = 25$ is $(\frac{4}{5}, \frac{3}{5})$. Similarly a unit vector corresponding to $\lambda = 50$ is $(-\frac{3}{5}, \frac{4}{5})$. Thus (see the argument following the statement of Theorem 9.1.2) if

$$\mathbf{H} = \begin{pmatrix} \frac{4}{5} & \frac{3}{5} \\ -\frac{3}{5} & \frac{4}{5} \end{pmatrix} = \frac{1}{5} \begin{pmatrix} 4 & 3 \\ -3 & 4 \end{pmatrix},$$

then $\mathbf{H}$ is orthogonal and

$$\mathbf{H}'\mathbf{A}\mathbf{H} = \begin{pmatrix} 25 & 0 \\ 0 & 50 \end{pmatrix}.$$

Thus the transformation $\mathbf{A} \to \mathbf{T}'\mathbf{A}\mathbf{T}$, where $\mathbf{T} = \mathbf{H}\mathbf{S}$ and

$$\mathbf{S} = \begin{pmatrix} \frac{1}{5} & 0 \\ 0 & \frac{1}{5\sqrt{2}} \end{pmatrix},$$

transforms $\mathbf{A}$ into $I(2, 0)$, and Q is of type $(2, 0)$. Moreover, we may choose orthonormal coordinate axes in $\mathbb{R}^2$ such that Q is given by $25x_1^2 + 50x_2^2$ (Proposition 9.1.3); and if we allow *any* basis in the vector space $\mathbb{R}^2$ then Q may be transformed into $x_1^2 + x_2^2$.

Notice that if we are only asked to determine the type of Q it suffices to 'complete the square'. That is, if $Q = ax^2 + 2hxy + by^2$, we compute the discriminant $d = h^2 - ab$ and then, assuming $a \neq 0$, we find

$$ax^2 + 2hxy + by^2 = a^{-1}((ax + hy)^2 - dy^2). \qquad (9.1.10)$$

Thus we see that if $a > 0$ and $d < 0$, then Q is of type $(2, 0)$ while if $a < 0$ and $d < 0$, then Q is of type $(0, 2)$. We will revert to this point of view in the next section.

Here we are content to note that the type of Q may be read off from (9.1.10) provided that $a \neq 0$. When $a = 0$ and $h \neq 0$, we have that

$$Q = 2hxy + by^2 = (2hx + by)y$$
$$= \tfrac{1}{4}(2hx + (b + 1)y)^2 - \tfrac{1}{4}(2hx + (b - 1)y)^2,$$

and we conclude that Q is of type $(1, 1)$. Finally, when $a = 0$, $h = 0$ and therefore $b \neq 0$, then Q reduces to

$$Q = by^2,$$

which is of type $(1, 0)$ when $b > 0$ and of type $(0, 1)$ when $b < 0$.

In Example 9.1.1 we have that $a = 34$, $d = -1250$, confirming that the quadratic form of that example is of type $(2, 0)$.

EXAMPLE 9.1.2. Plainly Q is of type $(0, 2)$ if and only if $-Q$ is of type $(2, 0)$. Thus the quadratic form $Q \equiv -34x^2 + 24xy - 41y^2$ is of type $(0, 2)$.

EXAMPLE 9.1.3. The remarks preceding Example 9.1.2 enable us easily to construct examples of quadratic forms of type $(1, 1)$. We will, however, take an example and analyse it—though not in detail—as in Example 9.1.1. Let $Q \equiv 119x^2 + 240xy - 119y^2$. The eigenvalues of the associated matrix $\mathbf{A}$ turn out to be ± 169. A unit eigenvector corresponding to the eigenvalue 169 is $(\tfrac{12}{13}, \tfrac{5}{13})$; and a unit vector corresponding to the eigenvalue -169 is $(-\tfrac{5}{13}, \tfrac{12}{13})$. Thus we may take

$$\mathbf{H} = \frac{1}{13} \begin{pmatrix} 12 & 5 \\ -5 & 12 \end{pmatrix}$$

and convert Q to $169(x_1^2 - x_2^2)$, showing that $\mathbf{A} \sim I(1, 1)$ and Q is of type $(1, 1)$.

It is clear that a quadratic form (in two or any number of variables) is of type $(1, 0)$ if and only if it is the square of a *linear form*, where a linear form in n variables is an inner product $(\mathbf{a}, \mathbf{x})$ where $\mathbf{a}$ is a constant vector in $\mathbb{R}^n$ and $\mathbf{x}$ a variable vector; in particular, a linear form in $\mathbb{R}^2$ is an expression $kx + ly$.

Thus if $Q = ax^2 + 2hxy + by^2$, then Q is of type $(1, 0)$ if and only if there are real numbers k, l, not both zero, such that

$$ax^2 + 2hxy + by^2 = (kx + ly)^2. \tag{9.1.11}$$

Similarly Q is of type $(0, 1)$ if and only if there are real numbers, k, l, not both zero, such that

$$ax^2 + 2hxy + by^2 = -(kx + ly)^2. \tag{9.1.12}$$

It follows by elementary algebra that *a necessary and sufficient condition for a non-zero quadratic form* $ax^2 + 2hxy + by^2$ *to be of type* $(1, 0)$ *or* $(0, 1)$ *is that the discriminant* $d = h^2 - ab = 0$.

EXAMPLE 9.1.4. Let $Q = 5x^2 - 30xy + 45y^2$. Since $Q = 5(x - 3y)^2$, Q is of type $(1, 0)$. If we seek an orthogonal transformation (change of coordinates) converting Q to λx_1^2, it suffices to take

$$\mathbf{H} = \frac{1}{\sqrt{10}} \begin{pmatrix} 1 & 3 \\ -3 & 1 \end{pmatrix}, \quad \mathbf{H}\begin{pmatrix} x \\ y \end{pmatrix} = \begin{pmatrix} x_1 \\ y_1 \end{pmatrix}.$$

Then Q becomes $50x_1^2$. Similarly, if $Q = -5x^2 + 30xy - 45y^2$, then Q is of type $(0, 1)$ and the same orthogonal transformation converts Q into $-50x_1^2$.

There are conclusions analogous to (9.1.8), (9.1.9) available for bilinear forms.

9.2. DEFINITENESS OF QUADRATIC FORMS

DEFINITION 9.2.1. We say that the quadratic form $Q(\mathbf{x})$ is

$$\begin{aligned}
&\textit{positive-definite} &&\text{if } \mathbf{x} \neq \mathbf{0} \Rightarrow Q(\mathbf{x}) > 0, \\
&\textit{negative-definite} &&\text{if } \mathbf{x} \neq \mathbf{0} \Rightarrow Q(\mathbf{x}) < 0, \\
&\textit{positive-semi-definite} &&\text{if } Q(\mathbf{x}) \geq 0 \text{ for all vectors } \mathbf{x}, \\
&\textit{negative-semi-definite} &&\text{if } Q(\mathbf{x}) \leq 0 \text{ for all vectors } \mathbf{x}.
\end{aligned}$$

It is plain that these properties are invariant under the congruence relation (9.1.6), since congruent quadratic forms take the same set of values. More precisely, one easily shows

THEOREM 9.2.1. *Let the quadratic form $Q(\mathbf{x})$ be associated with the real symmetric $(n \times n)$-matrix $\mathbf{A}$. Then [see Theorem 9.1.2]:*

$Q(\mathbf{x})$ *is positive-definite if and only if* $\mathbf{A} \sim \mathbf{I}(n, 0)$;
$Q(\mathbf{x})$ *is negative-definite if and only if* $\mathbf{A} \sim \mathbf{I}(0, n)$;
$Q(\mathbf{x})$ *is positive-semi-definite if and only if* $\mathbf{A} \sim \mathbf{I}(r, 0)$,

some $r \leq n$;

$Q(\mathbf{x})$ *is negative-semi-definite if and only if* $\mathbf{A} \sim \mathbf{I}(0, s)$,

some $s \leq n$.

(It is customary to use the terms 'positive-definite', positive-semi-definite' so that the latter excludes the former—that is, to exclude $r = n$ in the definition of 'positive-semi-definite', but this is mathematically inconvenient and we will not do so.)

We may equally well characterize these properties in terms of the eigenvalues of $\mathbf{A}$ [see § 7.9]. Thus

THEOREM 9.2.2. *In the terminology of Theorem 9.2.1,*
$Q(\mathbf{x})$ *is positive-definite if and only if the eigenvalues of $\mathbf{A}$ are all positive;*
$Q(\mathbf{x})$ *is negative-definite if and only if the eigenvalues of $\mathbf{A}$ are all negative;*
$Q(\mathbf{x})$ *is positive-semi-definite if and only if the eigenvalues of $\mathbf{A}$ are all non-negative;*

$Q(\mathbf{x})$ *is negative-semi-definite if and only if the eigenvalues of* $\mathbf{A}$ *are all non-positive.*

In particular we see from Theorem 9.2.1 and Proposition 9.1.3 that a positive-definite quadratic form (in n variables) may be reduced to a sum of n squares; and a positive-semi-definite quadratic form (in n variables) may be reduced to a sum of r squares, $r \leq n$.

Although Theorems 9.2.1 and 9.2.2 give valuable criteria for a quadratic form $Q(\mathbf{x})$ to be positive-definite, etc., they do not give effective ways of deciding such questions. It is no easy matter to calculate the eigenvalues of a matrix [see III § 4.6], so that Theorem 9.2.2 is not at first sight very useful. On the other hand, it is easy to decide whether the eigenvalues are all positive, etc., from just *looking at* the characteristic equation of the associated matrix $\mathbf{A}$ [see (7.3.9)–(7.3.10)]. Thus if

$$P(\lambda) \equiv \lambda^n - c_1\lambda^{n-1} + c_2\lambda^{n-2} - \ldots + (-1)^n c_n = 0 \qquad (9.2.1)$$

is the characteristic equation of $\mathbf{A}$, so that the eigenvalues of $\mathbf{A}$ are the roots of the equation (9.2.1) [see (7.2.3)], then we readily see that (referring to (9.2.1))

$Q(\mathbf{x})$ is positive-definite if and only if all the c_i are positive, for in that case the equation (9.2.1) cannot have a negative root;

$Q(\mathbf{x})$ is negative-definite if and only if all the c_i have the sign $(-1)^i$;

$Q(\mathbf{x})$ is positive-semi-definite if and only if, up to some $r \leq n$, all the c_i are positive, and the remainder are zero.

$Q(\mathbf{x})$ is negative-semi-definite if and only if, up to some $r \leq n$, all the c_i have the sign $(-1)^i$, and the remainder are zero.

The following criteria are even easier to apply. We will be content to state them just for positive-definite and negative-definite quadratic forms and will illustrate them by examples.

THEOREM 9.2.3.   *The quadratic form* $Q(\mathbf{x})$ *is positive-definite if and only if the leading minors* [*see* (6.13.6)] *of the associated matrix* $\mathbf{A}$ *are all positive; it is negative-definite if and only if, for all i, the leading* ($i \times i$)-*minor of the associated matrix A has the sign* $(-1)^i$.

EXAMPLE 9.2.1.   Consider the quadratic form in 3 variables

$$Q(\mathbf{x}) \equiv 5x_1^2 + 2x_2^2 + 2x_3^2 - 6x_1x_2 + 2x_1x_3 - 2x_2x_3.$$

The associated matrix is

$$\mathbf{A} = \begin{pmatrix} 5 & -3 & 1 \\ -3 & 2 & -1 \\ 1 & -1 & 2 \end{pmatrix}$$

and, if we compute the leading minors we obtain 5, 1, 1. Thus $Q(\mathbf{x})$ is positive-definite. The matrix associated with $-Q$ is, of course, $-\mathbf{A}$, and its leading minors are $-5, 1, -1$, showing that $-Q$ is negative-definite.

To explain the next (and last!) criterion which we will be giving, we revert to the general quadratic form (9.1.5) in n variables. Now it is plain that if $Q(\mathbf{x})$, given by (9.1.5), is positive-definite, then $a_{11} > 0$, since a_{11} is the value of the quadratic form Q at the unit vector $(1, 0, \ldots, 0)$ in the x_1-direction. Let us write $a = a_{11}$ so that

$$Q(\mathbf{x}) = \frac{1}{a}\left(a^2 x_1^2 + 2\sum_{j\geq 2} a a_{1j} x_1 x_j + \sum_{i\geq 2} a a_{ii} x_i^2 + 2\sum_{2\leq i<j} a a_{ij} x_i x_j\right)$$

$$= \frac{1}{a}\left(\left(a x_1 + \sum_{j\geq 2} a_{1j} x_j\right)^2 + \sum_{i\geq 2} b_{ii} x_i^2 + 2\sum_{2\leq i<j} b_{ij} x_i x_j\right), \quad (9.2.2)$$

where $b_{ii} = a_{ii} - a_{1i}^2$, $i \geq 2$; and $b_{ij} = a a_{ij} - a_{1i} a_{1j}$, $2 \leq i < j$.

We call this process '*completing the square*'—it generalizes the familiar process in solving quadratic equations. [See (9.1.10) where we carried out the process in the case $n = 2$.] We may now repeat the process with the quadratic form

$$Q' = \sum_{i\geq 2} b_{ii} x_i^2 + 2\sum_{2\leq i<j} b_{ij} x_i x_j$$

in the $(n-1)$ variables $x_2, \ldots, x_n$. We see that b_{22} is precisely the leading (2×2)-minor $a_{11}a_{22} - a_{12}^2$ of the matrix A associated with the original quadratic form Q; however, as we continue the process, the '*leading coefficients*' which emerge, $a_{11}, b_{22}, c_{33}, \ldots$, are *not* just the leading minors of $\mathbf{A}$, so that this criterion is strictly different from that given immediately previously. We illustrate the method by discussing again the quadratic form of Example 9.2.1.

EXAMPLE 9.2.1 (continued). We revert to the quadratic form in 3 variables

$$Q(\mathbf{x}) \equiv 5x_1^2 + 2x_2^2 + 2x_3^2 - 6x_1x_2 + 2x_1x_3 - 2x_2x_3.$$

Then

$$Q(\mathbf{x}) = \tfrac{1}{5}(25x_1^2 - 30x_1x_2 + 10x_1x_3 + 10x_2^2 + 10x_3^2 - 10x_2x_3)$$
$$= \tfrac{1}{5}((5x_1 - 3x_2 + x_3)^2 + x_2^2 + 9x_3^2 - 4x_2x_3)$$
$$= \tfrac{1}{5}((5x_1 - 3x_2 + x_3)^2 + (x_2 - 2x_3)^2 + 5x_3^2).$$

Thus the 'leading coefficients', when we have finished the process of 'completing the square' are 5, 1, 5. Since they are all positive, Q is positive-definite.

Now it is plain that, if we take $-Q$ instead of Q, then the first 'leading coefficient' is -5, but, after that, there is no difference between the leading coefficients arising from Q and $-Q$, since they produce the same Q'. Of course, as we have indicated several times, we do not need to test for negative-definiteness, since Q is negative-definite if and only if $-Q$ is positive-definite.

We give one more example to explain what we mean by 'leading coefficients'; we prefer to do this since we do not think a formal definition would be illuminating.

EXAMPLE 9.2.2. Let $Q(\mathbf{x}) = 3x_1^2 + 7x_2^2 + 6x_3^2 + 2x_1x_2 + 6x_1x_3 - 6x_2x_3.$

The first 'leading coefficient' is 3, the coefficient of x_1^2, and we proceed to complete the square:

$$Q(\mathbf{x}) = \tfrac{1}{3}(9x_1^2 + 6x_1x_2 + 18x_1x_3 + 21x_2^2 + 18x_3^2 - 18x_2x_3)$$
$$= \tfrac{1}{3}((3x_1 + x_2 + 3x_3)^2 + 20x_2^2 + 9x_3^2 - 24x_2x_3).$$

We now look at $Q' = 20x_2^2 + 9x_3^2 - 24x_2x_3$. Thus, the next 'leading coefficient' is 20 and we again complete the square:

$$Q' = \tfrac{1}{20}(400x_2^2 - 480x_2x_3 + 180x_3^2)$$
$$= \tfrac{1}{20}((20x_2 - 12x_3)^2 + 36x_3^2).$$

We see that the third 'leading coefficient' is 36. Since 3, 20, 36 are all positive, Q is positive-definite.

We point out, referring to this example, that it may be easier computationally to 'attack' the variables $x_1, x_2, x_3, \ldots$ in a different order. Thus, once we obtained Q' above, it is easier to proceed by observing that

$$Q' = (3x_3 - 4x_2)^2 + 4x_2^2,$$

showing that Q', and hence Q, is positive-definite. (This remark applies also to the 'leading minors' criterion of Theorem 9.2.3.)

Assuming the idea of leading coefficients is now clear, we may sum up our last discussion in a theorem.

THEOREM 9.2.4.   *The quadratic form $Q(\mathbf{x})$ is positive-definite if and only if the leading coefficients on carrying out the entire process of 'completing the square' are all positive.*

9.3.   HERMITIAN FORMS

There is a natural and useful generalization of the development above to the case of the complex vector space $\mathbb{C}^n$, whose elements are n-tuples of complex numbers [see Example 5.2.2]. The inner product of two vectors $\mathbf{z}, \mathbf{w}$ in $\mathbb{C}^n$ is then given by

$$(\mathbf{z}, \mathbf{w}) = z_1\bar{w}_1 + z_2\bar{w}_2 + \ldots + z_n\bar{w}_n \qquad (9.3.1)$$

where

$$\mathbf{z} = (z_1, z_2, \ldots, z_n)', \qquad \mathbf{w} = (w_1, w_2, \ldots, w_n)',$$

and $\bar{a}$ denotes the complex conjugate of the complex number a [see (2.7.10)]. This is called the *standard* inner product of $\mathbb{C}^n$. We again use the inner product to introduce a norm and a metric into $\mathbb{C}^n$; just as we did in section 9.1 in the case of the real vector space $\mathbb{R}^n$.

Now let $\mathbf{A}$ be a (complex) Hermitian $(n \times n)$-matrix [see (6.7.10)]. We define the associated Hermitian quadratic form Q as the function $Q: \mathbb{C}^n \to \mathbb{R}$ [see IV, § 5.3] given by

$$Q(\mathbf{z}) = (\mathbf{Az}, \mathbf{z}). \qquad (9.3.2)$$

Notice that this definition gives us $Q(\mathbf{z}) = \sum a_{ij}\bar{z}_i z_j$. To continue the extension of the theory to the complex case, we next point out that Q is, in fact, *real-valued*, as claimed above. To see this, notice first that if, for any complex matrix $\mathbf{A}$, $\mathbf{A}^*$ is the conjugate transpose of $\mathbf{A}$ [see (6.7.11)], then $\mathbf{A}$ is Hermitian if and only if $\mathbf{A} = \mathbf{A}^*$. To prove that Q is real-valued we first generalize Proposition 9.1.1 to read

PROPOSITION 9.3.1. *The inner product (9.3.1) satisfies the identity* $(\mathbf{Tz}, \mathbf{w}) = (\mathbf{z}, \mathbf{T}^*\mathbf{w})$ *for any* $\mathbf{z}, \mathbf{w} \in \mathbb{C}^n$ *and complex* $(n \times n)$-*matrix* $\mathbf{T}$, *where* $\mathbf{T}^* = \overline{\mathbf{T}}'$.

We then observe that $\overline{(z, w)} = (w, z)$ [cf. Example 2.7.7], so that, if $\mathbf{A}$ is Hermitian,

$$\overline{(\mathbf{Az}, \mathbf{z})} = (\mathbf{z}, \mathbf{Az}) = (\mathbf{z}, \mathbf{A}^*\mathbf{z}) = (\mathbf{Az}, \mathbf{z}),$$

and therefore $(\mathbf{Az}, \mathbf{z})$ is real. Next we recall that the eigenvalues of $\mathbf{A}$ are all real [see Theorem 7.8.1]. Finally, we point out that the congruence relation (9.1.6) naturally generalizes to the relation

$$\mathbf{A} \sim \mathbf{T}^*\mathbf{AT}, \quad \mathbf{T} \text{ non-singular complex } (n \times n)\text{-matrix.} \tag{9.3.3}$$

The remaining development is exactly as in the real case; in particular, a quadratic from Q is positive-definite if and only if it may be reduced to the form

$$Q(\mathbf{z}) = z_1\bar{z}_1 + z_2\bar{z}_2 + \ldots + z_n\bar{z}_n.$$

P.H.

BIBLIOGRAPHY

Finkbeiner, D. T. (1960). *Introduction to Matrices and Linear Transformations*, Freeman & Co.

Lowenthal, F. (1975). *Linear Algebra (with Linear Differential Equations)*, John Wiley & Sons.

Morris, A. O. (1978). *Linear Algebra*, Van Nostrand Reinhold Co.

Tropper, A. M. (1969). *Linear Algebra*, Nelson.

CHAPTER 10

Metric Vector Spaces

10.1. INNER PRODUCTS AND METRICS

We saw in the preceding chapter how to study quadratic and bilinear forms on $\mathbb{R}^n$ and $\mathbb{C}^n$, where $\mathbb{R}^n$ and $\mathbb{C}^n$ are endowed with their standard, Euclidean inner product. Here we consider a generalization to arbitrary vector spaces [see § 5.2]. We will assume the ground field to be the complex numbers $\mathbb{C}$—the adjustments necessary for vector spaces over the real numbers $\mathbb{R}$ should be obvious to the reader.

If we study the development in the preceding chapter, it readily becomes apparent what properties we used of the inner product (9.1.1) that we introduced. We are thus led to the following definition [cf. IV, § 11.6.3]:

DEFINITION 10.1.1.   Let V be a vector space over $\mathbb{C}$. Then an *inner product* on V is a complex-valued function, $(\mathbf{x}, \mathbf{y})$, of ordered pairs of vectors $\mathbf{x}, \mathbf{y}$ in V [see IV, § 5.3], subject to the conditions:

(a) $\overline{(\mathbf{x}, \mathbf{y})} = (\mathbf{y}, \mathbf{x})$ where $\bar{a}$ denotes the conjugate of the complex number a [see (2.7.10)]

(b) $(a_1\mathbf{x}_1 + a_2\mathbf{x}_2, \mathbf{y}) = a_1(\mathbf{x}_1, \mathbf{y}) + a_2(\mathbf{x}_2, \mathbf{y})$, $a_1, a_2 \in \mathbb{C}$

(c) $(\mathbf{x}, \mathbf{x}) \geq 0$

(d) $(\mathbf{x}, \mathbf{x}) = 0$ if and only if $\mathbf{x} = 0$.

We remark that (b) asserts the *linearity* of $(\mathbf{x}, \mathbf{y})$ in the first variable [cf. (5.11.1)]. From (a) and (b) we readily infer the *conjugate linearity* of $(\mathbf{x}, \mathbf{y})$ in the second variable, that is,

(b̄) $(\mathbf{x}, b_1\mathbf{y}_1 + b_2\mathbf{y}_2) = \bar{b}_1(\mathbf{x}, \mathbf{y}_1) + \bar{b}_2(\mathbf{x}, \mathbf{y}_2)$.

Just as for the standard inner product in $\mathbb{C}^n$ (or $\mathbb{R}^n$) (see (9.1.1) and (9.3.1)) we may define a *norm* and a *metric* in V in terms of the inner product, thus,

$$\|\mathbf{x}\| = \sqrt{(\mathbf{x}, \mathbf{x})}, \tag{10.1.1}$$

$$\rho(\mathbf{x}, \mathbf{y}) = \|\mathbf{x} - \mathbf{y}\|. \tag{10.1.2}$$

We now establish some properties of the norm [cf. IV, § 11.6.1] and the metric (including the defining properties of a metric [cf. IV, § 11.1]); however, it is of course crucial that *here* we only use properties (a)–(d) of Definition 10.1.1!

PROPOSITION 10.1.1.   *If* $x \in V, a \in \mathbb{C}$, *then* $\|ax\| = |a| \|x\|$ *where* $|a|$ *denotes the modulus of a* [*see* (2.7.13)].

Proof. $\|ax\|^2 = (ax, ax) = a\bar{a}(x, x)$, by (b) and $(\bar{b})$, $= |a|^2 \|x\|^2$.

PROPOSITION 10.1.2   (Schwarz's inequality). $|(x, y)| \le \|x\| \|y\|$, *equality occurring if and only if* x *and* y *are linearly dependent.*

Proof.   The proof given [IV, chapter 21] goes over essentially without change.

PROPOSITION 10.1.3.   *The function* (10.1.2) $\rho(x, y) = \|x - y\|$ *satisfies the conditions for a metric on* V, *namely,*
   (i) $\rho(x, y) = \rho(y, x)$
   (ii) $\rho(x, y) \ge 0$; *and* $\rho(x, y) = 0$ *if and only if* $x = y$
   (iii) (*the triangle inequality*) $\rho(x, y) + \rho(y, z) \ge \rho(x, z)$.
Proof.   Only (iii) need detain us. Since ρ is obviously *translation-invariant*, that is,

$$\rho(x, y) = \rho(x + v, y + v), \tag{10.1.3}$$

it suffices to show that $\|x\| + \|y\| \ge \|x + y\|$. Using conditions (b) and $(\bar{b})$ of Definition 10.1.1 and the fact that $(x, y) + (y, x) = \mathrm{Re}\,(x, y)$ we find that

$$\begin{aligned}
\|x + y\|^2 &= \|x\|^2 + 2\,\mathrm{Re}\,(x, y) + \|y\|^2 \\
&\le \|x\|^2 + 2|(x, y)| + \|y\|^2 \\
&\le \|x\|^2 + 2\|x\| \|y\| + \|y\|^2, \quad \text{by the Schwarz inequality} \\
&= (\|x\| + \|y\|)^2.
\end{aligned}$$

In this argument, $\mathrm{Re}\,z$ refers to the real part of the complex number z [see § 2.7.1].

We call a vector space V endowed with an inner product an *inner product space*. Then (10.1.1) gives us a definition of a *norm* in an inner product space and this norm satisfies the conditions

$$\begin{aligned}
\|ax\| &= |a| \|x\| \\
\|x\| &= 0 \quad \text{if and only if } x = 0 \qquad (10.1.4) \\
\|x\| + \|y\| &\ge \|x + y\|.
\end{aligned}$$

We define a *normed* vector space to be a vector space V endowed with a norm function $V \to \mathbb{R}^+$ (the non-negative reals), satisfying (10.1.4). Thus an inner product on V automatically yields a norm on V, but there are normed vector spaces whose norms do not arise from an inner product (see Remark (ii) below).

On the other hand, a norm on V automatically yields a translation-invariant metric on V by the rule (10.1.2) $\rho(\mathbf{x}, \mathbf{y}) = \|\mathbf{x} - \mathbf{y}\|$. If we wish only to emphasize the translation-invariant metric ρ obtained from the norm, we may call V a *metric vector space*, although there is no logical difference between this concept and that of a normed vector space.

Remarks. (i) The metric ρ obtained from a norm has further linearity properties beyond translation-invariance. Thus $\rho(a\mathbf{x}, a\mathbf{y}) = |a|\rho(\mathbf{x}, \mathbf{y})$.

(ii) A straightforward calculation shows that any norm arising from an inner product has the *Apollonius property*

$$\|\mathbf{x} + \mathbf{y}\|^2 + \|\mathbf{x} - \mathbf{y}\|^2 = 2(\|\mathbf{x}\|^2 + \|\mathbf{y}\|^2). \qquad (10.1.5)$$

A normed vector space satisfying (10.1.5), which is complete [see IV, § 11.3] is sometimes called a *Hilbert space* [see IV, § 11.6.3]. There are complete normed vector spaces, called *Banach spaces*, of great importance in functional analysis, some of which are not Hilbert spaces, that is, whose norms do not arise from an inner product [see IV, § 11.6.2].

Before proceeding further, we give some examples of inner product spaces. Of course, $\mathbb{R}^n$ and $\mathbb{C}^n$, with their standard inner products [see (9.1.1) and (9.3.1)], are inner product spaces, over $\mathbb{R}$ and $\mathbb{C}$ respectively.

EXAMPLE 10.1.1. We consider the set S of sequences of complex numbers

$$\mathbf{x} = (x_1, x_2, \ldots, x_n, \ldots)$$

such that the series $x_1\bar{x}_1 + x_2\bar{x}_2 + \ldots + x_n\bar{x}_n + \ldots$ converges [see IV, § 1.7]. This is plainly a complex vector space and we endow it with an inner product by the rule

$$(\mathbf{x}, \mathbf{y}) = x_1\bar{y}_1 + x_2\bar{y}_2 + \ldots + x_n\bar{y}_n + \ldots,$$

where $\mathbf{y} = (y_1, y_2, \ldots, y_n, \ldots)$. It may be shown that the series defining $(\mathbf{x}, \mathbf{y})$ does indeed converge (of course, the Schwarz inequality holds), and the verification of properties (a)–(d) of Definition 10.1.1 is immediate. Standard notation for S is $l^2(\mathbb{C})$; analogously we also have $l^2(\mathbb{R})$.

EXAMPLE 10.1.2. Let $P(\mathbb{C})$ be the vector space of polynomials in one variable over $\mathbb{C}$ [see Example 5.2.3]. We endow $P(\mathbb{C})$ with an inner product by the rule

$$(f, g) = \int_0^1 f(x)\overline{g(x)}dx.$$

It is again elementary to verify that this rule satisfies conditions (a)–(d) for an inner product. In fact we can embed $P(\mathbb{C})$ in the larger vector space $L^2(\mathbb{C})$ of square-integrable complex-valued functions [see IV, Example 11.6.4], that is, of functions $f(x)$ such that $\int_0^1 |f(x)|^2 dx$ exists; and the inner product rule given above then also extends. Of course, there is a similar story involving $P(\mathbb{R})$ and $L^2(\mathbb{R})$.

10.2. INNER PRODUCT SPACES

We now proceed to develop the theory of inner product spaces. Although we will deal explicitly with vector spaces over $\mathbb{C}$, the reader will readily adapt the treatment to the study of vector spaces over $\mathbb{R}$. Thus, let V be an inner product space, that is, a vector space (over $\mathbb{C}$) equipped with an inner product. Let us assume that V is n-dimensional [see Definition 5.4.1].

DEFINITION 10.2.1.   Two vectors **x** and **y** in V are called *orthogonal*, if their inner product is zero, that is if

$$(\mathbf{x}, \mathbf{y}) = 0.$$

The vector **x** is a *unit vector* if it has unit norm, that is

$$\|\mathbf{x}\| = 1, \quad \text{or} \quad (\mathbf{x}, \mathbf{x}) = 1.$$

Two unit vectors which are also orthogonal are called *orthonormal*.
    The next definition describes a particularly useful type of basis:

DEFINITION 10.2.2.   The basis $(\mathbf{u}_1, \mathbf{u}_2, \ldots, \mathbf{u}_n)$ for V is called *orthonormal* if each vector $\mathbf{u}_i$ is a unit vector and $(\mathbf{u}_i, \mathbf{u}_j) = 0$ if $i \neq j$.
    Notice that we may write the condition of orthonormality very succinctly using the *Kronecker delta* symbol [see (6.7.1)], thus

$$(\mathbf{u}_i, \mathbf{u}_j) = \delta_{ij}, \quad 1 \leq i, j \leq n. \tag{10.2.1}$$

Obviously the standard basis in $\mathbb{R}^n$ [see Example 5.4.1] (and in $\mathbb{C}^n$) is orthonormal. Thus the following theorem is very important in general.

THEOREM 10.2.1.   *Every n-dimensional inner product space V has an orthonormal basis.*

Proof.   We apply the *Gram-Schmidt* orthogonalization process [IV, § 11.7] to an arbitrary basis $(\mathbf{v}_1, \mathbf{v}_2, \ldots, \mathbf{v}_n)$ for the vector space V: we first define $\mathbf{u}_1 = \mathbf{v}_1/\|\mathbf{v}_1\|$, so that $\|\mathbf{u}_1\| = 1$. Now suppose that $(\mathbf{u}_1, \mathbf{u}_2, \ldots, \mathbf{u}_k, \mathbf{v}_{k+1}, \ldots, \mathbf{v}_n)$ is a basis for V with

$$(\mathbf{u}_i, \mathbf{u}_j) = \delta_{ij}, \quad 1 \leq i, j \leq k.$$

Set $\mathbf{v} = \mathbf{v}_{k+1} - \sum_{i=1}^{k} \lambda_i \mathbf{u}_i, \ \lambda_i \in \mathbb{C}$. Then $(\mathbf{u}_1, \mathbf{u}_2, \ldots, \mathbf{u}_k, \mathbf{v}, \mathbf{v}_{k+2}, \ldots, \mathbf{v}_n)$ is a basis for V and

$$(\mathbf{u}_i, \mathbf{v}) = (\mathbf{u}_i, \mathbf{v}_{k+1}) - \lambda_i, \quad 1 \leq i \leq k.$$

Set $\lambda_i = (\mathbf{u}_i, \mathbf{v}_{k+1})$ and define $\mathbf{u}_{k+1} = \mathbf{v}/\|\mathbf{v}\|$. Then

$$(\mathbf{u}_1, \mathbf{u}_2, \ldots, \mathbf{u}_k, \mathbf{u}_{k+1}, \mathbf{v}_{k+2}, \ldots, \mathbf{v}_n)$$

is a basis for V and

$$(\mathbf{u}_i, \mathbf{u}_j) = \delta_{ij}, \quad 1 \leq i, j \leq k + 1.$$

Thus we continue until we have constructed an orthonormal basis. Notice that any orthonormal set of vectors, that is, any set of vectors satisfying (10.2.1), is automatically linearly independent [see Definition 5.3.2].

Remark. This argument may readily be adapted to an inner product space with a *countable* basis, that is with a basis that can be arranged as an infinite sequence

$$(\mathbf{e}_1, \mathbf{e}_2, \ldots, \mathbf{e}_n, \ldots).$$

This generalization is very important in considering certain *orthonormal families of functions.*

When the elements of $\mathbb{C}^n$ are interpreted as column vectors the standard inner product of $\mathbf{x}$ and $\mathbf{y}$ is given by

$$(\mathbf{x}, \mathbf{y}) = \mathbf{y}^*\mathbf{x}, \tag{10.2.2}$$

where the asterisk denotes the conjugate complex transpose [see (6.7.11)]. Thus for an arbitrary matrix $\mathbf{K}$ we have that

$$\mathbf{K}^* = \overline{\mathbf{K}}', \qquad \mathbf{K}^{**} = \mathbf{K}.$$

Also $(\mathbf{KL})^* = \mathbf{L}^*\mathbf{K}^*$. Let $\mathbf{A}$ be an n by n matrix and apply (10.2.2) to the vectors $\mathbf{Ax}$ and $\mathbf{y}$. Then

$$(\mathbf{Ax}, \mathbf{y}) = \mathbf{y}^*\mathbf{Ax}. \tag{10.2.3}$$

The right-hand side of (10.2.3) can also be written as

$$(\mathbf{A}^*\mathbf{y})^*\mathbf{x} = (\mathbf{x}, \mathbf{A}^*\mathbf{y})$$

so that the identity

$$(\mathbf{Ax}, \mathbf{y}) = (\mathbf{x}, \mathbf{A}^*\mathbf{y})$$

holds for all $\mathbf{x}$ and $\mathbf{y}$ in $\mathbb{C}^n$.

In a more general setting, when A denotes a linear transformation of the vector space into itself [see § 5.13], we define the *adjoint* transformation A^* by stipulating that the relation

$$(A\mathbf{x}, \mathbf{y}) = (\mathbf{x}, A^*\mathbf{y}) \tag{10.2.4}$$

should hold for all $\mathbf{x}$ and $\mathbf{y}$.

Just as unitary matrices [see (6.7.13)] (and orthogonal matrices [see (6.7.9)]) are closely related to the considerations of orthonormal bases in $\mathbb{C}^n$ (and $\mathbb{R}^n$), so it is in our present more general setting.

DEFINITION 10.2.3.   A linear transformation U of the complex inner product space V is *unitary* if it preserves the inner product, that is, if $(U\mathbf{x}, U\mathbf{y}) = (\mathbf{x}, \mathbf{y})$.

THEOREM 10.2.2.   *Let* $U: V \to V$ *be a linear transformation of the vector space V. Then the following assertions are equivalent:*
  (i) *U is unitary;*
  (ii) *$\|U\mathbf{x}\| = \|\mathbf{x}\|$, for all $\mathbf{x}$;*
  (iii) *with respect to any orthonormal basis, $U^* = U^{-1}$.*

For real inner product spaces the corresponding statements are as follows:

DEFINITION 10.2.4.   A linear transformation R of the real inner product space V is *orthogonal* if it preserves the inner product, that is, if $(R\mathbf{x}, R\mathbf{y}) = (\mathbf{x}, \mathbf{y})$.

THEOREM 10.2.3.   *Let $R: V \to V$ be a linear transformation of the real inner product space V. Then the following assertions are equivalent:*
   (i) *R is orthogonal:*
   (ii) *$\|R\mathbf{x}\| = \|\mathbf{x}\|$ for all $\mathbf{x}$:*
   (iii) *with respect to any orthonormal basis, $R' = R^{-1}$.*

Proof.   We sketch the proof of Theorem 10.2.2 only. We first prove the equivalence of (i) and (iii). Now

$$(U\mathbf{x}, U\mathbf{y}) = (U^*U\mathbf{x}, \mathbf{y}).$$

Thus $(U\mathbf{x}, U\mathbf{y}) = (\mathbf{x}, \mathbf{y})$ for *all* $\mathbf{x}, \mathbf{y}$ if and only if $U^*U = 1$. The equivalence of (i) and (ii) follows from the identity

$$\|\mathbf{x} + \mathbf{y}\|^2 + \|\mathbf{x} - \mathbf{y}\|^2 + i\|\mathbf{x} + i\mathbf{y}\|^2 - i\|\mathbf{x} - i\mathbf{y}\|^2 = 4(\mathbf{x}, \mathbf{y}), \quad (10.2.5)$$

since the implication (i) $\Rightarrow$ (ii) is trivial. For (10.2.5) enables us to express the inner product in terms of the norm. The identity (10.2.5) is proved by straightforward computation, using the identity

$$\|\mathbf{x} + \mathbf{y}\|^2 = \|\mathbf{x}\|^2 + 2 \operatorname{Re}(\mathbf{x}, \mathbf{y}) + \|\mathbf{y}\|^2 \qquad (10.2.6)$$

exploited in the proof of Proposition 10.1.3(iii).

COROLLARY 10.2.4.   *Let $U: V \to V$ be a linear transformation of the vector space V. Then U is unitary if and only if it transforms an orthonormal basis for V into an orthonormal basis for V.*

Notice that if U transforms *one* orthonormal basis for V into an orthonormal basis for V then it is unitary and hence transforms *any* orthonormal basis for V into an orthonormal basis for V.

EXAMPLE 10.2.1.   Consider the real vector space $L^2(\mathbb{R})$ of real-valued square-integrable functions (see Example 10.1.2). Among such functions are the trigonometric functions $\sin 2\pi nx$, $n = 1, 2, 3, \ldots$ Now it may be shown [Example IV, 4.3.2(vii)] that, if m and n are any two positive integers, then

$$\int_0^1 \sin 2\pi nx \sin 2\pi mx\, dx = \begin{cases} 0 & \text{if } m \neq n \\ \tfrac{1}{2} & \text{if } m = n. \end{cases}$$

Thus, using the inner product in $L^2(\mathbb{R})$, as described in Example 10.1.2, we find that $\sin 2\pi n x$ and $\sin 2\pi m x$ are orthogonal if $m \neq n$, and that $\sqrt{2} \sin 2\pi n x$ is a unit vector. Thus the functions $\sqrt{2} \sin 2\pi n x$, $n = 1, 2, 3, \ldots$ form an orthonormal basis of the subspace of $L^2(\mathbb{R})$ which they span [see (5.5.6)].

EXAMPLE 10.2.2.   Here we again revert to Example 10.1.2 and consider, for any $n \geq 0$, the subspace $P_n(\mathbb{C})$ of $P(\mathbb{C})$ consisting of polynomials of degree not exceeding n. Then $P_n(\mathbb{C})$ admits the basis $(1, x, x^2, \ldots, x^n)$ [cf. Example 5.4.3] and we may apply the Gram-Schmidt process to obtain an orthonormal basis, as in the proof of Theorem 10.2.1. We start with $u_1 = 1$. We must compute the inner product of 1 and x; this is $\int_0^1 x\,dx$, which is $\frac{1}{2}$. Thus we set $v = x - \frac{1}{2}$ and then

$$\|v\|^2 = \int_0^1 (x^2 - x + \tfrac{1}{4})dx = \tfrac{1}{3} - \tfrac{1}{2} + \tfrac{1}{4} = \tfrac{1}{12}.$$

It follows that the next member of our orthonormal basis is $\sqrt{12}(x - \frac{1}{2}) = \sqrt{3}(2x - 1)$. Now we compute the inner product of 1 and x^2, which is $\frac{1}{3}$; and of $\sqrt{12}(x - \frac{1}{2})$ and x^2, which is $1/\sqrt{12}$. Thus we set $v = x^2 - \frac{1}{3} - (x - \frac{1}{2}) = x^2 - x + \frac{1}{6}$, and

$$\|v\|^2 = \int_0^1 (x^4 - 2x^3 + \tfrac{4}{3}x^2 - \tfrac{1}{3}x + \tfrac{1}{36})dx = \tfrac{1}{5} - \tfrac{1}{2} + \tfrac{4}{9} - \tfrac{1}{6} + \tfrac{1}{36} = \tfrac{1}{180}.$$

It follows that the next member of our orthonormal basis is $\sqrt{5}(6x^2 - 6x + 1)$. We may continue in this way to obtain an orthonormal basis for $P_n(\mathbb{C})$ for any value of n; indeed, if we proceed indefinitely in this way we obviously generate an orthonormal basis for $P(\mathbb{C})$ itself.

The following point should illustrate the utility of using orthonormal bases. Let $(\mathbf{u}_1, \mathbf{u}_2, \ldots, \mathbf{u}_n, \ldots)$ be a orthonormal basis for the inner product space V and let $\mathbf{v}$ be an arbitrary vector in V. Then [by (5.4.3)] for some n, $\mathbf{v}$ can be written uniquely as

$$\mathbf{v} = \sum_{i=1}^{n} \lambda_i \mathbf{u}_i, \quad \lambda_i \in \mathbb{C}. \tag{10.2.7}$$

PROPOSITION 10.2.5.   *In* (10.2.7) *we have*

$$\lambda_i = (\mathbf{v}, \mathbf{u}_i). \tag{10.2.8}$$

EXAMPLE 10.2.3.   We exemplify Proposition 10.2.5 by reverting to Example 10.2.2, with $n = 2$. Consider the polynomial $3x^2 - 5x - 1$. We wish to express it as a linear combination of the polynomials 1, $\sqrt{3}(2x - 1)$, $\sqrt{5}(6x^2 - 6x + 1)$, which form an orthonormal basis of $P_2(\mathbb{C})$. We find, computing inner products, that

$$(3x^2 - 5x - 1, 1)$$

$$= \int_0^1 (3x^2 - 5x - 1)dx = 1 - \tfrac{5}{2} - 1 = -\tfrac{5}{2},$$

$$(3x^2 - 5x - 1, \sqrt{3}(2x - 1))$$

$$= \sqrt{3} \int_0^1 (3x^2 - 5x - 1)(2x - 1)dx$$

$$= \sqrt{3} \int_0^1 (6x^3 - 13x^2 + 3x + 1)dx = -\frac{1}{\sqrt{3}},$$

$$(3x^2 - 5x - 1, \sqrt{5}(6x^2 - 6x + 1))$$

$$= \sqrt{5} \int_0^1 (3x^2 - 5x - 1)(6x^2 - 6x + 1)dx$$

$$= \sqrt{5} \int_0^1 (18x^4 - 48x^3 + 27x^2 + x - 1)dx = \frac{1}{2\sqrt{5}}.$$

The reader may confirm that

$$3x^2 - 5x - 1 = -\tfrac{5}{2} - \frac{1}{\sqrt{3}} \sqrt{3}(2x - 1) + \frac{1}{2\sqrt{5}} \sqrt{5}(6x^2 - 6x + 1).$$

Of course, we do not claim that the use of Proposition 10.2.5 necessarily provides the easiest way to obtain the coefficients λ_i in (10.2.7); we have simply been concerned to demonstrate that it works! Notice that the process may be easier if we are content to consider *orthogonal* bases instead of orthonormal bases, that is, if we retain the condition that $(u_i, u_j) = 0$ if $i \neq j$, but no longer insist that $\|u_i\| = 1$. Then we have to replace (10.2.8) by

$$\lambda_i = \frac{(v, u_i)}{\|u_i\|^2}.$$

10.3. COMPLEX INNER PRODUCT SPACES

We may now continue to imitate the development in Chapter 9 in our present more general setting. We introduce the congruence relation among (complex) $(n \times n)$-matrices A [cf. (9.1.6)]

$$A \sim T^*AT, \quad T \text{ non-singular } (n \times n)\text{-matrix} \tag{10.3.1}$$

and classify Hermitian $(n \times n)$-matrices [see (6.7.10)] under congruence exactly as before (note that in §9.1 we dealt *explicitly* with real symmetric matrices). The analogue of Theorem 9.1.2 holds, with the same proof. We will be content to expand on the first detail of that argument, in order to convince the reader that the proof may be executed, step by step, exactly as in that special case. Of course, a linear transformation A [see § 5.13] is *Hermitian* if and only if $A = A^*$, where A^* is defined in (10.2.4).

PROPOSITION 10.3.1.   *The eigenvalues of a Hermitian matrix are real.*

Proof.   Let $\mathbf{A}$ be Hermitian. Then, exactly as in (9.3.2), it follows that $(\mathbf{Az}, \mathbf{z})$ is real, $\mathbf{z} \in \mathbb{C}^n$; we need property (a) of Definition 10.1.1, of course. Thus if $\mathbf{Au} = \lambda\mathbf{u}$, with $\mathbf{u} \neq \mathbf{0}$, it follows that $(\mathbf{Au}, \mathbf{u}) = (\lambda\mathbf{u}, \mathbf{u}) = \lambda(\mathbf{u}, \mathbf{u})$, so that λ is real.

We may now continue just as in Chapter 9 to diagonalize a Hermitian matrix and obtain the obvious generalizations of Theorem 9.1.2 and Proposition 9.1.3; and we may similarly generalize the notions of positive-definite, negative-definite, positive-semi-definite, and negative-semi-definite quadratic forms, together with their characterizations.

P.H.

BIBLIOGRAPHY

Finkbeiner, D. T. (1960). *Introduction to Matrices and Linear Transformations*, Freeman & Co.

Lowenthal, F. (1975). *Linear Algebra (with Linear Differential Equations)*, John Wiley & Sons.

Morris, A. O. (1978). *Linear Algebra*, Van Nostrand Reinhold Co.

Tropper, A. M. (1969). *Linear Algebra*, Nelson.

CHAPTER 11

Linear Programming

11.1. INTRODUCTION

Linear programming can be dated from the year 1947, when G. B. Dantzig evolved an efficient technique, called the Simplex Method, for solving linear programming problems. The following decade saw the rapid development of both the theory and applications of linear programming, which were aided by the simultaneous introduction of the electronic computer.

One of the first problems to be solved by the simplex method was Stigler's diet problem (1945) and we shall use a simplified version of this problem to illustrate the linear programming problem.

11.1.1. The Diet Problem

Mrs. Jones feeds her husband on bread and cheese. Her husband has minimal daily calorie requirements of protein, fat and carbohydrate, which are shown in the Table 11.1.1. Also shown are the costs and calorific contents of bread and cheese.

	Protein	Fat	Carbohydrate	Cost
100 g bread	40	5	205	2.2p
100 g cheese	60	380	60	12p
Minimal daily requirements	300	790	1350	

Table 11.1.1

The problem is to determine how much bread and cheese Mrs. Jones should buy each day in order to minimize the cost of the diet, whilst fulfilling the calorie requirements. Suppose she buys $x_1 \times 100$ g of bread and $x_2 \times 100$ g of cheese, then the mathematical problem, known as a *linear programme*, is as follows.

309

Minimize $z = 2 \cdot 2x_1 + 12x_2,$ (cost of diet)
subject to $40x_1 + 60x_2 \geq 300,$ (at least 300 calories of protein)
                  $5x_1 + 380x_2 \geq 790,$ (at least 790 calories of fat)
              $205x_1 + 60x_2 \geq 1350,$ (at least 1350 calories of carbohydrates)
                  $x_1 \geq 0, x_2 \geq 0.$ (quantities must be non-negative).

The easiest and most illustrative method of solving problems in two unknowns is the *graphical method*. The values of x_1 and x_2 satisfying $40x_1 + 60x_2 \geq 300$ lie in the upper half-plane bounded by the straight line $40x_1 + 60x_2 = 300$, so the x_1 and x_2 satisfying all the above inequalities lie in the intersection of their respective half-planes, which is represented by the shaded region in Figure 11.1.1.

For constant z, the equation $z = 2 \cdot 2x_1 + 12x_2$ represents a straight line [see V, (1.2.1)], which moves parallel to itself away from the origin 0 as z increases from zero. The coordinates [see V, § 1.2.1] of the point P, in which this line first meets the shaded region, give the values of x_1 and x_2 which minimize z. These can be found by solving the simultaneous linear equations

$$5x_1 + 380x_2 = 790,$$
$$205x_1 + 60x_2 = 1350,$$

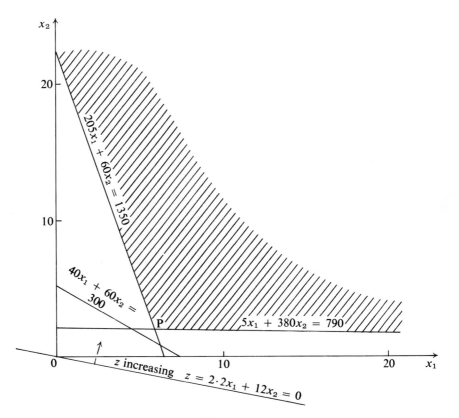

Figure 11.1.1

which have the solution $x_1 = 6$ and $x_2 = 2$ [see § 5.8]. In other words, Mrs. Jones solves her problem by buying each day 600 g of bread and 200 g of cheese at a cost of $z = 2 \cdot 2 \times 6 + 12 \times 2 = 37 \cdot 2$p.

11.1.2.   Nomenclature

We now introduce some nomenclature. The points (x_1, x_2) which satisfy the linear equalities and lie in the shaded region of Figure 11.1.1 are called *feasible solutions* of the linear programme, the point P is called an *optimal solution* and the value of z at P is known as the *value* of the linear programme. Two important properties to notice about the solution to the diet problem are that the region of feasible solutions is *polygonal* [see V, § 1.1.4] and that the optimal solution occurs at a *vertex* of the polygonal region. These properties are generally true for linear programmes.

11.2.   PRIMAL AND DUAL PROGRAMMES

11.2.1.   Dual of the Diet Problem

Given any linear programme, which we shall call the *primal* programme, another related linear programme, known as the *dual* programme, can be formulated. Consider the diet problem and suppose there is a manufacturer who makes synthetic protein, fat and carbohydrate. He will try to price his synthetic products so that they are competitive with the prices of bread and cheese. Let the prices per calorie of synthetic protein, fat, and carbohydrate be y_1, y_2 and y_3 pence, respectively. The minimal daily requirements of protein, fat and carbohydrate are shown in Table 11.1.1, so the mathematical problem is to

$$\text{maximize} \quad w = 300y_1 + 790y_2 + 1350y_3 \qquad \text{(the receipts)}$$

subject to
$$40y_1 + 5y_2 + 205y_3 \le 2 \cdot 2 \quad \text{(synthetic bread is not more expensive than bread)}$$
$$60y_1 + 380y_2 + 60y_3 \le 12 \quad \text{(synthetic cheese is not more expensive than cheese)}$$
$$y_1 \ge 0, y_2 \ge 0, y_3 \ge 0 \quad \text{(prices are non-negative)}.$$

In this dual programme the feasible points (y_1, y_2, y_3) lie in the three dimensional polyhedron bounded by the planes $40y_1 + 5y_2 + 205y_3 = 2 \cdot 2$, $60y_1 + 380y_2 + 60y_3 = 12$, $y_1 = 0$, $y_2 = 0$, $y_3 = 0$ [see V, § 2.2]. The receipts, w, will be maximized at a vertex of the polyhedron and it can be shown that the required vertex lies at the intersection of the planes

$$40y_1 + 5y_2 + 205y_3 = 2 \cdot 2, \qquad 60y_1 + 380y_2 + 60y_3 = 12, \quad y_1 = 0,$$

which is the point $y_1 = 0$, $y_2 = 0 \cdot 03$, $y_3 = 0 \cdot 01$. This gives receipts of $w = 300 \times 0 + 790 \times 0 \cdot 03 + 1350 \times 0 \cdot 01 = 37 \cdot 2$p.

There are two points to notice about this solution. Firstly, the maximum value of w equals the minimum value of z or, in other words, the values of the primal and dual programmes are equal, and secondly, the price of synthetic protein, y_1, is zero. If one looks at the optimal solution of the primal programme one finds that it provides $40 \times 6 + 60 \times 2 = 360$ calories of protein, which is more than the daily minimal requirement and so the marginal value* of protein is zero. In general, the variables of the dual programme can be interpreted as the marginal prices* of resources arising in the primal programme and are often called shadow prices.

11.2.2.   Matrix Formulation of Primal and Dual Programmes

Using Table 11.1.1 let

$$\mathbf{A} = \begin{pmatrix} 40 & 60 \\ 5 & 380 \\ 205 & 60 \end{pmatrix}, \qquad \mathbf{c} = \begin{pmatrix} 2 \cdot 2 \\ 12 \end{pmatrix}, \qquad \mathbf{b} = \begin{pmatrix} 300 \\ 790 \\ 1350 \end{pmatrix},$$

$$\mathbf{x} = \begin{pmatrix} x_1 \\ x_2 \end{pmatrix} \quad \text{and} \quad \mathbf{y} = \begin{pmatrix} y_1 \\ y_2 \\ y_3 \end{pmatrix},$$

so that the columns of $\mathbf{A}$ give the nutritional values of bread and cheese, $\mathbf{c}$ is the cost vector of bread and cheese, and $\mathbf{b}$ is the vector of daily nutritional requirements. So in matrix notation [see (5.7.4)], the primal programme is to

$$\begin{aligned} \text{minimize} \quad & z = \mathbf{c}'\mathbf{x}, \\ \text{subject to} \quad & \mathbf{Ax} \geq \mathbf{b}, \\ \text{and} \quad & \mathbf{x} \geq \mathbf{0}, \end{aligned}$$

and the dual programme is to

$$\begin{aligned} \text{maximize} \quad & w = \mathbf{b}'\mathbf{y}, \\ \text{subject to} \quad & \mathbf{A}'\mathbf{y} \leq \mathbf{c}, \\ \text{and} \quad & \mathbf{y} \geq \mathbf{0}, \end{aligned}$$

where the notation ''' is defined in section 6.5. The Matrix Algebra makes the relation between the primal and dual programmes much clearer and it can be shown that the dual of the dual programme is the primal programme.

EXAMPLE 11.2.1.   Find the dual of the programme:

$$\begin{aligned} \text{minimize} \quad & z = 2x_1 - x_2 + 3x_3, \\ \text{subject to} \quad & 3x_1 - 2x_2 + 5x_3 \geq -3, \\ & x_1 + x_2 + x_3 \geq 1, \\ \text{and} \quad & \mathbf{x} \geq \mathbf{0}. \end{aligned}$$

* The price or effective price of an additional unit of resource.

Solution. For this programme

$$A = \begin{pmatrix} 3 & -2 & 5 \\ 1 & 1 & 1 \end{pmatrix}, \qquad b = \begin{pmatrix} -3 \\ 1 \end{pmatrix}, \qquad c = \begin{pmatrix} 2 \\ -1 \\ 3 \end{pmatrix},$$

so

$$A' = \begin{pmatrix} 3 & 1 \\ -2 & 1 \\ 5 & 1 \end{pmatrix} \quad \text{and} \quad y = \begin{pmatrix} y_1 \\ y_2 \end{pmatrix}.$$

Hence the dual programme is to

$$\text{maximize} \quad w = -3y_1 + y_2,$$
$$\text{subject to} \quad 3y_1 + y_2 \leq 2,$$
$$-2y_1 + y_2 \leq -1,$$
$$5y_1 + y_2 \leq 3,$$
$$\text{and} \quad y_1 \geq 0, y_2 \geq 0.$$

Notice that in this example, the dual programme can easily be solved by the graphical method and it often happens that the dual programme is easier to solve than the primal.

11.2.3. Relations between Primal and Dual Programmes

From the matrix formulation of the primal and dual programmes it follows that for feasible x and y,

$$z = c'x \geq y'Ax \geq y'b = w,$$

and so the value of the primal is not smaller than the value of the dual. This leads to the important result that if feasible x and y are such that $z = w$, then x and y are optimal. Also the equation

$$z = c'x = y'Ax = y'b = w$$

implies that

$$\text{if} \quad (Ax)_i > b_i \quad \text{for some } i, \text{ then } y_i = 0,$$

and

$$\text{if} \quad (A'y)_i < c_i \quad \text{for some } i, \text{ then } x_i = 0.$$

These two relations are logically equivalent to

$$y_i > 0 \quad \text{implies} \quad (Ax)_i = b_i,$$

and

$$x_i > 0 \quad \text{implies} \quad (A'y)_i = c_i,$$

respectively.

We illustrate these results by returning once again to the diet problem for which

$$\text{minimum } z = 37 \cdot 2p = \text{maximum } w$$

and the optimal values of $\mathbf{x}$ and $\mathbf{y}$ are

$$\mathbf{x} = \begin{pmatrix} 6 \\ 2 \end{pmatrix} \quad \text{and} \quad \mathbf{y} = \begin{pmatrix} 0 \\ 0.03 \\ 0 \cdot 01 \end{pmatrix}.$$

It can easily be checked that these satisfy the relations

$$
\begin{aligned}
y_2 > 0 &\Rightarrow \quad 5x_1 + 380x_2 &&= 790 \\
y_3 > 0 &\Rightarrow 205x_1 + 60x_2 &&= 1350 \\
x_1 > 0 &\Rightarrow \quad 40y_1 + \quad 5y_2 + 205y_3 &&= 2 \cdot 2 \\
x_2 > 0 &\Rightarrow \quad 60y_1 + 380y_2 + 60y_3 &&= 12,
\end{aligned}
$$

where $\Rightarrow$ is the symbol for implies.

In the next section we bring together the major results of duality theory, which are useful for solving linear programmes. Some of the results are difficult to prove, so no attempt will be made to prove them in this chapter, but it is hoped that the foregoing discussion has made them plausible.

11.2.4.   Results of Duality Theory

1. If feasible solutions to both the primal and dual programmes exist, then there is an optimal solution to both programmes and

$$\text{minimum } z = \text{maximum } w.$$

2. If in the optimal solution of one programme the ith variable is positive, then the ith constraint in the dual programme is satisfied as an equation.
3. If either programme has no feasible solution, then neither has an optimal solution.

EXAMPLE 11.2.2.   The following example illustrates the third result of duality theory.

Consider the primal problem:

$$
\begin{aligned}
\text{minimize} \quad & z = x_1 - x_2, \\
\text{subject to} \quad & x_1 + 2x_2 \geq 4, \\
& x_1 \qquad\quad \geq 3, \\
\text{and} \quad & x_1 \geq 0, x_2 \geq 0.
\end{aligned}
$$

The feasible region is the shaded region in Figure 11.2.1 and it is clear that in this region the minimum value of z is $-\infty$.

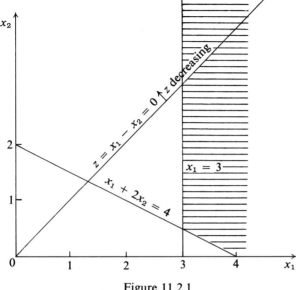

Figure 11.2.1

The dual problem is to

$$\text{maximize} \quad w = 4y_1 + 3y_2,$$
$$\text{subject to} \quad y_1 + y_2 \le 1,$$
$$2y_1 \qquad \le -1,$$
$$\text{and} \qquad y_1 \ge 0, y_2 \ge 0.$$

It is seen that the dual has no feasible solution, as the inequalities $2y_1 \le -1$, $y_1 \ge 0$ are incompatible, so this confirms the third duality result.

11.2.5. Alternative form for Primal and Dual Programmes

For the subsequent work in this chapter it will be better to consider the primal programme in the following form:

$$\text{maximize} \quad \mathbf{c'x},$$
$$\text{subject to} \quad \mathbf{Ax = b},$$
$$\text{and} \qquad \mathbf{x \ge 0},$$

for which the dual programme is to

$$\text{maximize} \quad \mathbf{b'y},$$
$$\text{subject to} \quad \mathbf{A'y \le c}.$$

In this form the inequalities in the primal programme have been replaced by equations and the non-negative condition on $\mathbf{y}$ has been dropped in the dual programme.

The above dual programme is consistent with our earlier definition, for we can write the primal programme as

$$\text{minimize} \quad \mathbf{c}'\mathbf{x},$$

$$\text{subject to} \quad \begin{pmatrix} \mathbf{A} \\ -\mathbf{A} \end{pmatrix}\mathbf{x} \geq \begin{pmatrix} \mathbf{b} \\ -\mathbf{b} \end{pmatrix},$$

$$\text{and} \qquad \mathbf{x} \geq \mathbf{0}.$$

The corresponding dual programme is to

$$\text{maximize} \quad (\mathbf{b}', -\mathbf{b}')\begin{pmatrix} \mathbf{y}_1 \\ \mathbf{y}_2 \end{pmatrix}$$

$$\text{subject to} \quad (\mathbf{A}', -\mathbf{A}')\begin{pmatrix} \mathbf{y}_1 \\ \mathbf{y}_2 \end{pmatrix} \leq \mathbf{c}'$$

$$\text{and} \qquad \mathbf{y}_1 \geq \mathbf{0}, \mathbf{y}_2 \geq \mathbf{0}.$$

On putting $\mathbf{y} = \mathbf{y}_1 - \mathbf{y}_2$, we recover the above dual programme.

The following example shows how any linear programme can be expressed in the form of the above primal programme and that the primal problem is not necessarily a minimizing problem.

EXAMPLE 11.2.3

$$\text{Maximize} \quad 2x_1 - 3x_2,$$
$$\text{Subject to} \quad 2x_1 + x_2 \leq 6,$$
$$x_1 - 3x_2 \geq 4,$$
$$x_1 \geq 0.$$

Since x_2 is unrestricted in sign, we put $x_2 = x_3 - x_4$, where $x_3 \geq 0$ and $x_4 \geq 0$, as any number can be expressed as the difference between two non-negative numbers. Maximize is changed into minimize on multiplying $2x_1 - 3x_2$ by -1 and the inequalities are transformed into equations by introducing x_5 and x_6, which are called *slack variables*. For example, the inequality $y \leq 2$ is equivalent to $y + z = 2$, $z \geq 0$, where z is a slack variable. Hence an equivalent linear programme in the required form is to

$$\text{minimize} \quad -2x_1 + 3x_3 - 3x_4,$$
$$\text{subject to} \quad 2x_1 + x_3 - x_4 + x_5 \qquad = 6,$$
$$x_1 - 3x_3 + 3x_4 \qquad - x_6 = 4,$$
$$x_1, x_3, x_4, x_5, x_6 \geq 0.$$

11.3.   THE TRANSPORTATION PROBLEM

11.3.1.   Formulation and Solution of the Transportation Problem

This problem, which was first formulated in 1941, is to find the cheapest scheme for sending ships from ports $P_1, P_2, \ldots, P_m$ to ports $Q_1, Q_2, \ldots, Q_n,$

when the cost of sending a ship from P_i to Q_j is known to be c_{ij}. Given that the supply of ships at P_i is s_i, the demand for ships at Q_j is d_j and that the total supply equals the total demand ($\sum_i s_i = \sum_j d_j$), then the problem is to determine the number of ships, x_{ij}, to send from P_i to Q_j so that the cost of the operation is minimized. The corresponding primal programme is to

minimize $\qquad z = \sum_i \sum_j c_{ij} x_{ij}$, $\qquad\qquad$ (the cost)

subject to $\qquad \sum_i x_{ij} = d_j \quad$ for $j = 1, 2, \ldots, n$, $\quad$ (demand at Q_j is met)

$\qquad\qquad\qquad -\sum_j x_{ij} = -s_i \quad$ for $i = 1, 2, \ldots, m$, $\quad$ (supply at P_i is used)

and $\qquad\qquad x_{ij} \geq 0$,

where the minus sign has been introduced in the supply equations, so that the following dual programme has an economic interpretation.

On putting $y' = (v_1, v_2, \ldots, v_n, u_1, u_2, \ldots, u_m)$, the dual programme becomes

$$\text{maximize} \quad w = \sum_j d_j v_j - \sum_i s_i u_i,$$

$$\text{subject to} \quad v_j - u_i \leq c_{ij} \quad \text{for all } i \text{ and } j. \qquad (11.3.1)$$

We can interpret this by supposing that a contractor offers to buy the ships at the port P_i for the price u_i per ship and to sell the ships at the ports Q_j for the price v_j per ship. His profits will equal w and (11.3.1) is the condition that his charges are competitive.

We shall now describe a method for solving the transportation problem using the second result of duality theory, which says that for optimal solutions

$$x_{ij} > 0 \Rightarrow v_j - u_i = c_{ij}. \qquad (11.3.2)$$

In this method we shall always take u_1 to be zero. This is allowable since, if v_j and u_i satisfy (11.3.1), so do $v_j + k$ and $u_i + k$ for any number k and k can be chosen to equal $-u_1$.

EXAMPLE 11.3.1. The costs, supplies and demands for a transportation problem are shown in the table below

		Q_1	Q_2	Q_3	
	d_j	4	2	2	
$\quad s_i$					
P_1	3	5	7	8	$= (c_{ij})$
P_2	5	1	2	3	

Solution. In the diet problem [§ 11.1.2] we noted that the optimal solution occurred at a vertex of the feasible region. Solutions corresponding to vertices are called *basic feasible solutions* and an optimal solution is to be found amongst them. A basic feasible solution of the transportation problem has at most $m + n - 1$ of the x_{ij} positive, which is the number of independent equations in the contraints [see § 5.8], as the following linear relation holds between the equations.

$$\sum_i \left(\sum_j x_{ij}\right) = \sum_i s_i = \sum_j d_j = \sum_j \left(\sum_i x_{ij}\right).$$

An initial basic feasible solution can be found by the *matrix minimum method*. This method first sends the maximum number of ships, compatible with the supplies and demands, along the cheapest route, i.e. 4 ships are sent from P_2 to Q_1. Then the next cheapest route, from P_2 to Q_2 is used and so on to give the following basic feasible solution:

$$(x_{ij}) = \begin{pmatrix} 0 & 1 & 2 \\ 4 & 1 & 0 \end{pmatrix}.$$

We now use condition (11.3.2) to calculate v_j and u_i.

$$x_{12} > 0 \Rightarrow v_2 - u_1 = c_{12} = 7.$$
$$x_{13} > 0 \Rightarrow v_3 - u_1 = c_{13} = 8.$$
$$x_{21} > 0 \Rightarrow v_1 - u_2 = c_{21} = 1.$$
$$x_{22} > 0 \Rightarrow v_2 - u_2 = c_{22} = 2.$$

Since there are only four equations to determine the five unknowns u_1, u_2, v_1, v_2, v_3, we put $u_1 = 0$, which has already been justified. Solving the equations gives in turn $v_2 = 7$, $v_3 = 8$, $u_2 = 5$ and $v_1 = 6$.

In the table below, the costs, c_{ij}, are shown in the left-hand side of each column and the initial basic feasible solution is shown on the right-hand side together with the circled entries, which are the values of $v_j - u_i$ when $x_{ij} = 0$.

d_j	4	2	2
v_j	6	7	8

s_i	u_i			
3	0	5 ⑥ θ	7 $1-\theta$	8 2
5	5	1 $4-\theta$	2 $1+\theta$	3 ③

Since $v_1 - u_1 = 6 > 5 = c_{11}$, the v_j and u_i are not feasible, so we have not arrived at an optimal solution. The next step is to send some ships along a route for which $v_j - u_i > c_{ij}$, so we put $x_{11} = \theta$ and adjust other *non-zero* x_i so that the supplies and demands are still satisfied. This leads to $x_{21} = 4 - \theta$, $x_{12} = 1 - \theta$, $x_{22} = 1 + \theta$ and the consequent change in cost is

$$\theta(c_{11} - c_{21} - c_{12} + c_{22})$$
$$= \theta\{c_{11} - (v_1 - u_2) - (v_2 - u_1) + (v_2 - u_2)\}$$
$$= \theta\{c_{11} - (v_1 - u_1)\} < 0$$

for $\theta > 0$. So provided we take $\theta > 0$, the cost is reduced. We choose $\theta = 1$, which is the maximum value of θ consistent with $x_{ij} \geq 0$. This procedure ensures that the new solution is both basic and feasible. On calculating the u_i and v_j as before, we obtain the next table.

v_j	5		6		8	
u_i						
0	5	$1+\theta$	7	⑥ 8	$2-\theta$	
4	1	$3-\theta$	2	2 3	④	θ

This time $v_3 - u_2 = 4 > 3 = c_{23}$, so we put $x_{23} = \theta$ and find the maximum value of θ is 2. Proceeding to the next table, we find that $v_j - u_i \leq c_{ij}$ for all i, j and so an optimal solution has been attained.

v_j	5		6		7	
u_i						
0	5	3	7	⑥ 8	⑦	
4	1	1	2	2 3	2	

The optimal solution is

$$(x_{ij}) = \begin{pmatrix} 3 & 0 & 0 \\ 1 & 2 & 2 \end{pmatrix}.$$

Integer Solutions

Provided the supplies and demands are positive integers, the matrix minimum method always leads to an optimal solution with integer values, as the method only involves operations on integers which results in integers. Obviously a non-integer optimal solution would be useless.

Uniqueness

It can happen that two or more different allocations of ships between ports give rise to the same minimum cost. However, if $v_j - u_i < c_{ij}$ for all $x_{ij} = 0$, as in the above example, then there is a unique optimal solution.

Degeneracy

Degeneracy occurs in a transportation problem when a partial sum of the supplies equals a partial sum of the demands, for example when $s_1 + s_4 = d_2 + d_3$. Under such circumstances, a basic feasible solution may be obtained in which less than $m + n - 1$ of the x_{ij} are positive, which results in too few equations to determine the u_i and v_j. This difficulty can be overcome by making the problem non-degenerate as follows. Let

$$\bar{s}_i = s_i + \varepsilon \qquad \text{for } i = 1, 2, \ldots, m$$
$$\bar{d}_j = d_j \qquad \text{for } j = 1, 2, \ldots, n - 1$$

and

$$\bar{d}_n = d_n + m\varepsilon,$$

where ε is a small quantity, which is chosen so that a partial sum of the $\bar{s}_i$ does not equal a partial sum of the $\bar{d}_j$. The optimal solution of the original problem is found by putting $\varepsilon = 0$ into the optimal solution of the perturbed problem.

EXAMPLE 11.3.2.   Solve the following transportation problem

		Q_1	Q_2	Q_3
	d_j	3	5	3
	s_i			
P_1	2	2	3	6
P_2	4	5	5	4
P_3	5	8	1	7

Solution

The problem is degenerate as $s_3 = d_2$ and the matrix minimum method gives

$$(x_{ij}) = \begin{pmatrix} 2 & 0 & 0 \\ 1 & 0 & 3 \\ 0 & 5 & 0 \end{pmatrix}.$$

This basic feasible solution only has 4 non-zero x_{ij} and so there are only four equations to determine u_2, u_3, v_1, v_2, v_3 ($u_1 = 0$). Consequently we make the problem non-degenerate, as described above, and then find an initial basic feasible solution by the matrix minimum method, which results in the following table.

s_i	u_i	d_j	3 2		5 −5		$3 + 3\varepsilon$ 1	
		v_j						
$2 + \varepsilon$	0	2	$2 + \varepsilon$	3	⟨−5⟩	6	①
$4 + \varepsilon$	−3	5	$1 - 2\varepsilon$	5	⟨−2⟩	4	$3 + 3\varepsilon$
$5 + \varepsilon$	−6	8	ε	1	5	7	⑦

Since $v_j - u_i \le c_{ij}$ for all $x_{ij} = 0$, we have an optimal solution to the perturbed problem. Putting $\varepsilon = 0$ gives

$$(x_{ij}) = \begin{pmatrix} 2 & 0 & 0 \\ 1 & 0 & 3 \\ 0 & 5 & 0 \end{pmatrix}$$

as an optimal solution to the original problem.

11.3.2. The Assignment Problem

In this problem, which is similar to a transportation problem, one is asked to allocate n candidates, $C_1, C_2, \ldots, C_n$, to n jobs, $J_1, J_2, \ldots, J_n$, given that the rating of C_i for J_j is a_{ij} and the sum of the ratings is to be maximized.

Let $x_{ij} = 1$ when C_i is assigned to J_j and zero otherwise, then the problem is to

maximize $\quad z = \sum_i \sum_j a_{ij} x_{ij} \qquad$ (the sum of the ratings)

subject to $\quad \sum_j x_{ij} = 1 \quad$ for all i (each candidate gets one job)

$\qquad\qquad \sum_i x_{ij} = 1 \quad$ for all j (each job is given to one candidate).

This problem can be solved as a degenerate transportation problem except that the dual variables, u_i and v_j, must satisfy

$$v_j - u_i \ge a_{ij}$$

for an optimal solution, as the primal programme is a maximization problem, whereas z is to be minimized in the transportation problem. The optimal solution produced by this method will have the x_{ij}'s equal to one or zero, as required, because of the nature of the constraints.

When there are more candidates than jobs, the shortfall in jobs can be made up by introducing fictitious jobs for which all the candidates have zero rating and a similar procedure can be used when there are more jobs than candidates.

11.3.3. The Transshipment Problem

In the transportation problem we assumed that the ships went directly from the ports P_i to the ports Q_j, whereas either type of port, as well as other ports, might be used as an intermediate stop. Such ports are known as ports of 'transshipment'. Also the shipping link between a P_i and Q_j may only be via a third port.

Consider the problem described by the network shown in Figure 11.3.1, where P_1, P_2 are ports with supplies of ships, Q_1, Q_2 are ports with demands for ships and R_1 is solely a transshipment port. The costs of shipping either way between two neighbouring ports is also shown.

Clearly one can easily reduce the above problem to the transportation problem, shown in the table below, by calculating the cost of the cheapest route

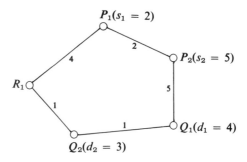

Figure 11.3.1

from a P_i to a Q_j. For example, the cheapest route from P_1 to Q_2 is via R_1 with a cost of 5.

	d_j	Q_1 4	Q_2 3
	s_i		
P_1	2	6	5
P_2	5	5	6

However, for larger problems the determination of the minimum costs between all pairs of ports could be tedious, so we now describe an alternative method. In the table below all the ports are listed as both source ports and demands ports and where there is no direct link between two ports, the square is left empty.

		P_1		P_2		Q_1		Q_2		R_1	
	d_j	0		0		4		3		0	
	s_i										
P_1	2	0	$-x_{11}$	2	x_{12}					4	x_{15}
P_2	5	2	x_{21}	0	$-x_{22}$	5	x_{23}				
Q_1	0			5	x_{32}	0	$-x_{33}$	1	x_{34}		
Q_2	0					1	x_{43}	0	$-x_{44}$	1	x_{45}
R_1	0	4	x_{51}					1	x_{54}	0	$-x_{55}$

The costs of transshipment are taken to be zero; x_{11}, x_{22}, etc. are the number of ships to be transshipped through P_1, P_2, etc., and x_{12}, x_{21} are the number of ships to be sent from P_1 to P_2 and from P_2 to P_1, respectively. We can now formulate the transshipment problem, which is to

$$\text{minimize} \quad z = \sum_i \sum_j c_{ij}x_{ij} \qquad \text{(the costs)}$$

$$\text{subject to} \quad \sum_{j \neq i} x_{ij} - x_{ii} = s_i \quad \text{for all } i, \quad \text{(satisfies supplies at } i\text{th port)}$$

$$\sum_{i \neq j} x_{ij} - x_{jj} = d_j \quad \text{for all } j, \quad \text{(satisfies demands at } j\text{th port)}.$$

$$\text{and} \qquad x_{ij} \geq 0.$$

The dual programme is to

$$\text{maximize} \qquad w = \sum_j d_j v_j - \sum_i s_i u_i,$$

$$\text{subject to} \qquad v_j - u_i \le c_{ij} \quad \text{for all } i, j \text{ and } i \ne j,$$

$$\text{and} \qquad -v_i + u_i \le 0 \qquad \text{for all } i.$$

It can be shown that there is an optimal solution to the dual programme with $v_i = u_i$ and an optimal solution to the primal problem with x_{ii}, for all i, occurring in the basic solution. These two properties are used in the method of solution, though as the problem is degenerate, difficulties can arise. A starting solution is to send 2 from P_1 to Q_2 via R_1, 1 from P_2 to Q_2 via P_1 and R_1 and 4 from P_2 to Q_1 and this leads to the table shown below in which $x_{11} = 1$ and $x_{55} = 3$ mean that one ship and three ships are transshipped through P_1 and R_1,

	S_i	u_i	P_1 d_j=0 u_j=0		P_2 d_j=0 u_j=-2		Q_1 d_j=4 u_j=3		Q_2 d_j=3 u_j=5		R_1 d_j=0 u_j=4	
P_1	2	0	0	$-1+\theta$	2	$\widehat{-2}$					4	$3-\theta$
P_2	5	-2	2	$1-\theta$	0	0	5	$4+\theta$				
Q_1	0	3			5	$\widehat{-5}$	0	$0-\theta$	1	$\widehat{2}$ θ		
Q_2	0	5					1	$\widehat{-2}$	0	0	1	$\widehat{-1}$
R_1	0	4	4	$\widehat{-4}$					1	$3-\theta$	0	$-3+\theta$

respectively. Since $u_4 - u_3 = 2 > 1 = c_{34}$, the solution is not optimal so we put $x_{34} = \theta$ and proceed as in the solution of the transportation problem, to find the maximum value for θ consistent with $x_{ij} \ge 0$, which is $\theta = 1$. Now the dual

	u_i	P_1 u_j=0		P_2 u_j=-1		Q_1 u_j=4		Q_2 u_j=5		R_1 u_j=4	
P_1	0	0	0	2	$\widehat{-1}$					4	2
P_2	-1	2	$\widehat{1}$	0	0	5	5				
Q_1	4			5	$\widehat{-5}$	0	-1	1	1		
Q_2	5					1	$\widehat{-1}$	0	0	1	$\widehat{-1}$
R_1	4	4	$\widehat{-4}$					1	2	0	-2

constraint $u_j - u_i \le c_{ij}$ is satisfied for all $i \ne j$ and hence the optimal solution is to send 2 from P_1 to Q_2 via R_1, 4 from P_2 to Q_1 and 1 from P_2 to Q_2 via Q_1.

Both the transportation and transshipment problems can be modified to deal with problems in which there are restrictions on the number of ships that can be sent between two ports.

11.4. THE SIMPLEX METHOD

11.4.1. Outline of Simplex Method

As we have already mentioned, in section 11.1.2, if an optimal solution to a linear programme exists, then there is an optimal solution which is a basic feasible solution. The simplex method searches through the basic feasible solutions in such a way that the value, z, of the linear function to be minimized is decreased by each choice of basic feasible solution. In fact the method we have just used for solving the transportation problem is a special case of the simplex method.

A basic feasible solution of a set of linear equations always has the number of non-zero x's equal to the number of equations, provided the system is non-degenerate and we shall discuss the degenerate case later.

11.4.2. Determination of Basic Feasible Solutions

We now show how to find the basic feasible solutions of the following linear programme:

$$\text{minimize} \quad c_1x_1 + c_2x_2 + c_3x_3 + c_4x_4,$$

$$\text{subject to} \quad x_1 \qquad + a_{13}x_3 + a_{14}x_4 = b_1, \qquad (11.4.1)$$

$$x_2 + a_{23}x_3 + a_{24}x_4 = b_2, \qquad (11.4.2)$$

$$x_1, x_2, x_3, x_4 \geq 0,$$

where we assume that b_1 and b_2 are positive. An immediate basic feasible solution is $x_1 = b_1$, $x_2 = b_2$, $x_3 = 0$, $x_4 = 0$, which can be represented by the following table. x_1 and x_2 are called the *basic variables*, as they are positive, and

		x_1	x_2	x_3	x_4		
basic	x_1	1	0	a_{13}	a_{14}	b_1	basic
variables	x_2	0	1	a_{23}	a_{24}	b_2	solution

Table 11.4.1

x_3 and x_4 are called the *non-basic variables*, as they are zero. Suppose we now look for a basic solution in which x_2 and x_3 are the basic variables. We can re-write (11.4.1) in the form

$$\frac{1}{a_{13}}x_1 + \qquad x_3 + \frac{a_{14}}{a_{13}}x_4 = \frac{b_1}{a_{13}}, \qquad (11.4.3)$$

and substituting for x_3 in (11.4.2) gives

$$-\frac{a_{23}}{a_{13}}x_1 + x_2 + \left(a_{24} - \frac{a_{23}a_{14}}{a_{13}}\right)x_4 = b_2 - \frac{a_{23}}{a_{13}}b_1, \qquad (11.4.4)$$

from which we obtain the basic solution

$$x_1 = 0, \qquad x_2 = b_2 - \frac{a_{23}}{a_{13}} b_1, \qquad x_3 = \frac{b_1}{a_{13}}, \qquad x_4 = 0,$$

and the corresponding table:

		x_1	x_2	x_3	x_4	
	x_3	$\dfrac{1}{a_{13}}$	0	1	$\dfrac{a_{14}}{a_{13}}$	$\dfrac{b_1}{a_{13}}$
	x_2	$-\dfrac{a_{23}}{a_{13}}$	1	0	$a_{24} - \dfrac{a_{23}}{a_{13}} a_{14}$	$b_2 - \dfrac{a_{23}}{a_{13}} b_1$

basic variables (left brace), basic solution (right brace)

Table 11.4.2

The circled element in Table 11.4.1, a_{13}, is called the *pivot*. It is at the intersection of the row of the basic variable to be replaced and the column of the new basic variable. We can now formulate the following rule for proceeding from one table to the next:

Replacement Rule

Divide the pivotal row by the pivot.
Subtract from each other row the multiple of the pivotal row, which leads to a zero in the pivotal column.

EXAMPLE 11.4.1.   A basic solution for

$$x_1 + 3x_2 + 8x_3 \qquad = 4,$$
$$x_2 + 12x_3 + x_4 = 5,$$

is

$$x_1 = 4, \, x_2 = 0, \, x_3 = 0, \, x_4 = 5.$$

	x_1	x_2	x_3	x_4	
x_1	1	3	8	0	4
x_4	0	1	(12)	1	5

Suppose we replace x_4 by x_3 as a basic variable, then the new table is found by dividing the second (pivotal) row by 12 and by subtracting $\frac{8}{12}$ of the second row from the first.

	x_1	x_2	x_3	x_4	
x_1	1	$\frac{7}{3}$	0	$-\frac{2}{3}$	$\frac{2}{3}$
x_3	0	$\frac{1}{12}$	1	$\frac{1}{12}$	$\frac{5}{12}$

The new basic solution is $x_1 = \frac{2}{3}$, $x_2 = 0$, $x_3 = \frac{5}{12}$, $x_4 = 0$.

11.4.3. Choice of Pivot

In solving equations (11.4.1) and (11.4.2), we are only interested in interchanges for which the new values of the basic variables are positive and the value of z is reduced. The values of x_i must be non-negative to satisfy the constraints. From Table 11.4.2 we see that the new basic variables are positive

$$\text{if} \quad \frac{b_1}{a_{13}} > 0 \quad \text{and} \quad b_2 - \frac{a_{23}}{a_{13}} b_1 > 0.$$

Since we assumed that b_1 and b_2 were positive, the condition is equivalent to

$$a_{13} > 0 \quad \text{and} \quad \frac{b_1}{a_{13}} < \frac{b_2}{a_{23}} \quad \text{if } a_{23} > 0.$$

We now derive the condition for the value of z to be reduced. For Table 11.4.1, $z_1 = c_1 b_1 + c_2 b_2$ and for Table 11.4.2,

$$z_2 = c_2 \left(b_2 - \frac{a_{23}}{a_{13}} b_1 \right) + c_3 \frac{b_1}{a_{13}}.$$

Hence for z to be reduced,

$$z_1 - z_2 = c_1 b_1 + c_2 \frac{a_{23}}{a_{13}} b_1 - c_3 \frac{b_1}{a_{13}} > 0,$$

or $a_{13} c_1 + a_{23} c_2 - c_3 > 0$, provided that $b_1 > 0$ and $a_{13} > 0$. These conditions can be summarized in the following rule:

Pivot Rule

For each column of the table calculate the value of $\sum_i a_{ij} c_i - c_j$. If this is positive for some j, say $j = l$, make x_l a basic variable. Next for $a_{il} > 0$, find the value of i, which minimizes b_i / a_{il}. Suppose the minimum occurs at $i = k$, then a_{kl} is the pivot.

11.4.4. Operation of Pivot Rule

To faciliate the use of the pivot rule, we extend Table 11.4.1, as shown below. The extreme right-hand element in the bottom row is the current value of z. If

		c_1	c_2	c_3	c_4	
		x_1	x_2	x_3	x_4	
c_1	x_1	1	0	a_{13}	a_{14}	b_1
c_2	x_2	0	1	a_{23}	a_{24}	b_2
		0	0	$a_{13}c_1 + a_{23}c_2 - c_3$	$a_{14}c_1 + a_{24}c_2 - c_4$	$c_1b_1 + c_2b_2$

any of the other elements in the bottom row are positive, then the pivotal column is chosen from amongst them. The pivotal row is then determined by finding the smallest of the ratios b_i/a_{il} for $a_{il} > 0$, given that the pivot lies in the lth column. Fortunately it can be shown that the bottom row transforms according to the replacement rule under a change of basic solution.

Finally, two things can go 'wrong' in the operation of the pivot rule. Firstly, no element in the bottom row may be positive, in which case an optimum solution has been achieved. Secondly, none of the a_{il} in the pivotal column may be positive. When this occurs, the linear programme has no optimal solution, as the value of z is unbounded below.

EXAMPLE 11.4.2

$$\text{Minimize} \quad z = 3x_1 + 2x_2 \qquad + 4x_4,$$
$$\text{subject to} \quad x_1 + 3x_2 + 8x_3 \qquad = 4,$$
$$x_2 + 12x_3 + x_4 = 5,$$
$$\text{and} \quad x_1, x_2, x_3, x_4 \geq 0.$$

Solution

We start the simplex method with the basic feasible solution $x_1 = 4$, $x_2 = 0$, $x_3 = 0$, $x_4 = 5$, as in the previous example. The elements in the bottom row

		3	2	0	4	
		x_1	x_2	x_3	x_4	
3	x_1	1	3	8	0	4
4	x_4	0	1	(12)	1	5
		0	11	72	0	32

under the basic variables are always zero, and the elements under x_2 and x_3 follow since $a_{12}c_1 + a_{22}c_4 - c_2 = 3 \times 3 + 1 \times 4 - 2 = 11$ and $a_{13}c_1 + a_{23}c_4 - c_3 =$

$8 \times 3 + 12 \times 4 - 0 = 72$. Let us make x_3 basic, then since $\frac{5}{12} < \frac{4}{8}$, 12 is the pivot and using the replacement rule we obtain the next table, where $\frac{72}{12}$ times the pivotal row has been subtracted from the bottom row.

	x_1	x_2	x_3	x_4	
x_1	1	$\frac{7}{3}$	0	$-\frac{2}{3}$	$\frac{2}{3}$
x_3	0	$\frac{1}{12}$	1	$\frac{1}{12}$	$\frac{5}{12}$
	0	5	0	-6	2

The c_i's can now be dropped from the second table, as they are no longer required in the calculations. This time the pivot is $\frac{7}{3}$ as $\frac{2}{7} < 5$. To save unnecessary computation, one should first calculate the bottom row of a table, for if these elements are non-positive, as in the table below, an optimum solution has been obtained, and there is no need to calculate all the remaining elements. The

	x_1	x_2	x_3	x_4	
x_2					$\frac{2}{7}$
x_3					$\frac{11}{28}$
	$-\frac{15}{7}$	0	0	$-\frac{32}{7}$	$\frac{4}{7}$

optimal solution is

$$x_1 = 0, \qquad x_2 = \tfrac{2}{7}, \qquad x_3 = \tfrac{11}{28}, \qquad x_4 = 0$$

and the value of the programme is $\frac{4}{7}$. It is a wise rule to check that your optimal solution does satisfy the equations.

EXAMPLE 11.4.3

$$
\begin{aligned}
\text{Minimize} \quad z = &\ x_1 + 2x_2 + x_3 \\
\text{subject to} \quad &\ 2x_1 + 3x_2 + 6x_3 = 8, \\
&\ -3x_1 - x_2 + 7x_3 = 4, \\
\text{and} \quad &\ x_1, x_2, x_3 \geq 0.
\end{aligned}
$$

Solution

In this example there is no obvious basic feasible solution with which to start the simplex method, so we introduce slack variables x_4 and x_5 and solve the following modified linear programme.

$$
\begin{aligned}
\text{Minimize} \quad z = &\ x_1 + 2x_2 + x_3 + M(x_4 + x_5), \\
\text{subject to} \quad &\ 2x_1 + 3x_2 + 6x_3 + x_4 = 8, \\
&\ -3x_1 - x_2 + 7x_3 + x_5 = 4, \\
\text{and} \quad &\ x_1, x_2, x_3, x_4, x_5 \geq 0,
\end{aligned}
$$

where M is a positive number, which is sufficiently large to ensure that both linear programmes have the same optimal solution, given that an optimal solution to the first linear programme exists.

We start the simplex method with x_4 and x_5 as the basic variables and obtain the following sequence of tables

		1	2	1	M	M	
		x_1	x_2	x_3	x_4	x_5	
M	x_4	2	3	6	1	0	8
M	x_5	-3	-1	⑦	0	1	4
		$-M-1$	$2M-2$	$13M-1$	0	0	$12M$
	x_4	$\left(\frac{32}{7}\right)$	$\frac{27}{7}$	0	1	$-\frac{6}{7}$	$\frac{32}{7}$
	x_3	$-\frac{3}{7}$	$-\frac{1}{7}$	1	0	$\frac{1}{7}$	$\frac{4}{7}$
		$\frac{32M-10}{7}$	$\frac{27M-15}{7}$	0	0	$\frac{-(13M-1)}{7}$	$\frac{32M+4}{7}$
	x_1				$\frac{7}{32}$	$-\frac{3}{16}$	1
	x_3				$\frac{3}{32}$	$\frac{1}{16}$	1
		0	$-\frac{15}{16}$	0	$-M+\frac{10}{32}$	$-M-\frac{1}{8}$	2

Since M is a suitably large positive number, the bottom row is non-positive and so the optimal solution to the original linear programme is $x_1 = 1$, $x_2 = 0$, $x_3 = 1$ with a value of 2.

11.4.5. Solution of Dual Programme

By extending the table, as in the above example, with a slack variable for each equation, we can find the solution of the dual programme simultaneously. The dual of the above primal programme is to

$$\text{maximize} \quad w = 8y_1 + 4y_2,$$
$$\text{subject to} \quad 2y_1 - 3y_2 \le 1,$$
$$3y_1 - y_2 \le 2,$$
$$6y_1 + 7y_2 \le 1,$$

and the optimal solution to this dual programme is given by the values at the bottom of the columns labelled by the slack variables, x_4 and x_5, when $M = 0$. This gives $y_1 = \frac{5}{16}$, $y_2 = -\frac{1}{8}$ as the optimal solution, which can easily be checked.

11.4.6. Inverse Matrix (Revised Simplex) Method

The inverse matrix method is used in many computer programmes, as it involves less computations when the number of columns in the constraint matrix **A** is much larger than the number of rows. It also uses less storage space and suffers less from rounding errors when **A** is sparse (contains mainly zeros). Consider the following linear programme:

$$\text{minimize} \quad z = c_1 x_1 + c_2 x_2 + c_3 x_3 + c_4 x_4,$$

$$\text{subject to} \quad a_{11} x_1 + a_{12} x_2 + a_{13} x_3 + a_{14} x_4 = b_1,$$

$$a_{21} x_1 + a_{22} x_2 + a_{23} x_3 + a_{24} x_4 = b_2,$$

$$\text{and} \quad x_1, x_2, x_3, x_4 \geq 0.$$

The simplex table extended by the slack variables, x_5 and x_6, is shown below.

		c_1	c_2	c_3	c_4	0	0	
		x_1	x_2	x_3	x_4	x_5	x_6	
0	x_5	a_{11}	a_{12}	a_{13}	a_{14}	1	0	b_1
0	x_6	a_{21}	a_{22}	a_{23}	a_{24}	0	1	b_2
		$-c_1$	$-c_2$	$-c_3$	$-c_4$	0	0	0

The inverse matrix method only uses the replacement rule on the last column and the columns corresponding to the slack variables. Suppose that after two replacements, the table takes the following form,

	x_1	x_2	x_3	x_4	x_5	x_6	
x_2	$\sum_i l_{1i} a_{i1}$				l_{11}	l_{12}	b_1^*
x_3	$\sum_i l_{2i} a_{i1}$				l_{21}	l_{22}	b_2^*
	$\sum_i y_i a_{i1} - c_1$	0	0	$\sum_i y_i a_{i4} - c_4$	y_1	y_2	z

where the values of l_{ij}, y_i, b_i^*, z are obtained by application of the replacement rule. The remaining non-zero elements in the bottom row, corresponding to the non-basic variables, are then computed using the original matrix **A** and the y_i. If the solution is not optimal, the pivotal column (the first) is chosen by the pivot rule and then the elements of the pivotal column are calculated using the matrices **L** and **A**, as shown in the table, where $\mathbf{L} = (l_{ij})$ is called the *inverse basis matrix*. Finally, the pivot for the next replacement is chosen using the second part of the pivot rule.

11.4.7. Dual Simplex Method

The dual simplex method uses a modified pivot rule and is a good method for problems of the form:

$$\text{minimize} \quad z \; = \; \mathbf{c}'\mathbf{x},$$
$$\text{subject to} \quad \mathbf{A}\mathbf{x} \geq \mathbf{b},$$
$$\text{and} \quad \mathbf{x} \geq \mathbf{0},$$

where $c_i \geq 0$ for all i and $\mathbf{b}$ contains both positive and negative elements, so that it is much easier to find a basic feasible solution for the dual programme than for the primal.

In this method the pivot rule is to choose some basic $x_k < 0$ and then for $a_{kj} < 0$, find the value j which minimizes

$$\frac{\sum_i a_{ij} c_i - c_j}{a_{kj}},$$

where $\sum_i a_{ij} c_i - c_j$ are the elements in the bottom row. If $j = l$, then a_{kl} is the pivot for the application of the replacement rule.

The dual simplex method works by determining a sequence of basic feasible solutions of the dual programme, which progressively increase w, the linear function to be maximized. The optimal solution is achieved when all the basic x_i are positive, so that the basic solution of the primal programme is also feasible (satisfies $\mathbf{x} \geq \mathbf{0}$ as well as $\mathbf{A}\mathbf{x} \geq \mathbf{b}$). If a situation occurs in which $x_k < 0$ and $a_{kj} \geq 0$ for all j, then the programme has no feasible solution.

11.4.8. Degeneracy

Degeneracy manifests itself in the simplex method when two or more elements in the same column can be chosen for the pivot. Often if one chooses one of the elements as the pivot and proceeds, an optimal solution is eventually obtained. However, it can happen that the method cycles through a set of basic feasible solutions for which z has the same value. In this case, the method breaks down, but the problem can be overcome by perturbing the value of $\mathbf{b}$ in the constraints, as is done in the transportation problem. The inverse matrix method is well adapted for doing this.

11.4.9. Sensitivity Analysis

Often in real problems the values of $\mathbf{A}$, $\mathbf{b}$ and $\mathbf{c}$ in the primal programme are only approximate or are subject to changes. In the case of variable $\mathbf{c}$, one can test how sensitive the optimal solution is to changes in the values of $\mathbf{c}$ by recomputing the bottom row of the optimal table with arbitrary c_i, and then finding the conditions on the c_i for the bottom row to remain non-positive. If $\mathbf{b}$ is changed to $\mathbf{b} + \mathbf{e}$, then the basic variables in the optimal solution remain unaltered provided $\mathbf{L}(\mathbf{b} + \mathbf{e}) \geq \mathbf{0}$, where $\mathbf{L}$ is the inverse basis matrix.

11.4.10.   Parametric Linear Programming

Parametric linear programming is concerned with problems in which the vectors c and/or b in the primal programme are linear functions of a parameter. As an example, we will solve the linear programme:

$$\text{minimize} \quad (1 + t)x_1 + (1 - 2t)x_2 + 3x_3,$$
$$\text{subject to} \quad x_1 + 3x_2 \qquad = 4 + s,$$
$$2x_2 + x_3 = 3,$$
$$\text{and} \qquad x_1, x_2, x_3 \geq 0,$$

where t is arbitrary and $s \geq -4$, so that the first constraint is not incompatible with $x_1, x_2 \geq 0$.

An initial basic feasible solution is $x_1 = 4 + s$, $x_3 = 3$ and the corresponding simplex table is shown below.

		$1 + t$	$1 - 2t$	3	
		x_1	x_2	x_3	
$1 + t$	x_1	1	3	0	$4 + s$
3	x_3	0	2	1	3
		0	$8 + 5t$	0	$(4 + s)(1 + t) + 9$

This is optimal for $8 + 5t \leq 0$ or $t \leq -\frac{8}{5}$. For $t > -\frac{8}{5}$ we pivot on 3 in the x_2 column if $4 + s/3 \leq \frac{3}{2}$ or $s \leq \frac{1}{2}$, and on 2 otherwise. Pivoting on 3 and 2, respectively, gives us the following two optimal tables:

	x_1	x_2	x_3	
x_2				$\dfrac{4 + s}{3}$
x_3				$\dfrac{1 - 2s}{3}$
	$\dfrac{-(8 + 5t)}{3}$	0	0	

	x_1	x_2	x_3	
x_1				$s - \frac{1}{2}$
x_2				$\frac{3}{2}$
	0	0	$\dfrac{-(8 + 5t)}{2}$	

Summarizing, the optimal solutions are

$$x_1 = 4 + s, \qquad x_2 = 0, \qquad x_3 = 3 \quad \text{for } t \leq -\tfrac{8}{5},$$
$$x_1 = 0, \qquad x_2 = (4 + s)/3, \qquad x_3 = (1 - 2s)/3 \quad \text{for } t \geq -\tfrac{8}{5}, s \leq \tfrac{1}{2},$$

and

$$x_1 = s - \tfrac{1}{2}, \qquad x_2 = \tfrac{3}{2}, \qquad x_3 = 0 \quad \text{for } t \geq -\tfrac{8}{5}, s \geq \tfrac{1}{2}.$$

11.4.11. Decomposition Method

The decomposition method is designed to deal with the following type of linear programme:

$$\text{maximize} \quad z = 3x_1 + 4x_2 + x_3 + 5x_4,$$
$$\text{subject to} \quad x_1 + 2x_2 + x_3 + 4x_4 \leq 10,$$
$$3x_1 + 2x_2 \qquad\qquad \leq 5,$$
$$x_1 + 4x_2 \qquad\qquad \leq 6,$$
$$5x_3 + x_4 \leq 7,$$
$$2x_3 + 3x_4 \leq 4,$$
$$\text{and} \quad x_1, x_2, x_3, x_4 \geq 0.$$

This sort of problem can arise in firms with several branches. Suppose in the above example, that x_1 and x_2 measure the outputs of one branch and x_3 and x_4, the outputs in a second branch, then the second and third inequalities represent the internal constraints in the first branch, whereas the fourth and fifth inequalities relate to the second branch. The first inequality represents a resource constraint for the whole firm. Clearly, for a firm with many branches, such problems can become very large and the decomposition method reduces the size of the problem by solving linear programmes for the individual branches and then checking whether the resulting solution is optimal for the overall linear programme. Details of the method can be found in the first four references listed at the end of the chapter.

11.5. APPLICATIONS

Linear programming methods can be used to solve a wide class of problems as exemplified by the following examples.

11.5.1. A Production Scheduling Problem

The demand for a particular make of jam is predicted to be d_1, d_2, d_3 and d_4 tonnes, respectively, during the next four months. The firm can produce up to s tonnes of jam per month and the costs of making one tonne of jam are c_1, c_2, c_3 and c_4, respectively, in the four months. If the storage cost is k per tonne per month, then the problem is to determine how much jam the firm should produce each month in order to minimize its production costs.

This problem can be formulated as a transportation problem by letting x_{ij} ($i, j = 1, 2, 3, 4$) be the number of tonnes produced in the ith month for delivery in the jth month and putting x_{i5} equal to the unused capacity in the ith month. The appropriate costs, supplies and demands are shown in the table below, where M represents a large cost which ensures that $x_{ij} = 0$ for $i > j$, as jam cannot be delivered before it is made. Assuming that the costs of producing no

	d_1	d_2	d_3	d_4	$4s - d_1 - d_2 - d_3 - d_4$
s	c_1	$c_1 + k$	$c_1 + 2k$	$c_1 + 3k$	0
s	M	c_2	$c_2 + k$	$c_2 + 2k$	0
s	M	M	c_3	$c_3 + k$	0
s	M	M	M	c_4	0

jam is zero, the last column consists of zeros, since x_{i5} is the unused capacity. Finally, the cost of producing one tonne of jam in the ith month for delivery in the jth month is $c_i + (j - i)k$ for $j \geq i$, where $(j - i)k$ is the storage cost.

Clearly, for a feasible solution we must have $s \geq d_1$, $2s \geq d_1 + d_2$, $3s \geq d_1 + d_2 + d_3$ and $4s \geq d_1 + d_2 + d_3 + d_4$, or otherwise the demands could not be met. The problem is now solved in the same way as a transportation problem.

11.5.2. A Cutting Problem

Wallpaper is supplied in rolls of length 9·9 m and 19 strips of length 2·1 m and 8 strips of length 0·9 m are required to paper a room. The problem is how to cut the rolls of paper so as to minimize the number of rolls used.

A roll of paper can be cut up into 2·1 m and 0·9 m lengths in the following ways:

2·1 m lengths	0·9 m lengths	Length wasted	Number of rolls used
4	1	0·6 m	x_1
3	4	0 m	x_2
2	6	0·3 m	x_3
1	8	0·6 m	x_4

The problem can now be formulated as the following linear programme:

minimize $z = x_1 + x_2 + x_3 + x_4,$ (number of rolls used)

subject to $4x_1 + 3x_2 + 2x_3 + x_4 \geq 19,$ (at least 19 strips of 2·1 m length)

           $x_1 + 4x_2 + 6x_3 + 8x_4 \geq 8,$ (at least 8 strips of 0·9 m length)

and $x_1, x_2, x_3, x_4 \geq 0.$

In order that the simplex method can be used, we re-write the constraints in the form

$$4x_1 + 3x_2 + 2x_3 + x_4 - x_5 + x_7 = 19$$
$$x_1 + 4x_2 + 6x_3 + 8x_4 - x_6 + x_8 = 8$$
$$x_1, x_2, \ldots, x_8 \geq 0,$$

and minimize

$$z = x_1 + x_2 + x_3 + x_4 + M(x_7 + x_8),$$

where x_5 and x_6 are introduced to transform the inequalities into equations, and x_7 and x_8 enable a basic feasible solution to be found. $M > 0$ is a large number chosen so that x_7 and x_8 are zero in the optimal solution. It can be seen that $M = 2$ is sufficiently large. Using the simplex method, we obtain the following table:

		1	1	1	1	0	0	$M = 2$	$M = 2$	
		x_1	x_2	x_3	x_4	x_5	x_6	x_7	x_8	
$M = 2$	x_7	4	3	2	1	-1	0	1	0	19
$M = 2$	x_8	1	④	6	8	0	-1	0	1	8
		9	13	15	17	-2	-2	0	0	54
	x_7	$(\tfrac{13}{4})$	0	$-\tfrac{5}{2}$	-5	-1	$\tfrac{3}{4}$	1	$-\tfrac{3}{4}$	13
	x_2	$\tfrac{1}{4}$	1	$\tfrac{3}{2}$	2	0	$-\tfrac{1}{4}$	0	$\tfrac{1}{4}$	2
		$\tfrac{23}{4}$	0	$-\tfrac{9}{2}$	-9	-2	$\tfrac{5}{4}$	0	$-\tfrac{13}{4}$	28
	x_1									4
	x_2									1
		0	0	$-\tfrac{1}{13}$	$-\tfrac{2}{13}$	$-\tfrac{3}{13}$	$-\tfrac{1}{13}$	$-\tfrac{23}{13}$	$-\tfrac{25}{13}$	5

Thus the optimal solution is to use 5 rolls of wallpaper, with 4 rolls cut into $4 \times 2 \cdot 1$ m $+ 1 \times 0 \cdot 9$ m lengths and 1 roll cut into $3 \times 2 \cdot 1$ m $+ 4 \times 0 \cdot 9$ m lengths. Only $2 \cdot 4$ m of paper is wasted.

The cutting problem does not always have an integer optimal solution, so integer programming methods may have to be used.

11.5.3. Revenue Maximization

Consider the following pair of dual programmes:

	Primal			Dual	
maximize	$w = \mathbf{b'y}$		minimize	$z = \mathbf{c'x}$	
subject to	$\mathbf{A'y} \le \mathbf{c}$		subject to	$\mathbf{Ax} \ge \mathbf{b}$	
and	$\mathbf{y} \ge \mathbf{0}$		and	$\mathbf{x} \ge \mathbf{0}.$	

Let c_j be the amount of the jth resource (e.g. coal, iron or manpower) available to a firm, y_i be the output of the ith manufactured product, b_i be the revenue from selling one unit of the ith output and a_{ij} be the amount of the jth resource required to manufacture one unit of the ith output. Then the primal programme is to maximize the revenue $w = \sum_i b_i y_i$, subject to the available resources not being exceeded.

If we interpret x_j as the value (accounting or shadow price) of one unit of the jth resource, then z is the value of the resources available to the firm and the ith constraint of the dual says that the value of the resources used in the manufacture of one unit of the ith output should not be less than the revenue from it. Hence the dual programme is to find the lowest valuation of the firm's resources which completely accounts for the revenue from each output.

With this interpretation, the first result of duality theory [see § 11.2.4] says that for optimal solutions, the total revenue from the outputs equals the value of the resources used. The second result of duality theory states that in an optimal solution, the firm will only manufacture those products ($y_i > 0$) for which the value of the resources used equals the revenue from the output ($\sum_j a_{ij}x_j = b_i$) and that only those resources that are completely consumed ($\sum_i a_{ij}y_i = c_j$) have a positive accounting value ($x_j > 0$).

A further result is that the value, x_j, assigned to one unit of the jth resource in the optimal solution, equals the marginal revenue of the jth resource. This follows since the marginal revenue of jth resource $= \partial w/\partial c_j = \partial z/\partial c_j = x_j$, provided the derivative exists. Thus the duality theory of linear programming is closely related to marginal analysis.

An an illustration of the foregoing theory we solve the following problem. A firm manufactures tables, desks and bookcases. These products are made from lengths of oak and plywood. The lengths of wood required for their manufacture, the revenues from their sale and the available resources of oak and plywood are summarized in the table below.

	One table	One desk	One bookcase	Amount available
oak	3 m	2 m	1 m	450 m
plywood	1 m	1·5 m	3 m	900 m
revenue	£64	£76	£56	

The problem is to find y_1, y_2 and y_3, the number of tables, desks and bookcases, respectively, the firm should manufacture so as to maximize its revenue. This reduces to the following linear programme:

$$\text{maximize} \quad w = 64y_1 + 76y_2 + 56y_3$$
$$\text{subject to} \quad 3y_1 + 2y_2 + y_3 \le 450$$
$$y_1 + 1{\cdot}5y_2 + 3y_3 \le 900$$
$$\text{and} \quad y_1, y_2, y_3 \ge 0,$$

and the dual programme is to

$$\text{minimize} \quad z = 450x_1 + 900x_2,$$
$$\text{subject to} \quad 3x_1 + x_2 \ge 64,$$
$$2x_1 + 1{\cdot}5x_2 \ge 76,$$
$$x_1 + 3x_2 \ge 56,$$
$$\text{and} \quad x_1, x_2 \ge 0,$$

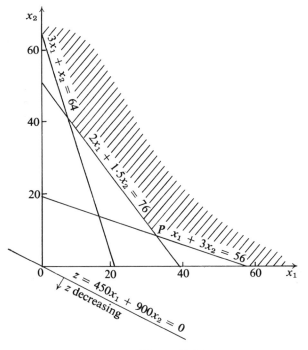

Figure 11.5.1

where x_1 and x_2 are interpreted as the accounting prices of the oak and plywood, respectively, and z is the value of the firm's resources of these woods. The dual programme can easily be solved by graphical methods (see Figure 11.5.1) and the optimal values are $x_1 = 32$ and $x_2 = 8$, the coordintes of the point P. This means that the accounting prices of one metre of oak and plywood are £32 and £8, respectively. To find the optimal solution of the primal programme we note that $3x_1 + x_2 > 64$, which says that the accounting price of a table is more than the revenue on it, consequently it is not worthwhile to manufacture tables and $y_1 = 0$. Since $x_1 > 0$ and $x_2 > 0$, it follows from the second result of duality theory that

$$3y_1 + 2y_2 + y_3 = 450,$$
$$y_1 + 1\cdot5y_2 + 3y_3 = 900,$$

and so putting $y_1 = 0$, we obtain $y_2 = 100$ and $y_3 = 250$ as the optimal solution to the primal programme. Hence the firm should manufacture no tables, 100 desks and 250 bookcases in order to maximize its revenue.

In general, the optimal solution to problems of this type will not be integer. Often a neighbouring feasible integer solution will be adequate, especially if the numbers are large. If not, then integer programming methods will have to be used.

This concludes the chapter on linear programming. Further applications of the subject can be seen in chapters 12 and 13, and the following references are recommended to readers who wish to follow up the methods and theory described in this chapter.

K.T.

REFERENCES

Beale, E. M. L. (1968). *Mathematical Programming in Practice*, Pitman.

Dantzig, G. B. (1963). *Linear Programming and Extensions*, Princeton University Press.

Hadley, G. (1962). *Linear Programming*, Addison-Wesley,

Krekó, B. (1968). *Linear Programming*, Pitman.

Trustrum, K. (1971). *Linear Programming*, Routledge Kegan Paul.

Vajda, S. (1961). *Mathematical Programming*, Addison-Wesley.

CHAPTER 12

Integer Programming

INTRODUCTION

Integer Programming is concerned with a class of optimization problems of which the following is an illustrative example.

(A)
$$
\begin{aligned}
\text{Maximize} \quad & x_1 + x_2 \\
\text{subject to} \quad & 10x_1 - 8x_2 \le 13 \\
& 2x_1 - 2x_2 \ge 1 \\
& x_1, x_2 \ge 0 \text{ and integer.}
\end{aligned}
$$

In this example we wish to find values for the variables x_1 and x_2 which make the expression $x_1 + x_2$ (the *objective function*) as large as possible but satisfy the *constraints* of the problem. These constraints are of two kinds. Firstly we have a number of inequality relations such as $10x_1 - 8x_2 \le 13$, $x_1 \ge 0$ etc. Secondly we have the stipulation that x_1 and x_2 must take *integer* values [see § 2.2.1]. Were it not for this second stipulation we would have a *Linear Programming* (LP) problem.

Superficially Integer Programming (IP) problems look very like LP problems. Any LP problem becomes an IP problem if we stipulate that some or all of the variables must take integer values. In the first case we obtain what is known as a *Mixed Integer Programming* (MIP) problem. In the latter case we have a *Pure Integer Programming* (PIP) problem. Example (A) above is clearly a PIP problem.

The integrality requirement on certain variables makes IP problems, in some ways, very different from LP problems in spite of their apparent similarity. IP problems are usually much more difficult to solve than correspondingly sized LP problems. Methods of solution are described in section 3. In compensation, however, it is possible to use IP to model a surprisingly wide range of practical problems. This is discussed in section 12.1. It is useful to make a distinction between a *problem* and a *model*. We will use the term 'problem' in a fairly general sense to indicate the practical situation we are dealing with. The term 'model' will be reserved for the mathematical representation of the problem. In section 12.2 we will be concerned with the transformation of practical problems into the particular type of model which concerns us in this chapter, i.e. an IP model.

In order to reinforce the distinction which must be made between LP and IP we present the optimal solution to example (A) below, firstly treating it as an LP and then as an IP.

LP Optimal Solution $x_1 = 4 \cdot 5$, $x_2 = 4$,   Objective $= 8 \cdot 5$

IP Optimal Solution $x_1 = 2$, $x_2 = 1$,   Objective $= 3$.

There is no obvious means of arriving at the IP Optimal Solution (the *integer optimum*) from the LP Optimal Solution (*the continuous optimum*). Rounding the fractional LP solution either up or down to the next integer produces a solution which violates the constraints (an infeasible solution). The deduction of the integer optimum requires a considerably more sophisticated approach than this. For general IP problems it often takes as much as ten times as long to find the IP optimum as the LP optimum using a computer. Our example has been purposely chosen to have only two variables so that we can illustrate the difficulty geometrically in Figure 12.1.1.

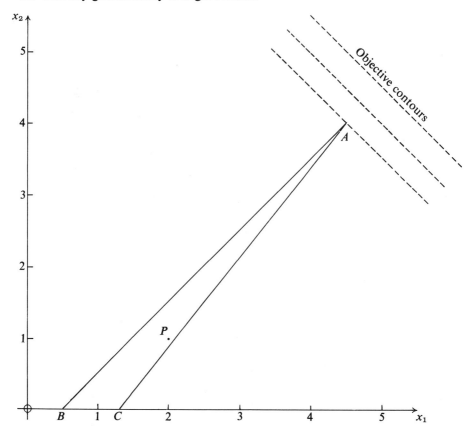

Figure 12.1.1

The values of variables satisfying the constraints of the LP problem are the coordinates of points in the region *ABC*. When optimizing a linear objective function we generally obtain a *vertex solution* for an LP model [see § 11.1.1]. In this case we arrive at point *A* by taking the highest of the objective contour lines which intersects the region *ABC*. For an IP model we are seeking integer solutions which may well lie in the interior of the region. In this example the optimal IP solution is represented by the point *P*.

In order to solve IP models we cannot, therefore, rely on any simple geometric property such as only looking at vertex solutions. Some, but not all, IP algorithms do, however, begin by solving the associated LP problem first. This is discussed in detail in section 12.3.

As in other branches of mathematics it seems surprising that integer problems should be more difficult to solve than problems involving real or rational numbers. This is indeed the case with Mathematical Programming. Kronecker's assertion that 'God invented the integers and man did the rest' seems to be of little avail.

12.1.   THE APPLICABILITY OF INTEGER PROGRAMMING

The range of practical problems which can be modelled using IP is very large. We briefly mention some of these applications below. The modelling of IP problems is often far from straightforward. Rather than describe how these practical problems are modelled in this section we devote section 12.2 to the subject of building IP models. Some of these applications will be used as illustrations there.

An alternative name for Integer Programming is *Discrete Programming*. In some ways this conveys the type of problem we are dealing with rather better. We are essentially dealing with quantities which change in discrete units rather than continuously.

It will be seen in the applications that follow that the majority of IP problems give rise to what are known as *Zero–One Integer Programming* models. These are models where some or all of the variables are restricted to two values 0 or 1. Such variables are used to indicate Yes/No type decisions. The use of such variables is described in detail in section 12.2. If the following inequalities are applied to a variable x:

$$0 \leq x \leq 1 \tag{12.1.1}$$

and the variable is restricted to take integer values, it can obviously only take the two values 0 or 1. 0–1 IP models can therefore be regarded as a special case of general IP models.

A description of areas where IP can be applied is now given.

12.1.1.   Problems Involving Discrete Activities

An obvious application of LP is to decide how much to make of various products subject to limited resources (machines, manpower, raw material etc.).

If these products can only come in integer quantities (such as cars, aeroplanes, bridges across a river etc.) we have to build an IP model. Although the most obvious application for IP, this is probably the least typical. In practice we would probably round the LP optimal solution off to the nearest integers. This will not guarantee us the IP optimal solution or even a feasible solution (as exemplified by example (A)) [see § 11.1.2]. For this type of problem, however, the values of the variables are usually sufficiently large (and approximate) to make rounding acceptable.

12.1.2.   Problems Involving Discrete Resources

Resources may be discrete (indivisible). Examples are manpower, machinery (half a machine is meaningless), storage capacity (we either use a whole tank or none of it) etc. In this case we must ensure that a model involves usage of discrete 'lumps' of the resources.

12.1.3.   Fixed Charge Problems

If a particular activity is carried out at all (such as making a product) it may involve a *fixed cost* (such as setting up a machine). This cost is quite independent of the level of the activity. The total cost of the activity therefore changes discretely between doing nothing (cost nothing) and doing something however small. This situation cannot be modelled by LP but can be dealt with by IP by using a 0–1 integer variable to represent whether the activity is, or is not, carried out at all. The formulation of this type of problem is described in section 12.2.

A special case of this sort of problem is the *Depot Location Problem*. This can be regarded as an extension of the *Transportation Problem* where we do not only want to determine a minimum cost distribution pattern but also decide where to build our depots (or warehouses) in the first place. The decision of whether to build a particular depot or not is represented by a 0–1 integer variable. This type of problem is again formulated in section 12.2.

Other particular applications which can be regarded as coming within this category are *Project Selection Problems* and *Capital Budgetting Problems*. Here it is necessary to decide whether or not to embark on particular projects or capital investments. The use of 0–1 integer variables again suggests itself.

Fixed charge problems account for a large number of the successful applications of IP.

12.1.4.   Problems with Logical Conditions

In addition to the fairly straightforward constraints which can be modelled by LP it is often necessary to impose extra 'logical' conditions. Such conditions can often be imposed by adding 0–1 integer variables and extra constraints into an otherwise LP model. One example of this situation is *limiting the number of ingredients* in a blending problem. The formulation of this type of problem is illustrated in section 12.2.

Other types of logical stipulation which we might wish to model in a practical situation are exemplified by the following:

If product A is made then we must make B and C as well. (12.1.2)

Either operation A must be finished before operation B or vice versa. (12.1.3)

If we use one of this class of ingredients then we cannot use any of the other class. (12.1.4)

We can use either this production process or that one. (12.1.5)

If this factory is kept open then we must retain these two depots. (12.1.6)

12.1.5. Nonlinear Problems

If non-linear functions occur in the constraints or objective function of a model then it is often possible to approximate these non-linearities by a series of linear functions. An extension of LP known as *Separable Programming* can be used to help solve the resulting model. In general, however, separable programming can only help one obtain a *local optimum*. An illustrative way of regarding this is to think of climbing a mountain range in a thick fog. A mountain peak is a local optimum which we should be able to recognize. We cannot, however, tell if there is another higher peak hidden in the fog. The highest such peak is known as a *global optimum*. An IP model can be used to obtain a global optimum, where separable programming finds only local optima. The formulation of such non-linearities using IP is described in section 12.2.

12.1.6. Combinatorial Problems

Many problems which arise in Operational Research are of a combinatorial character. This is a fairly loose classification and some of the problems mentioned above also exhibit combinatorial aspects. Most combinatorial problems can be formulated as IP models. The *Job-Shop Scheduling* problem is the problem of sequencing operations on machines so as to complete all operations as early as possible. Although IP is a superficially attractive way of modelling the problem such models are often very difficult to solve. Another objection to IP is that the dynamic nature of the situation often demands an adaptive approach to the solution rather than a once-and-for-only optimum. The *Travelling Salesman Problem* is famous because of its conceptual simplicity but difficulty in solution. It can be formulated as an IP model in a number of different ways. The problem is to find the minimum cost (distance) route of a salesman around a number of cities finally returning to his base. A mathematically equivalent version of the problem arises in sequencing jobs on machines in such an order as to minimize total set-up time. Set-up time depends on both the preceding job and its successor. It can be regarded as the 'distance' between jobs. An extension of the travelling salesman problem which again can be modelled using IP is the *Vehicle Scheduling Problem*. This involves deciding not just how to route lorries but also how to split different visits among different lorries.

A computationally very difficult problem which can be tackled through IP is the *Quadratic Assignment Problem*. This is an extension of the well-known assignment problem in LP. The cost of an assignment here depends not just on the individual assignment but on what other assignments have been made as well. If for example we were to locate a company division A in city P and division B in city Q the total cost might involve the degree of communication which takes place between the departments as well as the distances of the cities apart. The cost of assigning A to P is not therefore independent of the other assignments.

An important special class of problems gives rise to pure 0–1 IP models as well as having an extra special structure. These problems are known as *Set Covering Problems, Set Partitioning Problems* and *Set Packing Problems*. One of the famous problems of this type is the *Aircrew Scheduling Problem*. This involves assigning aircrews to 'rosters' so as to minimize either the total number of crews needed or their cost. The resultant problem is, in some cases, a Set Covering Problem or in other cases a Set Partitioning Problem. Such problems arise in scheduling other personnel. Another Set Partitioning Problem is the *Assembly Line Balancing Problem*. This is the problem of assigning workers to different stations on a production line in order to achieve a particular production rate. A slight extension of the Set Partitioning problem arises in *Political Districting*. This is the problem of splitting up an area into political constituencies so as to, as near as possible, equalize political representation.

Many combinatorial problems give rise to PIP models. Despite the seemingly large number of such applications it would be misleading to regard such problems as making up the bulk of IP applications. Most practical IP models arise from extensions of existing LP models. The extension usually (but not always) involves 0–1 integer variables which are used to model fixed costs, logical conditions or non-linearities. Clearly this will result in MIP models. Once integer variables are incorporated in the model it will be possible to use them to model, often complex situations, sometimes with a combinatorial flavour.

12.2. BUILDING INTEGER PROGRAMMING MODELS

The formulation of an IP model is often by no means straightforward. It is often possible to model a situation in more than one way. Some formulations may be more desirable than others. The aim of this section is to provide some systematization for the process of building a model. Model Building is still, however, to some extent an art and ingenuity can never be totally automated.

12.2.1.   The Uses of Integer Variables

Discrete Activities obviously give rise to integer, rather than continuous, variables and can be used to model indivisible quantities such as, for example, houses, cows or projects.

Discrete Resources can be represented by an expression involving 0–1 integer variables. For example the expression:

$$R_1\delta_1 + R_2\delta_2 + \ldots + R_n\delta_n$$

represents a resource which can come in quantities R_1, $R_1 + R_2$, $R_1 + R_2 + R_3$ etc. We adopt the convention of using the Greek symbol 'δ' to represent 0–1 integer variables.

Decision Variables are used in IP to represent different possible decisions. These are often 0–1 integer variables e.g.

$$\delta = 1 \quad \text{if the investment should be made,}$$
$$\delta = 0 \quad \text{if the investment should not be made.}$$

There are situations, however, in which we might use a more general integer variable as a decision variable e.g.

$$\gamma = 0 \quad \text{if we invest in a type 1 machine,}$$
$$\gamma = 1 \quad \text{if we invest in a type 2 machine,}$$
$$\gamma = 2 \quad \text{if we invest in a type 3 machine.}$$

Indicator Variables are used to indicate when a particular situation does or does not hold. Such variables are always 0–1 integer variables e.g.

$$\delta_i = 1 \quad \text{if ingredient } i \text{ is included in the blend,}$$
$$\delta_i = 0 \quad \text{if ingredient } i \text{ is not included in the blend.}$$

It is possible to 'link' such indicator variables to corresponding continuous variables by inequalities so that they have the desired effect. For example if x_i represents the amount of ingredient i in the blend we want δ_i to be forced to take the value 1 if x_i comes out greater than 0. In section 12.2.2 we describe how this may be done.

12.2.2.   The Uses of Integer Constraints

Resource Constraints

When activities can only take integer values the modelling of the conventional resource usage constraint found in LP is straightforward and identical in form to the LP case. If the resource itself is discrete we may have a constraint such as that below:

$$2x_1 + x_2 + 3x_3 \leq 10 + 2\delta_1 + 2\delta_2. \tag{12.2.1}$$

Here the conventional right-hand side constant of a resource constraint is replaced by an expression involving 0–1 integer variables. In this case the activities (represented by the x_i variables) use up different amounts of this discrete resource. Initially we have 10 units of the resource but this may be expanded to 12 or 14 if wished. The variables δ_1 and δ_2 would probably be given coefficients in the objective function representing the costs of expanding the resource availability. It is conventional to write both LP and IP models with variables on the left and constants on the right giving:

$$2x_1 + x_2 + 3x_3 - 2\delta_1 - 2\delta_2 \leq 10. \tag{12.2.2}$$

Linking Constraints

In, for example, *the fixed charge problem* of § 12.1.3, we wish to link a (probably continuous) variable x representing an activity level to an indicator variable δ. For example x might represent the quantity of a product to be made and δ indicate whether we make any (and incur a set up cost) or not. x and δ must be linked to establish the relation:

$$x > 0 \quad \text{implies} \quad \delta = 1. \tag{12.2.3}$$

This can (rather surprisingly) be done by the inequality

$$x - M\delta \leq 0. \tag{12.2.4}$$

M is a coefficient known to be as large or larger than x can ever be. The necessity to define such an 'upper bound' for x usually presents little practical difficulty.

Clearly if x comes out positive (we make some of the product) δ is forced to 1 (it cannot come out fractional). δ will probably have an objective coefficient representing the set up cost (the fixed charge).

It should be pointed out that constraint (12.2.4) does not force the reverse condition to hold i.e. if $\delta = 1$ x is not forced to be positive. In a situation such as this it is unnecessary to force this condition by a constraint. If δ were to be 1 and x had a value of 0 we could obtain a better (smaller cost) solution by changing δ to 0 and we would not violate the constraint (12.2.4). There are, however, situations where it is necessary to model the reverse condition by a constraint. This linking of 0–1 integer variables to continuous variables is at the heart of many IP formulations. We therefore give an example [see also § 12.1.3].

EXAMPLE 12.2.1 (The Depot Location Problem). Single depots can be located at m possible sites. If built, a depot at position i has a yearly capacity K_i and a yearly cost (in interest charges payable on the capital cost as well as running costs) of c_i. The depots will be used to supply n customers. Customer j has a yearly demand D_j. The cost of sending a unit quantity from depot i to customer j is c_{ij}. Where should depots be built and what is the distribution pattern which will minimize total cost?

In order to build an IP model the following variables are introduced:

$$x_{ij} = \text{quantity sent (per year) from depot } i \text{ to customer } j,$$
$$\delta_i = 1 \quad \text{if depot } i \text{ is built}$$
$$= 0 \quad \text{otherwise.}$$

x_{ij} is (generally) a continuous variable and δ_i a 0–1 integer variable.

The model is:

$$\text{minimize} \quad \sum_{ij} c_{ij} x_{ij} + \sum c_i \delta_i \qquad (12.2.5)$$

$$\text{subject to} \quad \sum_i x_{ij} \quad = D_j \quad \text{for all customers } j \qquad (12.2.6)$$

$$\sum_j x_{ij} \quad - K_i \delta_i \leq 0 \quad \text{for all depots } i. \qquad (12.2.7)$$

Note that the variable δ_i in constraints (12.2.7) is linked to the expression $\sum_j x_{ij}$. If $\delta_i = 0$ (depot i is not built) then $\sum_j x_{ij} = 0$ (nobody can be supplied from it).

Logical Constraints

Once 0–1 variables are introduced into a model they can be used to enforce logical conditions. For example $\delta_1, \delta_2, \ldots, \delta_n$ might be 0–1 indicator variables showing whether or not a particular continuous variable x_i takes a non-zero value. These variables will have been linked to the corresponding variables x_i by constraints of the form (12.2.4).

EXAMPLE 12.2.2 (Limiting the Number of Ingredients in a Blend). Let x_i be (continuous) variables representing quantities of ingredients in a blend. Suppose there are potentially n ingredients giving n such variables but we wish to limit the total number of ingredients to r ($r < n$).

Each x_i variable will be linked to a new 0–1 integer variable δ_i by a constraint of the form (12.2.4). M will be a coefficient representing some known upper limit to the quantity of ingredient i. It is then possible to limit the number of δ_i variables taking the value 1 (and therefore the number of x_i variables taking non-zero values) by the constraint:

$$\delta_1 + \delta_2 + \ldots + \delta_n \leq r. \qquad (12.2.8)$$

EXAMPLE 12.2.3 (Alternative Production Processes). Suppose we have three alternative production processes at our disposal for making a range of products. The capacities of the processes (in machine hours) are C_1, C_2 and C_3 respectively. Each unit of product j uses a_{ij} hours of process j. Each process imposes a constraint on the quantities which may be produced. Instead, however, of having to satisfy all three constraints we have only to satisfy at least one of them. This is sometimes referred to as a case of *disjunctive constraints*. If the variables x_j represent the quantities of products to be produced and δ_1, δ_2 and δ_3 the decisions of which process to use we have the following formulation:

$$\sum_j a_{ij} x_j + M_i \delta_i \leq M_i + C_i \quad \text{for } i = 1, 2, 3 \qquad (12.2.9)$$

$$\delta_1 + \delta_2 + \delta_3 \geq 1. \qquad (12.2.10)$$

M_i are coefficients representing upper bounds on the quantities $\sum_j a_{ij}$ in a similar manner to M in (12.2.4).

Any logical statement of the kind which might be made using Boolean Algebra (the Propositional Calculus, [see § 16.5]) can be modelled using 0–1 integer variables.

Non-Linearities

As was mentioned in section 12.1.5 non-linearities can sometimes be incorporated into an otherwise LP model by means of integer variables. We will demonstrate this by means of a simple example.

EXAMPLE 12.2.4 (A Non-Linear Function).   Suppose we wished to incorporate the non-linear expression $x^3 - 6x^2 + 8x$ into an otherwise LP model. This is a non-linear function of a single variable. Non-linear functions which can be expressed as the sum of non-linear functions of *single* variables are known as *separable functions* and amenable to an extension of LP known as *Separable Programming*. We will, however, describe how IP may be used instead. Figure 12.2.1 shows how the non-linear function $y = x^3 - 6x^2 + 8x$ is related to x.

Everywhere that this non-linear expression occurs in our model it will be replaced by the variable y. (In this example the variable y should not be constrained to be non-negative.) It is now necessary to relate y to x. We will assume that x always lies within the limits 0 to 5. It is necessary to make a piecewise linear approximation to the curve shown in Figure 12.2.1. For the purposes of illustration we will use the crude approximation given by $ABCDE$. New non-negative variables $\lambda_1, \lambda_2, \ldots, \lambda_5$ are introduced as 'weights' to be attached to the vertices $A, B, \ldots, E$. x and y can now be related by the expressions:

$$x = \lambda_2 + 3\lambda_3 + 4\lambda_4 + 5\lambda_5 \qquad (12.2.11)$$

$$y = 3\lambda_2 - 3\lambda_3 \qquad + 15\lambda_5. \qquad (12.2.12)$$

It is necessary to impose the following two conditions on the variables λ_i.

$$\lambda_1 + \lambda_2 + \lambda_3 + \lambda_4 + \lambda_5 = 1. \qquad (12.2.13)$$

$$\text{At most two adjacent } \lambda\text{'s can be non-zero.} \qquad (12.2.14)$$

Conditions (12.2.13) and (12.2.14) force x and y to be the coordinates of a point on one of the straight line segments AB, BC, CD or DE. For example if $\lambda_2 = 0.3$ and $\lambda_3 = 0.7$ we are at P where $x = 2.4$ and $y = -1.2$. The equations (12.2.11), (12.2.12) and (12.2.13) gives rise to conventional equality constraints among the variables.

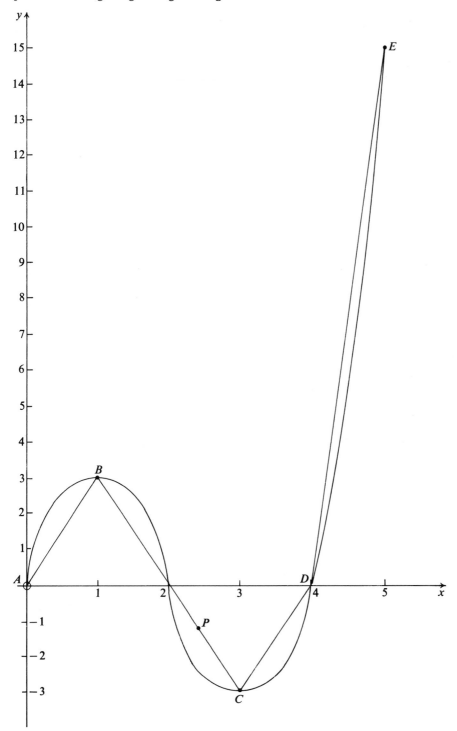

Figure 12.2.1

Condition (12.2.14) can only satisfactorily be dealt with by integer programming. Four integer 0–1 variables δ_1, δ_2, δ_3 and δ_4 are introduced into the model together with the extra constraints:

$$\lambda_1 \qquad\qquad\qquad\qquad - \delta_1 \qquad\qquad\qquad \leq 0 \qquad (12.2.15)$$

$$\lambda_2 \qquad\qquad\qquad - \delta_1 - \delta_2 \qquad\qquad \leq 0 \qquad (12.2.16)$$

$$\lambda_3 \qquad\qquad\quad - \delta_2 - \delta_3 \qquad \leq 0 \qquad (12.2.17)$$

$$\lambda_4 \qquad\qquad - \delta_3 - \delta_4 \leq 0 \qquad (12.2.18)$$

$$\lambda_5 \qquad\qquad - \delta_4 \leq 0 \qquad (12.2.19)$$

$$\delta_1 + \delta_2 + \delta_3 + \delta_4 = 1. \qquad (12.2.20)$$

It can be verified that these constraints have the effect of imposing conditions (12.2.13) and (12.2.14).

An alternative way of dealing with conditions (12.2.13) and (12.2.14) is to treat the set of variables λ_i as a so called *Special Ordered Set of Type* 2. Sets of variables arising with these conditions are sufficiently common to incorporate facilities for dealing with them into IP algorithms rather than model the conditions explicitly as we have done here.

Special Ordered Sets of Type 1 are an analogous (but slightly simpler) concept. A set of variables is said to form such a set if *exactly* one of the variables can be non-zero. Such a condition again could be coped with by an IP formulation. Many IP computer programs, however, also have special facilities for dealing with such sets. In this case it is not necessary to model the condition explicitly. A common example of a Special Ordered Set of Type 1 is when exactly one of a set of 0–1 IP variables is to take the value 1. Constraint (12.2.20) demonstrates one way of coping with the condition by modelling. The use of a Special Ordered Set of Type 1 avoids this.

Combinatorial Constraints

The formulation of combinatorial problems as IP models is often far from straightforward. A number of applications will be described by means of examples [see also § 12.1.6].

EXAMPLE 12.2.5 (Job-Shop Scheduling).   Suppose Operation A and Operation B both require the exclusive use of a particular machine. It is therefore necessary to delay one of the operations until the other has finished. Let the (known) durations of operations A and B be d_A and d_B respectively. Let

$$x_A = \text{time at which Operation } A \text{ starts,}$$
$$x_B = \text{time at which Operation } B \text{ starts.}$$

We have either

$$x_B \geq x_A + d_A \quad \text{or} \quad x_A \geq x_B + d_B.$$

This condition can be modelled by using a 0–1 integer variable δ and the constraints:

$$x_A - x_B - (M_A + d_A)\delta \leq -d_A \tag{12.2.21}$$

$$-x_A + x_B + (M_B + d_B)\delta \leq M_B; \tag{12.2.22}$$

M_A and M_B are known upper limits on the values of x_A and x_B.

In practice there will be many straightforward *sequencing constraints* among the variables x_i as well as the above *non-interference constraints* which involve 0–1 integer variables.

More complicated problems where the non-interference relations involve more than two operations and possibly more than two machines (e.g. three operations competing for the use of two similar machines) can still be modelled by IP. The formulation is, however, more complex.

EXAMPLE 12.2.6 (The Travelling Salesman Problem). A salesman wishes to set out from his base city (city 0) and visit n other cities (cities $1, 2, \ldots, n$) in some order finally returning to his base city. What order should he visit the cities in so as to minimize the total distance covered? The distance between cities i and j is c_{ij}. We do not necessarily assume that $c_{ij} = c_{ji}$. If this is done the model can be simplified somewhat.

The following variables are introduced:

$$x_{ij} = 1 \quad \text{if the salesman goes from city } i \text{ immediately to city } j$$
$$= 0 \quad \text{otherwise};$$

x_{ij} are 0–1 integer variables.

$$u_i = \text{the sequence number in which city } i \text{ is visited},$$

e.g. $u_0 = 0$, $u_3 = 1$, $u_6 = 2$ etc. if the cities are visited in the order $0, 3, 6, \ldots, u_i$ can be regarded as either general integer variables or continuous variables (they will take integer values in the optimal IP solution). The PIP (or MIP) model is:

$$\text{minimize} \quad \sum_{i,j} c_{ij} x_{ij}$$

$$\text{subject to} \quad \sum_i x_{ij} = 1 \quad \text{for all } j \tag{12.2.23}$$

$$\sum_j x_{ij} = 1 \quad \text{for all } i \tag{12.2.24}$$

$$u_i - u_j + n x_{ij} \leq n - 1 \quad \text{for all } i, j \neq 0 \text{ and } i \neq j. \tag{12.2.25}$$

Constraints (12.2.23) guarantee that each city is entered exactly once. Constraints (12.2.24) guarantee that each city is left exactly once.

Unfortunately only using constraints (12.2.23) and (12.2.24) could result in a solution involving subtours as illustrated by Figure 12.2.2 for a 7-city problem. Constraints (12.2.25) are needed to rule out subtours and provide a complete tour as illustrated in Figure 12.2.3.

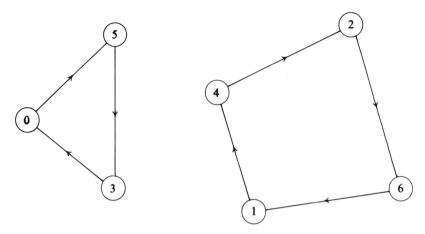

Figure 12.2.2

It is not immediately obvious why constraints (12.2.25) are sufficient to rule out subtours. If there are any subtours at least one of them will not include city 0. For example the right-hand subtour in Figure 12.2.2 does not include city 0. If we list all constraints (12.2.25) relating to the links in this subtour we get:

$$u_1 - u_4 + 6x_{14} \le 5 \qquad\qquad (12.2.26)$$

$$u_4 - u_2 + 6x_{42} \le 5 \qquad\qquad (12.2.27)$$

$$u_2 - u_6 + 6x_{26} \le 5 \qquad\qquad (12.2.28)$$

$$u_6 - u_1 + 6x_{61} \le 5. \qquad\qquad (12.2.29)$$

Since all the x_{ij} variables are 1 in these four constraints adding them all together gives:

$$24 \le 20 \qquad\qquad (12.2.30)$$

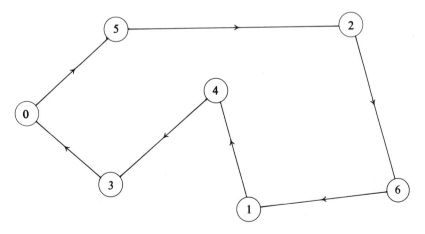

Figure 12.2.3

Since (12.2.30) is obviously false we have shown that the solution on figure 12.2.2 is not feasible according to constraints (12.2.25). Similarly it can be shown that all subtours not involving city 0 are ruled out.

It is, however, possible to have a complete tour since this must involve city 0 and constraints (12.2.25) do not apply when i or $j = 0$. The complete tour represented by Figure 12.2.3 does, for example, represent a feasible solution to our model. The values of the variables in this solution are:

$$x_{05} = x_{52} = x_{26} = x_{61} = x_{14} = x_{43} = x_{30} = 1,$$

all other $x_{ij} = 0$

$$u_0 = 0, \quad u_1 = 4, \quad u_2 = 2, \quad u_3 = 6, \quad u_4 = 5, \quad u_5 = 1, \quad u_6 = 3.$$

It should be noted that this model will become very large as the number of cities considered increases. If for example we have 100 cities the dimensions of the model are:

$$
\begin{array}{ll}
9900 & x \text{ variables} \\
100 & u \text{ variables} \\
100 & \text{constraints of type (12.2.23)} \\
100 & \text{constraints of type (12.2.24)} \\
9702 & \text{constraints of type (12.2.25)}
\end{array}
$$

There are other ways of formulating the Travelling Salesman Problem as an IP which can be found in the references.

EXAMPLE 12.2.7 (The Quadratic Assignment Problem).   In this type of model 0–1 integer variables with the following interpretations typically arise:

$$\delta_{ik} = 1 \text{ if department } i \text{ is located in city } k$$
$$= 0 \text{ otherwise,}$$

$$\delta_{jl} = 1 \text{ if department } j \text{ is located in city } l$$
$$= 0 \text{ otherwise.}$$

The objective function then involves the following types of term:

$$c_{ij} d_{kl} \delta_{ik} \delta_{jl}$$

where c_{ij} is the amount of communication between departments i and j; d_{kl} is the distance between cities k and l.

It is always possible to transform a model containing such non-linear terms into a linear IP model. Clearly the product $\delta_{ik}\delta_{jl}$ can only take the two values 0 or 1 depending on the values of δ_{ik} and δ_{jl}. $\delta_{ik}\delta_{jl}$ can be replaced by a single 0–1 variable δ_{ijkl} which can be related to δ_{ik} and δ_{jl} by two constraints:

$$\delta_{ik} + \delta_{jl} - \delta_{ijkl} \leq 1 \tag{12.2.31}$$

$$\delta_{ik} + \delta_{jl} - 2\delta_{ijkl} \geq 0. \tag{12.2.32}$$

Clearly this linearization of the original model results in a considerable increase
on its size. There is computational advantage to be gained from splitting constraints like (12.2.32) into two separate constraints. An instance of this is
described in Example 12.3.1.

EXAMPLE 12.2.8 (Aircrew Scheduling).   An airline is committed to flying the
following 'legs' over a particular period of time.

$$\{A,\ B,\ C,\ D,\ E,\ F,\ G\}.$$

The aircraft flying a leg has to be manned by an aircrew. Each aircrew is limited
to flying one 'roster' in the period of time. A roster is a set of legs which it is
possible for one crew to fly. The possible rosters are:

$$\{A,\ C\},\ \{B,\ D,\ E\},\ \{E,\ G\},\ \{A,\ F\},\ \{F\},\ \{C,\ E,\ F\},\ \{B,\ G\}.$$

Assigning an aircrew to roster j involves a cost c_j.
   The problem is to *cover* all legs with rosters (aircrews) in such a way as to
minimize total cost. As mentioned in section 12.1.6 this problem gives rise to a
type of PIP model known as a *Set Covering Problem*. Clearly this problem is a
case of a general abstract problem of covering a set of objects by members of
a class of subsets of those objects. Such problems arise in other contexts. 0–1
integer variables x_j are introduced so that

$$x_j = 1 \quad \text{if roster } j \text{ is included in the cover}$$
$$= 0 \quad \text{otherwise.}$$

The necessity to cover each leg gives rise to each constraint in the model.
For example the fact that leg A must be covered necessitates using at least one
of rosters 1 and 4. This gives rise to the first constraint in the model below.

$$
\begin{aligned}
\text{Minimize} \quad & c_1 x_1 + c_2 x_2 + c_3 x_3 + c_4 x_4 + c_5 x_5 + c_6 x_6 + c_7 x_7 \\
\text{subject to} \quad & x_1 \qquad\qquad\qquad + x_4 \qquad\qquad\qquad \geq 1 \\
& x_2 \qquad\qquad\qquad\qquad\qquad + x_7 \geq 1 \\
& x_1 \qquad\qquad\qquad\qquad + x_6 \qquad \geq 1 \\
& x_2 \qquad\qquad\qquad\qquad\qquad\qquad \geq 1 \\
& x_2 + x_3 \qquad\qquad\quad x_6 \qquad\quad \geq 1 \\
& \qquad\qquad x_4 + x_5 + x_6 \qquad \geq 1 \\
& x_3 \qquad\qquad\qquad\qquad\qquad x_7 \geq 1.
\end{aligned}
$$

Frequently the objective is simply to minimize the number of aircrews (rosters)
needed which leads to all the coefficients c_j being 1 in the above model.
   Clearly this model has a very special structure. Not only is it a PIP 0–1 model
but all the coefficients (apart from those in the objective function) are 0 or 1.
All constraints are of the '$\geq$' form. An extra restriction which might be imposed
on this problem is to prohibit aircrews flying as passengers on legs. If this is
the case then the above model is modified so that all the constraints become of

the '=' form. Such a problem is an instance of the *Set Partitioning Problem*. In addition to covering each element of our set we want to avoid overlap. Clearly we have to *partition* the original set into distinct members of the class of subsets.

The *Set Packing Problem* is another variant on the above type of model where all constraints are of the form '≤'. In fact this type of model is not essentially distinct from the *Set Partitioning Problem* since the introduction of slack 0–1 variables [see § 11.2.5] reduces it to this form.

In practice most crew scheduling problems give rise to models with relatively few constraints (legs) but a massive number of variables (rosters). Computational difficulties in solving such models usually arise not from the structure of the model (which is simple) but from the size.

Set Covering, Partitioning and Packing Problems give rise to models with important properties. For example an optimal IP solution can always be found which is a *basic* (vertex) solution in the LP sense [see Example 11.3.1]. This will, of course, not necessarily be the same basic solution as the LP optimum which can be fractional.

Special cases of the Set Covering, Packing and Partitioning Problems arise in graph theory. This is illustrated by Figure 12.2.4. The nodes of the graph can be regarded as elements of a set. Each edge defines a subset of the elements consisting of the two nodes which it connects. Each model which results will only have two non-zero coefficients per column. The minimum cover (each edge costed equally) for the graph in Figure 12.2.4 is represented by the bold edges. The corresponding Set Packing Problem on a graph is known as the *Matching*

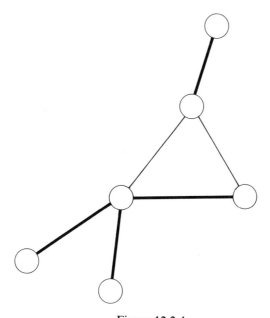

Figure 12.2.4

Problem [V, § 6.5]. It is the problem of finding the maximum subset of edges none of which have a node in common.

12.2.3. The Relationship between Integer Programming Models and Linear Programming Models

It has already been pointed out that in solving an IP model we are generally looking for integer points in the interior of the feasible region of the corresponding LP model. Suppose, for example, that we have the following constraints:

(B)

$$2x_1 + 2x_2 \geq 3$$
$$-2x_1 + 2x_2 \leq 3$$
$$4x_1 + 2x_2 \leq 19$$
$$x_1, \quad x_2 \geq 0.$$

The feasible region defined by these constraints is $ABCD$ in Figure 12.2.5.

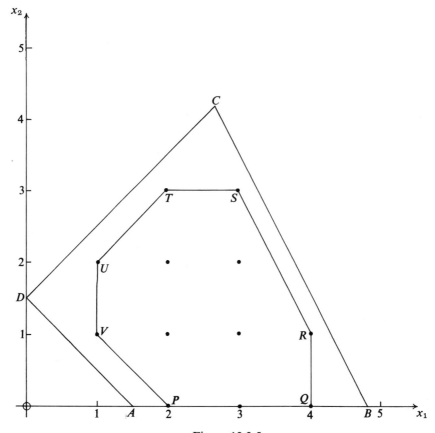

Figure 12.2.5

When an objective function is given with the constraints (B) we have an LP model whose optimal solution would lie on the boundary $ABCD$ of this feasible region.

If, however, x_1 and x_2 also have to take integer values we must confine our attention to the integer points marked by heavy dots inside this feasible region. One approach is to confine our attention to the region inside $PQRSTUV$. This is the so called *convex hull of feasible integer points*. It is the smallest convex region which contains all the feasible integer points. Such a convex region can itself be defined by linear inequalities among the variables. In this case we can use the following:

$$x_1 + x_2 \geq 2$$
$$x_1 \geq 1$$
$$-x_1 + x_2 \leq 1$$
(C) $$ x_2 \leq 3$$
$$2x_1 + x_2 \leq 9$$
$$x_1 \leq 4$$
$$ x_2 \geq 0.$$

These inequalities can be seen to define the boundaries of the region $PQRSTUV$. The edges of this region defined by these constraints are known as *facets*. This concept can obviously be generalized to models with more than two variables although it cannot be pictured geometrically for more than three variables. For any IP model there will exist facet constraints defining the convex hull of feasible integer points. If these facet constraints are found they can be used to replace those of the original model. This new version of the model can then be solved as an LP (which is computationally much easier) and we are guaranteed an integer solution.

Unfortunately it is often computationally very expensive to calculate all the facet constraints for an IP model. There are, however, practical situations in which the constraints of an IP model turn out to be facet constraints. In such circumstances we can ignore IP and treat the model as an LP. The optimal LP solution will yield integer values for the variables giving the IP optimum immediately. One well-known problem where this happens is the *Transportation Problem* [see § 11.3]. This is a special case of a wide class of problems concerned with *flows in networks*. Many problems in this class exhibit the property. For such problems there is no distinction to be made between IP and LP.

It will be seen in section 12.3 that a number of IP algorithms involve solving the associated LP model first. Even if we cannot guarantee an integer solution at this stage it is desirable that the LP optimal solution should be as 'close' to the IP optimum as possible. It is sometimes possible to reformulate a model to decrease the size of the feasible region of the associated LP model. In Figure 12.2.5 this amounts to defining new LP constraints which reduce the size of the region $ABCD$ even if it is not obvious how to reduce it right down to $PQRSTUV$.

There is another distinction between an IP model and its associated LP model which is illuminating and of practical importance. If an LP model is solvable (i.e. it is not infeasible or unbounded) there is an associated model, called the *dual model*, which yields the same optimal objective value as the original model [see § 11.2.1]. For example if we consider the LP model associated with (A) defined in the introduction we have the dual LP model:

$$\text{minimize} \quad 13y_1 - y_2$$
$$\text{subject to} \quad 10y_1 - 2y_2 \geq 1$$
(D)
$$-8y_1 + 2y_2 \geq 1$$
$$y_1, \quad y_1 \geq 0.$$

The optimal objective value of this model is 8·5; the same as that of the original LP version of model (A) (the *primal model*).

Suppose, however, we now consider model (A) with the integer restrictions. The optimal objective value is only 3. Clearly one would expect, with more restrictions (integrality), to do less well. The situation is represented diagramatically in Figure 12.2.6. There is a gap between the optimal solutions of the IP model and the dual LP model. This is known as a *duality gap*. It is of considerable computational significance and is referred to in section 12.3. If the variables

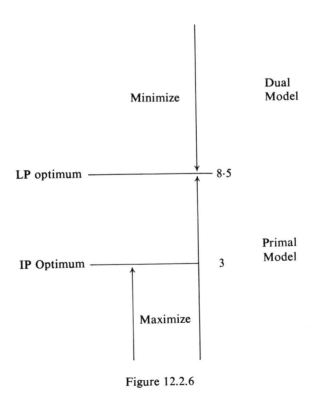

Figure 12.2.6

in the dual problem are also restricted to be integer there is, of course, an even bigger gap. In the example we would not then be able to make the objective on the dual model even as small as 8·5.

The purpose of this section has been to point out the differences and similarities between LP models and IP models. Their relationship to one another is important since LP algorithms play an important part in a number of IP solution methods. For some models (such as the transportation problem) the distinction disappears. For other models (particularly some representing combinatorial situations) the associated LP solution is so different as to be virtually useless.

12.3.   SOLVING INTEGER PROGRAMMING MODELS

It is possible to formulate a wide range of practical problems as IP models. Although IP is therefore a very attractive modelling tool it must be used with caution since IP models can be very difficult to solve. For LP models the Simplex Algorithm has proved to be a universally powerful method of solving surprisingly large problems (up to tens of thousands of constraints and variables) on a computer [see § 11.4]. No comparably powerful algorithm exists for IP models. Practical models are almost always solved using computer 'package' programs. As far as the author is aware all commercially available packages use a version of the *Branch and Bound Algorithm* for solving IP models. This algorithm is described in the next section. The Branch and Bound Algorithm has proved to be a more successful method than any other for solving a wide range of practical IP models. It is still very difficult to predict how long an IP model will take to solve on a computer. A very rough guide is provided by the number of integer variables. It is usually possible to solve problems involving a few hundred integer variables in a reasonable period of time. Larger problems involving thousands of variables can usually only be solved in a practical period of time if they have an especially simple structure (such as yielding an integer solution to the associated LP model). Nevertheless it is fairly easy to construct very small models which prove extremely difficult to solve by the Branch and Bound algorithm.

There is considerable scope in IP for building a model to suit an algorithm. Clearly a model which contains all the facet constraints defining the convex hull of feasible integer points is particularly easy to solve (as an LP). While it may not be possible to obtain the facet constraints easily it is sometimes possible to reformulate constraints in a 'tighter' form as far as the associated LP model is concerned. Clearly this is desirable if any IP algorithm is to be used which involves solving LP problems at stages in the solution. This is the case with both the Branch and Bound Algorithm and the Cutting Planes Algorithm described in section 12.3.1. In order to demonstrate this we give an example.

EXAMPLE 12.3.1 (Alternative Formulations for an Integer Programming Constraint).   Consider the logical condition (12.1.2). Suppose we have 0–1 integer variables with the following interpretations:

$$\delta_A = 1 \quad \text{if product } A \text{ is made}$$
$$= 0 \quad \text{otherwise};$$
$$\delta_B = 1 \quad \text{if product } B \text{ is made}$$
$$= 0 \quad \text{otherwise};$$
$$\delta_C = 1 \quad \text{if product } C \text{ is made}$$
$$= 0 \quad \text{otherwise}.$$

The condition can be modelled by:

$$2\delta_A - \delta_B - \delta_C \leq 0. \tag{12.3.1}$$

It can easily be verified that this constraint has the desired effect.

An alternative way of formulating the condition is to use the two constraints:

$$\delta_A - \delta_B \qquad \leq 0, \tag{12.3.2}$$
$$\delta_A \qquad - \delta_C \leq 0. \tag{12.3.3}$$

Again it can easily be seen that these two constraints have the desired effect.

Constraints (12.3.2) and (12.3.3), however, give a tighter associated LP problem than is the case with constraint (12.3.1). Clearly any LP (possibly fractional) solution which satisfies (12.3.2) and (12.3.3) must also satisfy (12.3.1) since (12.3.1) is simply the sum of constraints (12.3.2) and (12.3.3). The converse is not, however, true. There are fractions between 0 and 1 satisfying (12.3.1) but not satisfying (12.3.2) and (12.3.3) simultaneously. An example is:

$$\delta_A = \tfrac{1}{2}, \qquad \delta_B = \tfrac{1}{4}, \qquad \delta_C = \tfrac{3}{4}.$$

A model with constraint (12.3.1) could therefore yield a fractional solution to the associated LP problem which would not occur if it were replaced by (12.3.2) and (12.3.3).

One of the major attractions of the Branch and Bound Algorithm is that it is applicable to general MIP problems. This is also true of the Cutting Planes Algorithm which we describe in the next section although this approach has not had the same practical success and has not therefore been widely used to solve practical problems.

In view of the computational difficulty of solving general MIP models it seems sensible to consider *specialist algorithms* as well. These restrict themselves to a subclass of IP models but hopefully exploit the special structure of this subclass to good effect. One particular subclass of IP models worthy of special attention is the class of 0–1 PIP models. While such models do not arise anywhere near as frequently as general MIP models they are sufficiently common to merit special attention. It has already been mentioned in section 1.6 that Set Covering, Partitioning and Packing problems (such as crew scheduling) give rise to models of this sort. General PIP models can also be converted to 0–1 PIP models if bounds are known for the values of the variables (this is usually the case). In order to demonstrate this we present an example.

EXAMPLE 12.3.2 (Reducing a General Integer Variable to an Expression in 0–1 Integer Variables). Suppose u is an integer variable such that $0 \leq u \leq 12$. We may replace u by the expression:

$$\delta_0 + 2\delta_1 + 4\delta_2 + 8\delta_3 \qquad (12.3.4)$$

where δ_0, δ_1, δ_2 and δ_3 are 0–1 integer variables. It can easily be seen that different combinations of values for these 0–1 variables encompass all the possible values for u. (The unwanted values of u where $u > 12$ can be ruled out by an extra constraint if necessary.)

Such an expansion of a general PIP model will result in an increased number of variables in the 0–1 PIP form but if the integer variables can only take fairly small values such an expansion is not necessarily impractical.

In section 12.3.2 two specialist 0–1 PIP algorithms are described, *Implicit Enumeration* (Balas' Algorithm) and a *Pseudo-Boolean* method (Hammer's Algorithm). Specialist algorithms for IP abound and we only describe two of them. A few promising developments are described in section 12.3.3.

12.3.1. General Methods of Solving Integer Programming Models

(a) *The Branch and Bound Algorithm*

Strictly speaking 'Branch and Bound' is the name given to a general approach to solving a class of problems. This approach involves a tree search which will be demonstrated by an example below. We restrict our attention to a special version of the approach which is applicable to solving general MIP models. It makes use of LP at regular stages in the algorithm. Other IP algorithms (such as Balas' Algorithm described in section 12.3.2) also use a tree search and are sometimes referred to as Branch and Bound Methods. We prefer to reserve the name for the method described in this section.

The Branch and Bound Method begins by treating an IP model as an LP. This LP model is said to be a *relaxation* of the original model (we have 'relaxed' the integrality conditions). Clearly if the solution to this relaxed problem turns out to be integer (as would happen with the Transportation Problem) we need go on further. In general however the optimal solution to this LP model will give fractional values to some of the integer variables. When this happens one of these variables x, is chosen and a *separation* is performed on the model. This amounts to creating two new IP models. Thus the value of the variable x can be represented as $N + f$ where N is an integer and f is a fraction such that $0 < f < 1$. Since x must take an integer value in the IP model we can be sure one of the following two constraints must ultimately apply.

$$x \leq N \qquad (12.3.5)$$

$$x \geq N + 1. \qquad (12.3.6)$$

Two new IP models are therefore created. In the first we append the extra constraint (12.3.5) and in the second the extra constraint (12.3.6). These new

models are then treated in a similar manner to the original, i.e. they are relaxed
to become LP models and solved. If this procedure is repeated systematically
we can ultimately find the optimal IP solution if it exists. The method is best
demonstrated by a small numerical example.

$$\text{Maximize} \quad x_1 + x_2$$

$$\text{subject to} \quad 2x_1 + 2x_2 \geq 3$$

(E) $$\qquad\qquad\qquad -2x_1 + 2x_2 \leq 3$$

$$4x_1 + 2x_2 \leq 19$$

$$x_1, \quad x_2 \geq 0 \quad \text{and integer.}$$

We have purposely chosen an example with only two variables in order that we
may get some geometrical insight when we use the same example with another
algorithm (the cutting planes algorithm). Although the Branch and Bound
Algorithm solves general MIP models we are using a PIP model. There should
be no difficulty in seeing how the algorithm applies in the MIP case.

Strictly speaking the Branch and Bound Algorithm can only be guaranteed
to converge if the feasible region of the associated LP model [see § 11.1.2] is
closed. This is clearly the case in this example.

We begin by solving the *relaxation* of (E) obtained by ignoring the integrality
requirements. This LP model can be solved easily (the method of solution
need not concern us here. Normally the Simplex Algorithm would be used).
The solution of the LP relaxation (the *continuous optimum*) is:

$$x = 2\tfrac{2}{3}, \quad x = 4\tfrac{1}{6}, \quad \text{Objective} = 6\tfrac{5}{6}.$$

The next step is to choose an integer variable which has come out fractional
and create a *separation*. Which variable we choose is, to some extent, arbitrary
although in practical models this choice can be made in a sensible fashion in
order to speed the solution process. For this example we will simply choose x_1.
Since x_1 takes the value $2\tfrac{2}{3}$ we create two new models. In the first model we
append the constraint:

$$x_1 \leq 2 \qquad\qquad\qquad\qquad (12.3.7)$$

to the original model (E). In the second model we append the constraint:

$$x_1 \geq 3. \qquad\qquad\qquad\qquad (12.3.8)$$

The solution process is best illustrated by the solution tree shown in Figure
12.3.1. Each node of this tree represents an IP model. These models are numbered
in the order of their creation. Beside each node is the objective value of the
optimal solution to the LP relaxation. At each node (apart from 'terminal'
nodes) we do a separation and create the two new models drawn below the node.

In our small example we have solved the LP relaxation of the original model
(node 1) and created two new models (nodes 2 and 3). These are then relaxed
and solved yielding:

$$\text{Node 2.} \quad x_1 = 2, \quad x_2 = 3\tfrac{1}{2}, \quad \text{Objective} = 5\tfrac{1}{2}$$

$$\text{Node 3,} \quad x_1 = 3, \quad x_2 = 3\tfrac{1}{2}, \quad \text{Objective} = 6\tfrac{1}{2}.$$

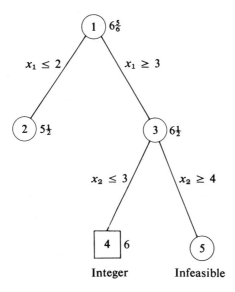

Figure 12.3.1

Nodes 2 and 3 now become *waiting nodes* (they are candidates for a further separation). One of them is chosen. Again this is, to some extent, an arbitrary choice but in practice the choice can be made with some foresight. For this example we will simply choose node 3 for immediate further development and create a separation using variable x_2. Nodes 4 and 5 are created and their relaxations solved yielding:

Node 4. $x_1 = 3$, $x_2 = 3$, Objective $= 6$

Node 5. Infeasible.

We now have an integer solution although we do not yet know if it is the optimal solution. Node 5 represents an infeasible model. This is quite likely to happen as we go down a branch of the tree since we are imposing progressively more severe constraints. The progress down a branch of the tree is known as *fathoming*. We are adding stronger and stronger bounds to the variables in the hope that we may eventually find an integer solution.

An important observation can now be made about the fathoming process. It can be seen that the optimal LP objective value associated with each node gets worse as we go down a branch of the tree. (For a maximization model this means getting smaller.) Clearly we would expect this to happen since the more we constrain a problem the less well we can do (strictly the objective gets no better since we might find an equally good solution at the next node). Once we have found an integer solution with a particular objective value we know that we need explore no other branches where the objective function has deteriorated as far as, or beyond, this value. This has happened in our example. Having obtained an integer solution with an objective value of 6 (node 4) we need not go back to explore beyond node 2 since the objective value is already worse than 6. If any integer solutions lie below node 2 they cannot be as good as the one we have already. We can therefore abandon node 2. Since there are no more waiting nodes to consider we can be sure that our best integer solution is the integer optimum.

For practically sized problems, of course, the solution tree will be much larger. Many integer solutions may be found on the way to the optimal integer solution. Even when the optimal integer solution is found it may be necessary to fathom other branches before being sure that it is optimal. Much 'backtracking' will probably take place when we have reached the end of branches. In this example we only backtracked once to node 2 when we had dispensed with nodes 4 and 5. No matter how large the model, however, the method consists of the two steps applied in turn repeatedly:

(a) *Relax* a model and solve it as an LP.
(b) *Separate* on an integer variable which has come out fractional in order to create two new models.

By applying these steps we fathom branches of the solution tree terminating branches for one of three reasons:

(i) An *integer* solution is reached.
(ii) A model is *infeasible*.
(iii) The objective value of the relaxation has become worse than the best integer solution obtained to date. In this case we say the submodel has been *bounded*, i.e. its best objective value can be no better than a known bound for the optimal value of the original model.

There is great virtue in terminating branches as quickly as possible since this rules out the need to explore large sections of the potential solution tree.

Two important choices are made at various stages in the algorithm. They are:

(i) Choice of separation variable.
(ii) Choice of waiting node.

These choices are important and affect the whole shape of the resultant solution tree. In practical problems when a computer package is used the user can give different rules for making these choices. A full discussion of this is beyond the scope of this exposition but is given in some of the references mentioned in section 5. One point is, however, worth making here. It is often possible to give an order of importance to the variables in an IP model related to their physical meaning. For example we might have the following two 0–1 integer variables.

$$\delta_1 = 1 \quad \text{if a depot is built}$$
$$\quad = 0 \quad \text{otherwise};$$

$$\delta_2 = 1 \quad \text{if a particular customer is supplied from the depot}$$
$$\quad = 0 \quad \text{otherwise}.$$

Then a constraint such as that below would probably be imposed.

$$\delta_1 - \delta_2 \geq 0; \qquad\qquad (12.3.9)$$

δ_1 would seem to be a more important variable to separate on than δ_2 if we had to choose between them.

(b) *The Cutting Planes Algorithm*

This algorithm can also be applied to general MIP models. For simplicity we will apply it to the PIP case and use model (E). Extending the algorithm to MIP models is fairly straightforward although not entirely obvious. The extension can be followed up through the references.

Like the Branch and Bound Algorithm the Cutting Planes Algorithm begins by solving the associated LP model. In this case it is necessary to use the Simplex Algorithm. We will again use model (E) as an example. Slack and surplus variables are added to the model [see § 11.2.5] in order to represent it in the form:

$$
\begin{array}{llll}
\text{maximize} & u \\
\text{subject to} & u - x_1 - x_2 & = 0 \\
& 2x_1 + 2x_2 - u_1 & = 3 \\
(F) & -2x_1 + 2x_2 + u_2 & = 3 \\
& 4x_1 + 2x_2 + u_3 = 19 \\
& x_1, \quad x_2, \quad u_1, \quad u_2, \quad u_3 \geq 0.
\end{array}
$$

We will express the model in an extended table which contains columns of coefficients of basic and non-basic variables [see § 11.4.2]. The initial table is:

u	x_1	x_2	u_1	u_2	u_3		
1	-1	-1	0	0	0		0
0	2	2	-1	0	0	$=$	3
0	-2	2	0	1	0		3
0	4	2	0	0	1		19

(F₁)

Solving the model by the Simplex Algorithm (including phase 1 of the algorithm in order to get a feasible basis to start with) yields the optimal LP table:

u	x_1	x_2	u_1	u_2	u_3		
1	0	0	0	$\frac{1}{6}$	$\frac{1}{3}$		$\frac{41}{6}$
0	1	0	0	$-\frac{1}{6}$	$\frac{1}{6}$	$=$	$\frac{8}{3}$
0	0	1	0	$\frac{1}{3}$	$\frac{1}{6}$		$\frac{25}{6}$
0	0	0	1	$\frac{1}{3}$	$\frac{2}{3}$		$\frac{32}{3}$

(F₂)

This gives us the continuous optimum (which we also obtained in the Branch and Bound Algorithm):

$$x_1 = 2\tfrac{2}{3}, \qquad x_2 = 4\tfrac{1}{6}, \qquad \text{Objective} = 6\tfrac{5}{6}.$$

One of the rows of the table must now be chosen in which the basic variable is an integer variable assuming a fractional value. In practice there is some flexibility about which row we choose but in order to be sure that the algorithm converges we should always choose the topmost row. In this case we consider the second row of the table. This corresponds (writing the basic variable in terms of the non-basic variables) to the equation:

$$x_1 = \tfrac{8}{3} + \tfrac{1}{6}u_2 - \tfrac{1}{6}u_3. \qquad (12.3.10)$$

We may separate out the integer and fractional parts of the coefficients to give:

$$x_1 = (2 + \tfrac{2}{3}) + (0 + \tfrac{1}{6})u_2 + (-1 + \tfrac{5}{6})u_3. \qquad (12.3.11)$$

Writing the integer quantities on the left gives:

$$(x_1 - 2 + u_3) = \tfrac{2}{3} + \tfrac{1}{6}u_2 + \tfrac{5}{6}u_3. \qquad (12.3.12)$$

We have purposely chosen the positive fractional portions of the original coefficients in (12.3.11) in order to yield a non-negative expression on the right of equation (12.3.12). This equation must hold in any feasible solution to our model. When x_1 does take an integer value the expression on the left-hand side of equation (12.3.12) will clearly represent an integer. Therefore the right-hand side expression must also represent an integer. Since it has non-negative coefficients (and some are positive) it must be strictly greater than 0. Any integer strictly greater than 0 must be greater than or equal to 1. We therefore have:

$$\tfrac{2}{3} + \tfrac{1}{6}u_2 + \tfrac{5}{6}u_3 \geq 1. \qquad (12.3.13)$$

This inequality is stronger than any that appeared in the original model. Our train of argument does, however, show it to be a valid inequality when x_1 is restricted to take integer values. The inequality is known as a *cut* since it cuts off part of the feasible region of the associated LP model. A geometrical representation of this cut is given after completing the solution to the example.

In order to incorporate the cut (12.3.13) as an extra constraint in our model it is convenient to rewrite it as:

$$-\tfrac{1}{3} + \tfrac{1}{6}u_2 + \tfrac{5}{6}u_3 \geq 0. \qquad (12.3.14)$$

The left-hand side of (12.3.14) can be represented by a new *non-negative* variable s_1 which we introduce into our model. This gives the equation:

$$s_1 = -\tfrac{1}{3} + \tfrac{1}{6}u_2 + \tfrac{5}{6}u_3. \qquad (12.3.15)$$

As a row of the table this becomes.

s_1	u	x_1	x_2	u_1	u_2	u_3		
1	0	0	0	0	$-\tfrac{1}{6}$	$-\tfrac{5}{6}$	$= -\tfrac{1}{3}$	(12.3.16)

Incorporating this row into the table F_2 (and adding a new column for s_1) and reoptimizing by the *Dual Simplex Algorithm* [see § 11.4.7] we obtain:

	s_1	u	x_1	x_2	u_1	u_2	u_3		
	$\frac{2}{5}$	1	0	0	0	$\frac{1}{10}$	0	=	$\frac{67}{10}$
	$\frac{1}{5}$	0	1	0	0	$-\frac{1}{5}$	0	=	$\frac{13}{5}$
(F₃)	$\frac{1}{5}$	0	0	1	0	$\frac{3}{10}$	0	=	$\frac{41}{10}$
	$\frac{4}{5}$	0	0	0	1	$\frac{1}{5}$	0	=	$\frac{52}{5}$
	$-\frac{6}{5}$	0	0	0	0	$\frac{1}{5}$	1		$\frac{2}{5}$

This gives a fractional solution. From the second row of the table we can obtain another cut:

$$\tfrac{4}{5}s_1 + \tfrac{1}{5}u_2 \geq \tfrac{2}{5}. \tag{12.3.17}$$

A new non-negative variable s_2 can be introduced into the model with the extra row below to represent this cut.

s_2	s_1	u	x_1	x_2	u_1	u_2	u_3		
1	$-\frac{4}{5}$	0	0	0	0	$-\frac{1}{5}$	0	=	$-\frac{2}{5}$	(12.3.18)

Incorporating this new row into the table F_3 and reoptimizing we obtain:

	s_2	s_1	u	x_1	x_2	u_1	u_2	u_3		
	$\frac{1}{2}$	0	1	0	0	0	0	0	=	$\frac{13}{2}$
	-1	1	0	1	0	0	0	0	=	3
(F₄)	$\frac{3}{2}$	-1	0	0	1	0	0	0	=	$\frac{7}{2}$
	1	0	0	0	0	1	0	0	=	10
	1	-2	0	0	0	0	0	1		0
	-5	4	0	0	0	0	1	0		2

We still have a fractional solution. From the third row of the table we obtain another cut:

$$\tfrac{1}{2}s_2 \geq \tfrac{1}{2}. \tag{12.3.19}$$

A new non-negative variable s_3 can be introduced into the model with the extra row below to represent this cut.

s_3	s_2	s_1	u	x_1	x_2	u_1	u_2	u_3		
1	$-\frac{1}{2}$	0	0	0	0	0	0	0	=	$-\frac{1}{2}$	(12.3.20)

Incorporating this new row into the table F_4 and reoptimizing we obtain:

	s_3	s_2	s_1	u	x_1	x_2	u_1	u_2	u_3	
	1	0	0	1	0	0	0	0	0	6
	-1	0	0	0	1	0	0	0	$\frac{1}{2}$	$\frac{7}{2}$
	2	0	0	0	0	1	0	0	$-\frac{1}{2}$	$\frac{5}{2}$
(F_5)	2	0	0	0	0	0	1	0	0	= 9
	-1	0	1	0	0	0	0	0	$-\frac{1}{2}$	$\frac{1}{2}$
	-6	0	0	0	0	0	0	1	2	5
	-2	1	0	0	0	0	0	0	0	1

Again we have a fractional solution. Both s_1 and s_2 have become basic in the last reoptimization. The cuts corresponding to these variables (rows 5 and 7 of the table F_5) may therefore be ignored. In fact it is strictly necessary to do this to guarantee convergence of the algorithm. From the second row of the table we can obtain a new cut:

$$\tfrac{1}{2}u_3 \geq \tfrac{1}{2}. \tag{12.3.21}$$

A new non-negative variable s can be introduced into the model with the extra row below to represent this cut:

s_4	s_3	u	x_1	x_2	u_1	u_2	u_3		
1	0	0	0	0	0	0	$-\frac{1}{2}$	$= -\frac{1}{2}$	(12.3.22)

Incorporating this new row into the table F_5 and reoptimizing we obtain:

	s_4	s_3	u	x_1	x_2	u_1	u_2	u_3	
	0	1	1	0	0	0	0	0	6
	1	-1	0	1	0	0	0	0	3
	-1	2	0	0	1	0	0	0	3
(F_6)	0	2	0	0	0	1	0	0	= 9
	4	-6	0	0	0	0	1	0	3
	-2	0	0	0	0	0	0	1	1

The optimal integer solution has now been obtained giving:

$$x_1 = 3, \qquad x_2 = 3, \qquad \text{Objective} = 6.$$

Four cuts were necessary to obtain this solution.

It is helpful to picture the above process geometrically. The original model (E) only contains two variables. These variables can therefore be represented by

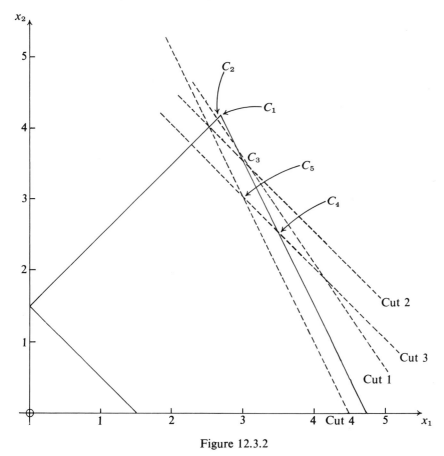

Figure 12.3.2

the coordinates of points in the plane. Figure 12.3.2 shows the feasible region of the associated LP model.

We begin by solving the model as an LP obtaining the solution represented by point C_1. Inequality (12.3.14) was then added as the first cut. The inequality must be expressed in terms of the variables x_1 and x_2 in order to represent it in Figure 12.3.2. u_2 and u_3 can be expressed in terms of x_1 and x_2 using the third and fourth rows of the form (E) of the model. We have:

$$u_2 = 3 + 2x_1 - 2x_2,$$

$$u_3 = 19 - 4x_1 - 2x_2.$$

Inequality 12.3.2 can therefore be rewritten substituting for u_2 and u_3 to give:

Cut 1. $\qquad\qquad\qquad 3x_1 + 2x_2 \le 16.$ $\qquad\qquad\qquad$ (12.3.23)

Inequalities (12.3.17), (12.3.19) and (12.3.21) give the other three cuts. These can also be reexpressed in terms of x_1 and x_2 (using where necessary the equations in which s_1 and s_2 are introduced). The resultant inequalities are:

Cut 2. $\qquad\qquad\qquad\qquad 2x_1 + 2x_2 \le 13 \qquad\qquad\qquad$ (12.3.24)

Cut 3. $\qquad\qquad\qquad\qquad\ \ x_1 +\ \ x_2 \le 6 \qquad\qquad\qquad$ (12.3.25)

Cut 4. $\qquad\qquad\qquad\qquad 4x_1 + 2x_2 \le 18. \qquad\qquad\qquad$ (12.3.26)

These four cuts are represented on Figure 12.3.2.

After imposing cut 1 a new fractional solution is obtained at C_2. After imposing cuts 2, 3 and 4 new fractional solutions are found at C_3 and C_4 before finding the optimal integer solution at C_5.

Although it is obviously mathematically more sophisticated than the Branch and Bound Algorithm the Cutting Planes Algorithm has not proved as efficient for solving practical problems.

One practical disadvantage of the Cutting Planes Algorithm is that all the solutions it obtains before obtaining the optimal integer solution are fractional and therefore of little practical use. This is in contrast to the Branch and Bound Algorithm where non-optimal integer solutions are often obtained on the way to optimality. It may often be worth stopping the solution process and contenting oneself with one of these non-optimal, yet practically useful, solutions.

Another practical consideration is that most computer packages for LP store the matrix of coefficients [see (5.7.3)] by *columns* [see § 6.1]. It is therefore very difficult to add extra *rows* to the matrix as is demanded by the *Cutting Planes Algorithm*. The Branch and Bound Algorithm imposes extra bounds on the variables in the course of optimization. Most computer packages handle bounds on variables in an especially simple manner which makes them easy to modify in the solution procedure.

It is finally worth pointing out that the constraints of the model which we have just solved by both the Branch and Bound and the Cutting Planes Algorithms are the set of constraints (B) discussed in section 12.2.3. We also considered there the constraints (C) defining the convex hull of feasible integer points. If these constraints had been used in place of the original constraints (shown by Figure 12.2.2) we could have solved our example as an LP. Unfortunately as we have already pointed out the derivation of such facet constraints is generally very expensive computationally. The Cutting Planes Algorithm could be regarded as going some way in this direction. We add 'tighter' constraints into the model in the course of optimization. Cut 4 which we added was in fact a facet constraint as can be seen by comparing Figure 12.3.2 with Figure 12.2.5.

The Cutting Planes Algorithm which we have described is only one algorithm using this general approach of adding tighter constraints into the model. There are other methods of deriving such cuts which have been incorporated into other algorithms. We mention one such method which can be applied to 0–1 PIP models in connection with our description of the Pseudo-Boolean approach in section 12.3.2.

12.3.2. Specialist Methods of Solving Integer Programming Models

Because of the computational difficulty of solving IP models there is often benefit to be gained from using specialist algorithms. By taking advantage of the special feature of a restricted class of IP models such an algorithm can often prove more efficient than one of the general methods discussed in section 12.3.1. We will restrict our attention to one particular specialist type of model. We consider 0–1 PIP models and describe two applicable algorithms, Implicit Enumeration (Balas' Algorithm) and a Pseudo-Boolean algorithm (Hammer and Granot's Algorithm).

In order to describe specialist algorithms for this class of problems we will use the following small numerical example (due to Balas).

(G)

$$\begin{aligned}
\text{Minimize} \quad & 5x_1 + 7x_2 + 10x_3 + 3x_4 + x_5 \\
\text{subject to} \quad & x_1 - 3x_2 + 5x_3 + x_4 - x_5 \geq 2 \\
& -2x_1 + 6x_2 - 3x_3 - 2x_4 + 2x_5 \geq 0 \\
& \quad\quad - x_2 + 2x_3 - x_4 - x_5 \geq 1 \\
& 0 \leq x_1, x_2, x_3, x_4, x_5 \leq 1 \text{ and integer.}
\end{aligned}$$

(a) *Implicit Enumeration*

This approach to solving 0–1 PIP models is similar to the Branch and Bound Algorithm in that it involves a tree search. The method does not, however, involve solving the associated LP model at any stage. One of the attractions of the method is that it can be done entirely in integer arithmetic (assuming all coefficients in the model are integers or can be made integers). Nowhere is it necessary to perform divisions and create fractional numbers. This feature of the algorithm makes it particularly suitable for automatic computation. The method is also known as *Balas' Additive Algorithm*.

In order to perform the algorithm it is convenient to express all models as minimizations where all objective coefficients are non-negative. It is also necessary to convert all constraints to the '$\leq$' form. In order to perform these transformations we:

(i) introduce complemented variables $\bar{x}_i$ into the model if necessary by the substitution $x_i = 1 - \bar{x}_i$;

(ii) convert '$=$' constraints to a '$\leq$' and a '$\geq$' form;

(iii) convert '$\geq$' constraints to '$\leq$'.

Performing those transformations necessary we convert (G) into the form:

(H)

$$\begin{aligned}
\text{Minimize} \quad & 5x_1 + 7x_2 + 10x_3 + 3x_4 + x_5 \\
\text{subject to} \quad & -x_1 + 3x_2 - 5x_3 - x_4 + x_5 \leq -2 \\
& 2x_1 - 6x_2 + 3x_3 + 2x_4 - 2x_5 \leq 0 \\
& \quad\quad x_2 - 2x_3 + x_4 + x_5 \leq -1 \\
& 0 \leq x_1, x_2, x_3, x_4, x_5 \leq 1 \text{ and integer.}
\end{aligned}$$

The variables in (H) are classified into 3 sets:

J_0 = set of indices of variables assigned the value 0;
J_1 = set of indices of variables assigned the value 1;
J_2 = set of indices of variables not assigned but temporarily given the value 0.

The algorithm proceeds by assigning variables from J_2 to J_0 or J_1 in a systematic fashion until a feasible solution is obtained. By constructing a 'tree' for the search the optimal solution can be ultimately obtained and verified.

In order to demonstrate the algorithm we will solve model (H). We begin by temporarily assigning all the variables in (H) to the value 0, i.e. placing their indices in J_2. Figure 12.3.3 represents the solution procedure. The following quantities are referred to in the solution process:

u = the current objective value;
z = the best feasible solution obtained to date (to begin with this can be taken as infinity);
v_i = the resulting total infeasibility when an unassigned variable x_i is assigned to J_1.

The steps of the procedure are summarized in Table 12.3.1 which should be read in conjunction with Figure 12.3.3. In Table 12.3.1, $\varnothing$ is used to denote the empty set [see § 1.2.1].

When all variables are assigned to J_2 at node 1 we have an infeasible solution

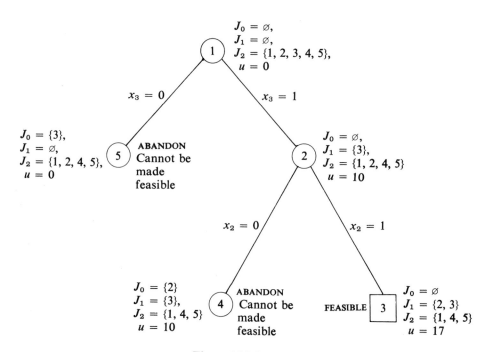

Figure 12.3.3

Node	u	z	J_0	J_1	J_2	v_1	v_2	v_3	v_4	v_5	Branching variable
1	0	∞	∅	∅	{1, 2, 3, 4, 5}	-4	-7	-3	-5	-5	3
2	10	∞	∅	{3}	{1, 2, 4, 5}	-5	0	—	-5	-1	2
3	17	17	∅	{2, 3}	{1, 4, 5}			Feasible solution			

Backtrack to node 2

Node	u	z	J_0	J_1	J_2	
4	10	17	{2}	{3}	{1, 4, 5}	Abandon branch—Impossible to make second constraint feasible

Backtrack to node 1

Node	u	z	J_0	J_1	J_2	
5	0	17	{3}	∅	{1, 2, 4, 5}	Abandon branch—Impossible to make third constraint feasible

No more unexplored branches. Best feasible solution found is optimal.

Table 12.3.1

since constraints 1 and 3 are unsatisfied. We consider the resultant total in-feasibility if we assign any one of the members of J_2 to J_1. These figures are given by the v_i columns in the top row of Table 12.3.1. The best effect on total infeasibility results from bringing in x_3. x_3 is therefore assigned to J_1 creating a right-hand branch from node 1 to node 2.

The procedure is then repeated by assigning x_2 to J_1 creating another right-hand branch to node 3. A feasible solution is obtained at node 3.

We then backtrack to examine the unexplored possibilities. At node 2 if we had assigned x_2 to J_0 we would have created a left-hand branch to node 4. We now take this route. At node 4 since variables x_2 and x_3 have been fixed at 0 and 1 respectively there is no way of making constraint 2 feasible. There is therefore no virtue in exploring beyond node 4.

We now backtrack to node 1 and consider the left-hand branch by assigning x_3 to J_0. There is now no way of making constraint 3 feasible.

Since all the possible branches have now been considered we can be sure that the only feasible solution found is the best. The optimal solution to (H) is therefore:

$$x_1 = 0, \qquad x_2 = x_3 = 1, \qquad x_4 = x_5 = 0, \qquad \text{Objective} = 17.$$

The choice, at each step, of the variable for branching was made according to which had the best effect on the sum of infeasibilities. This is not an enforced choice. Other rules for choosing the branching variable could be devised but would probably not work as well.

Two additional observations on the algorithm are important in practice although this particular example did not illustrate them. They are:

(i) The objective value down a branch may deteriorate beyond that of the best feasible solution so far obtained. Alternatively it may be possible to see that the potential value of the objective function down a branch cannot be better than this. In both these cases the branch may be abandoned.

(ii) It may be obvious that it is necessary to *fix* particular variables at specific values in order to achieve feasibility in a particular constraint. If this is so then these variables can be assigned to J_0 or J_1 immediately.

Any 0–1 PIP model could obviously be solved in theory by *complete enumeration*. We could, in an n variable problem, examine all 2^n possible settings for the variables and choose the best from the feasible ones. As n gets larger the number of possibilities becomes astronomic and such a method becomes impractical. The name 'implicit enumeration' indicates that the method described here is really little more than an enumeration but relies on the strategy of a tree search. Rules for abandoning branches enable one to avoid considering every possibility explicitly.

(b) *The Pseudo-Boolean Algorithm*

This method expresses the constraints of a 0–1 PIP model in terms of Boolean Algebra (see Chapter 16). The proposition '$x_i = 1$' is represented by the symbol X_i. Connectives are used to form compound propositions out of the 'atomic' propositions X_i. For the purposes of this application we can suffice with the following connectives:

'$\overline{}$' means 'not' e.g. '$\overline{A}$' negates proposition A,

'$\vee$' means 'or' e.g. '$A \vee B$' means 'A or B (or both)',

'$\wedge$' means 'and' e.g. '$A \wedge B$' means 'A and B'.

Each compound proposition has a truth value (true or false) depending on the truth values of the atomic propositions [see § 16.3].

It is possible to capture the import of each constraint in a 0–1 PIP model by a compound proposition involving the atomic propositions X_i. In order to do this it is convenient first to express the constraints of (G) in a form where all the coefficients are non-negative and all the constraints are of the '$\leq$' form. This gives:

$$\bar{x}_1 + 3x_2 + 5\bar{x}_3 + \bar{x}_4 + x_5 \leq 5 \qquad (12.3.27)$$

$$2x_1 + 6\bar{x}_2 + 3x_3 + 2x_4 + 2\bar{x}_5 \leq 8 \qquad (12.3.28)$$

$$x_2 + 2\bar{x}_3 + x_4 + x_5 \leq 1; \qquad (12.3.29)$$

$\bar{x}_i$ represents a new variable equal to $1 - x_i$.

Certain combinations of values for the variables in constraints (12.3.27), (12.3.28) and (12.3.29) are clearly impossible. We cannot, for example, have x_2 and $\bar{x}_3$ simultaneously taking the value 1 in constraint (12.3.27) since the sum of their coefficients exceeds the right-hand side value of 5. This observation can be summed up by saying that the proposition below must be FALSE:

$$X_2 \wedge \bar{X}_3 \qquad (12.3.30)$$

'$\bar{X}_i$' obviously stands for the proposition '$\bar{x}_i = 1$' which is the same as '$x_i = 0$'.

A set of indices of variables which cannot simultaneously be 1 in a constraint is known as a *cover*. Proposition (12.3.30) arises from the cover {2, 3}. We need only concern ourselves with *minimal covers*. If a set of indices is a cover this automatically implies that any larger set of indices containing it will also be a cover. In the Boolean Algebra representation [see § 16.3], for example, the falseness of Proposition (12.3.28) implies the falseness of:

$$\bar{X}_1 \wedge X_2 \wedge \bar{X}_3. \tag{12.3.31}$$

There is therefore no point in concerning ourselves with the non-minimal cover {1, 2, 3}.

In order to characterize a constraint completely we need to enumerate all the minimal covers. For constraint (12.3.27) we have:

{1, 3}	showing $\bar{X}_1 \wedge \bar{X}_3$ to be FALSE,
{2, 3}	showing $X_2 \wedge \bar{X}_3$ to be FALSE,
{3, 4}	showing $\bar{X}_3 \wedge \bar{X}_4$ to be FALSE,
{3, 5}	showing $\bar{X}_3 \wedge X_5$ to be FALSE,
{1, 2, 4, 5}	showing $\bar{X}_1 \wedge X_2 \wedge \bar{X}_4 \wedge X_5$ to be FALSE.

The falsity of the above four propositions can be summed up by the falsity of the single proposition:

$$(\bar{X}_1 \wedge \bar{X}_3) \vee (X_2 \wedge \bar{X}_3) \vee (\bar{X}_3 \wedge \bar{X}_4) \vee (\bar{X}_3 \wedge X_5) \vee (\bar{X}_1 \wedge X_2 \wedge \bar{X}_4 \wedge X_5). \tag{12.3.32}$$

Expression (12.3.32) is known as the *resolvent of the constraint* (12.3.27). There are technical advantages in working with FALSE propositions rather than TRUE ones although an analysis with TRUE propositions (negations of resolvents) would also be possible.

A resolvent can also be found for constraint (12.3.28) in a similar manner. This turns out to be:

$$(\bar{X}_2 \wedge X_3) \vee (X_1 \wedge \bar{X}_2 \wedge X_4) \vee (X_1 \wedge \bar{X}_2 \wedge \bar{X}_5) \vee (\bar{X}_2 \wedge X_4 \wedge \bar{X}_5)$$
$$\vee (X_1 \wedge X_3 \wedge X_4 \wedge \bar{X}_5) \tag{12.3.33}$$

For constraint (12.3.29) the resolvent is:

$$\bar{X}_3 \vee (X_2 \wedge X_4) \vee (X_2 \wedge X_5) \vee (X_4 \wedge X_5). \tag{12.3.34}$$

In order to have a feasible solution to the original model (H) we must have (12.3.32), (12.3.33) and (12.3.24) all FALSE. This happens if the following proposition, which results from combining them, is also FALSE.

$$(\bar{X}_1 \wedge \bar{X}_3) \vee (X_2 \wedge \bar{X}_3) \vee (\bar{X}_3 \wedge \bar{X}_4) \vee (\bar{X}_3 \wedge X_5)$$
$$\vee (\bar{X}_1 \wedge X_2 \wedge \bar{X}_4 \wedge X_5) \vee (\bar{X}_2 \wedge X_3) \vee (X_1 \wedge \bar{X}_2 \wedge X_4)$$
$$\vee (X_1 \wedge \bar{X}_2 \wedge \bar{X}_5) \vee (\bar{X}_2 \wedge X_4 \wedge \bar{X}_5)$$
$$\vee (X_1 \wedge X_3 \wedge X_4 \wedge \bar{X}_5) \vee \bar{X}_3 \vee (X_2 \wedge X_4)$$
$$\vee (X_2 \wedge X_5) \vee (X_4 \wedge X_5) \tag{12.3.35}$$

(12.3.35) is known as the *resolvent of the system*.

Simplification procedures can now be applied to the expression (12.3.35). It is beyond the scope of this description to give full details of the procedures but these may be followed up through the references. We outline the simplification. Expressions involving only '$\wedge$' to combine a number of propositions X_i and $\bar{X}_i$ will be referred to as 'clauses'. If in (12.3.35) any clause completely contains another clause we may remove the containing clause as redundant. Clause eleven for example, in (12.3.35) allows us to remove clauses one, two, three and four. Applying this procedure where possible reduces (12.3.35) to:

$$(\bar{X}_2 \wedge X_3) \vee (X_1 \wedge \bar{X}_2 \wedge X_4) \vee (X_1 \wedge \bar{X}_2 \wedge \bar{X}_5) \vee (\bar{X}_2 \wedge X_4 \wedge \bar{X}_5)$$
$$\vee (X_1 \wedge X_3 \wedge X_4 \wedge \bar{X}_5) \vee \bar{X}_3 \vee (X_2 \wedge X_4)$$
$$\vee (X_2 \wedge X_5) \vee (X_4 \wedge X_5) \tag{12.3.36}$$

If one clause contains, or is the same as, another apart from a particular X_i being negated in one but unnegated in the other this X_i (or $\bar{X}_i$) may be removed from the larger clause. When the clauses are the same size it may be removed from either and the other (duplicate) clause entirely removed. Applying this procedure and then the previous one when it becomes applicable reduces (12.3.36) to:

$$\bar{X}_2 \vee \bar{X}_3 \vee X_4 \vee X_5. \tag{12.3.37}$$

Since (12.3.37) must be FALSE its negation must be TRUE. The negation of (12.3.27) is:

$$X_2 \wedge X_3 \wedge \bar{X}_4 \wedge \bar{X}_5. \tag{12.3.38}$$

Negating (12.3.37) requires the use of the so called De Morgan's Law in Boolean Algebra [see § 16.1].

The truth of (12.3.38) characterizes all feasible solutions to (H). In order for (H) to be feasible we must therefore have:

$$x_2 = x_3 = 1 \quad \text{and} \quad x_4 = x_5 = 0.$$

x_1 is undetermined but must clearly be 0 from considerations of optimality. We therefore have the optimal solution:

$$x_1 = 0, \qquad x_2 = x_3 = 1, \qquad x_4 = x_5 = 0, \qquad \text{Objective} = 17.$$

The characterization of the feasible solutions will not generally be as simple as (12.3.38). It can, however, always be expressed in a standard form which allows one to deduce the optimal solution from these feasible solutions.

This method of solving 0–1 PIP models has not, as yet, had much success in solving practical models partly through lack of experience. There are circumstances in which the Boolean Algebra representation of a constraint can be unwieldy. For example the 0–1 constraint below would require an enormous number of clauses to represent it.

$$x_1 + x_2 + \ldots + x_{100} \le 1. \tag{12.3.39}$$

The method does have the great virtue, however, of revealing all sorts of logical implications hidden within a model. By making these implications explicit it may then be much easier to solve the model. For example the second clause of (12.3.32) indicates that the following constraint is implicit in the model:

$$x_2 - x_3 \le 0. \tag{12.3.40}$$

This is not obvious from a cursory examination of the model. One approach would be to add (12.3.40) or similarly deduced constraints as *cutting planes* and then to solve the model by another method.

12.3.3. Other Approaches to Solving Integer Programming Models

In this section we will discuss some other approaches to solving both general and special types of IP model. Some of these approaches are highly technical and we will only indicate their nature. Full descriptions can be followed up through the references in section 12.5.

(a) *Partitioning a MIP model (Benders' Decomposition)*

This method involves partitioning the matrix of an MIP model into itsi nteger and continuous portions [see § 6.6]. If the integer variables are then given particular values we are left with an LP model. The dual of this LP model [see § 11.2.1] has constraints which are all independent of the values assigned to the integer variables. Only the objective function depends on the values given by these integer variables. If we could enumerate all the vertices of the feasible region defined by the constraints of this dual LP model [see § 11.1.2] we could find which vertex provided its optimal solution for each setting of the integer variables. In this way the optimal settings for the integer variables and consequently the optimal solution to the original model could be deduced. Unfortunately there will generally be a very large number of vertices of the dual model. We therefore proceed to derive them iteratively as follows: a setting for the integer variables leads to a vertex (basic) solution for the dual model. This vertex solution is then added to those already obtained and new settings obtained for the integer variables using these dual vertex solutions. The problem of deducing the 'best' new settings for these integer variables gives rise to a PIP model. Once this is solved the cycle can be repeated to find another dual vertex solution.

This method therefore involves successively solving an LP model and then a PIP model. One of the attractions of the method is that it is possible to use specialist algorithms for the PIP model. Considerable success has been achieved in solving some very large MIP models by this method. Full details of the method and some reported results of using it can be found in the references.

(b) *Dynamic Programming*

Chapter 16 of Volume IV is devoted to this method of solving particular types of optimization problems. It is theoretically possible to solve any IP (or LP) model by dynamic programming. This would be an extremely inefficient method of solving most practical models. There is, however, one special type of IP model which is efficiently solved in this way. This is the so called *Knapsack Problem*. A knapsack problem is a PIP model with *one* constraint. The name 'knapsack' arises from an improbable practical application. A hiker wishes to fill his knapsack with a number of items of differing value. The weight limitation on his knapsack provides the single constraint and the maximization of the total value of the contents provides the objective function. We give a general knapsack model below:

$$\text{maximize} \quad p_1 x_1 + p_2 x_2 + \ldots + p_n x_n$$

(I)   subject to $a_1 x_1 + a_2 x_2 + \ldots + a_n x_n \leq b$

$$x_1, x_2, \ldots, x_n \geq 0 \text{ and integer.}$$

There are variants and extensions of such a model which are also efficiently solved by dynamic programming. For example, each variable might have a simple upper bound constraint. A common instance of this is the 0–1 knapsack problem where each variable has a simple upper bound of 1.

The knapsack problem does not arise directly very frequently. It does, however, arise as a subsidiary problem to be solved in larger IP (or LP) models. In particular the *Cutting Stock* problem requires the generation of columns for an IP (or LP) model at stages in the optimization. Each column of coefficients generated comes from the solution of a knapsack problem. The cutting stock problem is the problem of cutting up standard lengths of material (lengths of wood, widths of paper etc.) to meet specified orders but so as to minimize wastage. Each possible 'pattern of orders' which fits into the standard length gives rise to a potential column of the IP model. There are usually a vast number of potential patterns only a few of which will be ultimately used. It is therefore desirable only to generate them (and so solve a knapsack model) when they appear worth considering in the course of optimization.

Dynamic programming approaches the solution to a problem in *stages*. In order to solve model (I) we define the *r*th stage model as:

$$\text{maximize} \quad f_r(y) = p_1 x_1 + p_2 x_2 + \ldots + p_r x_r$$

(J)   subject to $a_1 x_1 + a_2 x_2 + \ldots + a_r x_r \leq y$

$$x_1, x_2, \ldots, x_r \geq 0.$$

Notice that model (J) is defined in terms of a parameter y. $f_1(y)$ is easy to solve. Clearly

$$f_1(y) = p_1 \left[\frac{y}{a_1} \right] \quad \text{and} \quad x_1(y) = \left[\frac{y}{a_1} \right];$$

'[ ]' indicates the next integer less than or equal to the expression inside the brackets. '$x_1(y)$' indicates that x_1 depends on the value of the parameter y.

It is possible to calculate $f_{r+1}(y)$ when $f_r(y)$ is known for all values of y. This can be done by means of the following *recursion*

$$f_{r+1}(y) = \max \left[p_{r+1} x_{r+1} + f_r(y - a_{r+1} x_{r+1}) \right]. \tag{12.3.41}$$

$$x_{r+1} \leq \left[\frac{y}{a_{r+1}} \right]$$

This relationship is fairly easy to justify. It depends on a fairly obvious principle which is central to dynamic programming known as the *Principle of Optimality*. This is discussed in volume IV, § 16.2.1.

Once $f_1(y)$ is calculated for each y it is possible to calculate $f_2(y)$, $f_3(y)$ etc. up to $f_n(y)$. $f_n(b)$ provides the optimal solution to model (I). Such a recursive procedure lends itself well to automatic computation.

There is an alternative recursion to (12.3.41) which is more efficient to execute. This is equally easily justified. It is:

$$f_{r+1}(y) = \max [f_r(y), p_{r+1} + f_{r+1}(y - a_{r+1})]. \qquad (12.3.42)$$

(c) *Group Theory*

A *group* is an abstract mathematical concept which is widely studied in pure mathematics and has many important applications in applied mathematics. Should the reader be unfamiliar with the concept he is referred to Chapter 8. In order to show how group theory can be applied to solving IP we will consider model (E) of section 12.3.1. This is a PIP model. The method can be extended (with some difficulty) to deal with MIP models but we will confine ourselves here to an outline of how PIP models are treated.

Once slack variables are added [see § 11.2.5] and the associated LP model is solved by the simplex algorithm we have the following representation of the model (derived from tableau (F_2) of section 12.3.1).

(K)

$$\text{Maximize} \qquad \tfrac{41}{6} - \tfrac{1}{6}u_2 - \tfrac{1}{3}u_3$$

$$\text{subject to} \quad x_1 = \tfrac{8}{3} + \tfrac{1}{6}u_2 - \tfrac{1}{6}u_3$$

$$x_2 = \tfrac{25}{6} - \tfrac{1}{3}u_2 - \tfrac{1}{6}u_3$$

$$u_1 = \tfrac{32}{3} - \tfrac{1}{3}u_2 - \tfrac{2}{3}u_3.$$

The three constraint equations in (K) can be written entirely in integers as:

$$6x_1 = 16 + u_2 - u_3 \qquad (12.3.43)$$

$$6x_2 = 25 - 2u_2 - u_3 \qquad (12.3.44)$$

$$3u_1 = 32 - u_2 - 2u_3. \qquad (12.3.45)$$

Any integer solution must satisfy the above three equations. These conditions can be restated in congruence relations as [see § 4.3]:

$$-u_2 + u_3 \equiv 16 \pmod 6 \qquad (12.3.46)$$

$$2u_2 + u_3 \equiv 25 \pmod 6 \qquad (12.3.47)$$

$$u_2 + 2u_3 \equiv 32 \pmod 3. \qquad (12.3.48)$$

In addition, in order that the variables be non-negative we must have:

$$16 + u_2 - u_3 \geq 0 \qquad (12.3.49)$$

$$25 - 2u_2 - u_3 \geq 0 \qquad (12.3.50)$$

$$32 - u_2 - 2u_3 \geq 0 \qquad (12.3.51)$$

$$u_2, u_3 \geq 0. \qquad (12.3.52)$$

From the definition of congruence it follows that (12.3.46), (12.3.47) and (12.3.48) can be written more simply as:

$$5u_2 + u_3 \equiv 4 \quad (\text{mod } 6) \qquad\qquad (12.3.53)$$

$$2u_2 + u_3 \equiv 1 \quad (\text{mod } 6) \qquad\qquad (12.3.54)$$

$$u_2 + 2u_3 \equiv 2 \quad (\text{mod } 3). \qquad\qquad (12.3.55)$$

The vectors

$$\begin{pmatrix} 5 \\ 2 \\ 1 \end{pmatrix}, \begin{pmatrix} 1 \\ 1 \\ 2 \end{pmatrix}, \begin{pmatrix} 4 \\ 1 \\ 2 \end{pmatrix}$$

can be regarded as members of a group under the appropriate modular addition in each of the elements [see Example 8.2.7]. If this addition between elements in the group is denoted by $\oplus$ and the three elements by the symbols g_A, g_B and g_C we can express (12.3.53), (12.3.54) and (12.3.55) as:

$$g_A u_2 \oplus g_B u_3 = g_C. \qquad\qquad (12.3.56)$$

It is possible to reexpress the equations in (K) in a standard form known as *Smith Normal Form*. This enables us to determine the group which we are dealing with in its simplest presentation [Theorem 8.2.13]. We have not done this here since we are only sketching the approach.

Having reexpressed the constraints of the original model in the form (12.3.56) we wish to choose non-negative u_2, u_3 so as to maximize the objective of (K) subject to these constraints. This gives us a *group knapsack model* analogous to the knapsack problem considered in the last subsection. The single constraint now applies to elements of a group rather than integers in arithmetic. Dynamic programming is again an efficient way of solving this knapsack problem.

The group knapsack model is not equivalent to the original model since we have ignored the conditions (12.3.49), (12.3.50) and (12.3.51) implying the non-negativity of the LP basic variables x_1, x_2 and u_1.

Our model is a relaxation of the original model. Should it yield non-negative values for x_1, x_2 and u_1 we will have the optimal solution to the original model as well. Otherwise we will have to perform a tree search using this and other relaxations to fathom the tree in a similar manner to the Branch and Bound Algorithm.

The application of group theory to solving IP models is very deep and further discussion is beyond the scope of this description. It can be followed further through the references. Although it represents perhaps the most sophisticated approach to solving IP models it has met with limited success to date in solving practical models.

(d) *Lagrangian Relaxation*

It has already been pointed out [in § 12.3.1(a) and § 12.3.3(c)] that both the LP model associated with an IP model and the group knapsack model are *relaxations* of the original model. Another relaxation arises if we subtract suitable multiples of some of the constraints from the objective function and optimize relative to the other constraints. The resulting relaxed model may have a particularly simple structure if the constraints to be subtracted from the objective function and their multiples are suitably chosen. If the model is regarded as a maximization and the constraints of the form '≤' we must subtract *non-negative* multiples of these constraints from the objective function. The multipliers play an analogous role to dual variables in LP [see § 11.2.1]. It may be computationally relatively easy to obtain an optimal solution to the relaxed model. The optimal objective value associated with this solution will provide a bound on the optimal objective value of the original model. Such a bound can be put to good use in a Branch and Bound method. Considerable computational success has been achieved with this approach to the *Travelling Salesman Problem*. References are given in section 12.5.

We will demonstrate the approach by a Lagrangian Relaxation of model (E) from section 12.3.2. It is convenient to think of all constraints in the '≤' form. This gives:

$$\text{maximize} \quad x_1 + x_2$$

(L)
$$\begin{aligned} \text{subject to} \quad -2x_1 - 2x_2 &\leq -3 \\ -2x_1 + 2x_2 &\leq 3 \\ 4x_1 + 2x_2 &\leq 19 \\ x_1, x_2 &\geq 0 \quad \text{and integer.} \end{aligned}$$

Suppose we subtract λ_1 times the first constraint and λ_2 times the second constraint from the objective function. This gives the Lagrangian relaxation:

$$\text{maximize} \quad (1 + 2\lambda_1 + 2\lambda_2)x_1 + (1 + 2\lambda_1 - 2\lambda_2)x_2 - 3\lambda_1 + 3\lambda_2$$

(M)
$$\begin{aligned} \text{subject to} \quad 4x_1 + 2x_2 &\leq 19 \\ x_1, x_2 &\geq 0 \quad \text{and integer.} \end{aligned}$$

It can be shown that whatever non-negative values of λ_1 and λ_2 are chosen the optimal objective of (M) will not be less than that of (L).

For example if we take $\lambda_1 = 0$ and $\lambda_2 = \frac{1}{6}$ we obtain:

$$\text{maximize} \quad \tfrac{4}{3}x_1 + \tfrac{2}{3}x_2 + \tfrac{1}{2}$$

(N)
$$\begin{aligned} \text{subject to} \quad 4x_1 + 2x_2 &\leq 19 \\ x_1, x_2 &\geq 0 \quad \text{and integer.} \end{aligned}$$

The optimal objective value of (N) is $6\frac{1}{2}$. We can therefore be sure that the optimal objective value of (L) will never be greater than $6\frac{1}{2}$. Such a bound on the optimal objective value could be put to good use in the Branch and Bound Algorithm. Notice that this is a tighter (and therefore more useful) bound than that provided by the optimal solution to the LP model associated with (L). This was shown to be $6\frac{5}{6}$ in section 12.3.1.

Lagrangian relaxation is clearly an attempt to exploit duality concepts in IP.

For an LP model the optimal solution to the dual model provides a set of prices (Lagrange multipliers) which can be applied as 'weights' to the constraints in order to yield the optimal solution. With IP models it can be shown that no such multipliers necessarily exist. (This subject is mentioned further in section 12.4.) The λ's in (M) go some way towards making up for this. It is instructive to picture the situation for model (L) in Figure 12.3.4.

For the associated LP model the optimal solution to its dual leads to the optimal objective value of $6\frac{5}{6}$. For the IP model this provides no more than an upper bound on the optimal objective value. The gap between the IP optimum (6) and the LP optimum ($6\frac{5}{6}$) was referred to in section 2.3 as a *duality gap*. Lagrangian relaxation goes some way towards closing this gap. The above example has enabled us to close up the gap from $\frac{5}{6}$ to $\frac{1}{2}$. In general it is never possible to close the gap completely. With this example for instance there is no 'better' choice of λ_1 and λ_2 in (M) which will close the gap any further.

We have discussed Lagrangian relaxation because of its considerable computational success in helping to solve IP models. In practice the method depends very much on a suitable partitioning of the constraints of a model into those which are incorporated into the objective function (and their multipliers) and

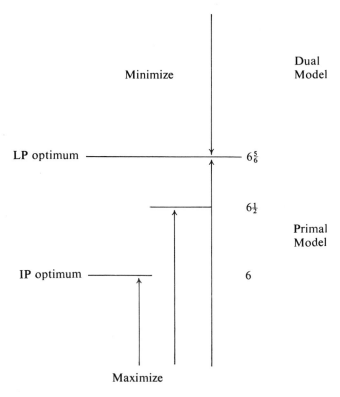

Figure 12.3.4

those which are not. This must be dictated by the physical meaning of the problem which has been modelled.

(e) *Approximate Methods*

Since IP models are often very difficult to solve there is attraction in looking at approximate (sometimes called heuristic) methods of solution which may not be guaranteed to produce the optimal solution but often produce a useful solution fairly quickly. In many practical problems a reasonably good feasible solution is all that is required. The expenditure in finding both the optimal solution and then proving it to be optimal is often not justified.

Unfortunately it is sometimes difficult to find even a feasible solution to an IP model. Once a feasible solution has been found it is possible to examine neighbouring solutions to see if they yield improvements by, for example, altering values of some of the integer variables up or down. We therefore confine our attention to the problem of finding a feasible solution. It may of course happen that a model has no feasible solution in which case any approach must fail.

We describe a method due to Hillier and apply it to model (E). It is convenient to regard the method as applying to maximization models with '$\leq$' constraints. Model (E) is rewritten in this form below.

$$
\begin{aligned}
\text{Maximize} \quad & x_1 + x_2 \\
\text{subject to} \quad & -2x_1 - 2x_2 \leq -3 \\
& -2x_1 + 2x_2 \leq 3 \\
& 4x_1 + 2x_2 \leq 19 \\
& x_1, x_2 \geq 0 \quad \text{and integer.}
\end{aligned}
$$

(O)

We apply the following steps and relate them to Figure 12.3.5.

Step 1. Solve the model as an LP.
This yields $x_1 = 2\frac{2}{3}$, $x_2 = 4\frac{1}{6}$ represented by vertex P of the LP feasible region in Figure 12.3.4.

Step 2. Ignore any constraints which are not 'binding' in the optimal LP solution. Consider only non-slack variables which take non-zero values in the optimal solution.
We therefore consider only the second and third constraints in (O) (represented by the boundaries AP and BP of the feasible region). Only variables x_1 and x_2 are considered.

Step 3. Calculate $\frac{1}{2}$ times the sum of absolute coefficients corresponding to each variable considered for each non-ignored constraint. Subtract these quantities from the corresponding right-hand side coefficients for the constraints.
The right-hand side coefficients for constraints two and three now become 1 and 16. In effect the edges AP and BP have been moved back to CQ and DQ.

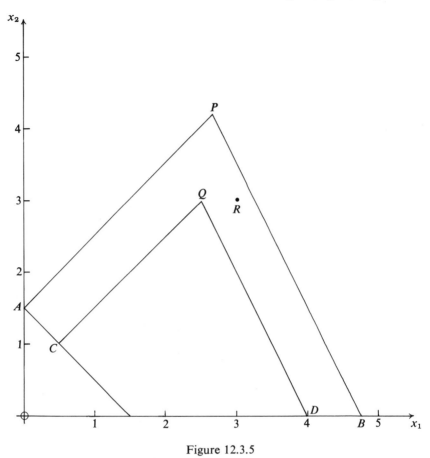

Figure 12.3.5

Step 4.   Solve the new constraints as if they were equations.
          This yields $x_1 = 2\frac{1}{2}$, $x_2 = 3$ represented by point Q.
Step 5.   Add $\frac{1}{2}$ to each of these new values of the variables being considered
          and set these variables to the integer portions of these values.
          This gives $x_1 = 3$, $x_2 = 3$ represented by point R.
For this particular example the procedure has resulted in both a feasible
solution and the optimal solution. Clearly the motive behind the method is to
seek a feasible integer solution inside the feasible region but close to the LP
optimum. The method does not always yield a feasible solution (it fails for model
(A) in the introduction) but works sufficiently often to be worth trying. Should
it not yield a feasible solution additional procedures can be used to carry the
search further. Full details and further approximate methods can be obtained
from the references.

An additional motive for seeking good feasible solutions through computationally quick approximate methods is that they can be used as *bounds* for
subproblems in the Branch and Bound Algorithm (or Implicit Enumeration).

It has already been pointed out how good bounds on the optimal objective value speed the tree search.

Finally it should be mentioned that the physical nature of many practical problems suggests approximate methods. Rather than build an IP model it may be worth considering such a method first. An obvious case is the *Travelling Salesman Problem*. Numerous heuristic algorithms exist for this, some of which are cited in the references. While such methods may yield good solutions one cannot *prove* optimality by such an approach.

12.4.   AREAS FOR FUTURE DEVELOPMENT

Mathematical Programming has been one of the fastest growing branches of Applied Mathematics in recent years. This growth has been motivated by its widespread economic use. Integer Programming is probably the most important extension of Linear Programming. As has been seen in section 12.1, IP opens up the possibility of modelling a wide range of practical problems. Unfortunately its power as a modelling tool has not been matched by the appearance of any universally powerful general algorithm. If such an algorithm could be devised its economic importance would be immense. Much research is going on into devising new IP algorithms. While many ingenious algorithms have been published most of them only prove efficient for solving very small models or specialized classes of model. In addition ways are still being devised of improving existing algorithms. This, coupled with the increasing efficiency of package programs as well as computer hardware, is constantly making it economically possible to solve larger IP models.

Although there is some hope that some day an algorithm will be devised for solving large IP models in a reasonable period of time this hope must be tempered by some recent results in the *Theory of Computational Complexity*. It seems almost certain that no IP algorithm can ever be devised for which the solution time will never be more than a polynomial function of the size of the data. In other words for any algorithm it will almost always be possible to find models of increasing size which result in correspondingly enormous increases in computer time to solve. This does not necessarily mean that no efficient algorithm can ever be found since practical models tend to have very special structures and may not represent the 'worst cases' which cause an algorithm to take its maximum possible time. References to the Theory of Computational Complexity and journals where new algorithms for IP are often published are given in section 12.5.

It was pointed out in section 12.2 that the way in which an IP model is built can affect the time it takes to solve it. As a result of the relative newness of IP as a technique the modelling aspect has not yet been fully developed. New insights into how to build an IP model of particular practical situations are still to be gained. It is often possible to model the same situation in radically different ways. The resultant models may take considerably different times to solve. Some references to the modelling aspect of IP are given in section 12.5.

In LP there are well-known results concerning *duality* [see § 11.2.4]. To any LP model there corresponds a dual model. The solution to this dual model provides important subsidiary economic information in a practical situation. The dual solution can also be used, in conjunction with easily found extra information, to do a *sensitivity analysis* on an LP model [see § 11.4.9]. No really satisfactory duality theory exists for IP. A number of attempts have been made to construct an analogous theory but all have serious shortcomings. The absence of a satisfactory duality theory presents a serious gap in the theory of IP which, hopefully, may one day be filled.

12.5. REFERENCES AND FURTHER READING

Articles on the Theory of Integer Programming as well as descriptions of Applications and Case Studies may be found in the following journals.

Mathematical Programming (*Math. Prog.*)

Mathematical Programming Studies (*Math. Prog. Studies*)

Operations Research (*Opns. Res.*)

Journal of the Operational Research Society (*JORS*)

Management Science (*Mgmt. Sci.*)

Methods of Operations Research (*Meth. O.R.*)

Discrete Mathematics (*Disc. Math.*)

Naval Research Logistics Quarterly (*Nav. Res. Log. Quart.*)

The following books, given in section 12.6, are devoted in whole, or in part, to Integer Programming; (Garfinkel and Nemhauser, 1972), (Hu, 1969), (Greenberg, 1971), (Plane and McMillan, 1971), (Zionts, 1974) and (Salkin, 1974).

Survey papers on Integer Programming are given by Dantzig, (1957), Beale, (1965) Balinski, (1970) and Geoffrion and Marsten, (1972). A survey of business applications is given by Cushing, (1970).

There are a large number of papers on Integer Programming. A selection from these is given below.

Section 12.1. Applications involving *Discrete Activities and Resources* tend to be similar to standard applications of Linear Programming and can be pursued through the literature on LP.

The *Fixed Charge Problem* is discussed by Hirsh and Dantzig, (1968).

The *Depot Location Problem* is described by Baumol and Wolfe, (1958) as well as many other authors.

Project Selection and *Capital Budgeting problems* are discussed by Weingartner, (1974).

A number of problems with *Logical Conditions* are described by Williams (1974 and 1978). A systematic approach to the formulation of logical problems is given by Williams (1977).

The *Job-Shop Scheduling* problem is discussed by Conway, Maxwell and Miller, (1967). One way of formulating the problem as an IP is described by Manne, (1960).

There are many papers on the *Travelling Salesman Problem.* (Bellmore and Nemhauser, 1968) give a survey paper. Held and Karp, (1970 and 1971) describe one of the most successful methods to date of solving the problem using Lagrangian

Relaxation. This problem together with the related *Vehicle Scheduling Problem* is discussed in Eilon, Watson-Gandy and Christofides, (1971).

The *Quadratic Assignment Problem* is discussed by (Koopmans and Beckmann, 1957). A useful survey with applications is given by Lawler, (1974). A practical application is described by Beale and Tomlin, (1972).

Set Covering, Partitioning and Packing Problems, together with a comprehensive list of applications are discussed by Balas and Padberg, (1972). A specialist algorithm for the set partitioning problem is given by Marsten, (1972). The *Assembly Line Balancing* problem is described by Roberts and Villa, (1970). *Political Districting* by IP is described by Garfinkel and Nemhauser, (1970). The *Matching Problem* is discussed by Edmonds and Johnson, (1970).

Section 12.2. The formulation of IP models is discussed at length by Williams, (1977). Dantzig, (1963) devotes a section to the subject. The formulations of specific applications as IP models can be obtained from most of the references for section 1.

Methods of obtaining some of the facets of the convex hull of feasible integer points for 0–1 IP models are given by Balas, (1975a), Hammer, Johnson and Peled, (1975) and Wolsey, (1975). Reformulations of two particular applications to yield integer LP optimal solutions are given by Rhys, (1970) and Williams, (1974).

Section 12.3. Methods of solving IP models are discussed in some of the survey papers mentioned above.

Special Ordered Sets of variables are a concept due to Beale and Tomlin, (1969).

The *Branch and Bound Algorithm* described here is due to Dakin, (1965). Another form of Branch and Bound Algorithm is given by Land and Doig, (1960). A useful framework within which to view Branch and Bound and *Implicit Enumeration* methods is given by Geoffrion and Marsten, (1972). A practical discussion on solving real life IP models is given by Forrest, Hirst and Tomlin, (1974).

The *Cutting Planes Algorithm* described here is due to Gomory, (1963). An extension to deal with MIP models is described by Beale, (1958). Totally different approaches to generating cuts are given by Balas, (1971) and Balas, (1975b).

The method of *Implicit Enumeration* described here known as *Balas' Additive Algorithm* was proposed by Balas, (1965). Another description is given by Glover, (1965).

The *Pseudo-Boolean Algorithm* is due to Hammer and Granot, (1972). A survey of the use of Boolean Algebra for solving 0–1 PIP models is given by Hammer, (1974). The generation of cuts through the Boolean approach is described by Hammer and Nguyen, (1972).

The method of *Partitioning* an MIP model known as *Benders' Decomposition* is due to Benders, (1962).

The way in which the *Knapsack Problem* arises from the *Cutting Stock Problem* is described in Gilmore and Gomory, (1961 and 1963). Dynamic Programming as a means of solving the knapsack problem is described by Bellman, (1957).

Group Theory is applied to IP models by Gomory, (1965). Further papers on this approach are Wolsey, (1971), Gorry, Shapiro and Woolsey, (1972) and Johnson, (1974).

Lagrangian Relaxation is very fully discussed by Geoffrion, (1974). Its use in solving the *Travelling Salesman Problem* is described by Held and Karp, (1970 and 1971). Shapiro, (1971) uses the approach in combination with group theory.

The *Approximate Method* described is due to Hillier, (1969). Other approximate methods of solving IP models are described by Senju and Toyoda, (1968), Trauth and Woolsey, (1968) and Ibaraki, Ohashi and Mine, (1974).

Section 12.4. Current developments in Integer Programming can be followed by reading articles appearing in the journals mentioned above.

The *Theory of Computational Complexity* is described by Aho *et al.*, (1974).

An attempt to provide an economically useful dual for IP models was made by Gomory and Baumol, (1960). A survey of more recent attempts is given by Williams, (1976).

12.6. BIBLIOGRAPHY

Aho, A. V., Hopcroft, J. E. and Ullman, J. O. (1974). *The Design and Analysis of Computer Algorithms*, Addison-Wesley.

Balas, E. (1965). An Additive Algorithm for Solving Linear Programs with Zero–One Variables, *Opns. Res.* **13**, 517–546.

Balas, E. (1971). Intersection Cuts—A New Type of Cutting Planes for Integer Programming, *Opns. Res.* **19**, 19–39.

Balas, E. (1975a). Facets of the Knapsack Polytope, *Math. Prog.* **8**, 146–164.

Balas, E. (1975b). Disjunctive Programming: Cutting Planes from Logical Conditions, in *Nonlinear Programming 2* (Ed. O. L. Mangasarian), Academic Press, 279–312.

Balas, E. and Padberg, M. W. (1972). On the Set Covering Problem II, *Opns. Res.* **20**, 1152–1161.

Balinski, M. L. (1970). On Recent Developments in Integer Programming, in *Proceedings of the Princeton Symposium on Mathematical Programming*, Ed. H. W. Kuhn, Princeton University Press, 267–302.

Baumol, W. J. and Wolfe, P. (1958). A Warehouse Location Problem, *Opns. Res.* **6**, 252–263.

Beale, E. M. L. (1958). A Method for Solving Linear Programming Problems When Some But Not All of The Variables Must Take Integral Values, *Stat. Tech. Rep. No. 19*, Princeton University.

Beale, E. M. L. (1965). Survey of Integer Programming, *ORQ.* **16**, 219–228.

Beale, E. M. L. and Tomlin, J. A. (1969). Special Facilities in a General Mathematical Programming System for Non-Convex Problems Using Ordered sets of Variables, in *Proceedings of the Fifth International Conference of O.R.*, Ed. J. Lawrence, Tavistock.

Beale, E. M. L. and Tomlin, J. A. (1972). An Integer Programming Approach to a Class of Combinatorial Problems, *Math. Prog.* **1**, 339–344.

Bellman, R. (1957). *Dynamic Programming*, Princeton University Press.

Bellmore, M. and Nemhauser, G. L. (1968). The Travelling Salesman Problem, *Opns. Res.* **16**, 538–558.

Benders, J. F. (1962). Partitioning Procedures for Solving Mixed-Variables Programming Problems, *Numerische Mathematik* **4**, 238–252.

Conway, R. W., Maxwell, W. L. and Miller, L. W. (1967). *Theory of Scheduling*, Addison-Wesley.

Cushing, B. E. (1970). The Application Potential of Integer Programming, *J. of Bus.* **43**, 457–467.

Dakin, R. J. (1965). A Tree Search Algorithm for Mixed Integer Programming, *Computer J.* **8**, 250–255.

Dantzig, G. B. (1957). Discrete Variable Extremum Problems, *Opns. Res.* **5**, 266–277.

Dantzig, G. B. (1963). *Linear Programming and Extensions*, Princeton University Press, 535–550.

Edmonds, J. and Johnson, E. L. (1970). Matching: A Well-Solved Class of Integer Linear Programs, *Proc. of the Calgary Int. Conf. on Comb. Structures and Their Application*, Gordon and Breach, 89–92.

Eilon, S., Watson-Gandy, C. D. T. and Christofides, N. (1971). *Distribution Management: Mathematical Modelling*, Griffin.

Forrest, J. J. H., Hirst, J. P. H. and Tomlin, J. A. (1974). Practical Solution of Large Mixed Integer Programming Problems with UMPIRE, *Mgmt. Sci.* **20**, 733–773.

Garfinkel, R. S. and Nemhauser, G. L. (1970). Optimal Political Districting by Implicit Enumeration Techniques, *Mgmt. Sci.* B495–B508.

Garfinkel, R. S. and Nemhauser, G. L. (1972). *Integer Programming*, Wiley.

Geoffrion, A. M. (1974). Lagrangian Relaxation for Integer Programming, *Math. Prog. Studies* **2**, 82–114.

Geoffrion, A. M. and Marsten, R. E. (1972). Integer Programming Algorithms: A Framework and State of the Art Survey, *Mgmt. Sci.* **18**, 465–491.

Gilmore, P. C. and Gomory, R. E. (1961). A Linear Programming Approach to the Cutting Stock Problem, *Opns. Res.* **9**, 849–859.

Gilmore, R. C. and Gomory, R. E. (1963). A Linear Programming Approach to the Cutting Stock Problem, Part II, *Opns. Res.* **11**, 863–888.

Glover, F. (1965). A Multiphase-Dual Algorithm for the Zero–One Integer Programming Problem, *Opns. Res.* **13**, 879–919.

Gomory, R. E. (1963). An Algorithm for Integer Solutions to Linear Programs, in *Recent Advances in Mathematical Programming*, Eds. R. L. Graves and P. Wolfe, McGraw Hill, 269–302.

Gomory, R. E. (1965). On the Relation between Integer and Non-Integer Solutions to Linear Programs, *Proc. Nat. Acad. Sci.* **53**, 260–265.

Gomory, R. E. and Baumol, W. J. (1960). Integer Programming and Pricing, *Econometrica* **28**, 521–550.

Gorry, G. A., Shapiro, J. F. and Wolsey, L. A. (1972). Relaxation Methods for Pure and Mixed Integer Programming Problems, *Mgmt. Sci.* **18**, 229–239.

Greenberg, H. (1971). *Integer Programming*, Academic Press.

Hammer, P. L. (1974). Boolean Elements in Combinatorial Optimisation, paper presented at the Advanced Study Institute in Combinatorial Optimisation, Versailles, France, September 1974.

Hammer, P. L. and Granot, F. (1972). On the Use of Boolean Functions in 0–1 Programming, *Meth. O. R.* **12**, 154–184.

Hammer, P. L., Johnson, E. L. and Peled, U. N. (1975). Facets of Regular 0–1 Polytopes, *Math. Prog.* **8**, 179–206.

Hammer, P. L. and Nguyen, S. (1972). APOSS, A Partial Order in the Solution Space of Bivalent Programs, 41st meeting ORSA, New Orleans, Indiana.

Held, M. and Karp, R. M. (1970). The Travelling Salesman Problem and Minimum Spanning Trees, *Opns. Res.* **18**, 1138–1162.

Held, M. and Karp, R. M. (1971). The Travelling Salesman Problem and Minimum Spanning Trees: Part II, *Math. Prog.* **1**, 6–25.

Hillier, F. S. (1969). Efficient Heuristic Procedures for Integer Linear Programs with General Variables, *Opns. Res.* **17**, 600–637.

Hirsh, W. M. and Dantzig, G. B. (1968). The Fixed Charge Problem, *Nav. Res. Log. Quart.* **15**, 413–424.

Hu, T. C. (1969). *Integer Programming and Network Flows*, Addison-Wesley.

Ibaraki, T., Chashi, T. and Mine, H. (1974). A Heuristic Algorithm for Mixed Integer Programming Problems, *Math. Prog. Studies* **2**, 115–136.

Johnson, E. L. (1974). The Group Problem for Mixed Integer Programming, *Math. Prog. Studies* **2**, 137–179.

Koopmans, T. C. and Beckmann, M. J. (1957). Assignment Problems and the Location of Economic Activities, *Econometrica* **25**, 53–76.

Land, A. H. and Doig, A. G. (1960). An Automatic Method for Solving Discrete Programming Problems, *Econometrica* **28**, 497–520.

Lawler, E. L. (1974). The Quadratic Assignment Problem; A Brief Review, paper presented at Advanced Study Institute on Combinatorial Programming, Versailles, France, September 1974.

Manne, A. S. (1960). On the Job-Shop Scheduling Problem, *Opns. Res.* **8**, 219–223.

Marsten, R. E. (1972). An Algorithm for Large Set Partitioning Problems, Center for Mathematical Studies in Economics and Management Science, Northwestern University, Discussion Paper No. 8.

Plane, D. R. and McMillan, C. (1971). *Discrete Optimisation: Integer Programming and Network Analysis for Management Decisions*, Prentice-Hall.

Rhys, J. M. W. (1970). A Selection Problem of Shared Fixed Costs and Network Flows, *Mgmt. Sci.* **17**, 200–207.

Roberts, S. D. and Villa, C. D. (1970). On a Multiproduct Assembly Line Balancing Problem, *AIIE Trans.* **II**, 361–364.

Salkin, H. M. (1974). *Integer Programming*, Addison Wesley.

Senju, S. and Toyoda, Y. (1968). An Approach to Linear Programming with 0–1 Variables, *Mgmt. Sci.* **15**, B196–B207.

Shapiro, J. F. (1971). Generalised Lagrange Multipliers in Integer Programming, *Opns. Res.* **19**, 68–76.

Trauth, C. A. and Woolsey, R. E. (1968). MESA; A Heuristic Integer Linear Programming Technique, Res. Rep. SC-RR-68-299, Sandia Labs., Albuquerque, New Mexico.

Weingartner, H. M. (1974). *Mathematical Programming and the Analysis of Capital Budgetting Problems*, Academic Press.

Williams, H. P. (1974). Experiments in the Formulation of Integer Programming Problems, *Math. Prog. Studies* **2**, 180–197.

Williams, H. P. (1976). The Economic Interpretation of Duality for Practical Mixed Integer Programming Problems, *Proceedings of IX Symposium on Mathematical Programming*, Budapest, Hungary, August 1976. Ed. by A. Prèkopa, 569–588.

Williams, H. P. (1977). Logical Problems and Integer Programming, *Bulletin of the Institute of Mathematics and its Applications*, **13**, 18–20.

Williams, H. P. (1977). *Model Building in Mathematical Programming*, Wiley.

Williams, H. P. (1978). The Reformulation of Two Integer Programming Problems, *Math. Prog.*, **14**, 325–331.

Wolsey, L. A. (1971). Group-Theoretical Results in Mixed Integer Programming, *Opns. Res.* **19**, 1691–1697.

Wolsey, L. A. (1975). Faces for a Linear Inequality in 0–1 Variables, *Math. Prog.* **8**, 165–178.

Zionts, S. (1974). *Linear and Integer Programming*, Prentice Hall.

CHAPTER 13

The Theory of Games

13.1. INTRODUCTION

13.1.1. Background

The theory of games is concerned with decisions, and in particular with choosing from a number of possible decisions one which is, in some sense to be defined, best. This assumes, to begin with, that we have some scale of priorities which enables us to say which of two decisions is to be preferred. If the consequences of possible decisions are precisely known, then no difficulty arises; however, this is only rarely the case in real life situations. In economic or military competition, and in many other sociological, political, or diplomatic circumstances, it is necessary to make decisions, the consequences of which are not within the control of the decision maker.

The consequences may depend on actions and decisions of others, be they partners or opponents, and frequently also on chance, or on acts of nature, such as for instance the weather.

To analyse logically such situations it is convenient to use a model which typifies at least some of the features mentioned. For this purpose parlour games are an excellent example, without the unpleasant connotations which some of those features have. When we decide which moves to make in a game of chess, say, then we cannot say with certainty that we shall win, because this depends also on the moves of our oppoent. When we play a game of cards, then the outcome depends also on chance, such as the deal.

The title of John von Neumann's seminal paper of 1928 was, significantly, 'On the theory of parlour games'. It may be assumed that this great mathematician was attracted to this topic by the intricacies of its logical structure. On the other hand, the classical text of von Neumann and Morgenstern of 1944 was inspired by the challenge of applying abstract theory to economic behaviour. Sociology as well as branches of Operational Research have benefited from applying concepts which have originated in the theory of games.

The theory of games will not teach anybody how to be proficient in a particular game, say chess, or football. Nor is it concerned with computing probabilities for games of chance. It is, rather, an abstract theoretical construction for

391

elucidating the essential logical features of a competitive environment. It has been useful in suggesting approaches to debates and to arbitration, and within its framework suggestions have been made to such topics as bidding for contracts, or the discovery of fraud.

13.1.2. Concepts and Definitions

We start by introducing some concepts on which the theory is based.

By a *game* we mean a set of rules which define the actions which a set of 'players' can take, and the resulting gains or losses. This set of rules is given before a 'play' begins, i.e. a particular instance of performing the game. Every game can be performed in a number (though not necessarily an infinite number) of possible plays, every one of these consisting of a series of 'moves', made in accordance with the rules of the game. Some of the moves might depend on chance as well. At the end of the play the players receive or pay amounts dependent on the moves made.

It is customary to imagine that the players decide on their moves as the play develops. However, this is not essential for the theory to be applicable. We might imagine that each player decides, before the play starts, on what moves they would make in every conceivable situation during the play, as it developed from the moves that have been made. Thereby the outcome, and if chance moves are involved, the expected outcome, is already determined. In effect, the game has been reduced to one of a single move by each player, made unknown to the other players. Such a reduction is called *normalizing* the game.

Of course, such a reduction is practically impossible for any game of intellectual interest. It can, in effect, be done for Noughts and Crosses, for instance, and thereby this game has become trivial. But lovers of chess need not fear that it will ever be normalized.

A game played by two players is referred to as a 'two-person-game', and one played by more than two players as an *n-person-game*. A *zero-sum-game* is a game where payments are only made between the players, so that the sum of all their gains is zero. A *finite* game is one where each player has only a finite number of possible actions to choose from.

A finite zero-sum two-person-game is also called a *matrix* game, or a *rectangular* game. This name is due to the form in which the results, the *pay-offs* can be displayed, when the game is normalized.

In the following sections we consider, first, rectangular games. Then we deal with games played by more than two players. This model is particularly suitable for economic situations, where the interests of the participants are not altogether opposed to one another. *Non-zero-sum games*, where the payments, i.e. gains and losses, do not add up to zero, can be given similar significance.

This is followed by a brief description of *extensive* games in a not normalized form. Finally, we treat *infinite* games, and introduce in this context statistical considerations, though not on a very high level.

The bulk of our presentation concerns rectangular games. The theory of these games can be said to be fairly complete. The detailed treatment of the other types, and a description of their state of development would require a more sophisticated treatment, which is beyond the scope of these pages.

13.2. MATRIX GAMES

13.2.1. Pure Strategies

Consider a specific example.

EXAMPLE 13.2.1.   Assume that one player, R (for row) chooses one of the numbers 2, 3, or 5, and the other player C (for column) chooses one of the numbers 1, 2, or 4. The player who has chosen the higher number receives from his opponent the difference between the chosen numbers, in some agreed monetary units.

We obtain an overall view from the following *pay-off table* (Table 13.2.1), which lists the amount received by R from C, dependent on the pair of choices of the two players:

		Choice of C		
		1	2	4
	2	1	0	−2
Choice of R	3	2	1	−1
	5	4	3	1

Table 13.2.1

A negative pay-off is, of course, a payment from R to C. Because R receives the pay-off listed, he wants it to be as high as possible; we call him the *maximizing* player, and C is the *minimizing* player.

The respective choices of the players will depend on their attitudes to risk. The theory of games is concerned with decisions made by cautious, safety-first players, and we shall now try to find some test by which such a player can call a decision the best choice he could have made under the circumstances: consider the pay-off Table 13.2.1. If R chooses 2 (or 3, or 5), then he can be sure that he will gain at least −2 (or −1, or 1). He will choose 5, to make his smallest gain as large as possible. Similarly, if C chooses 1 (or 2, or 4), he might have to pay as much as 4 (or 3, or 1). He will choose 4, to make his largest loss as small as possible.

If each player takes this attitude, then the pay-off, i.e. the payment from C to R, will be 1. The player making some other choice would risk to be worse off. (In this rather trivial game the choices of the players were obvious to begin with: each chooses the highest number at his disposal.)

We can describe the manner in which we have found the outcome in Example 13.2.1 by saying that we have determined the minima of the rows, and then we have chosen the largest of these, the *maximin*. We have also determined the maxima of the columns, and then we chose the smallest of these, the *minimax*. In this case it turned out that the maximin equalled the minimax. A game with a pay-off table where this is so is called *strictly determined*. Such a game has some comfortable features which we shall now describe.

We can say that when R chose 5, he was guaranteed to win at least 1, and C, choosing 4, had a guarantee of losing not more than 1. Any other choice would have given them a worse guarantee. Because the two guarantees were equal (though a gain for one, a loss for the other) we can now say that each player has made the best possible choice, that is, the strategies chosen were *optimal*. Indeed, since R knew that with his choice he must win at least 1, whatever C's choice may be, the best that C could do was to restrict R's gain to this amount. Conversely, since C's choice ensured that he would not lose more than 1, whatever R's choice may be, the best R could do was to make C lose precisely that much. Thus both have made the choice which was the best answer to the choice of the opponent. This result depended on the game being strictly determined.

The equality of the two guarantees has a further consequence. Each player would have made the same choice if he had known his opponent's strategy in advance. Although such knowledge is excluded by the rules of the game, such consideration could be relevant in realistic situations. It follows that when a strictly determined game is played repeatedly, the players would repeat their respective choices.

The entry in the pay-off table corresponding to the minimax and the maximin in a strictly determined game (in Example 13.2.1: 1 in the third row and third column) is called, by a geometrical analogy, a *saddle point entry*. It must be understood, though, that the saddle point must have the correct orientation. Any entry which is largest in its row and at the same time smallest in its column would also be in a saddle point, but it would not have the significance for which we are looking.

A pay-off table (Table 13.2.2) can have more than one saddle point, for instance:

$$
\begin{array}{rrrr}
0 & 2 & 1 & 0 \\
-1 & 5 & -3 & -2 \\
0 & 4 & 2 & 0
\end{array}
$$

Table 13.2.2

where all the 0's are saddle point entries. We shall return to this later, in section 13.2.3. Here we merely point out that in such a game the players could change their strategies when the game is repeated aiming for different saddle points, without thereby making any difference to the outcome.

The satisfactory results for strictly determined games suggest the question of whether something similar can be found for those games which are not strictly determined. That such games exist is evident from such games as *matching pennies*, with the pay-off table:

	H	T
H	1	−1
T	−1	1

Table 13.2.3

or the game of *Scissors-Paper-Stone* (also called *Criminals' Baccarat, Japanese Morra, Amchara*, etc.) with pay-off table (Table 13.2.4).

	Sc	P	St
Sc	0	1	−1
P	−1	0	1
St	1	−1	0

Table 13.2.4

We shall analyse a less trivial, but still very simple game:

EXAMPLE 13.2.2. Let R choose two different numbers from the set $(1, 2, 4)$, and let C choose one number from the same set. If R chooses the two numbers different from the choice of C, then R receives from C the sum of the values of the numbers chosen by R. But if one number has been chosen by both players, then R receives the value of the number which only he has chosen, less the value of the number which both have chosen. The pay-off table (Table 13.2.5) is then as follows:

		C chooses			
		1	2	4	row minima
	1, 2	1	−1	3	−1
R chooses	1, 4	3	5	−3	−3
	2, 4	6	2	−2	−2
column maxima		6	5	3	

Table 13.2.5

The minimax 3 differs from the maximin -1. There is no entry in the table which is, at the same time, the smallest in its row and also the largest in its column: there is no saddle point.

In the above example the maximin is smaller than the minimax. In fact, it can never be larger, as we now show:

Denote the entry in row i and column j by a_{ij}. Let, for all i,

$$\min_j a_{ij} = a_i, \quad \text{say, so that} \quad a_i \le a_{ij} \quad \text{for all } j$$

and let, for all j

$$\max_i a_{ij} = A_j, \quad \text{say, so that} \quad A_j \ge a_{ij} \quad \text{for all } i.$$

Let

$$\max_i \min_j a_{ij} = \max_i a_i = a_{i_0}, a_{i_0} \le a_{i_0 j} \quad \text{for all } j$$

and

$$\min_j \max_i a_{ij} = \min_j A_j = A_{j_0}, A_{j_0} \ge a_{i j_0} \quad \text{for all } i.$$

Then

$$a_{i_0} \le a_{i_0 j_0} \le A_{j_0}, \quad \text{q.e.d.}$$

The question now arises whether in a game with no saddle point there is some sense in which a strategy can be called better, or at least not worse, than any other strategy. To answer this question, we take our clue from the fact that in a strictly determined game no player need change his choice when the game is repeated, and that it does not help any player to know the choice of the other player in advance. If there is no saddle point, such advance knowledge might help, and a player would be ill advised to show that he will make always the same choice. Therefore, when playing such a game repeatedly, we shall change our strategies, and concentrate at obtaining as much as possible, or losing as little as possible, in the long run.

13.2.2. Mixed Strategies

Guided by such considerations as outlined above, we shall now look at the consequences of choosing strategies at random with given long-term relative frequencies [see II, § 2.1]. We call such a procedure a *mixed strategy*, to distinguish it from the 'pure' strategies which we have considered so far. The latter are, of course, a special case of the former, with relative frequency 1 for just one of the possible choices.

EXAMPLE 13.2.3.   We take as the basis for our investigation the pay-off table of Example 13.2.2, and assume now that R chooses, in the long run, the pair $(1, 2)$ with relative frequency $\frac{2}{3}$, and the pair $(1, 4)$ with relative frequency $\frac{1}{3}$.

He never chooses (2, 4). We write this mixed strategy $(\frac{2}{3}, \frac{1}{3}, 0)$. In the long run his gain will still depend on the strategy of C:

> if C chooses 1 all the time, R will gain $1 \times \frac{2}{3} + 3 \times \frac{1}{3} = \frac{5}{3}$,
> if C chooses 2 all the time, R will gain $(-1) \times \frac{2}{3} + 5 \times \frac{1}{3} = 1$

and

> if C chooses 4 all the time, R will gain $3 \times \frac{2}{3} + (-3) \times \frac{1}{3} = 1$.

Thus, if C chooses one of his pure strategies, R is guaranteed a gain of at least 1. Also, even if C chooses a mixed strategy, the average pay-off to R cannot be smaller than 1. Therefore, if R chooses the mixed strategy $(\frac{2}{3}, \frac{1}{3}, 0)$, his guarantee is a gain of 1.

Now let C choose the mixed strategy $(0, \frac{1}{2}, \frac{1}{2})$, i.e. never 1, and the other two numbers at random, with equal relative frequencies.

If R chooses	then C loses on average
(1, 2)	$(-1) \times \frac{1}{2} + 3 \times \frac{1}{2} = 1$
(1, 4)	$5 \times \frac{1}{2} + (-3) \times \frac{1}{2} = 1$
(2, 4)	$2 \times \frac{1}{2} + (-2) \times \frac{1}{2} = 0.$

Thus when C chooses $(0, \frac{1}{2}, \frac{1}{2})$, he has a guarantee that he will not lose more than 1, whichever strategy R chooses, pure or mixed.

With the two strategies chosen, the two guarantees are now equal. By the same argument as that in the case of a strictly determined game, and in the same sense, we conclude that the chosen strategies are optimal, for either player.

We notice that, compared with the maximin and minimax obtained when both players employed the pure strategies of Example 13.2.2, the mixed strategies of Example 13.2.3 obtained a better guarantee for each player: a gain of 1 is better than a gain of -1, and a loss of 1 is better than a loss of 3. However, the more important observation is this: *the two guarantees are numerically equal.*

13.2.3. Solutions and Equilibrium Points

We are now interested in knowing whether we can always obtain this equality by appropriate, possibly mixed strategies. The answer is affirmative. This is the *Main Theorem*, or *Minimax Theorem*, of the Theory of Games. It was first proved, after ealier partial proofs by E. Borel, by John von Neumann in a paper read to the Göttingen Mathematical Society on 7 December 1926, and published in von Neumann (1928). He proved an existence theorem, and used for this purpose the fixed point theorem of Kakutani [V, § 5.5]. Later J. Ville gave an algebraic proof. By using Linear Programming a constructive proof can be given, i.e. a proof which contains also a procedure for finding the optimal

strategies (see, for instance, Vajda (1975), p. 39). A pair of optimal strategies
which result in equal guarantees for each player, is called a *solution* of the game,
and the resulting pay-off is its *value*.

We have mentioned in section 13.2.1 that a pay-off table can have more than
one saddle point. Even if there is no saddle point, there can be more than one
'solution', as defined above. Also, if there is just one saddle point, the game can
have an additional solution in mixed strategies.

Let the pure strategies of the maximizing player be chosen with relative
frequencies $x_1, \ldots, x_n$, and those of the minimizing player with relative frequencies $y_1, \ldots, y_m$. These relative frequencies are non-negative, and the x_i add
up to 1, as do the y_j. We describe the set of the x_i briefly as (x), and similarly the
set of the y_j by (y).

Now assume that there are two solutions, $((x'), (y'))$, and $((x''), (y''))$. Then

$$\sum_i \sum_j a_{ij} x_i'' y_j' \geq \sum_i \sum_j a_{ij} x_i'' y_j'' \geq \sum_i \sum_j a_{ij} x_i' y_j''$$

because in answering (x''), the minimizing player does at least as well with (y'')
as he would with (y'), and in answering (y''), the maximizing player does at
least as well with (x''), as he would with (x')—this is, after all, the meaning of
the pair $((x''), (y''))$ being a solution.

Similarly, because $((x'), (y'))$ is a solution,

$$\sum_i \sum_j a_{ij} x_j' y_j'' \geq \sum_i \sum_j a_{ij} x_i' y_j' \geq \sum_i \sum_j a_{ij} x_i'' y_j'.$$

If we now look at the extreme terms in both sequences of inequalities, we see
that they are the same, but in reversed order. Therefore all these double sums
must be equal, and hence $((x''), (y'))$, and $((x'), (y''))$ are also solutions. This
result generalizes an observation which we have made when we gave an example
of more than one saddle point. There were, then, not merely two, but four, in
the corners of a rectangle.

Mixed strategies were introduced in the context of repeated plays of the same
game, to produce the best long-term average pay-off. However, mixed strategies
may have relevance to single plays as well.

For instance, pure strategies might mean investing a given capital in various
ventures. If the yield of each depends in a known manner on various potential
circumstances, then we may wish to invest so that the smallest realizable yield
is as high as possible. In this example there is no conscious opponent. Nevertheless, the concept of an optimal opposing strategy has still some meaning. It is
the most pessimistic assumption about the probabilities of the various circumstances which could occur. If these probabilities describe correctly the future
relative frequencies of occurrence, then our own strategy ensures the highest
average yield.

With a view to later developments we introduce here the concept of an
equilibrium point. This is a pair of strategies (x) and (y), such that even if a
player knew his opponent's strategy, he would have no reason to change his

own. It is easily seen that an equilibrium point is a solution of a rectangular (matrix) game. However, this concept can also be applied to other games, as we shall see.

Because the definition of an equilibrium point does not refer to any payments, it is also applicable when no common unit for measuring utility is assumed, or when the players attach different values to the same outcome (perhaps because one of the opponents knows something about circumstances which are unknown to others).

13.2.4.   Methods for Solving Small Rectangular Games

We mention here a few methods which are suitable for solving games with small pay-off tables, or pay-off tables which can be simplified.

(a) If only one of the players has only two pure strategies at his disposal, and of course their mixtures, then a geometric representation is instructive. We illustrate this by means of Example 13.2.3, where the last line of the pay-off table (Table 13.2.5) is omitted since it is not used in the optimal strategy of R which we obtained in Example 13.2.3:

$$\begin{array}{rrr} 1 & -1 & 3 \\ 3 & 5 & -3 \end{array}$$

Table 13.2.6

We draw a diagram with baseline $[0, 1]$ [see (2.6.10)] and indicate on it, by the distance from 0, the relative frequency of the second row in the strategy of R. The two pay-offs for R, by his first or his second strategy, dependent on C's choosing one of his pure strategies, are drawn as ordinates above 0 and 1 respectively [see V, § 1.2.1]. We connect the end-points of the ordinates belonging to the same pure strategy of C, and call these lines the *strategy lines* of C.

Suppose we draw a vertical from the point at the base-line with distance $\frac{3}{5}$ from 0 and $\frac{2}{5}$ from 1. This means that we ask what happens if R chooses his first row with relative frequency $\frac{2}{5}$, and his second row with relative frequency $\frac{3}{5}$. The intersection with the three strategy lines of C will have heights

$$1 \times \left(\tfrac{2}{5}\right) + 3 \times \left(\tfrac{3}{5}\right) = \tfrac{11}{5}$$
$$(-1) \times \left(\tfrac{2}{5}\right) + 5 \times \left(\tfrac{3}{5}\right) = \tfrac{13}{5}$$
$$3 \times \left(\tfrac{2}{5}\right) + (-3) \times \left(\tfrac{3}{5}\right) = -\tfrac{3}{5}.$$

The lowest of these—the guarantee for R's strategy $\left(\tfrac{2}{5}, \tfrac{3}{5}\right)$—has height $-\tfrac{3}{5}$. But this is not R's best strategy. His best strategy is that for which the lowest intersection of the corresponding vertical with the strategy lines is highest. This is the case at distance $\frac{1}{3}$ from 0, indicating the strategy $\left(\tfrac{2}{3}, \tfrac{1}{3}\right)$, our previous result.

If it is the minimizing player who has two pure strategies, then an analogous diagram can be constructed and interpreted. In this case we shall draw strategy

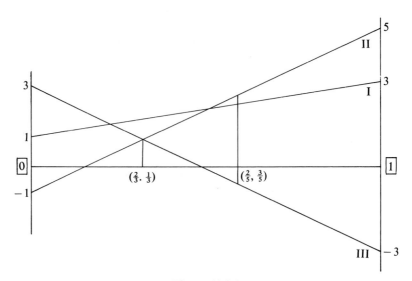

Figure 13.2.1

lines of R, and it is then the lowest of the uppermost intersections of verticals with the strategy lines which gives the optimal strategy. For instance, if the pay-off table (Table 13.2.7) were

$$
\begin{array}{cc}
-1 & 5 \\
2 & 3 \\
4 & 1 \\
\end{array}
$$

Table 13.2.7

then the best strategy for C would be $(\frac{1}{2}, \frac{1}{2})$, with value $\frac{5}{2}$. (R would use $(0, \frac{3}{4}, \frac{1}{4})$).

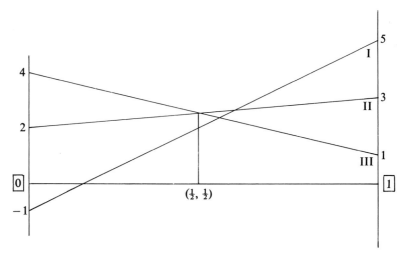

Figure 13.2.2

A representation which superimposes the strategies of both players on the same graph, still assuming that one of them has only two pure strategies at his disposal, is explained in S. Vajda (1967), on pages 14–17.

(b) If both players have only two pure strategies, then we look first whether there is a saddle point, which gives the solution as well as the value of the game. If there is none, then both players will have to choose a mixed strategy. The best result is obtained if the pay-off is independent of the opponent's strategy. Otherwise the opponent could choose that pure strategy which makes the pay-off more attractive to him. Therefore we have to solve [see § 5.8]

for R	for C

$$a_{11}x_1 + a_{21}x_2 = a_{12}x_1 + a_{22}x_2 \qquad a_{11}y_1 + a_{12}y_2 = a_{21}y_1 + a_{22}y_2$$

$$x_1 + x_2 = 1 \qquad\qquad\qquad y_1 + y_2 = 1$$

$$x_1, x_2 \geq 0 \qquad\qquad\qquad y_1, y_2 \geq 0.$$

Solution: mixed strategy (x_1, x_2) $\qquad$ Solution: mixed strategy (y_1, y_2)

$$x_1 = (a_{22} - a_{21})/(a_{11} + a_{22} - a_{12} - a_{21}) \qquad y_1 = (a_{22} - a_{12})/(a_{11} + a_{22} - a_{12} - a_{21})$$

$$x_2 = (a_{11} - a_{12})/(a_{11} + a_{22} - a_{12} - a_{21}) \qquad y_2 = (a_{11} - a_{21})/(a_{11} + a_{22} - a_{12} - a_{21})$$

These solutions are valid only if the values are non-negative, and this is so if $a_{11} - a_{12}$ and $a_{21} - a_{22}$ have different signs and so have $a_{11} - a_{21}$ and $a_{12} - a_{22}$. If this is not so, then the pay-off table has a saddle point. Hence the solutions given are valid if there is no saddle point. The computation of the optimal mixed strategies is extremely simple. In words: R computes the absolute values of the differences in his rows [see (2.2.2)], and mixes the rows in the inverse proportions. C acts analogously on his columns. For instance, if the pay-off table were

$$\begin{matrix} -1 & 3 \\ 5 & -3 \end{matrix}$$

the optimal mixed strategies for the maximizing and minimizing players would be:

$$R: \quad |-1 \quad -3| = 4$$
$$|5 - (-3)| = 8 \quad \text{mixture 8 to 4, i.e. } (\tfrac{2}{3}, \tfrac{1}{3});$$

and

$$C: \quad |-1 \quad -5| = 6$$
$$|3 - (-3)| = 6 \quad \text{mixture 1 to 1, i.e. } (\tfrac{1}{2}, \tfrac{1}{2}).$$

13.2.5.   Dominance

The concept which we shall now introduce is often useful as a device to simplify the pay-off table. We call a row (or a column) dominated by another, if no entry in the former is larger (smaller) then the parallel entry in the latter row (column). No dominated row or column need be considered as part of a mixture of an optimal strategy.

We can also compare mixed strategies. For instance, we could have excluded the first column of Table 13.2.5:

$$
\begin{array}{rrr}
1 & -1 & 3 \\
3 & 5 & -3 \\
6 & 2 & -2
\end{array}
$$

by observing that

$$\tfrac{1}{2}(-1 + 3) = 1$$
$$\tfrac{1}{2}(5 - 3) < 3$$
$$\tfrac{1}{2}(2 - 2) < 6.$$

If no (mixed) strategy is preferable to a given strategy, whatever the opponent chooses, then the latter is called *admissible*, and a set S of strategies such that to any possible strategy there is one uniformly better or equally good in S, is called *complete*. There is clearly no point in considering any strategy not within a complete set (if such a set exists).

We add a few more examples, to illustrate the versatility of the concepts introduced.

EXAMPLE 13.2.4.   The rules of the Bluffing Game are as follows:

In a pack of cards there are high cards and low cards in equal numbers. A draws a card. His probability of drawing a high card is $\tfrac{1}{2}$, and so is his probability of drawing a low card. If he draws a high card, he bids 2. If he draws a low card, he can pay 1 (strategy A_1), or he can bid 2 (strategy A_2, Bluffing). If A has bid, then B can either challenge him (strategy B_1), in which case B either gains 2 or loses 2, according as A has a low or a high card, or B can pay 1 (strategy B_2).

In this game the pay-off depends also on chance (the card A has drawn). The expected pay-offs to A are:

	B_1	B_2
A_1	$\tfrac{1}{2} \times 2 + \tfrac{1}{2} \times (-1) = \tfrac{1}{2}$	$\tfrac{1}{2} \times 1 + \tfrac{1}{2} \times (-1) = 0$
A_2	$\tfrac{1}{2} \times 2 + \tfrac{1}{2} \times (-2) = 0$	$\tfrac{1}{2} \times 1 + \tfrac{1}{2} \times 1 = 1$

Using the method (b) of section 13.1.3 we find that the solution is $(\tfrac{2}{3}, \tfrac{1}{3})$ for A, $(\tfrac{2}{3}, \tfrac{1}{3})$ for B, with value $\tfrac{1}{3}$.

The optimal mixed strategy for A implies that once on three occasions he will bid, although he has drawn the low card: he will be bluffing.

EXAMPLE 13.2.5.   The rules of two-finger Morra, a game dating back to antiquity, are as follows.

Each of two players shows, simultaneously, one or two fingers and calls a number he guesses is the number of fingers which the other shows. If just one player has guessed correctly, then he wins the sum of the number of fingers shown. Otherwise no payment is made.

In the following pay-off table (x, y) indicates x $(= 1$ or $2)$ fingers shown, y $(= 1$ or $2)$ fingers guessed:

		C plays:			
		(1, 1)	(1, 2)	(2, 1)	(2, 2)
	(1, 1)	0	2	-3	0
	(1, 2)	-2	0	0	3
R plays:	(2, 1)	3	0	0	-4
	(2, 2)	0	-3	4	0

This table is skew-symmetric [cf. (6.7.8)] and the value of the game will be 0, the game is 'fair'. This is obvious, since both players are in the same position.

Computation shows that each player may choose any mixture $(0, a, 1 - a, 0)$ where $\frac{4}{7} \leq a \leq \frac{3}{5}$ and the two players need not, of course, choose the same value for a. R's pay-off is 0, because of the inner square of zeros in the table. Values outside the given range for a must be avoided: otherwise the opponent could make the pay-off worse, by using the first, or the fourth row or column.

It is also easily seen that no appropriate mixtures of the form $(b, 0, 0, 1 - b)$ exist, though one might be tempted by the zeros in the corners to try and find one.

13.2.6. Games Against Nature

We have mentioned in our introductory remarks that the consequences of any of our decisions might depend also on natural phenomena like the weather, which are not within our control, but not within the control of any conscious opponent either. We then speak of *games against nature*. We can still use concepts of the theory of games, and to give an example, we discuss the problem of betting on a race course.

EXAMPLE 13.2.6. A bookmaker offers you odds $a_i: 1$ on horse $H_i.(i = 1, \ldots, n)$.

The possible strategies of the horses, as it were, are for horse H_1, or for horse $H_2, \ldots,$ or for horse H_n to win. (We leave it to the reader's imagination to say if there are also conceivable mixed strategies.) Your pure strategies are betting on one horse, with all your stake, and the mixed strategies are partitions of your stake, on a number of horses. You stake the portion x_i of the total on horse H_i, and $x_1 + \ldots + x_n = 1$. If you bet 1 on H_i, and this horse wins, then you receive

a_i, and your stake back. Otherwise you lose 1. This leads to the following pay-off table:

Winning horse

		H_1	H_3	...	H_n
	H_1	a_1	-1	...	-1
Your bet	H_2	-1	a_2	...	-1
	$\vdots$	$\vdots$	$\vdots$	$\ddots$	
	H_n	-1	-1	...	a_n

Table 13.2.8

We want to find non-negative $x_1, \ldots, x_n$, adding up to 1 and such that the smallest of the values

$$a_1 x_1 - x_2 - \ldots - x_n$$
$$-x_1 + a_2 x_2 - \ldots - x_n$$
$$\vdots \qquad \vdots \qquad \ddots \qquad \vdots$$
$$-x_1 - x_2 - \ldots + a_n x_n$$

i.e. of $(a_1 + 1)x_1 - 1, (a_2 + 1)x_2 - 1, \ldots, (a_n + 1)x_n - 1$ is maximized.

Now if the x_i are chosen so as to make all these values equal i.e. if $x_i(a_i + 1) = x_j(a_j + 1)$ and hence, since $\sum_i x_i = 1$,

$$x_i = \frac{1}{a_i + 1} \Big/ \sum_i \frac{1}{a_i + 1} \quad \text{for all } i,$$

then the guarantee is any one of $(a_i + 1)x_i - 1$, and this is the gain, whichever horse wins.

We show now that this is, in fact, the highest guarantee, and hence the value of the game.

Suppose you chose, instead of x_i, ratios $x_i + d_i$, where of course $\sum_i d_i = 0$. Then $(a_i + 1)x_i - 1$ will be replaced by

$$(a_i + 1)(x_i + d_i) - 1 = (a_i + 1)x_i - 1 + (a_i + 1)d_i.$$

But the additional terms cannot all be positive, because they must add up to zero. Therefore the smallest value, and this is the guarantee, will not be smaller than before.

Thus you have solved the problem of how to partition your total stake, in the spirit of the theory of games. You know that whatever the outcome of the race, your gain will be $(a_i + 1)x_i - 1$, which is equal to

$$\left[1 \Big/ \sum_i \frac{1}{a_i + 1} \right] - 1.$$

But it would be foolish to bet at all, if this gain were negative. The system should only be used when $\sum_i 1/(a_i + 1) \leq 1$. Now every experienced punter knows that this is unlikely to be the case. However, it has been recorded that at the call-over prices on the eve of the 1954 Grand National in Aintree (England) it would have been possible to win 4%, not enough to make a poor man rich, but it can make a rich man richer (if bets of any amount are accepted).

It might be of interest to look now at the bookmaker's position. Which odds does he offer? Suppose the total bets offered on H_i amount to b_i. Then the highest odds he can offer are $a_i = \sum_{j \neq i} b_j/b_i$, hence

$$a_i + 1 = \sum_j b_j/b_i, \quad \text{and} \quad \sum_i 1/(a_i + 1) = 1.$$

He will, of course, offer less, and you should not use the system, unless you want to lose.

13.2.7.   Fixed Points

We have mentioned in section 13.2.3 that the first proof of the Minimax Theorem, by von Neumann (1928), used a fixed point theorem of topology. We do not enter into this field here, but we shall show how the idea of a fixed point of a transformation [V, § 5.1] can be relevant to our topic.

EXAMPLE 13.2.7.   If in Table 13.2.6 of section 13.2.4;

$$\begin{array}{rrr} 1 & -1 & 3 \\ 3 & 5 & -3 \end{array}$$

R chooses his first strategy R_1 with a relative frequency larger than $\frac{2}{3}$, and hence his second strategy R_2 with a relative frequency smaller than $\frac{1}{3}$, then C should choose his second column, i.e. his strategy C_2. If the upper and lower bounds are reversed, then he should choose his third column, C_3. If R chooses R_1 and R_2 with relative frequencies precisely $\frac{2}{3}$ and $\frac{1}{3}$ respectively, then any pure or mixed strategy of C which uses only the second and third column produces the same result. We illustrate these facts as shown in Figure 13.2.3.

For the analogous representation of R's best strategies we have Figure 13.2.4.

Any point within the triangle, or on its boundary, will be the centre of gravity when appropriate weights, adding up to 1, are attached to the three vertices. These weights represent the relative frequencies with which the corresponding columns of C are chosen. Thus every point of the triangle represents one of C's (mixed) strategies.

Figure 13.2.3

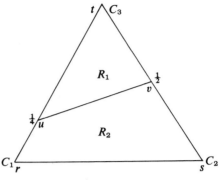

Figure 13.2.4

We have also indicated the best strategy of R, as an answer to any of C's chosen strategies. For instance, if C chooses C_1 with relative frequency larger than $\frac{3}{4}$, and C_3 with relative frequency smaller than $\frac{1}{4}$, but C_2 not at all, (we are then on the straight line segment ru), then R's best strategy would be R_2, his second row. On the other hand, if C chooses a point in the triangle uvt, then R should choose R_1.

We can indicate all this as follows:

$$\text{line } ac \rightarrow s, \qquad \text{line } cb \rightarrow t, \qquad \text{point } c \rightarrow \text{line } st.$$

Also

$$\text{quadrilateral } ruvs \rightarrow b, \qquad \text{triangle } uvt \rightarrow a, \qquad \text{line } uv \rightarrow \text{line } ab.$$

For instance, the best response to a is s, and the best response to s is b. We say a has been *transformed*, via s, into b. Also, the best response to c is anywhere on the line st, and the best response to some point on this line is somewhere on the line ab. Now c is on the line ab, and is transformed via the line st into anywhere on the line ab, for instance into itself. We say it is a *fixed point* of the transformation of the line ab, via the triangle rst, into ab itself. This fixed point defines R's optimal strategy: $(\frac{2}{3}, \frac{1}{3})$.

In the same way we note, for player C, that

$$\text{quadrilateral } ruvs \rightarrow b \rightarrow t, \qquad \text{triangle } uvt \rightarrow a \rightarrow s,$$

$$\text{line } uv \rightarrow \text{line } ab \rightarrow \text{line } st.$$

Now v is on uv and also on st. It is a fixed point of the transformation of the triangle rst, via the line ab, into itself. It indicates C's optimal strategy: $(0, \frac{1}{2}, \frac{1}{2})$.

The crucial point in von Neumann's proof was the demonstration that fixed points always exist in configurations occurring in connection with rectangular games.

13.3.   MORE-PERSON GAMES

Until now we have been dealing only with two-person games, although obviously some of the concepts can be applied to more general games as well. We shall now consider one of these generalizations, viz. the case of more than two players, but to begin with we shall still assume that payments are only made from players to players, i.e. we shall still be dealing with *zero-sum games*. The total of all payments, positive if made, negative if received, is zero.

13.3.1.   Non-cooperative Games

The first type of more-person games to be considered is that of 'non-co-operative' games (Nash (1951)).

If the game is played by two players, then their respective interests are entirely contradictory. This is not necessarily so when there are more than two players, and it is possible for them to cooperate to some extent. It is natural to think that such games will more readily represent real life situations, since in real life we are compelled to cooperate nearly all the time. However, to begin with, we shall assume that no cooperation is possible, or allowed, for legal, or psychological, or perhaps for physical reasons.

We have seen in section 13.2.3 that every rectangular game has a solution, and that such a solution is an equilibrium point. It is this concept which remains useful in the case of more-person games, and we repeat it here, for convenience, and for emphasis:

An *equilibrium point* is a set of strategies, pure or mixed, one for each player, such that no player can improve his pay-off by changing his own strategy, as long as each of the other players keeps his own strategy unchanged.

Nash has proved that every non-cooperative game has an equilibrium point. His method of proof is outside the scope of this chapter, but we give illustrative examples.

EXAMPLE 13.3.1   In the game of 'Odd Man Out' there are three players. Each of them chooses one of the numbers 1 or 2. If all three choose the same number, then no payment is made. Otherwise if one player chooses a number which no other has chosen, the first pays to each of the other two their choice. Table 13.3.1 gives the pay-offs to A, B, and C, in this order.

		C chooses 1		C chooses 2	
		B chooses		B chooses	
		1	2	1	2
A chooses	1	0, 0, 0	1, −2, 1	1, 1, −2	−4, 2, 2
	2	−2, 1, 1	2, 2, −4	2, −4, 2	0, 0, 0

Table 13.3.1

This game has two equilibrium points: each player chooses 1, or each player chooses 2.

We shall give another example later, in connection with non-zero-sum games. We shall then also see that different equilibrium points may lead to different values for the participants. In such a case it is not possible to argue that the equilibrium point is optimal, in the sense in which we could say it for rectangular games. Investigations concerning reasons why any particular equilibrium point should emerge in preference to others are being pursued by psychologists and sociologists, so far without any generally persuasive results.

13.3.2.   Cooperative Games

The structure of the pay-off arrangements may be such that there are good reasons why players should combine, to their mutual advantage, in *coalitions*. This leads to the consideration of cooperative games. (We have here reversed the historical sequence. Cooperative games were already considered by von Neumann and Morgenstern (1944), before Nash (1951) introduced non-cooperative games.) We imagine that some players form a coalition, and play a matrix game against the coalition of the remaining players. The pay-off is then divided between the partners in some pre-arranged manner.

EXAMPLE 13.3.2.   We illustrate this again by the game 'Odd-Man-Out' (cf. Example 13.3.1). In a three-person game, only a coalition of two against the third is conceivable, and there are three such possible coalitions. The pay-off table (Table 13.3.2), showing the pay-off to the lone player, is as follows:

	Choice of the partners			
	1, 1	1, 2	2, 1	2, 2
Choice of 1	0	1	1	−4
the lone				
player 2	−2	2	2	0

Table 13.3.2

The second and the third columns are dominated, and the solution is $(\frac{1}{3}, \frac{2}{3})$ for the lone player, and $(\frac{2}{3}, 0, 0, \frac{1}{3})$ for the partners. The value of the game is $-\frac{4}{3}$, i.e. the coalition wins $\frac{4}{3}$, in the long run. This is more than the pay-off 0 in the non-cooperative game, so that each player is encouraged to form a coalition with one of the other players. But which coalition is formed, depends on psychological, sociological, and other incidentals, and we cannot study these here. There remains also the problem of how the winning partners should divide the joint gain between them, though in this particular instance it will, presumably, be agreed generally that each should get half of it.

The results concerning the value of a cooperative game, having a 'coalition' vs. the 'complementary coalition', are described by the *characteristic function* $v(.)$. This is a function [see § 1.4] whose values depend on the set of players in a coalition, including the case of a single member playing against the others combined and, formally, that of all players together. This last value is 0 in a zero-sum game.

EXAMPLE 13.3.3. For Odd-Man-Out we have the following characteristic function:

$$v(A) = v(B) = v(C) = -\tfrac{4}{3}, \text{ each player playing alone loses } \tfrac{4}{3},$$

$$v(AB) = v(AC) = v(BC) = \tfrac{4}{3}, \text{ a coalition wins } \tfrac{4}{3},$$

$$v(ABC) = 0, \text{ a zero-sum-game.}$$

In general, the values of a characteristic function will depend on which players have formed a coalition, while in this simple case any coalition will fare equally well.

Also, in this simple case we shall call $\tfrac{4}{3}$ the 'value of the game', but in a case where a variety of coalitions is possible, with varying outcomes, we need a more sophisticated definition of what we can reasonably call a solution, and a value of the game.

13.3.3. von Neumann–Morgenstern Solution

The von Neumann–Morgenstern definition is based on the following concepts:

An *imputation* is a sharing out of a total available (zero for zero-sum games) amongst the members of a coalition, such that each player receives at least as much as he would receive if he had remained alone against all others. Formally, if the players are A_i ($i = 1, \ldots, n$), then an imputation is a partition $\alpha = (\alpha_1, \ldots, \alpha_n)$, $\alpha_i \geq v(A_i)$ for all i [see § 3.9].

An imputation α *dominates* an imputation $\beta = (\beta_1, \ldots, \beta_n)$, if there exists a set $S = (i_1, \ldots, i_m)$ such that $\sum_{i \in S} \alpha_i \leq v(S)$, $\sum_{i \in S} \beta_i \leq v(S)$, and $\alpha_i > \beta_i$ for all i in S. It should be noticed that the last inequality is strict, and that m is less than n.

For instance, for Odd-Man-Out, let $\alpha = (0, \tfrac{4}{3}, -\tfrac{4}{3})$, $\beta = (\tfrac{4}{3}, 0, -\tfrac{4}{3})$, and $\gamma = (\tfrac{4}{3}, -\tfrac{4}{3}, 1)$. Then β dominates γ (for $S = (1, 2)$), and γ dominates α (for $S = (1, 3)$), but β does not dominate α: dominance is not transitive. Note also that it is possible that in a pair of imputations each dominates the other (obviously for different sets S).

Having introduced these concepts, we are now ready to define what we mean by a *solution*: it is a set of imputations, such that any imputation not in the set is dominated by one inside the set, and no imputation in the set is dominated by another in the set. In general, there will be more than one such set. We mention also, to avoid any misconception, that an equilibrium point need not be part of any solution set. For instance, in Odd-Man-Out the imputation $(0, 0, 0)$ is dominated by any other imputation, but marks an equilibrium point.

The multiplicity of solution sets does not make this concept inapplicable as a useful model in economics. But the definition has other features which might be considered unsatisfactory. To begin with, it has been proved by W. F. Lucas in an (as yet?) unpublished memorandum that not every game has such a solution set. Also, there may be imputations which appear in more than one of the solution sets, and there exist even games where every imputation is a member of some solution set. Moreover, because two imputations might dominate one another, it is quite possible that an imputation in a solution set is dominated by one outside the set.

It might appear natural to define a solution as a set of imputations, none of which is dominated by any other. However, many games, and in particular zero-sum games have no such imputations at all.

13.3.4.   Another Concept of Solution: the Shapley Value

All this has made it tempting to look for other definitions of the solution of a more-person game. We mention just one of them, the 'Shapley value' (Shapley, 1953). It is, in fact, an imputation, but because it is unique, the portions of the players may be called the *values* of the game for them, individually. These values are defined as being the average values by which a player increases the value of a characteristic function when he joins one or more others to form a coalition. The precise meaning of this will become clear from an example.

EXAMPLE 13.3.4.   Consider a three-person game with characteristic function

$$v(A) = -\tfrac{1}{2}, \qquad v(B) = -\tfrac{1}{3}, \qquad v(C) = -\tfrac{1}{4}$$
$$v(BC) = \tfrac{1}{2}, \qquad v(AC) = \tfrac{1}{3}, \qquad v(AB) = \tfrac{1}{4}$$
$$v(ABC) = 0.$$

The three players can be ordered in six different ways [see § 3.7]:

$$ABC \quad ACB \quad BAC \quad BCA \quad CAB \quad CBA.$$

In the first permutation, A is alone to begin with, $v(A)$ $= -\tfrac{1}{2}$

This is also the case in the second permutation, $v(A)$ $= -\tfrac{1}{2}$

According to the third permutation, A joins B, and $v(AB) - v(B)$ $= \tfrac{7}{12}$

According to the fourth permutation, A joins the coalition of BC, and
$v(ABC) - v(BC)$ $= -\tfrac{1}{2}$

According to the fifth permutation, A joins C, and $v(AC) - v(C)$ $= \tfrac{7}{12}$

According to the sixth permutation, A joins the coalition of BC, and
$v(ABC) - v(BC)$ $= -\tfrac{1}{2}$

Total	$-\tfrac{5}{6}$
Mean	$-\tfrac{5}{36}.$

Similarly, we have

	for B			for C	
$v(AB) - v(A)$	$\frac{3}{4}$		$v(ABC) - v(AB)$	$-\frac{1}{4}$	
$v(ABC) - v(AC)$	$-\frac{1}{3}$		$v(AC) - v(A)$	$\frac{5}{6}$	
$v(B)$	$-\frac{1}{3}$		$v(ABC) - v(AB)$	$-\frac{1}{4}$	
$v(B)$	$-\frac{1}{3}$		$v(BC) - v(B)$	$\frac{5}{6}$	
$v(ABC) - v(AC)$	$-\frac{1}{3}$		$v(C)$	$-\frac{1}{4}$	
$v(BC) - v(C)$	$\frac{3}{4}$		$v(C)$	$-\frac{1}{4}$	

$\frac{1}{6}$, mean $\frac{1}{36}$. $\frac{2}{3}$, mean $\frac{1}{9}$.

The Shapley values are $-\frac{5}{36}$ for A, $\frac{1}{36}$ for B, $\frac{1}{9}$ for C. They add up to zero. It can be shown that in all cases they add up to the value of the coalition of all players together, and hence in a zero-sum game always to zero.

13.4. NON-ZERO-SUM GAMES

It is possible to think of situations where the pay-offs to the players do not add up to zero, but where the 'game' produces, in fact, an increase of total utility. These games will frequently be models for economic activities, where cooperation between potential competitors will be beneficial for both of them.

EXAMPLE 13.4.1. Consider a two-person, non-zero-sum game, with the pay-off table as shown in Table 13.4.1. The first entry in each cell is the pay-off

		C	
		c	d
R	a	50, 75	200, 225
	b	150, 300	100, 150

Table 13.4.1

to R, the second is the pay-off to C. Because this is not a zero-sum game, we must enter two values in each cell, and the second is not necessarily the negative of the first. Such games are called *b-matrix* games, and methods have been devised for computing equilibrium points. One such method is as follows:

A non-zero-sum game can be transformed into a zero-sum game if we add one more player, D. Admittedly, this is a dummy with only one strategy: to wait for the choices of the real players, and then to balance the resulting payments to zero. In the present example he would have to pay

D

125	425
450	250.

Although this is now a rather peculiar zero-sum three-person game, it has nevertheless at least one equilibrium point, like all such games. They are, in this case, (a, d), and (b, c), and also, with mixed strategies, $(\frac{1}{2}a + \frac{1}{2}b, \frac{1}{2}c + \frac{1}{2}d)$. In the latter case the pay-off is 125 to R, and 187·5 to C.

No player gets the same pay-off in all three equilibrium points, and they will therefore prefer different ones. Thus, if somehow the equilibrium point (a, d) had been reached, the players would, considering the total payments, have a joint interest to move to (b, c). However, if then R were expected to be satisfied with 150, he would feel this to be unjust, and the question arises how he should be induced, perhaps by a side-payment, to consent to the move.

The pay-offs which are possible with either pure or mixed strategies are exhibited in Figure 13.4.1, where all points within the shaded area indicate a pair of pay-offs to R (abscissa) and to C (ordinate). In general, such an area is not convex for non-zero-sum games, but its shape can differ from that in the diagram of our example.

EXAMPLE 13.4.2.   In this context we might mention the so-called *Prisoners' Dilemma*: this is a non-cooperative game, called so in view of an anecdote which relates the question of whether two prisoners, held incommunicado, should or should not give evidence damaging to themselves, but more to the other. H has pay-off table as shown in Table 13.4.2.

Here the only equilibrium point is (a, c). Also, if each player considers only

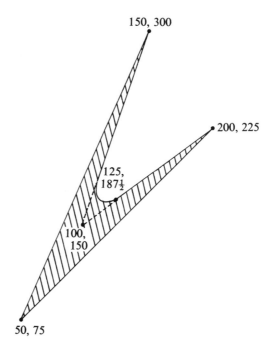

Figure 13.4.1

$$
\begin{array}{c|cc}
 & c & d \\
\hline
a & -1, -1 & 1, -2 \\
b & -2, \ 1 & 0, \ 0 \\
\end{array}
$$

Table 13.4.2

the pay-offs to himself, he would choose a, and c, respectively. However, considered as a cooperative game—against the dummy player—this is the worst they can do. Obviously (b, d) would be preferable.

This fact has led to quite extensive speculation about how the two would, in fact, act. Would they trust each other to be reasonable and to decide b, or d? Or would one of them rather hope that the other will trust him, and will he himself double-cross? Interesting questions, but hardly addressed to a mathematician.

EXAMPLE 13.4.3.   The situation is quite different in the following pay-off table (Table 13.4.3):

$$
\begin{array}{c|cc}
 & c & d \\
\hline
a & -1, -1 & -2, \ \ \ 2 \\
b & 2, -2 & -100, -100 \\
\end{array}
$$

Table 13.4.3

(the game is 'chicken': each player dares the other to stay longest in a dangerous position which, if he stays too long, will kill him).

Here we have two equilibrium points, (a, d) and (b, c), with equal total pay-off, but with different imputations. In terms of the interpretation as 'chicken', this result is trivially obvious.

13.5.   EXTENSIVE GAMES

We have described games where each player makes one single move. If we think now of a game consisting of a succession of moves, then we call it *extensive*. We shall consider here, in particular, games with *perfect information*, i.e. games where each player knows at every stage which moves have been made by himself and by his opponent. This is so, for instance, in chess, but not in card games where a player can discard a card without telling which one it is. It can be shown that a game with perfect information is strictly determined after it has been transformed into an equivalent rectangular game, by normalization (see § 13.1.2). The proof was given by E. Zermelo in the *Transactions of the 5th International Congress of Mathematicians* (Cambridge 1912, vol. II, p. 501).

Although this proves that there exists in chess an optimal pure strategy (and possibly more than one), it is fortunately—for lovers of the game—practically impossible to determine it. Also, we do not know if the result would be a win for White, or for Black, or if it would be a draw.

It can also be shown that a normalized game of perfect information can always be solved by applying the concept of dominance to reduce the pay-off table to a single element. (However, not every strictly determined game can be solved in this manner:

$$
\begin{array}{ccc}
2 & 6 & 1 \\
3 & 2 & 5 \\
3 & 4 & 6
\end{array}
$$

disproves such an assumption.)

EXAMPLE 13.5.1.   Let *A* and *B* play the following game:

First, *A* chooses 4 or 3, and *B*, who knows what *A* has chosen, chooses now 2 or 6. Then a coin is spun and if it falls Head, then *A* wins the amount that *B* has chosen. If the coin falls Tail, then he loses the amount he himself had chosen.

We have here introduced a chance effect, and take as pay-off the expected value of the result.

The tree shown in Figure 13.5.1 gives an overall view of the possible plays. The pay-off table, for the normalized game, is as follows:

B's choice

		if 4, then 2 if 3, then 2	if 4, then 6 if 3, then 2	if 4, then 2 if 3, then 6	if 4, then 6 if 3, then 6
A's choice	4	-1	1	-1	1
	3	$-\frac{1}{2}$	$-\frac{1}{2}$	$\frac{3}{2}$	$\frac{3}{2}$

Table 13.5.1

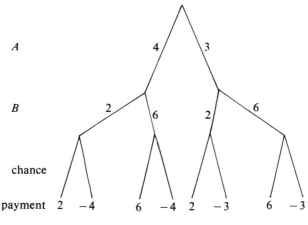

Figure 13.5.1

For instance, B may have decided before the game starts (this is what we mean by normalizing) that if A chooses 4, he will choose 6, but if A chooses 3, then he will choose 2 (the second of his strategies of the table above).

Then, if A chooses 4 (the first strategy of A in the table), and hence B chooses 6, the expected pay-off will be $6 \times \frac{1}{2} + (-4) \times \frac{1}{2} = 1$. On the other hand, if A chooses 3, (the second of his strategies), and hence B chooses 2, then the expected pay-off will be $2 \times \frac{1}{2} + (-3) \times \frac{1}{2} = -\frac{1}{2}$. These are the entries in the second column of the pay-off table.

Omission of dominated columns reduces the table to

$$-1$$
$$-\tfrac{1}{2}$$

and omitting the dominated first row gives $-\frac{1}{2}$ as the value of the game. The cell in the second row and first column is a saddle point.

EXAMPLE 13.5.2.   On the network of roads shown in Figure 13.5.2 the numbers indicate the time it takes, in hours, to pass through a link. A traveller wants to get to T from S as quickly as possible. One of the intermediate links has been blocked, and he will notice it if he gets to it. If he still uses it, it delays him by 2 hours.

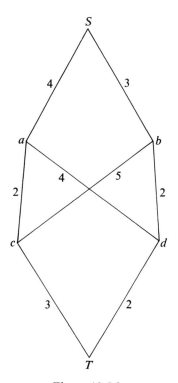

Figure 13.5.2

In which direction should he start off? Which link should a potential adversary block, who wishes to delay the traveller?

The pay-off table is shown in Table 13.5.2. (The traveller is the minimizing player.)

Use link

		ac	ad	bc	bd	row minima
	ac	11	10	11	7	7
	ad	9	12	11	7	7
Block link	bc	9	10	13	7	7
	bd	9	10	11	9	9
column maxima		11	12	13	9	

Table 13.5.2

There is a saddle point at (bd, bd), which means that the traveller should move first to b, and then to d, even if that link is blocked. The travelling time is at most 9.

As in Example 13.5.1 we could find the saddle point by eliminating, step by step, dominated strategies.

There exists a developed theory for extensive games without perfect information, but we do not describe it here.

The concept of dominance is also applicable to infinite games, to which we now turn.

13.6. INFINITE GAMES

Until now we have been dealing with finite games. We turn now to games where at least one player has an infinite set of pure strategies at his disposal. These strategies might be continuous, for instance they might consist of choosing some point in time, or enumerable, for instance choosing a positive number. Mixtures of these two types are also possible.

We define a *solution* again as a set of strategies, one for each player, which form an equilibrium point.

EXAMPLE 13.6.1.   To show that the idea of an equilibrium point is an obvious one in conflicts of interest, and to introduce a homely example, we mention the story of two children who want to divide a cake 'fairly'. They agree that one of them should divide the cake into two portions, and the other will choose the portion he wants. The first child divides into parts x and $1 - x$. The second child has the dominating strategy of taking the larger portion, if there is one, and otherwise any of the two. The first child makes $x = \frac{1}{2}$, and will know that he could not have done better, even had he known the other's strategy (which he can of course guess).

We do not have for infinite games a comprehensive and elegant theory such as we showed to exist for rectangular games. However, some facts concerning solutions are known.

13.6.1. Pure Strategies

A *pure strategy* is defined as a right-hand continuous distribution function $F(x)$ [see II, § 10.3.1], and a game *on the unit square* is one where $F(0) = 0$, $F(1) = 1$. It can then be shown that for two players, with pure strategies $F(x)$ and $G(y)$,

$$\max_F \min_G \int_0^1 \int_0^1 M(x, y)dF(x)dG(y) = \min_G \max_F \int_0^1 \int_0^1 M(x, y)dF(x)dG(y) \quad (13.6.1)$$

where $M(x, y)$ is a continuous function of x and y [see IV, § 5.3].

If we interpret $M(x, y)$ to mean the pay-off from the minimizing player (choosing $G(y)$) to the maximizing player (choosing $F(x)$), then we have here a Minimax Theorem for a special type of infinite games. We illustrate this with an example

EXAMPLE 13.6.2. If $M(x, y) = |x - y|$, then we can guess that the best strategy for the maximizing player is to choose $F(x)$ as a step function [see IV, § 2.3] with step $\frac{1}{2}$ at 0, and step $\frac{1}{2}$ at 1 (note that this is a pure strategy), and for the other player to choose $G(y)$ with a single step of height 1, at $\frac{1}{2}$. It is easily confirmed that these two strategies form an equilibrium point.

This procedure, guessing and then confirming the solution, is in many cases the most efficient method of obtaining a solution. Nevertheless, theory can sometimes offer some guidance:

If the pay-off function $M(x, y)$ is strictly convex in y for all x [see V, Definition 4.3.1], then the minimizing player has a unique optimal strategy with a finite number of steps; if $M(x, y)$ is strictly concave in x for all y, then the maximizing player has such an optimal strategy. (Luce and Raffa (1957)).

If $M(x, y)$ has a saddle point [see IV, Figure 5.6.4] with the right orientation to have both these conditions satisfied, then the optimal strategy for both players is to aim at that saddle point.

EXAMPLE 13.6.3. Let

$$M = (y - 0.8)^2 - (x - 0.6)^2 + 2(x - 0.6)(y - 0.8) + 10.$$

This quadratic surface has a saddle point at $(0.6, 0.8)$. If the maximizing player chooses 0.6, then $M = (y - 0.8)^2 + 10$ is minimized by $y = 0.8$. Conversely, when $y = 0.8$, then $M = -(x - 0.6)^2 + 10$ is maximized by $x = 0.6$. The value of the game is 10.

On the other hand, if y were 0.5, say, then $x = 0.3$ would be best, and the pay-off would be increased, to the minimizing player's disadvantage, to 10.18.

We add another example, with continuous, but not strictly convex or concave pay-off. This is an example of Colonel Blotto games, after a character from Caliban's Weekend Problems book.

EXAMPLE 13.6.4.   Blotto wants to cross a mountain range with his un-divided force, and he can choose one of n roads across. The Enemy can divide his force, putting a proportion of y_i on road i ($= 1, \ldots, n$). Any of these partitions is one of his infinitely many pure strategies, while choosing one of them by chance is a mixed strategy.

If Blotto chooses road i, then the pay-off to him is $p(y_i)$. This assumes that the enemy forces on other roads than the one which Blotto uses are irrelevant.

We make two alternate assumptions about the form of $p(y_i)$. First we assume that

$$\sum_i p(y_i)/n \geq p\left(\sum_i y_i/n\right) = p(1/n).$$

This is the case, for instance, if the function $p(y_i)$ is convex [see V, § 4.3.1]. The solution consists then of

Blotto's mixed strategy: use each road with equal frequency
The Enemy's pure strategy: $y_1 = y_2 = \ldots = y_n = 1/n$.

Proof.   Given Blotto's strategy, he obtains, if the Enemy uses $(y_1, \ldots, y_n)$, $\sum_i p(y_i)/n$, and in particular $p(1/n)$, if $y_i = 1/n$. Conversely, given the Enemy's strategy $y_i = 1/n$, the pay-off to Blotto is $p(1/n)$, and is independent of i. Hence $p(1/n)$ is the value of the game.

Our second assumption is

$$y_i p(1) + (1 - y_i)p(0) \leq p(y_i) \quad \text{for all } y_i.$$

This is the case, for instance, if $p(y_i)$ is concave [again, see V, § 4.3.1].

Now the solution consists of

Blotto's mixed strategy: use each road with equal probability
The Enemy's mixed strategy: choose just one road, each with equal prob-
ability (i.e. choose $(0, \ldots, 1, \ldots, 0)$, the 1 being in the ith position with relative frequency $1/n$).

Proof.   Given Blotto's strategy, he obtains, if the Enemy uses road i with probability z_i,

$$\sum_i [z_i p(1) + (1 - z_i)p(0)]/n = p(1)/n + (1 - 1/n)p(0),$$

which is independent of i.

Conversely, if the enemy uses relative frequencies z_i for road i, then the pay-off to Blotto is

$$\sum_i p(z_i)/n \quad \text{which is larger than}$$

$$\sum_i [z_i p(1) + (1 - z_i)p(0)]/n = p(1)/n + (1 - 1/n)p(0)$$

unless one of the z_i equals one, and all others are zero.

A game with pay-off of the form

$$\sum_s \sum_t a_{st} u_s(x) v_t(y)$$

on the unit square $0 \le x \le 1, 0 \le y \le 1$ is called *separable*. Such games have optimal strategies composed of a finite number of pure strategies. In fact, rectangular games are a special case of separable games, and we can equate pure strategies of the latter to mixed strategies of the former.

EXAMPLE 13.6.5. There are infinite games with no solution (equilibrium point). We mention here one on the unit square, but whose pay-off is not continuous. No strategies make the minimax equal to the maximin. For the proof, see Dresher (1961), p. 115. Its pay-off is defined as

$$0 \text{ when } x = y,$$
$$-1 \text{ when } x = 1, y < 1, \text{ or when } x < y < 1$$
$$1 \text{ when } y = 1, x < 1, \text{ or when } y < x < 1$$

represented in Figure 13.6.1.

13.6.2. Duels

Another group of games extensively studied are games of timing. Dresher (1961) analyses such games, called *duels*, where each player has reason to delay his action as long as possible, but might lose by delaying too long.

Two duellists, each with one bullet, approach one another. $P(x)$ is the probability of hitting the opponent when fired at a distance x [see II, § 10.3.2], and this probability increases monotonically with decreasing x [see IV, Definition 2.7.1]. The duellist who hits his opponent puts him out of action. We assume, for the sake of simplicity, that the accuracy of hitting is the same for both.

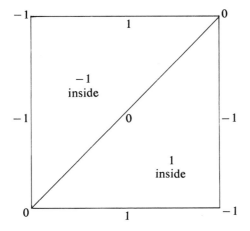

Figure 13.6.1

We have to consider two cases: (i) the 'noisy duel', when each knows if the other has fired, with or without a hit, and (ii) the 'silent duel', when neither knows if the opponent has fired without hitting. The point of this distinction is that if A knows that B has fired, but missed, then he can wait until he is near enough to be sure not to miss as well. Our model assumes that, in any case, they carry on approaching one another.

Let 1 be the pay-off for the winning duellist, and 0 if none or both are put out of action.

It turns out that this is a game with a minimax-maximin solution, but the solutions are different in the two cases. We quote them, to give an idea of what sort of solutions can occur: In the noisy duel, both should fire when their distance x satisfies $P(x) = \frac{1}{2}$. This is, for both of them, a pure strategy and the value of the game is, of course, 0.

In the silent duel the solution is more complicated, as we would expect. The optimal strategies are mixed. They should not fire before $P(x) = \frac{1}{3}$, and afterwards, they should fire when their distance is x, with probability density $\frac{1}{4}x^3$, as long as they are active.

13.6.3. Statistical Decision Procedures

The following considerations form part of a theory initiated by A. Wald (1950). We shall give an abbreviated and simplified account of it.

We consider the decision procedures of a statistician to be a game against nature N. The statistician must make decisions, not knowing the actual strategies of N, i.e. the probability distributions of relevant parameters which affect the consequences of his decisions. However, let us assume that he does know that they belong to a given set of distribution functions $F(x_1, \ldots, x_n)$, i.e. the probabilities that $X_i \le x_i$ for the random variables X_i ($i = 1, \ldots, n$) [see II, § 10.3.1].

The pay-off is measured by the importance of making the correct guess about which distribution function holds, i.e. 'being the strategy of N'. It equals some given function $W(F, d)$ which indicates the expected loss to the statistician if he makes the decision d, while the true distribution function is F. We illustrate this with the following example.

EXAMPLE 13.6.6.   Assume that the statistician has to decide, on the basis of 9 independent observations [see II, Definition 3.5.1], whether the mean of a normal population with variance 9 is or is not -1 [see II, § 11.4.3]. The mean has, consequently, variance 1 [cf. II, § 8.1 and § 9.2.1].

Let there be three possibilities: the mean could be -1, 0.5, or $+1$. Let us also assume that the statistician has two pure strategies: he can accept that the mean equals -1 if and only if the mean of the observations is larger than 0.5, or he can accept that the mean is -1 if and only if the mean of the observations is less than -0.5. The pay-off is $-p$ if p is the probability of rejecting -1 incorrectly, and 0 if that rejection is correct, i.e. if the mean of the population is, in fact, 0.5 or 1.

Thus, if the mean is actually -1, then we shall reject it, incorrectly, by the first mentioned strategy if the average of the observations is larger than 0.5, and the probability of this is [cf. II, (11.4.5)]

$$\frac{1}{\sqrt{2\pi}} \int_{0.5}^{\infty} \exp\left[-(x+1)^2/2\right]dx = \frac{1}{\sqrt{2\pi}} \int_{1.5}^{\infty} \exp\left[-x^2/2\right]dx.$$

Similarly, if the population mean is 0.5, and we use, say, the second strategy, then we accept that the mean is -1, incorrectly, with probability

$$\frac{1}{\sqrt{2\pi}} \int_{-\infty}^{-0.5} \exp\left[-(x-0.5)^2/2\right]dx = \frac{1}{\sqrt{2\pi}} \int_{-\infty}^{0} \exp\left[-x^2/2\right]dx.$$

The complete pay-off table will be

		Actual value of the population mean		
		-1	0.5	1
Strategy: reject if the	< 0.5	$-(1.5, \infty)$	$-(-\infty, 0)$	$-(-\infty, -0.5)$
mean of observations is	< -0.5	$-(0.5, \infty)$	$-(-\infty, -1)$	$-(-\infty, -1.5)$

Table 13.6.1

where (a, b) stands for

$$\frac{1}{\sqrt{2\pi}} \int_{a}^{b} \exp\left[-x^2/2\right]dx.$$

From tables of the normal distribution we obtain:

$$-0.07 \quad -0.50 \quad -0.31$$
$$-0.31 \quad -0.16 \quad -0.07.$$

The third column is dominated by the second so that the statistician's optimal strategy emerges from

$$-0.07 \quad -0.50$$
$$-0.31 \quad -0.16$$

as $(\frac{15}{58}, \frac{43}{58})$, and the average long term loss is approximately 0.25.

Nature's optimal mixed strategy is $(\frac{34}{58}, \frac{24}{58})$; we interpret this as the 'most unfavourable' distribution which the statistician has to expect.

More about this will be found in Blackwell and Girshick (1954).

S.V.

REFERENCES

Blackwell, D. and Girshick, M. A. (1954). *Theory of Games and Statistical Decisions*, John Wiley and Sons.

Dresher, M. (1961). *Games of Strategy. Theory and Applications*, Prentice Hall.

Kuhn, H. W. and Tucker, A. W. (eds.) (1953). *Contributions to the Theory of Games*, vol. II, Princeton Univ. Press.

Luce, R. D. and Raiffa, H. (1957). *Games and Decisions*, John Wiley and Sons.

Nash, J. (1951). Non-cooperative games, *Ann. of Math.* **54**, 286–295.

von Neumann, J. (1928). Zur Theorie der Gesellschaftsspiele, *Math. Ann.* **100**, 295–320.

von Neumann, J. and Morgenstern, O. (1944, 2nd ed. 1947). *Theory of Games and Economic Behaviour*, Princeton Univ. Press.

Raiffa, H. (1953). Arbitration schemes for generalized two-person games, in Kuhn and Tucker (eds.) 361–387.

Shapley, S. L. (1953). The value of *N*-person games, in Kuhn and Tucker (eds.) 307–317.

Vajda, S. (1967). *The Theory of Games and Linear Programming*, Methuen and Co.

Vajda, S. (1975). *Problems in Linear and Non-linear Programming*, Charles Griffin and Co.

Wald, A. (1950). *Statistical Decision Functions*, John Wiley and Sons.

CHAPTER 14

Polynomials and Rational Functions

14.1. POLYNOMIALS ON A SINGLE VARIABLE

A polynomial in a *variable* or *indeterminate* x is an expression of the form

$$p(x) = c_0 + c_1 x + c_2 x^2 + \ldots + c_n x^n, \tag{14.1.1}$$

where the *coefficients* $c_0, c_1, \ldots, c_n$ are given numbers or, more generally, elements of a specified field [see § 2.4.2] or ring [see § 2.3].

When the coefficients are real [see § 2.6.1], we say that $p(x)$ is a *real polynomial*. The set of all real polynomials is denoted by $\mathbb{R}[x]$ and we write $p(x) \in \mathbb{R}[x]$. Similarly, the symbols $\mathbb{Q}[x]$, $\mathbb{Z}[x]$ and $\mathbb{C}[x]$ describes the set of all polynomials whose coefficients are rational numbers, integers and complex numbers, respectively. We have the inclusions

$$\mathbb{Z}[x] \subset \mathbb{Q}[x] \subset \mathbb{R}[x] \subset \mathbb{C}[x]. \tag{14.1.2}$$

EXAMPLE 14.1.1. If $f(x) = \pi x^2 - (\sqrt{2})x + e^2$, then $f(x) \in \mathbb{R}[x]$. It would be correct, but less informative, to say that $f(x)$ belongs to $\mathbb{C}[x]$.

EXAMPLE 14.1.2. $g(x) = 5x^3 + 2x - 7 \in \mathbb{Z}[x]$.

EXAMPLE 14.1.3. $h(x) = ix^2 + 1 \in \mathbb{C}[x]$.

In the great majority of cases we are concerned with polynomials that lie in one of the sets mentioned in (14.1.2). For an arbitrary field $\mathbb{F}$ we denote by $\mathbb{F}[x]$ the set of polynomials with coefficients in $\mathbb{F}$.

When we wish to indicate that the coefficients of $p(x)$ lie in a domain D, we write $p(x) \in D[x]$. It is, however, essential to assume that D is an integral domain [see § 2.3], so that zero divisors are excluded.

The highest power of x that occurs in $p(x)$ with a non-zero coefficient is called the *degree* of $p(x)$, abbreviated $\deg p(x)$. Thus, if $c_n \neq 0$ in (14.1.1) we have that

$$\deg p(x) = n.$$

A polynomial of degree zero does not involve x and reduces to a non-zero constant. If all coefficients are zero, the polynomial is called the *zero polynomial*;

it is denoted by 0, although this symbol is also used for the number zero. The above definition does not assign a degree to the zero polynomial, but it is convenient to put

$$\deg 0 = -\infty,$$

and to declare that, for any real number k,

$$(-\infty) + k = (-\infty),$$
$$(-\infty) + (-\infty) = (-\infty).$$
(14.1.3)

It is worthwhile having a short name for a rather special, but important, type of polynomial:

DEFINITION 14.1.1. A polynomial whose highest or *leading* coefficient is equal to unity, that is

$$p(x) = c_0 + c_1 x + \ldots + c_{n-1} x^{n-1} + x^n,$$

is said to be *monic*.

Polynomials of degrees unity, two or three are called *linear*, *quadratic* or *cubic* respectively. When the coefficients lie in a field, so that division by constants is possible, then every non-zero polynomial $f(x)$ can be written in the form

$$f(x) = c_n f_0(x),$$
(14.1.4)

where $c_n \neq 0$ and $f_0(x)$ is a monic polynomial.

EXAMPLE 14.1.4. In $\mathbb{Q}[x]$ we can write

$$5x^3 + 2x - 7 = 5(x^3 + \tfrac{2}{5}x - \tfrac{7}{5}).$$

But this equation is not available in $\mathbb{Z}[x]$, because the coefficients $\tfrac{2}{5}$ and $-\tfrac{7}{5}$ do not lie in $\mathbb{Z}$.

Throughout this chapter we assume that the domain D of coefficients consists of numbers, that is $D \subseteq \mathbb{C}$. Then if α is any number, we may substitute α for x and obtain the *value* of f at α, namely

$$f(\alpha) = c_0 + c_1\alpha + c_2\alpha^2 + \ldots + c_n\alpha^n.$$
(14.1.5)

Finally, it is appropriate to draw attention to the difference between the algebraic notion of a polynomial and the function, in the sense of analysis, which this polynomial describes. Thus in algebra we stipulate that two polynomials are *equal* if and only if they have the same coefficients. However, in analysis we say that the functions $f(x)$ and $g(x)$ are equal if and only if

$$f(\alpha) = g(\alpha)$$

for all admissible values α of x. When the coefficients are numbers, as we shall always assume, then the two definitions of equality coincide. But this need no longer be true when the coefficients are taken from a finite field [see § 4.3.3].

14.2. ALGEBRA OF POLYNOMIALS

Two polynomials can be added and subtracted in the obvious way. Thus if

$$f(x) = c_0 + c_1 x + \ldots + c_m x^m \quad (c_m \neq 0)$$

and

$$g(x) = d_0 + d_1 x + \ldots + d_n x^n \quad (d_n \neq 0)$$

are polynomials of degrees m and n respectively, then

$$f(x) \pm g(x) = (c_0 \pm d_0) + (c_1 \pm d_1)x + (c_2 \pm d_2)x^2 + \ldots$$

and

$$\deg (f(x) \pm g(x)) \leq \max (m, n).$$

When $m \neq n$, the equality holds; but, if $m = n$, the coefficients c_m, d_m may cancel to give a polynomial of degree less than m.

Multiplication of polynomials is also natural; each term of $f(x)$ is multiplied by each term of $g(x)$, and the terms involving the same power of x are collected together. So, for example,

$$(2 + x - x^2)(1 - 3x + 2x^2)$$
$$= (2 - 6x + 4x^2) + (x - 3x^2 + 2x^3) + (-x^2 + 3x^3 - 2x^4)$$
$$= 2 - 5x + 5x^3 - 2x^4.$$

The degrees satisfy

$$\deg (f(x)g(x)) = \deg f(x) + \deg g(x). \tag{14.2.1}$$

For if $f(x)$ and $g(x)$ are both non-zero, this is true because the leading coefficient of $f(x)g(x)$ is equal to $c_m d_n$ and is therefore non-zero; if either is the zero polynomial, then (14.2.1) holds in virtue of the conventional addition rules for $-\infty$ given in (14.1.3).

If D is an integral domain [see § 2.3], it can be verified that the set of polynomials $D[x]$ forms a ring with respect to the above rules for addition, subtraction and multiplication.

14.3. THE DIVISION ALGORITHM

Polynomials in a single variable have many properties that are analogous to those of integers. It is well known that if m and n are positive integers, then n can be uniquely expressed as

$$n = qm + r,$$

where q and r are non-negative integers and $0 \leq r < m$ [see (4.1.10)]. There is a strikingly similar result for polynomials:

THEOREM 14.3.1 (*Division algorithm for polynomials*). *Let* $\mathbb{F}$ *be a field and let* $f(x)$ *and* $g(x)$ *be polynomials in* $\mathbb{F}(x)$, *and suppose that* $f(x) \neq 0$. *Then there exists a unique pair of polynomials* $q(x)$ *and* $r(x)$ *in* $\mathbb{F}[x]$, *such that*

$$g(x) = q(x)f(x) + r(x) \tag{14.3.1}$$

and

$$\deg r(x) < \deg f(x). \tag{14.3.2}$$

The polynomials $q(x)$ *and* $r(x)$ *are called the* quotient polynomial *and* remainder polynomial *respectively.*

When $\deg f(x) > \deg g(x)$, the equation is trivially satisfied if we put $q(x) = 0$ and $r(x) = g(x)$. Henceforth, we shall assume that $\deg f(x) \leq \deg g(x)$. The method of obtaining $q(x)$ and $r(x)$ is best explained by working through a typical case.

EXAMPLE 14.3.1. Let

$$g(x) = 2x^4 + 3x^2 - 5x - 2$$

and

$$f(x) = 3x^2 + x - 5.$$

The idea is to remove the leading term of $g(x)$, that is $2x^4$, by subtracting or adding to $g(x)$ a polynomial of the form $cx^k f(x)$. The first step in the reduction is given by

$$g(x) - \tfrac{2}{3}x^2 f(x) = (2x^4 + 3x^2 - 5x - 2) - \tfrac{2}{3}x^2(3x^2 + x - 5)$$
$$= -\tfrac{2}{3}x^3 + \tfrac{19}{3}x^2 - 5x - 2.$$

Next we remove the leading term $-\tfrac{2}{3}x^3$ by constructing the polynomial

$$g(x) - \tfrac{2}{3}x^2 f(x) + \tfrac{2}{9}xf(x) = \tfrac{59}{9}x^2 - \tfrac{55}{9}x - 2.$$

Finally, the leading term $\tfrac{59}{9}x^2$ is deleted by forming

$$g(x) - \tfrac{2}{3}x^2 f(x) + \tfrac{2}{9}xf(x) - \tfrac{59}{27}f(x) = -\tfrac{224}{27}x + \tfrac{241}{27}.$$

The process now terminates since we have reached a polynomial whose degree is less than the degree of $f(x)$. Hence we write the last equation as

$$g(x) - q(x)f(x) = r(x),$$

where

$$q(x) = \tfrac{2}{3}x^2 - \tfrac{2}{9}x + \tfrac{59}{27},$$
$$r(x) = -\tfrac{224}{27}x + \tfrac{241}{27},$$

as required in (14.3.1).

The complete division algorithm is set out in the scheme below:

$$\tfrac{2}{3}x^2 - \tfrac{2}{9}x + \tfrac{59}{27}$$

$$3x^2 + x - 5\overline{)2x^4 + 0x^3 + 3x^2 - 5x - 2}$$
$$2x^4 + \tfrac{2}{3}x^3 - \tfrac{10}{3}x^2$$

$$-\tfrac{2}{3}x^3 + \tfrac{19}{3}x^2 - 5x - 2$$
$$-\tfrac{2}{3}x^3 - \tfrac{2}{9}x^2 + \tfrac{10}{9}x$$

$$\tfrac{59}{9}x^2 - \tfrac{55}{9}x - 2$$
$$\tfrac{59}{9}x^2 + \tfrac{59}{27}x - \tfrac{295}{27}$$

$$-\tfrac{224}{27}x + \tfrac{241}{27}$$

The quotient $q(x)$ appears at the top as $\tfrac{2}{3}x^2 - \tfrac{2}{9}x + \tfrac{59}{27}$ and $r(x)$ at the bottom as $-\tfrac{224}{27}x + \tfrac{241}{27}$.

A consequence of the division algorithm is the following important result:

THEOREM 14.3.2 (*Remainder theorem*). *If $x - k$ is divided into the polynomial $g(x)$, the remainder is the constant $g(k)$, thus*

$$g(x) = q(x)(x - k) + g(k). \tag{14.3.3}$$

EXAMPLE 14.3.2. Let $g(x) = 2x^4 + 3x^2 - 5x - 2$ and $k = -1$. Then $g(-1) = 8$. Hence there exists a polynomial $q(x)$ such that

$$g(x) = q(x)(x + 1) + 8.$$

In fact,

$$q(x) = 2x^3 - 2x^2 + 5x - 10.$$

If the remainder $r(x)$ in (14.3.1) is equal to zero we say that $f(x)$ *divides* $g(x)$ and we write

$$f(x) \mid g(x). \tag{14.3.4}$$

Thus (14.3.4) means that there exists a polynomial $q(x)$ such that

$$g(x) = q(x)f(x).$$

For the rest of this section we consider polynomials whose coefficients lie in a given field $\mathbb{F}$. It is one of the most remarkable consequences of the division algorithm that any two polynomials $f(x)$ and $g(x)$ in $\mathbb{F}[x]$, which are not both zero, possess a unique *greatest common division* (G.C.D.), also known as *highest common factor* (H.C.F.). This fact is described more fully in the following

THEOREM 14.3.3. *If $\mathbb{F}$ is a field, any two polynomials $f(x)$ and $g(x)$ in $\mathbb{F}[x]$, which are not both zero, possess a unique highest common factor $d(x)$, written*

$$d(x) = (f(x), g(x)). \tag{14.3.5}$$

It has the following properties:

    (i) $d(x)$ *is monic,*

    (ii) $d(x)|f(x)$ *and* $d(x)|g(x)$,

    (iii) *if* $t(x)|f(x)$ *and* $t(x)|g(x)$, *then* $t(x)|d(x)$,

    (iv) $d(x)$ *is the monic polynomial of lowest degree that can be expressed in the form*

$$d(x) = a(x)f(x) + b(x)g(x), \qquad (14.3.6)$$

where $a(x)$ *and* $b(x)$ *are suitable polynomials in* $F[x]$.

The following observations are obvious, always on the understanding that the domain of coefficients is a field.

1. $(f(x), g(x)) = (g(x), f(x))$.

2. If $f(x) = a_0x^n + a_1x^{n-1} + \ldots + a_n \ (a_0 \neq 0)$, then

$$(f(x), 0) = a_0{}^{-1}f(x), \qquad (14.3.7)$$

this being the unique monic polynomial which is proportional to $f(x)$.

3. If k is a non-zero constant,

$$(kf(x), g(x)) = (f(x), g(x)). \qquad (14.3.8)$$

4. If $f(x)$ and $g(x)$ have no factor in common except non-zero constants, then

$$(f(x), g(x)) = 1 \qquad (14.3.9)$$

and (14.3.6) reduces to

$$1 = a(x)f(x) + b(x)g(x). \qquad (14.3.10)$$

Conversely, if $f(x)$ and $g(x)$ satisfy (14.3.10) for suitable $a(x)$ and $b(x)$, then (14.3.9) must hold. Two polynomials $f(x)$ and $g(x)$ for which (14.3.10) or (14.3.9) is true, are said to be *relatively prime*.

EXAMPLE 14.3.3.   If $f(x) = x^4 + x - 1$ and $g(x) = x^5 + x^2 - x + 2$, then $(f(x), g(x)) = 1$ because

$$1 = \tfrac{1}{2}g(x) - \tfrac{1}{2}xf(x).$$

We shall not prove Theorem 14.3.3 in detail. Instead we shall describe a systematic method, known as *Euclid's Algorithm*, for computing the highest common factor of any pair of polynomials. The procedure is based on the following idea: if $\deg g(x) \geq \deg f(x)$, let

$$g(x) = q(x)f(x) + r(x), \quad \deg r(x) < \deg f(x),$$

as in (14.3.1). Then it is easy to show that

$$d(x) = (f(x), g(x)) = (f(x), r(x)).$$

If $r(x) \neq 0$ we can repeat the division algorithm and write

$$f(x) = q_1(x)r(x) + r_1(x), \quad \deg r_1(x) < \deg r(x)$$

and we obtain that

$$d(x) = (r_1(x), r(x)).$$

Thus at each stage we diminish the degree of one or the other of the polynomials without altering the highest common factor. The process terminates either if one of the polynomials becomes zero, in which case we apply (14.3.7), or else if one of the polynomials is reduced to a non-zero whence constant $d(x) = 1$.

EXAMPLE 14.3.4. We use the Euclidean algorithm to show that the highest common factor of $g(x) = x^4 - 1$ and $f(x) = x^3 - 3x + 2$ is equal to $x - 1$.
First we divide $g(x)$ by $f(x)$, thus

$$g(x) = xf(x) + r(x), \qquad r(x) = 3x^2 - 2x - 1.$$

Next we divide $f(x)$ by $r(x)$, and we obtain that

$$f(x) = (\tfrac{1}{3}x + \tfrac{2}{9})r(x) + r_1(x), \qquad r_1(x) = -\tfrac{20}{9}(x - 1).$$

Finally, division of $r(x)$ by $r_1(x)$ yields

$$r(x) = (-\tfrac{27}{20}x - \tfrac{9}{20})r_1(x)$$

with a zero remainder. Hence $d(x) = -\tfrac{9}{20}r_1(x)$.
In the general case, Euclid's algorithm may be set out as follows: assume that $\deg f(x) \leq \deg g(x)$ and that the process terminates at the hth step. We then have the following chain of division formulae:

$$
\begin{aligned}
g(x) &= q(x)f(x) + r(x), & \deg r(x) &< \deg f(x) \\
f(x) &= q_1(x)r(x) + r_1(x), & \deg r_1(x) &< \deg r(x) \\
r(x) &= q_2(x)r_1(x) + r_2(x), & \deg r_2(x) &< \deg r_1(x)
\end{aligned}
$$
$$\cdots \qquad\qquad\qquad \cdots$$
$$
\begin{aligned}
r_{h-2}(x) &= q_h(x)r_{h-1}(x) + r_h(x), & \deg r_h(x) &< \deg x_{h-1}(x) \\
r_{h-1}(x) &= q_{h+1}(x)r_h(x)
\end{aligned}
$$

(14.3.11)

At the final stage the remainder would be the zero polynomial, which we have not written explicitly.
Generally, we have the rule that the last non-zero remainder in Euclid's algorithm is proportional to the highest common factor, that is

$$d(x) = kr_h(x), \tag{14.3.12}$$

where k is chosen in such a way that $d(x)$ is rendered monic.
Euclid's algorithm may also be used to compute polynomials $a(x)$ and $b(x)$

which occur in (14.3.6). To this end we write the penultimate equation of (14.3.11) in the form

$$r_h(x) = -q_h(x)r_{h-1}(x) + r_{h-2}(x) = a_1(x)r_{h-1}(x) + b_1(x)r_{h-2}(x),$$

say. The preceding equation in the scheme (14.3.11) can be recast as

$$r_{h-1}(x) = -q_{h-1}(x)r_{h-2}(x) + r_{h-3}(x)$$

and may therefore serve to eliminate $r_{h-1}(x)$ from the above expression for $r_h(x)$, thus

$$r_h(x) = a_1(x)(-q_{h-1}(x)r_{h-2}(x) + r_{h-3}(x)) + b_1(x)r_{h-2}(x)$$
$$= a_2(x)r_{h-2}(x) + b_2(x)r_{h-3}(x),$$

say. On eliminating the remainders $r_{h-1}(x), r_{h-2}(x), \ldots, r(x)$ in turn we arrive at an equation of the form

$$r_h(x) = a_{h+1}(x)f(x) + b_{h+1}g(x).$$

Finally, we refer to (14.3.12): after multiplying the last equation by k we obtain the equation (14.3.6).

EXAMPLE 14.3.5.   As in Example 14.3.4, let $f(x) = x^3 - 3x + 2$ and $g(x) = x^4 - 1$. Using the same notation as before we have that

$$r_1(x) = f(x) - (\tfrac{1}{3}x + \tfrac{2}{9})r(x)$$
$$= f(x) - (\tfrac{1}{3}x + \tfrac{2}{9})(g(x) - xf(x))$$
$$= (\tfrac{1}{3}x^2 + \tfrac{2}{9}x + 1)f(x) - (\tfrac{1}{3}x + \tfrac{2}{9})g(x).$$

Since $d(x) = -\tfrac{9}{20}r_1(x) = x - 1$ we finally obtain that

$$x - 1 = (-\tfrac{3}{20}x^2 - \tfrac{1}{10}x - \tfrac{9}{20})f(x) + (\tfrac{3}{20}x + \tfrac{1}{10})g(x).$$

Incidentally, the polynomials $a(x)$ and $b(x)$ in (14.3.6) are not uniquely determined; for if we put

$$A(x) = a(x) - p(x)g(x)$$
$$B(x) = b(x) + p(x)f(x),$$

where $p(x)$ is an arbitrary polynomial, then

$$A(x)f(x) + B(x)g(x) = a(x)f(x) + b(x)g(x) = d(x).$$

14.4.   FACTORIZATION

In this section we discuss the way in which a polynomial can be split into 'simpler' polynomials. The idea is analogous to the decomposition of a positive integer, greater than unity, into a product of primes [see the Fundamental Theorem of Arithmetic in section 4.2]. However, when we are dealing with polynomials, it is essential to state what domain of coefficients is involved.

DEFINITION 14.4.1.   The polynomial $f(x)$ in $L[x]$ is said to be *reducible* in $L[x]$ (or over L) if there exist polynomials $g(x)$ and $h(x)$ in $L[x]$, both of positive degree, such that

$$f(x) = g(x)h(x). \tag{14.4.1}$$

The non-constant polynomial $f(x)$ in $L[x]$ is said to be *irreducible* in $L[x]$ (or over L) if every equation of the form (14.4.1) implies that either $g(x)$ or $h(x)$ is a constant.

Notice that a constant polynomial is not regarded as an irreducible polynomial just as the integer unity is not considered to be a prime number.

EXAMPLE 14.4.1.   The polynomial $x^4 + 1$ is irreducible over $\mathbb{Z}$ and over $\mathbb{Q}$, but it is reducible over $\mathbb{R}$, because

$$x^4 + 1 = (x^2 + x\sqrt{2} + 1)(x^2 - x\sqrt{2} + 1).$$

The following facts, which are analogous to those about primes, are fairly easy consequences of the notion of irreducibility.

PROPOSITION 14.4.1.   *If $p(x)$ is irreducible and monic over L and if $f(x)$ is an arbitrary polynomial over L, then either $(f(x), p(x)) = p(x)$ or $(f(x), p(x)) = 1$.*

PROPOSITION 14.4.2.   *If $p(x)$ and $q(x)$ are distinct irreducible monic polynomials then $(p(x), q(x)) = 1$.*

PROPOSITION 14.4.3.   *If the irreducible polynomial $p(x)$ divides the product $f(x)g(x)$, then either $p(x)|f(x)$ or $p(x)|g(x)$ (or both).*

Of course, a linear polynomial in $L[x]$ is always irreducible, whatever the domain L is.

When the domain is a field, we apply (14.1.4) and confine ourselves to the decomposition of monic polynomials into monic irreducible factors. As in the case of integers, this factorization is unique with the obvious proviso that we do not distinguish between decompositions that differ merely in the arrangements of the factors. Thus we quote the following fundamental fact.

THEOREM 14.4.4.   *Any monic polynomial of positive degree over a field $\mathbb{F}$ can be uniquely factorized into a product of monic irreducible polynomials in $\mathbb{F}[x]$.*

14.5.   ROOTS OF POLYNOMIALS

Let $f(x)$ be a polynomial over the field $\mathbb{F}$ and suppose that $\mathbb{K}$ is a field containing $\mathbb{F}$, that is $\mathbb{F} \subseteq \mathbb{K}$. If k is an element of $\mathbb{K}$ such that $f(k) = 0$, we say that k is a *zero* of $f(x)$ or that k is a *root* of the equation $f(x) = 0$. Using equation (14.3.3) we can enunciate the

PROPOSITION 14.5.1.   *The element k is a root of* $f(x) = 0$ *if and only if* $f(x)$
is divisible by $x - k$, *that is*

$$f(x) = (x - k)g(x).$$

A linear polynomial

$$ax + b \quad (a \neq 0)$$

has precisely one zero, namely

$$k = -b/a,$$

which always lies in the field of coefficients.

The roots of the quadratic equation

$$q(x) = ax^2 + bx + c = 0 \quad (a \neq 0)$$

are [cf. (2.7.9)]

$$\alpha = \frac{-b + \sqrt{b^2 - 4ac}}{2a}, \qquad \beta = \frac{-b - \sqrt{b^2 - 4ac}}{2a}. \tag{14.5.1}$$

But these do not necessarily lie in the field, as root extraction cannot always be
carried out in a given field. For example

$$x^2 - 3$$

lies in $\mathbb{Q}[x]$, but its roots $\pm\sqrt{3}$ do not lie in $\mathbb{Q}$. Notice that

$$\alpha + \beta = -b/a \quad \text{and that} \quad \alpha\beta = c/a.$$

Evidently, if a quadratic polynomial $q(x)$ in $\mathbb{F}[x]$ is reducible over $\mathbb{F}$, it can
only break up into two linear factors, that is

$$ax^2 + bx + c = a(x - k)(x - l),$$

where k and l lie in $\mathbb{F}$. Conversely, if $q(x)$ has a zero in $\mathbb{F}$, say $q(k) = 0$, where
$k \in F$, than the other zero is

$$l = \frac{-b}{a} - k$$

and also lies in $\mathbb{F}$. Thus we can state that $q(x)$ is reducible over $\mathbb{F}$ if and only if
it has a zero in $\mathbb{F}$.

For the remainder of this section we confine ourselves to polynomials with
coefficients in $\mathbb{R}$ or $\mathbb{C}$ or, as we shall briefly say, to real or complex polynomials.
The case of $\mathbb{C}[x]$ is settled, at least theoretically, by a celebrated and deep
result:

THEOREM 14.5.2 (*Fundamental Theorem of Algebra*).   *Every complex poly-*
nomial $f(x)$ *of degree greater than zero has a root in* $\mathbb{C}$, *that is, there exists a*
complex number α *such that* $f(\alpha) = 0$.

It should be noted that this theorem only guarantees the existence of a root. The numerical determination of a root may involve considerable labour and is treated elsewhere [see III, Chapter 5].

In order to study the immediate consequences of this theorem consider a monic polynomial $f(x)$ over $\mathbb{C}$ of degree n (≥ 1). By virtue of Proposition 14.5.1 we can assert that there exists a complex number α_1 such that

$$f(x) = (x - \alpha_1)f_1(x),$$

where $f_1(x)$ is a polynomial of degree $n - 1$. If $n \geq 2$ we can apply the same argument to $f_1(x)$: thus there exists a complex number α_2, not necessarily distinct from α_1, such that $f_1(x) = (x - \alpha_2)f_2(x)$ and hence

$$f(x) = (x - \alpha_1)(x - \alpha_2)f_2(x).$$

Continuing in this manner we find ultimately that $f(x)$ is split into a product of linear factor in $\mathbb{C}[x]$, that is

$$f(x) = (x - \alpha_1)(x - \alpha_2)\ldots(x - \alpha_n). \tag{14.5.2}$$

The phenomenon of repeated factors will be studied in section 14.6.

It has now become clear that in $\mathbb{C}[x]$ the only irreducible polynomials are the linear polynomials.

Next, we turn to a discussion of $\mathbb{R}[x]$. Let

$$f(x) = a_0 x^n + a_1 x^{n-1} + \ldots + a_{n-1}x + a_n \quad (a_0 \neq 0)$$

be an arbitrary real polynomial and suppose that α is a complex root of the equation $f(x) = 0$, that is

$$a_0\alpha^n + a_1\alpha^{n-1} + \ldots + a_n = 0. \tag{14.5.3}$$

An equation between complex numbers remains valid if each term is replaced by its conjugate complex [see (2.7.10)]. Since the coefficients of $f(x)$ are real, we have that

$$\bar{a}_m = a_m \quad (m = 0, 1, \ldots, n).$$

Hence we deduce from (14.5.3) that

$$a_0\bar{\alpha}^n + a_1\bar{\alpha}^{n-1} + \ldots + a_n = 0,$$

or, more briefly,

$$f(\bar{\alpha}) = 0.$$

Thus we have

PROPOSITION 14.5.3. *The complex zeros of a real polynomial always occur in pairs of conjugate complex numbers.*

At this point it is convenient to summarize the properties of real quadratics:

PROPOSITION 14.5.4. *Let*

$$q(x) = ax^2 + bx + c \quad (a \neq 0)$$

be a real quadratic. Three cases arise:

(i) *If $b^2 < 4ac$, then $q(x)$ is irreducible over $\mathbb{R}$ and has a pair of conjugate complex zeros:*

(ii) *If $b^2 > 4ac$, then $q(x)$ is reducible over $\mathbb{R}$ and has two distinct real roots:*

(iii) *If $b^2 = 4ac$, then*

$$q(x) = a\left(x + \frac{b}{2a}\right)^2$$

and both zeros are equal to $-b/2a$.

Resuming the general discussion let

$$f(x) = x^n + a_1 x^{n-1} + a_2 x^{n-2} + \ldots + a_n \tag{14.5.4}$$

be a monic real polynomial of degree n. When $f(x)$ is considered as a polynomial in $\mathbb{C}[x]$, it splits into n linear factors as shown in (14.5.2). By virtue of Proposition 14.5.3 we may pair off conjugate complex factors. If we put

$$(x - \alpha)(x - \bar{\alpha}) = x^2 - bx + c,$$

than b and c are real numbers such that $b^2 < 4c$. Thus if $f(x)$ has r pairs of conjugate complex roots and s real roots, the decomposition of $f(x)$ into irreducible factors over $\mathbb{R}$ is as follows:

$$f(x) = (x^2 - b_1 x + c_1) \ldots (x^2 - b_r x + c_r)(x - k_1) \ldots (x - k_s). \tag{14.5.5}$$

Of course, in particular cases the integers r or s may be zero and the quadratic or the linear factors may be absent.

EXAMPLE 14.5.1. Let $f(x) = x^6 - 3x^4 + 3x^2 - 2$. The decomposition into irreducible factors in $\mathbb{R}[x]$ is given by

$$f(x) = (x^2 - \sqrt{3}x + 1)(x^2 + \sqrt{3}x + 1)(x - \sqrt{2})(x + \sqrt{2}).$$

14.6.   REPEATED FACTORS

Let $f(x)$ be a monic polynomial in $\mathbb{C}[x]$ and assume that its distinct zeros are

$$\alpha_1, \alpha_2, \ldots, \alpha_k. \tag{14.6.1}$$

Gathering together equal factors in (14.5.2) we obtain the factorization

$$f(x) = (x - \alpha_1)^{r_1}(x - \alpha_2)^{r_2} \ldots (x - \alpha_k)^{r_k}, \tag{14.6.2}$$

where $r_j \geq 1$ ($j = 1, 2, \ldots, k$). When $r_j \geq 2$, we say that α_j is a *repeated root* of $f(x) = 0$; in particular, when $r_j = 2$, then α_j is called a *double root*.

Let $f(x)$ be given by (14.5.4). We associate with it the *derivative* [cf. IV, Example 3.1.1]

$$f'(x) = nx^{n-1} + (n - 1)a_1 x^{n-2} + \ldots + a_{n-1}. \tag{14.6.3}$$

If the coefficients of $f(x)$ lie in the field $\mathbb{F}$ ($\subseteq \mathbb{C}$), so do the coefficients of $f'(x)$. Now let

$$d(x) = (f(x), f'(x))$$

be the highest common factor of $f(x)$ and $f'(x)$. By Euclid's algorithm, or otherwise, the polynomial $d(x)$ can be computed entirely by operations in $\mathbb{F}$ and for
this purpose it is unnecessary to know the zeros of $f(x)$. On the other hand, with
the aid of the differential calculus it can be shown that

$$d(x) = (x - \alpha_1)^{r_1-1}(x - \alpha_2)^{r_2-1} \ldots (x - \alpha_k)^{r_k-1}.$$

Hence

$$g(x) = \frac{f(x)}{d(x)} = (x - \alpha_1)(x - \alpha_2) \ldots (x - \alpha_k) \qquad (14.6.4)$$

is a polynomial in $\mathbb{F}[x]$ whose zeros are precisely the distinct zeros of $f(x)$.

When $r_j = 1$, then α_j is an unrepeated or *simple* root, and we mention the
following

DEFINITION 14.6.1.   A polynomial is called *separable* if all its zeros are
simple.

If $h(x)$ is an irreducible polynomial in $\mathbb{F}[x]$ which divides both $f(x)$ and $f'(x)$,
that is if $h(x)|f(x)$, then

$$(h(x))^2|f(x).$$

Every polynomial $f(x) \in \mathbb{F}[x]$ can be factorized as

$$f(x) = f_1(x)f_2(x),$$

where $f_1(x)$ and $f_2(x)$ are polynomials in $\mathbb{F}[x]$ such that $f_1(x)$ consists of all the
repeated factors of $f(x)$ and $f_2(x)$ is separable. Indeed, we have that

$$f_1(x) = (f(x), (d(x))^2). \qquad (14.6.5)$$

EXAMPLE 14.6.1.   Let $f(x) = x^3 - 3x + 2$. Then

$$f'(x) = 3x^2 - 3 \quad \text{and} \quad (f(x), f'(x)) = x - 1.$$

Also

$$\frac{f(x)}{x - 1} = x^2 + x - 2,$$

which is a separable polynomial.

14.7.   LOCATING THE REAL ZEROS

There is a crude but useful way of finding whereabout the real zeros of a real
polynomial are: it is find, if possible, two values a and b with the property that
$f(a)$ and $f(b)$ have opposite signs. There must then be at least one zero between
a and b; there may, of course, be more. In fact, this method applies to any
continuous real function [see IV, § 2.1.2]. The information can then be refined
to any degree of accuracy by taking a mesh of points between a and b and finding
the signs of $f(x)$ at these points.

EXAMPLE 14.7.1.   Let $f(x) = x^4 + 5x - 4$. Since $f(0) = -4 < 0$ and $f(1) = 2 > 0$, we can infer that $f(x)$ has a zero between 0 and 1. On searching further we find that $f(\frac{1}{2}) < 0$ and $f(\frac{4}{5}) > 0$. Hence we know that there is a zero between $\frac{1}{2}$ and $\frac{4}{5}$.

There are more efficient ways of approximating to a root which depend on the differential calculus, as we shall see in the next section.

14.8.   FINDING THE NUMBER OF REAL ROOTS OF A POLYNOMIAL (EQUATION)

Before embarking on any method of locating the real roots, it is desirable to know how many there are; if, for instance, there are none, any method will be time-consuming and unprofitable. There is one very simple test that gives an upper bound to the number of real zeros. If $f(x) = x^n + a_{n-1}x^{n-1} + \ldots + a_0$, we say that the sequence of real coefficients $1, a_{n-1}, a_{n-2}, \ldots, a_0$ has a *change of sign* for each pair of successive coefficients of opposite signs after coefficients that are zero have been omitted. So $x^8 + x^6 - x^5 - x^4 - x^3 + x - 1$ has three changes of sign.

THEOREM 14.8.1 (*Descartes' rule of signs*).   *A real polynomial $f(x)$ cannot have more positive real roots than there are changes of sign in its sequence of coefficients: nor more negative real roots than there are changes of sign in the sequence of coefficients of $f(-x)$.*

Here, if b is a k-fold root, the root b counts as k roots.

EXAMPLE 14.8.1.   The polynomial $x^8 + x^6 - x^5 - x^4 - x^3 + x - 1$ cannot have more than three positive roots; and since

$$f(-x) = x^8 + x^6 + x^5 - x^4 + x^3 - x - 1,$$

the original polynomial cannot have more than three negative roots.

A little more can be said; *the number of positive or negative roots will be even if there is an even number of changes of sign, and odd if there is an odd number of changes.*

Although Descartes' rule of signs is very quick to apply, it only gives an upper bound to the number of roots. There is a much deeper method due to Sturm which gives the exact number and can also help to locate them.

First we must define a *Sturm Sequence* for a given polynomial. This is a sequence of polynomials

$$h_1(x), \qquad h_2(x), \qquad h_3(x), \ldots \tag{14.8.1}$$

which will be defined step by step. We begin by putting

$$h_1(x) = f(x) \quad \text{and} \quad h_2(x) = f'(x) \tag{14.8.2}$$

where $f'(x)$ is the derivative of $f(x)$. The rest of the sequence is determined by saying how, for $r \geq 1$, the polynomial h_{r+2} is derived from h_{r+1} and h_r. These three polynomials are related by the Division Algorithm [Theorem 14.3.1]: $h_{r+2}(x)$ is the negative of the remainder when $h_r(x)$ is divided by $h_{r+1}(x)$, that is

$$h_r(x) = q_r(x)h_{r+1}(x) - h_{r+2}(x). \qquad (14.8.3)$$

It is clear that the degrees of the polynomials in the sequence decrease. The last term is the one whose successor would be the zero polynomial; it is in fact some constant multiplied by $(f(x), f'(x))$. But (14.8.3) is only one of the possible ways of constructing a Sturm sequence. The procedure described above is liable to involve alarming fractions, even when the coefficients of $f(x)$ are moderately small integers. It will presently become clear that we require to know only the signs of the polynomials (14.8.1) when we substitute for x some real values. Therefore the results will not be affected if each term of the sequence is multiplied by an arbitrary positive number. Accordingly, it is prudent to modify (14.8.3) and to write instead

$$u_r h_r(x) = q_r(x)h_{r+1}(x) - v_r h_{r+2}(x) \qquad (14.8.4)$$

where u_r and v_r are positive numbers at our disposal. An example will make the point.

EXAMPLE 14.8.2.   Let $f(x) = x^5 + x^4 - x^3 + 3x^2 - 4 = h_1(x)$. Then

$$f'(x) = 5x^4 + 4x^3 - 3x^2 + 6x = h_2(x).$$

To avoid fractions we multiply $h_1(x)$ by 25 before carrying out the division by $h_2(x)$, giving:

$$25h_1(x) = (5x + 1)h_2(x) + (-14x^3 + 48x^2 - 6x - 100).$$

In order to obtain $h_3(x)$ we have to reverse the sign of the polynomial in the last bracket. At the same time it is convenient to multiply by $\frac{1}{2}$. Thus we put

$$h_3(x) = 7x^3 - 24x^2 + 3x + 50.$$

Now we apply the division algorithm to $49h_2(x)$ and $h_3(x)$:

$$49h_2(x) = (35x + 148)h_3(x) + (3300x^2 - 1900x - 7400).$$

Reversing the sign of the remainder and dropping the factor 100 we obtain

$$h_4(x) = -33x^2 + 19x + 74.$$

Next we apply the algorithm to $1089h_3(x)$ and $h_4(x)$, thus

$$1089h_3(x) = (-231x + 659)h_4(x) + (7840x + 5684).$$

We observe that $7840x + 5684 = 196(40x + 29)$. On dropping the factor 196 and changing the sign we obtain that

$$h_5(x) = -40x - 29.$$

Finally, we divide $1600h_4(x)$ by $h_5(x)$, giving

$$1600(-33x^2 + 19x + 74) = (1320x - 1717) + 68{,}647.$$

The remainder is a positive constant. In accordance with our rule we reverse its sign and we might just as well replace it by -1. Hence

$$h_6(x) = -1.$$

Summarizing the computations we list the terms of the Sturm sequence as follows:

$$
\begin{aligned}
h_1(x) &= x^5 + x^4 - x^3 + 3x^2 && - 4\\
h_2(x) &= \qquad\quad 5x^4 + 4x^3 - 3x^2 + 6x\\
h_3(x) &= \qquad\qquad\quad\ 7x^3 - 24x^2 + 3x + 50\\
h_4(x) &= \qquad\qquad\qquad\ -33x^2 + 19x + 74\\
h_5(x) &= \qquad\qquad\qquad\qquad\ -40x - 29\\
h_6(x) &= \qquad\qquad\qquad\qquad\qquad\ -1.
\end{aligned}
$$

Returning now to the general case, for any value a we can calculate $C(a)$, the number of changes of sign in the sequence $h_1(a)$, $h_2(a)$, $h_3(a)$, $\ldots$, after any zeros in the sequence have been omitted. We can now state:

THEOREM 14.8.2 (*Sturm's Theorem*).   *For any real polynomial, if $C(a)$ is as just defined, then, provided that $a < b$, and neither a nor b is a zero of the polynomial, the number of real zeros between a and b is $C(a) - C(b)$. In this any repeated zero is only counted once.*

Once the Sturm sequence has been computed this theorem is particularly easy to apply if we only want to know the number of positive and negative roots of the polynomial equation $f(x) = 0$. Take $a = -K$, $b = +K$ where K is so large that the signs in the sequence are determined only by the signs of the highest powers of x; then $C(a)$ and $C(b)$ can be found from the polynomials in the Sturm sequence at a glance. In much the same way $C(0)$ can be found by looking only at the sequence of signs of the constants in the sequence. In our example the signs are:

	$-K$	0	$+K$
h_1	$-$	$-$	$+$
h_2	$+$	0	$+$
h_3	$-$	$+$	$+$
h_4	$-$	$+$	$-$
h_5	$+$	$-$	$-$
h_6	$-$	$-$	$-$
C	4	2	1

So the number of negative roots is $C(-K) - C(0) = 2$; and of positive roots is $C(0) - C(+K) = 1$. Notice that Descartes' rule of signs tells us that there are at most 3 positive roots and at most 2 negative roots, whereas Sturm's method gives us the exact numbers.

There is a third method for getting information about the number and positions of the roots of a real polynomial equation; it is known as Budan's method but is due to Fourier. It stands between Descartes' and Sturm's both in scope and in simplicity, being less informative than Sturm's and rather more trouble to apply than Descartes'. It uses a sequence of functions that is easy to write down, $f(x), f'(x), f''(x), \ldots$, each being the derivative of its predecessor. Let $B(a)$ be the number of changes of sign in the sequence $f(a), f'(a), f''(a), \ldots$, after any zeros have been omitted.

THEOREM 14.8.3 (*Budan-Fourier*). *For a real polynomial $f(x)$, if neither a nor b is a root of $f(x) = 0$, and $a < b$, then the number of roots between a and b cannot exceed $B(a) - B(b)$, a k-fold root counting as k roots.*

If we take a to be 0 and b to be some large enough number, this gives Descartes' rule for positive roots; his rule for negative roots follows from the method of Budan-Fourier applied in the same way to $f(-x)$.

14.9. REAL RATIONAL FUNCTIONS

A real *rational function* of x is the ratio $f(x)/g(x)$ of two real polynomials $f(x)$ and $g(x)$. (Complex rational polynomials are similarly defined.) If for some polynomials $h(x), f_0(x), g_0(x)$, $f(x) = h(x)f_0(x)$ and $g(x) = h(x)g_0(x)$, then $f(x)/g(x) = f_0(x)/g_0(x)$; it is (as with rational numbers) generally sensible to reduce any rational function to its simplest terms by dividing the top (*numerator*) and the bottom (*denominator*) by any common factor of degree greater than or equal to unity. If $\deg f(x) > \deg g(x)$, we can apply the division algorithm to express $f(x)$ as

$$f(x) = q(x)g(x) + r(x), \qquad \deg r(x) < \deg g(x),$$

and we recall that the polynomials $q(x)$ and $r(x)$ are uniquely determined. It follows that

$$\frac{f(x)}{g(x)} = q(x) + \frac{r(x)}{g(x)}. \tag{14.9.1}$$

So any rational function can be expressed as the sum of a polynomial (perhaps zero) and a *proper* rational function in which the numerator has degree lower than that of the denominator.

This is like writing a vulgar fraction as a mixed fraction, except that in the latter situation uniqueness does not obtain if negative fractions are allowed; for example

$$-\tfrac{3}{2} = -2 + \tfrac{1}{2} = -1 - \tfrac{1}{2}.$$

But the following proposition has no analogue for fractions.

PROPOSITION 14.9.1.   *The sum of two proper rational functions is again a proper rational function.*

For if

$$\deg f_1(x) < \deg g_1(x) \quad \text{and} \quad \deg f_2(x) < \deg g_2(x),$$

then

$$\frac{f_1(x)}{g_1(x)} + \frac{f_2(x)}{g_2(x)} = \frac{f_1(x)g_2(x) + f_2(x)g_2(x)}{g_1(x)g_2(x)},$$

where

$$\deg\{f_1(x)g_2(x) + f_2(x)g_1(x)\} < \deg\{g_1(x)g_2(x)\}.$$

14.10. PARTIAL FRACTIONS

Suppose we have a proper rational function $f(x)/g(x)h(x)$, whose denominator can be factorized into factors $g(x)$ and $h(x)$ that have themselves no factor in common, that is

$$\deg f(x) < \deg\{g(x)h(x)\}, \quad (g(x), h(x)) = 1.$$

It is always possible to split the original functions into the sum of two proper rational functions, thus

$$\frac{f(x)}{g(x)h(x)} = \frac{a(x)}{g(x)} + \frac{b(x)}{h(x)}, \tag{14.10.1}$$

where $\deg a(x) < \deg g(x)$ and $\deg b(x) < \deg h(x)$. The proof of this important fact is as follows: since $(g(x), h(x)) = 1$, we can apply (14.3.10) and find polynomials $a_0(x)$ and $b_0(x)$ such that

$$1 = a_0(x)h(x) + b_0(x)g(x).$$

On multiplying throughout by $f(x)$ we obtain that

$$f(x) = a_1(x)h(x) + b_1(x)g(x), \tag{14.10.2}$$

where $a_1(x) = f(x)a_0(x)$, $b_1(x) = f(x)b_0(x)$. Next, we divide (14.10.2) by $g(x)h(x)$, thus

$$\frac{f(x)}{g(x)h(x)} = \frac{a_1(x)}{g(x)} + \frac{b_1(x)}{h(x)}. \tag{14.10.3}$$

It is possible that the two terms on the right-hand side of (14.10.3) are not proper rational functions, in which case we shall write

$$\frac{a_1(x)}{g(x)} = G(x) + \frac{a(x)}{g(x)}, \quad \frac{b_1(x)}{h(x)} = H(x) + \frac{b(x)}{h(x)},$$

where $G(x)$ and $H(x)$ are polynomials and $\deg a(x) < \deg g(x)$, $\deg b(x) < \deg h(x)$· Hence

$$\frac{f(x)}{g(x)h(x)} - \frac{a(x)}{g(x)} - \frac{b(x)}{h(x)} = G(x) + H(x).$$

The expression on the left-hand side of this equation is a proper rational function by virtue of Proposition 14.9.1. Therefore a contradiction would arise, unless $G(x) + H(x) = 0$. This proves (14.10.1). The process described in this equation is called splitting a rational function into *partial fractions*.

One method of carrying out the splitting involves the solution of a set of linear equations for the (initially unknown) coefficients of $a(x)$ and $b(x)$. Let us illustrate it with an example.

EXAMPLE 14.10.1.   We apply the method to

$$\frac{x - 1}{(x + 1)^2(x^3 + 2)}.$$

Since the numerators are to have lower degrees than the denominators, we can write

$$\frac{x - 1}{(x + 1)^2(x^3 + 2)} = \frac{a_1 x + a_0}{(x + 1)^2} + \frac{b_2 x^2 + b_1 x + b_0}{(x^3 + 2)}.$$

We have now to find a_1, a_0, b_2, b_1, b_0. To do this we multiply through by $(x + 1)^2(x^3 + 2)$ to get the polynomial identity

$$x - 1 = (a_1 x + a_0)(x^3 + 2) + (b_2 x^2 + b_1 x + b_0)(x^2 + 2x + 1)$$

or

$$x - 1 = (a_1 + b_2)x^4 + (a_0 + b_1 + 2b_2)x^3 + (b_0 + 2b_1 + b_2)x^2$$
$$+ (2a_1 + 2b_0 + b_1)x + (2a_0 + b_0).$$

Since these polynomials are to be identically equal, the coefficients of any power of x must be the same on each side of the equation [see § 14.1]; so, comparing coefficients, we obtain that

$$a_1 + b_2 = 0$$
$$a_0 + b_1 + 2b_2 = 0$$
$$b_0 + 2b_1 + b_2 = 0$$
$$2a_1 + 2b_0 + b_1 = 1$$
$$2a_0 + b_0 = -1.$$

The solution of these equations is [see § 5.10]:

$$a_0 = 5, \quad a_1 = 7, \quad b_0 = -11, \quad b_1 = 9, \quad b_2 = -7.$$

Therefore

$$\frac{x - 1}{(x + 1)^2(x^3 + 2)} = \frac{7x + 5}{(x + 1)^2} + \frac{-7x^2 + 9x - 11}{(x^3 + 2)}.$$

This straightforward example illustrates the two main features of the method: it is simple in principle and rather tedious to carry out in practice. It can be extended to the case in which the denominator can be factorized into k factors, no two having a common factor; the rational function can then be expressed as the sum of k partial fractions.

There is, fortunately, another method that can be used instead of, or in conjunction with, the method of comparing coefficients. If two polynomials are identical, they have the same values for any value of x; by suitable choices of value for x it is often possible to get very simple equations for the unknown coefficients.

EXAMPLE 14.10.2.   Let

$$\frac{3x - 4}{(x - 2)^2(x - 1)(x + 1)} = \frac{a_1 x + a_0}{(x - 2)^2} + \frac{b_0}{(x - 1)} + \frac{c_0}{(x + 1)}.$$

Multiplying out, as before, we find that

$$3x - 4 = (a_1 x + a_0)(x - 1)(x + 1)$$
$$+ b_0(x - 2)^2(x + 1) + c_0(x - 2)^2(x - 1).$$

Let us put $x = 1$ on both sides, since then the first and the third terms on the right-hand side become 0; we are left with:

$$-1 = 2b_0.$$

Similarly putting $x = -1$, we get:

$$-7 = -18c_0.$$

So

$$b_0 = -\tfrac{1}{2} \quad \text{and} \quad c_0 = \tfrac{7}{18}.$$

We can get two equations relating a_1 and a_0 by putting $x = 2$ and $x = 0$; they are:

$$2 = 3(2a_1 + a_0)$$

and

$$-4 = -a_0 + 4b_0 - 4c_0 = -a_0 - 2 - \tfrac{14}{9}.$$

From these we find that $a_0 = \tfrac{4}{9}$ and $a_1 = \tfrac{1}{9}$. The result, therefore, is:

$$\frac{3x - 4}{(x - 2)^2(x - 1)(x + 1)} = \frac{1}{9}\frac{x + 4}{(x - 2)^2} - \frac{1}{2}\frac{1}{x - 1} + \frac{7}{18}\frac{1}{x + 1}.$$

The values b_0 and c_0 can be found even more directly. For instance, b_0 is the value of

$$\frac{3x - 4}{(x - 2)^2(x + 1)} \quad \text{when } x = 1;$$

similarly c_0 is the value of

$$\frac{3x - 4}{(x - 2)^2(x - 1)} \quad \text{when } x = -1.$$

There is in fact a general rule for finding the constant numerator for any unrepeated linear factor of the original denominator: if

$$\frac{f(x)}{(x - \alpha)g(x)}$$

is a proper rational function and $(x - \alpha)$ is not a factor of $g(x)$, then

$$\frac{f(x)}{(x - \alpha)g(x)} = \frac{f(\alpha)}{g(\alpha)} \frac{1}{x - \alpha} + \frac{h(x)}{g(x)}.$$

When the denominator has a repeated factor, there is another and more useful form for the decomposition into partial fractions. For instance, in the previous example we can express

$$\frac{1}{9} \frac{x + 4}{(x - 2)^2}$$

in the form

$$\frac{d}{(x - 2)^2} + \frac{e}{x - 2}.$$

After multiplication the requirement is seen to be that

$$\tfrac{1}{9}(x + 4) = d + e(x - 2).$$

So d must be $\tfrac{2}{3}$ and $e = \tfrac{1}{9}$. The alternative decomposition is then

$$\frac{3x - 4}{(x - 2)^2(x - 1)(x + 1)}$$

$$= \frac{2}{3} \frac{1}{(x - 2)^2} + \frac{1}{9} \frac{1}{x - 2} - \frac{1}{2} \frac{1}{x - 1} + \frac{7}{18} \frac{1}{x + 1}. \quad (14.10.4)$$

This alternative form for the partial fractions can be extended both to the case in which the repeated factor is not linear and to the case in which it is raised to a power > 2. To give an example of both extensions, suppose that

$$\phi(x) = \frac{f(x)}{(x^2 + 1)^3 g(x)}$$

is a proper rational function, in which $g(x)$ and $x^2 + 1$ share no common factor. Then $\phi(x)$ can be split into partial fractions as follows:

$$\frac{a_1x + a_0}{(x^2 + 1)^3} + \frac{b_1x + b_0}{(x^2 + 1)^2} + \frac{c_1x + c_0}{x^2 + 1} + \frac{h(x)}{g(x)}.$$

14.11.   APPLICATIONS OF PARTIAL FRACTIONS

The partial fractions decomposition is extremely useful in integrating rational functions [IV, Ex. 4.2.4]. It is often easy to integrate each separate term in the decomposition when the original function would be very hard to deal with as it stands.

The decomposition can also be used to express a rational function as a power-series [see IV, § 1.10]. This may be of value in itself and it has an application to the study of linear recursions [see § 14.12].

Consider Example 14.10.2. The first term on the right-hand side of (14.10.4) is [see (3.10.3) and IV, § 1.10].

$$\frac{2}{3}\frac{1}{(x-2)^2} = \frac{2}{3}\frac{1}{4(1-\frac{1}{2}x)^2} = \frac{1}{6}(1 + 2(\tfrac{1}{2}x) + 3(\tfrac{1}{2}x)^2 + 4(\tfrac{1}{2}x)^3 + \ldots).$$

The second term is:

$$\frac{1}{9}\frac{1}{x-2} = -\frac{1}{18}\frac{1}{1-\frac{1}{2}x} = -\frac{1}{18}(1 + \tfrac{1}{2}x + (\tfrac{1}{2}x)^2 + (\tfrac{1}{2}x)^3 + \ldots).$$

The third term is $\frac{1}{2}(1 + x + x^2 + x^3 + \ldots)$ and the fourth is $\frac{7}{18}(1 - x + x^2 - \ldots)$. So

$$\frac{3x-4}{(x-2)^2(x-1)(x+1)} = \sum_{n=0}^{\infty} \{\tfrac{1}{6}(n+1)(\tfrac{1}{2})^n - \tfrac{1}{18}(\tfrac{1}{2})^n + \tfrac{1}{2} + \tfrac{7}{18}(-1)^n\}x^n,$$

provided that each of the expansions converges [also see (3.10.3)]. In this case the restrictions are that $|\frac{1}{2}x| < 1$ and $|x| < 1$; the expansion is therefore valid if $|x| < 1$.

This method can always be applied when the denominator can be factorized into real linear factors, none of which is the factor x. If the denominator has a quadratic factor that has no real root, it can be factorized into two linear terms involving a pair of complex conjugate numbers [see Proposition 14.5.3]. The partial fractions decomposition can be performed using complex numbers, and the power-series expansions can be written down. The corresponding terms from the two power-series will have as coefficients pairs of conjugate complex numbers, so that they can be added together to give a real power-series.

Let us, for example suppose that the real quadratic polynomial $x^2 - bx + c$ has complex roots α, $\bar{\alpha}$ [see § 14.5], and that $\alpha = r(\cos \theta + i \sin \theta)$ [see § 2.7.3]; then $\bar{\alpha} = r(\cos \theta - i \sin \theta)$ [see (2.7.24)]. By Demoivre's theorem [see (2.7.29)]

$$\alpha^n = r^n(\cos n\theta + i \sin n\theta).$$

We can express

$$\frac{1}{x^2 - bx + c} = \frac{1}{(x-\alpha)(x-\bar{\alpha})}$$

in partial fractions as

$$\frac{1}{\alpha - \bar{\alpha}}\frac{1}{x - \alpha} + \frac{1}{\bar{\alpha} - \alpha}\frac{1}{x - \bar{\alpha}}$$

$$= \frac{1}{2ri\sin\theta}\frac{1}{x - \alpha} - \frac{1}{2ri\sin\theta}\frac{1}{x - \bar{\alpha}}$$

$$= \frac{i}{2r\sin\theta}\left(\frac{1}{x - \bar{\alpha}} - \frac{1}{x - \alpha}\right)$$

$$= \frac{i}{2r\sin\theta}\left(\frac{1}{\alpha}\left(\frac{1}{1 - x/\alpha}\right) - \frac{1}{\bar{\alpha}}\left(\frac{1}{1 - x/\bar{\alpha}}\right)\right)$$

$$= \frac{i}{2r\sin\theta}\left(\frac{1}{\alpha}\left(1 + \frac{x}{\alpha} + \frac{x^2}{\alpha^2} + \dots\right) - \frac{1}{\bar{\alpha}}\left(1 + \frac{x}{\bar{\alpha}} + \frac{x^2}{\bar{\alpha}^2} + \dots\right)\right)$$

$$= \frac{i}{2r\sin\theta}\{r^{-1}(\cos\theta - i\sin\theta) + r^{-2}(\cos 2\theta - i\sin 2\theta)x + \dots$$

$$- r^{-1}(\cos\theta + i\sin\theta) - r^{-2}(\cos 2\theta + i\sin 2\theta)x - \dots\}$$

$$= \frac{1}{r\sin\theta}(r^{-1}\sin\theta + r^{-2}\sin 2\theta x + r^{-3}\sin 3\theta x^2 + \dots)$$

$$= \frac{1}{r\sin\theta}\sum_{n=0}^{\infty}\frac{x^n\sin(n+1)\theta}{r^{n+1}}.$$

This expansion converges provided that $|x| < r$.

We can express the left-hand side in terms of r and θ instead of b and c, since

$$b = \alpha + \bar{\alpha} = 2r\cos\theta$$

and

$$c = \alpha\bar{\alpha} = r^2.$$

The expansion then reads:

$$\frac{1}{x^2 - 2xr\cos\theta + r^2} = \frac{1}{r\sin\theta}\sum_{n=0}^{\infty}\frac{x^n\sin(n+1)\theta}{r^{n+1}}.$$

It is instructive, particularly in view of the methods of the next section, to check the formula by multiplying both sides by $x^2 - 2rx\cos\theta + r^2$. The left-hand side, of course, becomes 1. The right-hand side becomes

$$\frac{x^2 - 2xr\cos\theta + r^2}{r\sin\theta}\sum_{n=0}^{\infty}\frac{x^n\sin(n+1)\theta}{r^{n+1}}.$$

We collect terms in equal powers of x. The constant term is obtained by putting $x = 0$, which gives

$$\frac{r^2}{r\sin\theta}\left(\frac{\sin\theta}{r}\right) = 1,$$

in agreement with the left-hand side. When $n \geq 1$, the term in x^n is equal to

$$\frac{1}{r^n \sin \theta} \{\sin (n + 1)\theta - 2 \cos \theta \sin n\theta + \sin (n - 1)\theta\} x^n.$$

Now, since [see V, (1.2.20)]

$$\sin (n + 1)\theta + \sin (n - 1)\theta = \sin (n\theta + \theta) + \sin (n\theta - \theta)$$
$$= 2 \sin n\theta \cos \theta,$$

the coefficient of x^n is (as required) 0 for $n \geq 1$.

14.12. LINEAR RECURSIVE SEQUENCES

A sequence of real numbers $u_0, u_1, u_2, \ldots$ is said to be defined *recursively* if u_n is defined in terms of the preceding u's. Given the initial terms of the sequence all later terms can then be calculated. A recursive definition of importance in Numerical Analysis is:

$$u_n = \frac{1}{2} \left(u_{n-1} + \frac{a}{u_{n-1}} \right).$$

Whatever positive number is chosen for u_0, the sequence of u's converges rapidly to $\sqrt{a}$. Here u_n is given in terms of just its immediate predecessor, and only one initial term is needed to determine the sequence. Although this is a recursive sequence it is not a linear one.

The sequence $u_0, u_1, u_2, \ldots$ is called a *homogeneous linear recursive sequence* if each u_n is a fixed linear combination of a finite number of its predecessors:

$$u_n = a_1 u_{n-1} + a_2 u_{n-2} + \ldots + a_k u_{n-k}. \tag{14.12.1}$$

Unless otherwise stated, it will always be assumed that $a_1, a_2, \ldots, a_k$ are given (usually real) numbers, which do not depend on n. Equation (14.12.1) is also called a homogeneous *linear recurrence relation* (with constant coefficients) of *order* k, where k is the number of predecessors of u_n involved in the relation (it is assumed that $a_k \neq 0$). Yet another name for (14.12.1) is a *recursion of order k*. Sometimes it is convenient to replace n by $n + k$ and to write (14.12.1) in the form

$$u_{n+k} = a_1 u_{n+k-1} + a_2 u_{n+k-2} + \ldots + a_k u_n. \tag{14.12.2}$$

Adopting a different point of view we may interpret (14.12.2) as a *difference equation*. For this purpose it is helpful to think of a sequence as a function defined on the set of all non-negative integers [see § 1.4.1], and we use the notation

$$u_n = u(n) \quad (n = 0, 1, 2, \ldots)$$

in analogy with a function $f(x)$ defined, say, on the set of real numbers. Corresponding to the first derivative $f'(x)$ [see IV, § 3.1.1] we introduce the *first difference*

$$\Delta u(n) = u(n + 1) - u(n). \tag{14.12.3}$$

The *second difference* is defined as

$$\Delta^2 u(n) = \Delta(\Delta u(n)) = (u(n + 2) - u(n + 1)) - (u(n + 1) - u(n)),$$

that is

$$\Delta^2 u(n) = u(n + 2) - 2u(n + 1) + u(n). \tag{14.12.4}$$

Higher differences are defined similarly, but we confine ourselves here to a second order equation. Consider the recurrence relation

$$u(n + 2) + b_1 u(n + 1) + b_2 u(n) = 0. \tag{14.12.5}$$

We express $u(n + 2)$ and $u(n + 1)$ in terms of $\Delta u(n)$ and $\Delta^2 u(n)$:

$$u(n + 1) = \Delta u(n) + u(n),$$
$$u(n + 2) = \Delta^2 u(n) + 2u(n + 1) - u(n)$$
$$= \Delta^2 u(n) + 2\Delta u(n) + u(n).$$

Hence (14.12.5) is equivalent to

$$\Delta^2 u(n) + (2 + b_1)\Delta u(n) + (1 + b_1 + b_2)u(n) = 0$$

or, after renaming the constants,

$$\Delta^2 u(n) + c_1 \Delta u(n) + c_2 u(n) = 0. \tag{14.12.6}$$

This is a *linear difference equation* of the second order. Conversely, any difference equation can be turned into a recurrence relation.

A famous example is the *Fibonacci sequence*

$$0, 1, 1, 2, 3, 5, 8, 13, 21, \ldots \tag{14.12.7}$$

which is generated by the recursion

$$u_n = u_{n-1} + u_{n+2} \tag{14.12.8}$$

from initial values $u_0 = 0$, $u_1 = 1$.

In the general case, if u_n depends on its k previous members, then the k initial values determine the rest of the sequence. It is natural to look for a formula that gives u_n in terms of n, the a's, and constants determined by the initial values. This problem will be discussed in the next section.

14.13. SOLUTION OF RECURRENCE RELATIONS

In this section we are going to discuss methods for solving equations of the type (14.12.1).

(i) One fruitful approach to this problem is to form the *generating function*, that is the power series

$$U(x) = u_0 + u_1 x + u_2 x^2 + \ldots + u_n x^n + \ldots, \tag{14.13.1}$$

without being concerned whether it converges or not [see IV, § 1.10]. We now multiply (14.13.1) by $1 - a_1x - a_2x^2 - \ldots - a_kx^k$, where the a's are those of the recurrence relation (14.12.1). Thus

$$(1 - a_1x - a_2x^2 - \ldots - a_kx^k)U(x)$$

$$= u_0 + u_1x + \quad u_2x^2 + \ldots + \quad u_kx^k + \ldots + \quad u_nx^n + \ldots$$

$$\quad - a_1u_0x - a_1u_1x^2 - \ldots - a_1u_{k-1}x^k - \ldots - a_1u_{n-1}x^n - \ldots$$

$$\quad - a_2u_0x^2 - \ldots - a_2u_{k-2}x^k - \ldots - a_2u_{n-2}x^n - \ldots$$

$$\cdots\cdots\cdots\cdots\cdots\cdots\cdots\cdots\cdots\cdots\cdots\cdots$$

$$\quad - a_ku_0x^k - \ldots - a_ku_{n-k}x^n - \ldots$$

$$= p_{k-1}(x) - \sum_{n=k}^{\infty} (u_n - a_1u_{n-1} - \ldots - a_ku_{n-k})x^n$$

$$= p_{k-1}(x),$$

since $u_n = a_1u_{n-1} + a_2u_{n-2} + \ldots + a_ku_{n-k}$. In this, $p_{k-1}(x)$ is the polynomial of degree not exceeding $k - 1$ that arises from collecting terms at the start of the series where there will be fewer than $k + 1$ terms to collect, explicitly

$$p_{k-1}(x) = u_0 + (u_1 - a_1u_0)x + (u_2 - a_1u_1 - a_2u_0)x^2 + \ldots$$

$$+ (u_k - a_1u_{k-1} - a_2u_{k-2} - \ldots - a_{k-1}u_0)x^{k-1}.$$

This implies that

$$U(x) = \frac{p_{k-1}(x)}{1 - a_1x - a_2x^2 - \ldots - a_kx^k}. \tag{14.13.2}$$

In fact (14.13.1) is the power-series expansion of the proper rational function (14.13.2).

The methods of section 14.11 now allow us to decompose this rational function into partial fractions and thereby to express it as a power-series in x. By equating this with $U(x)$ and then equating coefficients, we can derive the required formula for u_n.

The denominator can be factorized into (possibly complex) linear factors. Let us suppose that these are all distinct and that

$$1 - a_1x - a_2x^2 - \ldots - a_kx^k = (1 - \beta_1x)(1 - \beta_2x)\ldots(1 - \beta_kx).$$

The β's are in fact the roots of the '*reverse*' polynomial

$$y^k - a_1y^{k-1} - \ldots - a_k. \tag{14.13.3}$$

Then, for some values of the constants $b_1, b_2, \ldots, b_k$,

$$U(x) = \frac{b_1}{1 - \beta_1x} + \frac{b_2}{1 - \beta_2k} + \ldots + \frac{b_k}{1 - \beta_kx}$$

$$= b_1 \sum_{n=0}^{\infty} (\beta_1x)^n + \ldots + b_k \sum_{n=0}^{\infty} (\beta_kx)^n$$

$$= \sum_{0=n}^{\infty} (b_1\beta_1^n + b_2\beta_2^n + \ldots + b_k\beta_k^n)x^n.$$

Equating coefficients we deduce that

$$u_n = b_1\beta_1{}^n + b_2\beta_2{}^n + \ldots + b_k\beta_k{}^n.$$

The values of $b_1, b_2, \ldots, b_k$ can be found from the initial values in the sequence $u_0, u_1, \ldots$ by solving the equations

$$
\begin{aligned}
b_1 \quad + b_2 \quad + b_3 \quad + \ldots + b_k &= u_0 \\
b_1\beta_1 \quad + b_2\beta_2 \quad + b_3\beta_3 + \ldots + b_k\beta_k &= u_1 \\
\vdots \\
b_1\beta_1^{k-1} + b_2\beta_2^{k-1} \quad\quad + \ldots + b_k\beta_k^{k-1} &= u_{k-1}.
\end{aligned}
$$

The determinant of these equations for the b's, namely

$$
\begin{vmatrix}
1 & 1 & \cdots & 1 \\
\beta_1 & \beta_2 & \cdots & \beta_k \\
\beta_1{}^2 & \beta_2{}^2 & \cdots & \beta_k{}^2 \\
\vdots & \vdots & & \vdots \\
\beta_1^{k-1} & \beta_2^{k-1} & \cdots & \beta_k^{k-1}
\end{vmatrix}
$$

is a Vandermonde determinant, whose value is the product of the terms $(\beta_r - \beta_s)$, for $r > s$ [cf. (6.14.3)]; since we have assumed that all β's are distinct, this does not vanish and the equations have a unique solution [see § 6.8 and § 5.8].

EXAMPLE 14.13.1.   We consider the Fibonacci series, (14.12.7). The recursion is $u_n = u_{n-1} + u_{n-2}$; the reverse polynomial is $y^2 - y - 1$, whose roots are $\frac{1}{2} \pm \frac{1}{2}\sqrt{5}$ [see (14.5.1)]. So, for some constants b_1 and b_2,

$$u_n = b_1(\tfrac{1}{2} + \tfrac{1}{2}\sqrt{5})^n + b_2(\tfrac{1}{2} - \tfrac{1}{2}\sqrt{5})^n. \tag{14.13.4}$$

To find b_1 and b_2, we use the initial values $u_0 = 0$, $u_1 = 1$. So

$$b_1 + b_2 = 0$$
$$(\tfrac{1}{2} + \tfrac{1}{2}\sqrt{5})b_1 + (\tfrac{1}{2} - \tfrac{1}{2}\sqrt{5})b_2 = 1.$$

The solution of these equations is $b_1 = 1/\sqrt{5}$, $b_2 = -1/\sqrt{5}$ [cf. Example 5.8.1]; this gives the final result that, for the Fibonacci sequence,

$$u_n = \frac{1}{\sqrt{5}}\{(\tfrac{1}{2} + \tfrac{1}{2}\sqrt{5})^n - (\tfrac{1}{2} - \tfrac{1}{2}\sqrt{5})^n\}. \tag{14.13.5}$$

Despite appearances such a term is always an integer.

When the reverse polynomial has a pair of complex conjugate roots, it is generally best to express them in the form $r\cos\theta \pm ir\sin\theta$.

EXAMPLE 14.13.2.   The recursion

$$u_n = 2u_{n-1} - 4u_{n-2}, \qquad (14.13.6)$$

has reverse polynomial $y^2 - 2y + 4$; its roots are

$$1 \pm i\sqrt{3} = 2\left(\cos\frac{\pi}{3} \pm i\sin\frac{\pi}{3}\right).$$

[cf. (2.7.9) and (2.7.22)]. The general real solution is therefore

$$u_n = 2^n\left\{a\left(\cos\frac{\pi}{3} + i\sin\frac{\pi}{3}\right)^n + \bar{a}\left(\cos\frac{\pi}{3} - i\sin\frac{\pi}{3}\right)^n\right\}$$

$$= 2^n\left\{a\left(\cos n\frac{\pi}{3} + i\sin n\frac{\pi}{3}\right) + \bar{a}\left(\cos n\frac{\pi}{3} - i\sin n\frac{\pi}{3}\right)\right\}$$

[see (2.7.29)]

$$= 2^n\left(b_1\cos n\frac{\pi}{3} + b_2\sin n\frac{\pi}{3}\right). \qquad (14.13.7)$$

If the reverse polynomial has a repeated root, the form of the general solution can still be obtained by using partial fractions and expanding in a power-series. If, say, $(y - \beta)^r$ is a factor of the polynomial, the corresponding general term is $(b_0 + b_1 n + b_2 n^2 + \ldots + b_{r-1} n^{r-1})\beta^n$.

EXAMPLE 14.13.3.   Consider the recursion $u_0, u_1, \ldots$, where

$$u_n = 3u_{n-2} - 2u_{n-3}, \qquad (n \geqslant 3) \qquad (14.13.8)$$

The reverse polynomial is $y^3 - 3y + 2 = (y - 1)^2(y + 2)$. The general solution is [see Proposition 14.13.1]:

$$u_n = (b_0 + b_1 n)(1)^n + b_2(-2)^n$$
$$= b_0 + b_1 n + b_2(-2)^n. \qquad (14.13.9)$$

If, for instance, the sequence starts $0, 0, 1, 0, 3, -2, 9, -12, 31, \ldots$, then $u_n = \frac{1}{9}(-1 + 3n + (-2)^n)$.

   (ii) The second approach is based on the *principle of superposition*. Although the argument applies in general, we shall explain it only for second order equations. The computations ultimately turn out to be much the same as in the first method. But the point of view is different; it resembles the treatment of linear differential equations [see IV, § 7.4].

   We wish to solve the recurrence relation

$$u_{n+2} = a_1 u_{n+1} + a_2 u_n \quad (a_2 \neq 0) \qquad (14.13.10)$$

on the assumption that the *initial values*

$$u_0 \quad \text{and} \quad u_1 \qquad (14.13.11)$$

are prescribed. Disregarding the initial conditions for the moment, we shall show that if s_n and t_n are solutions of (14.13.10) then so is any sequence w_n obtained by superposition, that is

$$w_n = ps_n + qt_n,$$

where p and q are arbitrary constants. Indeed, suppose that

$$s_{n+2} = a_1 s_{n+1} + a_2 s_n$$

and

$$t_{n+2} = a_1 t_{n+1} + a_2 t_n.$$

On multiplying these equations by p and q respectively and then adding we obtain that

$$ps_{n+2} + qt_{n+2} = a_1(ps_{n+1} + qt_{n+1}) + a_2(ps_n + qt_n),$$

that is

$$w_{n+2} = a_1 w_{n+1} + a_2 w_n,$$

as claimed. Using the language of linear algebra [see Definition 5.2.1], we can state that the set of solutions of (14.13.10) forms a vector space. Since w_n involves two arbitrary constants, it may be expected that the initial conditions can be satisfied by a suitable choice of p and q.

The problem has now been reduced to finding two *particular* solutions s_n and t_n of (14.13.10). However, these solutions must be linearly independent, that is a relation of the type

$$as_n + bt_n = 0 \quad \text{for all } n$$

can hold only if $a = b = 0$. This implies that neither s_n nor t_n can be the trivial solution ($u_n = 0$), nor can we allow proportional solutions ($t_n = ks_n$ for all n). We consider a *trial solution* of the form

$$u_n = x^n \quad (x \neq 0),$$

where x remains to be determined. Substituting in (14.13.10) we find that

$$x^{n+2} = a_1 x^{n+1} + a_2 x^n;$$

since $x \neq 0$ we may cancel a factor x^n and obtain that

$$x^2 - a_1 x - a_2 = 0. \tag{14.13.12}$$

This is called the *auxiliary equation* associated with the recurrence relation. Here it is a quadratic equation; had we been concerned with a recurrence relation of order k, then the auxiliary equation would have been of degree k.

As we are assuming that a_1 and a_2 are real, we have to distinguish between three types of quadratic equation (14.13.12) [see Proposition 14.5.4].

1st case: $a_1{}^2 + 4a_2 > 0$. There are two distinct real roots, say

$$\lambda = \alpha, \qquad \lambda = \beta$$

and the general solution is

$$w_n = p\alpha^n + q\beta^n.$$

In order to satisfy the initial condition we put $n = 0$ and $n = 1$ and solve the equations

$$u_0 = p + q, \qquad u_1 = \alpha p + \beta q.$$

Thus

$$p = \frac{u_1 - \beta u_0}{\alpha - \beta}, \qquad q = \frac{\alpha u_0 - u_1}{\alpha - \beta},$$

and the sequence with the appropriate initial values is given by

$$u_n = \left(\frac{u_1 - \beta u_0}{\alpha - \beta}\right)\alpha^n + \left(\frac{\alpha u_0 - u_1}{\alpha - \beta}\right)\beta^n \quad (n = 0, 1, 2, \ldots). \quad (14.13.13)$$

2nd case: $a_1{}^2 + 4a_2 < 0$. The solution of (14.13.12) is a pair of conjugate complex numbers, α and $\bar{\alpha}$. In this situation it is often convenient to use polar coordinates [see § 2.7.3]. Thus put

$$\alpha = r(\cos \phi + i \sin \phi) \quad (\sin \phi \neq 0).$$

The particular solutions of (14.13.10) may be taken to be the sequences α^n and $\bar{\alpha}^n$, as in the real case. But it is more advantageous to use instead the sequences [see (2.7.29)]:

$$\tfrac{1}{2}(\alpha^n + \bar{\alpha}^n) = r^n \cos n\phi, \qquad \frac{1}{2i}(\alpha^n - \bar{\alpha}^n) = r^n \sin \phi,$$

which are also solutions by virtue of the principle of superposition. The general solution now takes the form

$$w_n = pr^n \cos n\phi + qr^n \sin n\phi.$$

In order to satisfy the initial condition we have to put

$$p = u_0$$

$$pr \cos \phi + qr \sin \phi = u_1.$$

Hence the required solution is

$$u_n = u_0 r^n \cos n\phi + \left(\frac{u_1 - u_0 r \cos \phi}{r \sin \phi}\right) r^n \sin n\phi. \qquad (14.13.14)$$

3rd case: $a_1{}^2 + 4a_2 = 0$. The auxiliary equation (14.13.12) becomes

$$(\lambda - \tfrac{1}{2}a_1)^2 = 0, \qquad a_1 \neq 0.$$

There is now a double root, and the previous method yields only one particular solution, namely

$$s_n = (\tfrac{1}{2}a_1)^n.$$

However, it can be verified that in this case the sequence

$$t_n = n(\tfrac{1}{2}a_1)^n$$

is an independent particular solution; indeed,

$$t_{n+2} - a_1 t_{n+1} + \tfrac{1}{4} a_1^2 t_n$$
$$= (n + 2)(\tfrac{1}{2} a_1)^{n+2} - a_1(n + 1)(\tfrac{1}{2} a_1)^{n+1} + \tfrac{1}{4} a_1^2 n(\tfrac{1}{2} a_1)^n = 0.$$

The general solution is

$$w_n = p(\tfrac{1}{2} a_1)^n + qn(\tfrac{1}{2} a_1)^n,$$

where p and q are determined by the initial conditions

$$u_0 = p$$
$$u_1 = p\tfrac{1}{2} a_1 + q\tfrac{1}{2} a_1.$$

Thus the solution we seek is given by

$$u_n = u_0(\tfrac{1}{2} a_1)^n + \left(\frac{2u_1}{a_1} - u_0\right) n(\tfrac{1}{2} a_1)^n. \qquad (14.13.15)$$

In general, the phenomenon of multiple roots is as follows:

PROPOSITION 14.13.1. *If a linear recurrence relation has an auxiliary equation with an m-fold root* α, *then each of the sequences*

$$\alpha^n, n\alpha^n, n^2\alpha^n, \ldots, n^{m-1}\alpha^n$$

is a particular solution.

We conclude this section by mentioning some linear recurrence relations which are either non-homogeneous, that is contain a term that is independent of the sequence u_n, or have coefficients which depend on n.

EXAMPLE 14.13.4. The recurrence relation

$$u_{n+1} - nu_n = 0, \qquad u_1 = 1$$

has solution $u_n = (n - 1)!$

EXAMPLE 14.13.5. The recurrence relation

$$u_{n+1} - u_n = n, \quad u_0 = 0$$

has solution

$$u_n = \tfrac{1}{2} n(n - 1).$$

EXAMPLE 14.13.6. The recurrence relation

$$u_n = (n - 1)(u_{n-1} + u_{n-2}), \quad u_0 = 1, u_1 = 0$$

has solution

$$u_n = n! \sum_{r=0}^{n} \frac{(-1)^k}{r!} \quad (n \geq 0).$$

14.14.   SYSTEMS OF RECURRENCE RELATIONS

The first order recurrence relation

$$u_{n+1} = au_n \quad (a \neq 0) \tag{14.14.1}$$

is solved by

$$u_n = a^n u_0, \tag{14.14.2}$$

where u_0 is the initial value.

Generalizing this concept we shall study a system of simultaneous recurrence relations involving several unknown sequences $x_n, y_n, z_n, \ldots$. We shall discuss in detail only the case of two relations for two sequences. Thus we wish to solve the equations

$$
\begin{aligned}
x_{n+1} &= a_1 x_n + a_2 y_n \\
y_{n+1} &= b_1 x_n + b_2 y_n
\end{aligned}
\tag{14.14.3}
$$

on the assumption that the initial values x_0 and y_0 are prescribed. An instantaneous solution is at hand if the system is written in matrix form. For let

$$
\mathbf{w}_n = \begin{pmatrix} x_n \\ y_n \end{pmatrix}, \qquad \mathbf{A} = \begin{pmatrix} a_1 & a_2 \\ b_1 & b_2 \end{pmatrix}.
$$

Then (14.14.3) becomes

$$\mathbf{w}_{n+1} = \mathbf{A}\mathbf{w}_n,$$

and by analogy with (14.14.2) the solution is

$$\mathbf{w}_n = \mathbf{A}^n \mathbf{w}_0. \tag{14.14.4}$$

Although this formula is concise and theoretically satisfactory, there remains the practical problem of finding a tractable expression for an arbitrary power of $\mathbf{A}$. In order to simplify the argument we make the additional hypothesis that $\mathbf{A}$ has *distinct eigenvalues*, which we shall denote by α and β. We recall that by the Cayley-Hamilton Theorem [see Theorem 7.5.1] the matrix $\mathbf{A}$ satisfies its own characteristic equation, namely [see § 7.3: (7.3.9) and (7.3.10) in particular]:

$$\mathbf{A}^2 - (\alpha + \beta)\mathbf{A} + \alpha\beta\mathbf{I} = 0,$$

which we rewrite as

$$\mathbf{A}^2 = p_2\mathbf{I} + q_2\mathbf{A}.$$

Multiplying both sides by $\mathbf{A}$ we obtain that

$$\mathbf{A}^3 = p_2\mathbf{A} + q_2\mathbf{A}^2 = p_2\mathbf{I} + q_2(p_2\mathbf{I} + q_2\mathbf{A}) = p_3\mathbf{I} + q_3\mathbf{A},$$

say. Continuing the argument we can easily prove by induction [see § 2.1] that there are scalars p_n, q_n ($n = 0, 1, 2, \ldots$) such that

$$\mathbf{A}^n = p_n\mathbf{I} + q_n\mathbf{A}. \tag{14.14.5}$$

Again, by induction, it can be shown that

$$p_n = \frac{\alpha\beta^n - \beta\alpha^n}{\alpha - \beta}, \qquad q_n = \frac{\alpha^n - \beta^n}{\alpha - \beta} \quad (n \geq 0). \tag{14.14.6}$$

Substituting in (14.14.5) and (14.14.4) we obtain an explicit formula for the solution of (14.14.4), thus

$$\mathbf{w}_n = \left(\frac{\alpha\beta^n - \alpha^n\beta}{\alpha - \beta}\right)\mathbf{w}_0 + \left(\frac{\alpha^n - \beta^n}{\alpha - \beta}\right)A\mathbf{w}_0. \tag{14.14.7}$$

EXAMPLE 14.14.1. Consider the system

$$x_{n+1} = 4x_n - 2y_n$$
$$y_{n+1} = 3x_n - y_n. \tag{14.14.8}$$

The eigenvalues of the coefficient matrix are given by [see Theorem 7.2.3]

$$\begin{vmatrix} t - 4 & 2 \\ -3 & t + 1 \end{vmatrix} = t^2 - 3t + 2,$$

that is [see (14.5.1)]

$$\alpha = 2, \qquad \beta = 1.$$

Substituting in (14.14.7) we obtain the solution of (14.14.8):

$$\binom{x_n}{y_n} = (2 - 2^n)\binom{x_0}{y_0} + (2^n - 1)\begin{pmatrix} 4 & -2 \\ 3 & -1 \end{pmatrix}\binom{x_0}{y_0}$$

or, after a short calculation,

$$x_n = (3 \cdot 2^n - 2)x_0 + (-2^{n+1} + 2)y_0$$
$$y_n = (3 \cdot 2^n - 3)x_0 + (3 - 2^{n+1})y_0.$$

It is interesting to observe that the theory of first order systems can be adapted to solve recurrence relations of higher order. We illustrate the technique by considering the second order equation

$$u_{n+2} = a_1 u_{n+1} + a_2 u_n \tag{14.14.9}$$

with initial values u_0 and u_1 already studied in section 14.13.
    The idea is to introduce a new sequence

$$v_n = u_{n+1} \quad (n \geq 0). \tag{14.14.10}$$

The single equation (14.14.9) is equivalent to the system

$$u_{n+1} = v_n$$
$$v_{n+1} = a_1 v_n + a_2 u_n \tag{14.14.10}$$

with the initial values u_0 and $v_0 = u_1$. We write the system in vector form, thus

$$\mathbf{w}_{n+1} = A\mathbf{w}_n, \tag{14.14.11}$$

where

$$\mathbf{w}_n = \begin{pmatrix} u_n \\ v_n \end{pmatrix}, \qquad \mathbf{A} = \begin{pmatrix} 0 & 1 \\ a_2 & a_1 \end{pmatrix}, \qquad \mathbf{w}_0 = \begin{pmatrix} u_0 \\ u_1 \end{pmatrix}. \qquad (14.14.12)$$

The characteristic polynomial [see § 7.3] is given by

$$|t\mathbf{I} - \mathbf{A}| = \begin{vmatrix} t & -1 \\ -a_2 & t - a_1 \end{vmatrix} = t^2 - a_1 t - a_2,$$

which is identical with the reversed polynomial (14.13.3) and, when equated to zero, with the auxiliary equation (14.13.12). Provided that the eigenvalues of $\mathbf{A}$ are distinct, the vector $\mathbf{w}_n$ is given by (14.14.7).

EXAMPLE 14.14.2.   We consider once again the Fibonacci sequence defined by

$$u_{n+2} = u_{n+1} + u_n, \qquad u_0 = 0, \qquad u_1 = 1.$$

The equivalent first order system is described in (14.14.11) and (14.14.12). Its matrix and initial vector are

$$\mathbf{A} = \begin{pmatrix} 0 & 1 \\ 1 & 1 \end{pmatrix} \quad \text{and} \quad \mathbf{w}_0 = \begin{pmatrix} 0 \\ 1 \end{pmatrix}.$$

The characteristic equation for $\mathbf{A}$ is

$$t^2 - t - 1 = 0,$$

and the eigenvalues are given by [see Theorem 7.2.3 and (14.5.1)]:

$$\alpha = \tfrac{1}{2}(1 + \sqrt{5}), \qquad \beta = \tfrac{1}{2}(1 - \sqrt{5}).$$

Hence we may use (14.14.7) to obtain the solution, thus

$$\begin{pmatrix} u_n \\ v_n \end{pmatrix} = \left(\frac{\alpha\beta^n - \beta\alpha^n}{\alpha - \beta} \right) \begin{pmatrix} 0 \\ 1 \end{pmatrix} + \frac{\alpha^n - \beta^n}{\alpha - \beta} \begin{pmatrix} 0 & 1 \\ 1 & 1 \end{pmatrix} \begin{pmatrix} 0 \\ 1 \end{pmatrix}.$$

For our purpose it suffices to compare the first components on both sides of this vector equation, giving

$$u_n = \frac{\alpha^n - \beta^n}{\alpha - \beta} = \frac{1}{\sqrt{5}} \left\{ \left(\frac{1 + \sqrt{5}}{2} \right)^n - \left(\frac{1 - \sqrt{5}}{2} \right)^n \right\},$$

in agreement with (14.13.5).

14.15.   POLYNOMIALS IN SEVERAL INDETERMINATES

A polynomial $f(x, y)$ over D in the *indeterminates* (variables) x and y is an expression of the form

$$a + (bx + cy) + (dx^2 + exy + gy^2) + (hx^3 + jxy^2 + kxy^2 + my^3) + \cdots, \qquad (14.15.1)$$

where the coefficients $a, b, c, \ldots$ are elements of the domain D, and there are only finitely many terms. The domain of coefficients can be an integral domain [see § 2.3], like the integers [see § 2.2.1]; but unless otherwise indicated it will be assumed that D is a field $\mathbb{F}$ [see § 2.4.2], usually $F = \mathbb{R}$ [see § 2.6.1] or $F = \mathbb{C}$ [see § 2.7.1].

The terms in (14.15.1) have been bracketed into the component *homogeneous* polynomials: a polynomial in x and y is said to be homogeneous of degree d if it is of the form

$$a_0 x^d + a_1 x^{d-1} y + a_2 x^{d-2} y^2 + \ldots + a_r x^{d-r} y^r + \ldots + a_d y^d$$

each term having *total degree* d when the degrees of x and y have been added together. In an arbitrary polynomial the degree of the term or terms of highest total degree is called the *degree* of the polynomial.

It will be clear how these definitions are extended to polynomials over D in any finite number of indeterminates

$$\mathbf{x} = (x_1, x_2, \ldots, x_n).$$

Such a polynomial will be denoted by $f(x_1, x_2, \ldots, x_n)$ or more briefly by $f(\mathbf{x})$, and it is equal to a finite sum

$$f(\mathbf{x}) = \sum_\alpha a_\alpha x_1{}^{\alpha_1} x_2{}^{\alpha_2} \ldots x_n{}^{\alpha_n},$$

where $\alpha = (\alpha_1, \alpha_2, \ldots, \alpha_n)$ is an n-tuple of non-negative integers and $a_\alpha \in D$. Two polynomials are equal if and only if they consist of the same terms. A single term, that is

$$x_1{}^{\alpha_1} x_2{}^{\alpha_2} \ldots x_n{}^{\alpha_n} \tag{14.5.2}$$

is called a *monomial* in $x_1, x_2, \ldots, x_n$. It is sometimes desirable to list the monomials in a systematic manner. For this purpose we use the *lexical* (*lexicographical* or *dictionary*) order: thus

$$x_1{}^{\alpha_1} x_2{}^{\alpha_2} \ldots x_n{}^{\alpha_n} \quad \text{precedes} \quad x_1{}^{\beta_1} x_2{}^{\beta_2} \ldots x_n{}^{\beta_n}$$

if in the sequence of differences

$$\alpha_1 - \beta_1, \alpha_2 - \beta_2, \ldots, \alpha_n - \beta_n$$

the first non-zero term is positive. For example, $x_1{}^2 x_2{}^2 x_3$ precedes $x_1{}^2 x_2 x_3 x_4$, and $x_1 x_2 x_3$ precedes $x_2{}^3$.

The polynomial $f(x_1, x_2, \ldots, x_n)$ is homogeneous of degree d if, with a further indeterminate t, we have that

$$f(t x_1, t x_2, \ldots, t x_n) = t^d f(x_1, x_2, \ldots, x_n). \tag{14.15.3}$$

The definition is in fact used as the definition of a homogeneous function of degree d, whether it is a polynomial or not. For instance, the rational function

$$\frac{x}{y^2 - 2z^2}$$

in the variables x, y, z is homogeneous of degree -1; but is not, of course, a polynomial.

As with polynomials in a single indeterminate, two polynomials in the same set of indeterminates can be added, subtracted, or multiplied to give polynomials in the same set of indeterminates. For this it must be accepted that a polynomial in x, y, z may happen to have zero coefficients for all the terms involving, say, y; it will then look like a polynomial in only x, z, but it is to be regarded as being one in x, y, z. This is not unfamiliar; the zero polynomial is regarded as a polynomial in x. Since polynomials can be combined in the ways described, and since the usual laws obtain, we have that the polynomials over D in any set of indeterminates form a ring, which we denote by

$$D[x_1, x_2, \ldots, x_n]. \tag{14.15.4}$$

It sometimes happens that D is itself a ring of polynomials, for example, $D = \mathbb{F}[x]$, the ring of polynomials in a single indeterminate x over a field $\mathbb{F}$ [see § 14.2]. If we adjoin a second indeterminate y to form $D[y]$, we can assert that

$$(D[x])[y] = D[x, y].$$

This means simply that a polynomial in x and y can either be expressed as a sum of terms $ax^\alpha y^\beta$ as in (14.15.1), or as a polynomial

$$p_0(x) + p_1(x)y + p_2(x)y^2 + \ldots + p_r(x)y^r$$

in y whose coefficients are polynomials in x.

The polynomial $f(x_1, x_2, \ldots, x_n)$ of $D[x_1, x_2, \ldots, x_n]$ is said to be *reducible* over D if

$$f(x_1, \ldots, x_n) = g(x_1, \ldots, x_n)h(x_1, \ldots, x_n), \tag{14.15.5}$$

where $g(x_1, \ldots, x_n)$ and $h(x_1, \ldots, x_n)$ are non-constant polynomials over D. If an equation like (14.15.5) is not possible, then we say that $f(x_1, \ldots, x_n)$ is *irreducible*.

When D is a field, we have the analogue of Theorem 14.4.1:

THEOREM 14.15.1.   *If $\mathbb{F}$ is a field, then every polynomial in $\mathbb{F}[x_1, \ldots, x_n]$ can be written as a product of irreducible polynomials over $\mathbb{F}$. The factorization is unique, except for the order of the factors and for the insertion of non-zero constant factors.*

14.16.   SYMMETRIC POLYNOMIALS

A polynomial in several indeterminates is said to be *symmetric* if it is unchanged by any permutation [see § 3.7] of the indeterminates. If, for instance, the indeterminates are x, y, z, then the polynomial $f(x, y, z)$ is symmetric if and only if

$$f(x, y, z) = f(y, x, z) = f(z, y, x) = f(x, z, y) = f(y, z, x) = f(z, x, y).$$

For example,

$$3x + 3y + 3z + x^2y + x^2z + xy^2 + xz^2 + y^2z + yz^2$$

is symmetric in x, y, z; whereas

$$2x - y + z$$

is not.

There is a set of symmetric polynomials in the indeterminates $x_1, x_2, \ldots, x_n$ that are called the *elementary symmetric functions*; they are:

$$
\begin{aligned}
c_1 &= x_1 + x_2 + \ldots + x_n \\
&= \sum x_i \qquad (1 \le i \le n) \\
c_2 &= x_1x_2 + x_1x_3 + \ldots + x_ix_j + \ldots + x_{n-1}x_n \\
&= \sum x_ix_j \qquad (1 \le i < j \le n) \qquad\qquad (14.16.1) \\
c_3 &= x_1x_2x_3 + \ldots + x_ix_jx_k + \ldots + x_{n-2}x_{n-1}x_n \\
&= \sum x_ix_jx_k \qquad (1 \le i < j < k \le n) \\
&\vdots \\
c_n &= x_1x_2 \ldots x_n.
\end{aligned}
$$

In general, c_r is the sum of all the different products of r distinct x's from among $x_1, x_2, \ldots, x_n$; thus c_r is the sum of $\binom{n}{r}$ terms [see § 3.8].

There is one immediate result: let t be an additional indeterminate. Then we have the important identity

$$
\begin{aligned}
(t - x_1)(t - x_2)(t - x_3) &\ldots (t - x_n) \\
&= t^n - c_1t^{n-1} + c_2t^{n-2} - \ldots + (-1)^nc_n. \quad (14.16.2)
\end{aligned}
$$

Suppose that the right-hand side of (14.16.2) is a real polynomial with known coefficients, and that the roots $x_1, x_2, \ldots, x_n$ are not known; it is often sufficient for an immediate purpose to know, for instance, that the sum of the roots is equal to c_1 the negative of the coefficient of t^{n-1}; or that their product is c_n, which is $(-1)^n$ times the constant term in (14.16.2) [cf. (14.5.1)]. As an illustration we discuss the following geometric problem.

EXAMPLE 14.16.1.   Under what conditions will four points on a parabola be concyclic (lie on a circle)? The parabola has equation $y^2 = 4ax$ for which the standard parametric form is $(at^2, 2at)$ [see V, § 1.3.2], that is the point (x, y) lies in the parabola, if and only if there exists a real number t such that

$$x = at^2, \qquad y = 2at. \qquad\qquad (14.16.3)$$

We shall simply refer to 'the point t' of the parabola, if the equations (14.16.3) are satisfied.

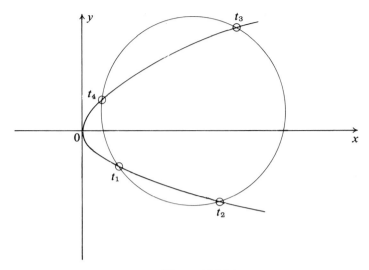

Figure 14.16.1

Suppose now that the four points with parametric values t_1, t_2, t_3, t_4 are, in fact concyclic. The general equation for a circle is [see V, § 1.1.8]

$$x^2 + y^2 + 2gx + 2fy + c = 0. \qquad (14.16.4)$$

Hence if the four points lie on this circle, then t_1, t_2, t_3, t_4 are the roots of the equation formed by substituting (14.16.3) in (14.16.4), giving

$$a^2t^4 + 4a^2t^2 + 2gat^2 + 4fat + c = 0$$

or

$$a^2t^4 + (4a^2 + 2ga)t^2 + 4fat + c = 0.$$

The relevant feature of this equation is that, whatever the values of g, f, c, the coefficient of t^3 is zero. Therefore, by (14.16.2)

$$t_1 + t_2 + t_3 + t_4 = 0. \qquad (14.16.5)$$

It is not hard to prove the converse, and so to establish that the four points are concyclic if and only if (14.16.5) is true.

The elementary symmetric functions are so called because of the following result:

THEOREM 14.16.1 (*Fundamental theorem on symmetric functions*). *Let* $S(x_1, x_2, \ldots, x_n)$ *be a symmetric polynomial in* $x_1, x_2, \ldots, x_n$. *Then there exists a polynomial G in n indeterminates such that*

$$S(x_1, x_2, \ldots, x_n) = G(c_1, c_2, \ldots, c_n). \qquad (14.16.6)$$

This is an identity when the functions $c_1, c_2, \ldots, c_n$ *are expressed in terms of* $x_1, x_2, \ldots, x_n$ *in accordance with* (14.16.1). *The expression* (14.16.6) *is unique in the sense that*

$$G(c_1, c_2, \ldots, c_n) = H(c_1, c_2, \ldots, c_n)$$

if and only if G and H are the same polynomial. Finally, if S has integral coefficients, so has G.

EXAMPLE 14.16.2.   Consider the polynomial

$$S(x_1, x_2, x_3) = x_1^2 x_2 + x_1^2 x_3 + x_2^2 x_1 + x_2^2 x_3 + x_3^2 x_1 + x_3^2 x_2$$

which is symmetrical in x_1, x_2, x_3. By expanding we find that

$$c_1 c_2 = (x_1 + x_2 + x_3)(x_1 x_2 + x_1 x_3 + x_2 x_3) = S(x_1, x_2, x_3) + 3 x_1 x_2 x_3.$$

Therefore

$$S(x_1, x_2, x_3) = c_1 c_2 - 3 c_3.$$

There is an infinite sequence of symmetric polynomials in $x_1, x_2, \ldots, x_n$, namely the *sums of powers*:

$$s_r = x_1^r + x_2^r + \ldots + x_n^r \quad (r = 0, 1, 2, \ldots). \tag{14.16.7}$$

When $n = 0$, each term is equal to unity, so that $s_0 = n$. By Theorem 14.16.1 each s_r can be expressed as a polynomial in $c_1, c_2, \ldots, c_n$. Now s_r is of degree r in $x_1, x_2, \ldots, x_n$. Hence on the right-hand side only $c_1, c_2, \ldots, c_r$ can occur, but not $c_{r+1}, \ldots, c_n$, whose degrees are greater than r. Hence there exist polynomials G_r such that

$$s_r = G_r(c_1, c_2, \ldots, c_r) \quad (r = 0, 1, 2, \ldots). \tag{14.16.8}$$

The expressions on the right-hand side of (14.16.8) can be found successively by using a set of relations between the power sums and the elementary symmetric functions which were discovered by Newton.

THEOREM 14.16.2 (*Newton's identities*).   *If* $1 \le r \le n$,

$$s_r - c_1 s_{r-1} + c_2 s_{r-2} - \ldots + (-1)^{r-1} c_{r-1} s_1 + (-1)^r r c_r = 0. \tag{14.16.9}$$

If $r > n$,

$$s_r - c_1 s_{r-1} + c_2 s_{r-2} - \ldots + (-1)^{n-1} c_{n-1} s_{r-n+1} + (-1)^n c_n s_{r-n} = 0. \tag{14.16.10}$$

When $n \ge 4$, the first 4 equations in (14.16.9) are

$$s_1 - c_1 = 0$$
$$s_2 - c_1 s_1 + 2 c_2 = 0$$
$$s_3 - c_1 s_2 + c_2 s_1 - 3 c_3 = 0$$
$$s_4 - c_1 s_3 + c_2 s_2 - c_3 s_1 + 4 c_4 = 0.$$

Solving for the s's in terms of the c's we obtain that

$$s_1 = c_1 \tag{14.16.11}$$
$$s_2 = c_1 s_1 - 2 c_2 = c_1^2 - 2 c_2 \tag{14.16.12}$$
$$s_3 = c_1 s_2 - c_2 s_1 + 3 c_3$$

whence by substituting for s_2 and s_1

$$s_3 = c_1{}^3 - 3c_1c_2 + 3c_3. \qquad (14.16.13)$$

Similarly,

$$s_4 = c_1{}^4 - 4c_1{}^2c_2 + 4c_1c_3 + 2c_2{}^2 - 4c_4. \qquad (14.16.14)$$

It will be observed that, in accordance with the last part of Theorem 14.16.1, no fractional coefficients appear in these formulae.

Conversely, Newton's identities can also be used to express $c_1, c_2, \ldots, c_n$ in terms of $s_1, s_2, \ldots, s_n$. In this case, however, the polynomials no longer have integral coefficients; for instance,

$$c_2 = \tfrac{1}{2}(s_1{}^2 - s_2),$$
$$c_3 = \tfrac{1}{6}(s_1{}^3 - 3s_1s_2 + 2s_3).$$

EXAMPLE 14.16.3.   Let p, q and r be arbitrary real numbers. Denote the roots of

$$t^5 - 2t^4 + 3t^3 - pt^2 + qt - r = 0$$

by x_1, x_2, x_3, x_4, x_5. Since $c_1 = 2$ and $c_2 = 3$ we find that

$$s_2 = \sum_{i=1}^{5} x_i{}^2 = c_1{}^2 - 2c_2 = -2.$$

It follows that the roots cannot all be real, because the sum of squares of real numbers is not negative.

BIBLIOGRAPHY

Fraleigh, J. B. (1967). *A First Course on Abstract Algebra*, Addison-Wesley, Chapter 3.

Jacobson, N. (1974). *Basic Algebra I*, Freeman & Co., Chapter 2.

MacLane, S. and Birkhoff, G. (1967). *Algebra*, Macmillan, Chapter 3.

Marcus, M. (1977). *Introduction to Modern Algebra*, Dekker, Chapter 3.

van der Waerden, B. L. (1949). *Modern Algebra I*, Unger, Chapters 3 and 4.

CHAPTER 15

Financial Mathematics

15.1. INVESTMENTS, ANNUITIES AND LOANS

15.1.1. Interest

Suppose we invest a capital of S pounds at an annual rate of interest i. This means that at the end of one year the value of that sum will have increased to $S(1 + i)$ pounds. If this capital remains invested, at the same rate of interest, then after n years it will have increased to $S(1 + i)^n$ pounds, and we say accumulation has taken place at *compound interest*.

Conversely, if we know that we shall receive a payment of S at the end of n years from now, then with an interest rate of i p.a. the present value of that payment is $S(1 + i)^{-n}$, since this is the amount which, if invested now, would have increased, with compound interest at an annual rate of i, to S at the time when that payment is due. We say S, payable after n years, is *discounted* at a rate of interest i.

We shall write v for $(1 + i)^{-1}$, and call it the *present value* of 1 due a year hence. We shall also write d for $iv = 1 - v$, the *discount* on 1 due a year hence (i.e. the difference between the value of 1 payable one year hence, and of 1 payable now).

In financial practice it is usual to pay interest at the end of fractions of one year. If after one mth of a year an amount of $i_{(m)}/m$ is paid for a unit sum, so that after one year altogether $i_{(m)}$ has been paid as interest, then $i_{(m)}$ is called the *nominal rate* of interest, *convertible* m times a year. It corresponds to the *effective* rate of interest i, when

$$\left(1 + \frac{i_{(m)}}{m}\right)^m = 1 + i, \quad \text{i.e. } i_{(m)} = m[(1 + i)^{1/m} - 1].$$

We notice that i is larger than $i_{(m)}$, which is also obvious from the meaning of these two symbols.

Imagine now that m increases, so that the fraction of a year, at the end of which interest $i_{(m)}/m$ or a unit capital is due, decreases. We have then [using IV (2.11.6) and (2.11.4)]:

$$\lim_{m \to \infty} i_{(m)} = \lim_{m \to \infty} \frac{(1 + i)^{1/m} - 1}{1/m} = \log_e (1 + i) = -\log_e (1 - d). \quad (15.1.1)$$

We denote this limit by δ, the *force of interest*, or *force of discount*. It follows [from IV (2.11.6)] that $(1 + i) = e^{\delta}$.

15.1.2.   Annuities

When payments are made at regular intervals, fixed in advance, and independent of circumstances such as survival of a recipient or other conditions, then we speak of an *annuity certain*.

The value of an annuity certain of 1 p.a. for n years, payments being made at the end of each year, is [IV, Example 1.7.1]

$$v + v^2 + \ldots + v^n = v(1 - v^n)/(1 - v) = (1 - v^n)/i \qquad (15.1.2)$$

and is denoted by $a_{\overline{n}|}$.

The value of an annuity certain of 1 p.a. for n years, payments being made at the beginning of each year, is

$$1 + v + \ldots + v^{n-1} = (1 - v^n)/(1 - v) = (1 - v^n)/d \qquad (15.1.3)$$

and is denoted by $\ddot{a}_{\overline{n}|}$.

We have

$$a_{\overline{n}|} = \ddot{a}_{\overline{n+1}|} - 1.$$

If the number of payments is not limited, then we obtain

$$a_{\overline{\infty}|} = 1/i, \quad \text{and} \quad \ddot{a}_{\overline{\infty}|} = 1/d.$$

At the time of the last payment of an annuity certain for n years the value of the annuity is:

$$1 + (1 + i) + (1 + i)^2 + \ldots + (1 + i)^{n-1} = \frac{(1 + i)^n - 1}{i}. \qquad (15.1.4)$$

If an annuity is paid in intervals of $1/m$ years, during n years altogether, in instalments of $1/m$, the first instalment being due now, then its present value is

$$\frac{1}{m}\left[\sum_{j=0}^{nm-1}\left(1 + \frac{i_{(m)}}{m}\right)^{-j/m}\right] = \frac{1}{m}\frac{1 - v_{(m)}^n}{1 - v_{(m)}^{1/m}}, \quad \text{where } v_{(m)} = \left(1 + \frac{i_{(m)}}{m}\right)^{-1}, \qquad (15.1.5)$$

and when m increases without limit, we have the *continuous annuity*

$$\bar{a}_{\overline{n}|} = \frac{1 - e^{-\delta n}}{\delta}. \qquad (15.1.6)$$

We mention also an annuity whose yearly payments are 1 (after one year), $\ldots$, k (after k years), $\ldots$, n (after n years). Its present value is [IV, Example 1.7.1]

$$v(1 + 2v + 3v^2 + \ldots + nv^{n-1}) = v\frac{1 - v^n}{d^2} - \frac{nv^{n+1}}{d}. \qquad (15.1.7)$$

At the time of the last payment the accumulated value of such an annuity is:

$$\frac{(1 + i)^n - 1}{vi^2} - \frac{n}{i}. \qquad (15.1.8)$$

Similarly, a decreasing annuity of n (now), decreasing by 1 each year until 1 (after $n - 1$ years) has present value:

$$\frac{n}{d} - \frac{v(1 - v^n)}{d^2} \qquad (15.1.9)$$

and its value at the time of the last payment is:

$$\frac{n(1 + i)^n}{i} - \frac{(1 + i)^n - 1}{i^2}. \qquad (15.1.10)$$

With increasing rate of interest the value of an annuity decreases. A formula which gives the precise relationship between annuities calculated at different rates of interest is [see IV, § 1.11]:

$$a_{\overline{n}|} - a'_{\overline{n}|} = (i' - i)v \sum_{t=1}^{n} \sum_{m=t}^{n} v^{m-t} v'^t$$

$$= (i' - i)v' \sum_{t=1}^{n} \sum_{m=t}^{n} v'^{m-t} v^t, \qquad (15.1.11)$$

where primed symbols refer to rate i' and not primed to rate i of annual interest. (Note that the 'roof' $\overline{}|$ on the subscript of symbols such as $a_{\overline{n}|}$, $\breve{a}_{\overline{n}|}$, etc. is an essential part of the notation.)

15.1.3. Repayment of Loans

Let us imagine that a loan of 1000, given now, is to be repaid in some way which takes account of a yield of i p.a. We consider various ways of repayment. In each case, the present value of the repayments will have to equal 1000.

If the loan is to be repaid by a lump-sum after n years, then obviously that lump-sum must be $1000(1 + i)^n$.

Another simple method of repayment consists of paying 1000 after n years, and paying meanwhile at the end of each year an amount of $1000i$. The present value of such payment is $1000ia_{\overline{n}|} + 1000v^n = 1000(1 - v^n) + 1000v^n = 1000$, as it clearly must be. When this scheme operates, then after the payment of the interest due at the end of a year, the outstanding loan is still 1000.

In certain circumstances it might be considered more appropriate to repay the loan by equal annual payments p, at the end of each of the next n years. The value of such payments is $pa_{\overline{n}|}$, so that p will equal $1000/a_{\overline{n}|}$.

After t years, when the tth instalment is just due, the still outstanding fraction of the loan will be $p\breve{a}_{\overline{n-t+1}|}$. Each payment of p can be considered as the payment of interest on the loan still outstanding, and partial repayment. The two parts form different portions of p as time progresses, because the part representing interest will gradually diminish. We illustrate this fact by an example.

EXAMPLE 15.1.1. Let $n = 10$, $i = 0.05$. Then $(1 - v^n)/i = (1 - 0.61391)/0.05 = 7.7218$, and $1000/7.7218 = 129.50$. The repayment plan is as follows:

End of year	Loan outstanding before payment of annuity	Interest on outstanding loan	Repayment
1	1000·00	50·00	79·50
2	920·50	46·02	83·48
3	837·02	41·85	87·65
4	749·37	37·45	92·05
5	657·32	32·87	96·63
6	560·69	28·03	101·47
7	459·22	22·96	106·54
8	352·68	17·62	111·88
9	240·80	12·04	117·46
10	123·34	6·16	123·34

The loan outstanding after 6 years, for instance, before payment of the instalment due, will be

$$129 \cdot 50 \ddot{a}_{\overline{5}|} = 129 \cdot 50 \times 4 \cdot 3296 = 560 \cdot 69$$

as in the table above.

If the repayment is to be made throughout an unlimited time, then the instalment of the annuity is $1000i$, which is obvious from first principles: the payments are simply interest on the loan, and no repayment takes place at all.

Consider now the case when during the time of repayment of a loan (e.g. a mortgage) the rate of interest is increased. There are then (at least) two possible arrangements. If the outstanding part of the loan is S', say, and the remaining number of instalments—the first is to be paid at once—is t, then the yearly amount of the instalment can be increased to $S'/\ddot{a}'_{\overline{t}|}$, where $\ddot{a}'_{\overline{t}|}$ is the value of an annuity at the higher rate of interest. This will result in an increase, because $\ddot{a}'_{\overline{t}|} < \ddot{a}_{\overline{t}|}$.

An alternative arrangement is possible, by retaining the amount of the yearly instalment, but extending the number of future payments. If the amount to be paid annually is p, then a time t' will have to be found such that $S' = p\ddot{a}'_{\overline{t'}|}$ where $a'_{\overline{t'}|}$ is again calculated at the higher rate of interest. It could be that t' turns out to be an integer plus a fraction. This would mean that the last payment would be smaller than p.

If no tables are readily available for the new rate of interest, then a simple computation can serve. For instance, if in the example above after 8 years, just before the payment of 129·50, the rate of interest is increased from 5% to 7%, the calculation would proceed as follows:

End of year	Loan outstanding	Interest	Repayment
8	352·68	24·69	104·81
9	247·87	17·33	112·17
10	135·70	9·50	120·00
11	15·70	1·10	15·70

so that the last payment, after 11 years, would be $1 \cdot 10 + 15 \cdot 70 = 16 \cdot 80$.

Now let us think of an organization which invites the public to subscribe to a loan under the following conditions: The amount of the loan is, say £1,000,000, in 1000 portions of £1000 each. The time of repayment is not fixed in advance, but at n different specified numbers of years, $t_1, \ldots, t_n$ one nth of the 1000 portions is being repaid, to holders chosen at random.

While a portion of the loan is not repaid, the holder of that portion receives interest at a rate of i p.a. From his point of view he has acquired the right to repayment after t_1, or $t_2, \ldots$, or t_n years, with probability $1/n$ for each of these periods [see II, § 3.3], with the right to interest until repayment.

It is clear that the value of all payments by the borrowing organization, discounted to the time of the launching of the loan, and computed at a rate of interest i, equals the total capital borrowed, since the arrangement can be viewed as comprising n different loans, repayable after different numbers of years, with interest being paid before repayment. For each of these portions we have seen above that the result holds.

If no interest is being paid before redemption of a capital S, and the probability of repayment after $1, 2, \ldots, n$ years is $1/n$ is each case, then the present, discount value of the capital is [see (15.1.2)]

$$S(v + v^2 + \ldots + v^n)/n = Sv(1 - v^n)/nd. \qquad (15.1.12)$$

15.1.4. Intrinsic Rate of Interest

Consider again an investment which guarantees the receipt of 1 after n years, with a yield of i at the end of the 1st, 2nd, $\ldots$, nth year from now. We know that the present value of this investment, discounted at a rate of interest i, equals 1.

The right of the investor to the stipulated receipts can be sold, but the price will not be necessarily equal to 1, because it will depend on assumptions about future rates of interest on the market. Let the price be P. By paying this price, the buyer acquires the right to the stipulated payments. He will then want to know which rate of interest, say i_0, would make the discounted value of all those payments equal to P. That rate i_0 is called the rate of interest *intrinsic* to the transaction.

Formally, we have to solve

$$P = v_0^n + iv_0(1 - v_0^n)/(1 - v_0) \qquad (15.1.13)$$

for v_0 and then find $i_0 = (1/v_0) - 1$.

In practice no great accuracy is required for the solution of this equation, and various methods of approximate solution are used, assuming that v_0 equals $(1 + x)v$, where v is, of course, $1/(1 + i)$ and x is small enough for higher powers than the first to be ignored.

More generally, let us assume that we pay P for the right to receive, after t ($= 1, 2, \ldots, n$) amounts p_t, and that we wish to compute the intrinsic rate of interest in such an arrangement. The payments cease altogether after n years, and therefore they can be considered as covering not only interest, but also part repayment of the invested capital, P. In practice, though, p_n might be much larger than the other p_i. This was so in the last example.

To solve the problem, we have to solve

$$P = p_1v + p_2v^2 + \ldots + p_nv^n$$

for v, and the intrinsic rate of interest will be $i = (1/v) - 1$.

If P and all p_i are positive, then only one positive value of v satisfies the equation, and all other solutions will be irrelevant.

The situation is different, though, if the contract provides for payments to be made during the time of its validity by either party to the other, which is formally equivalent to some of the p_i being negative.

As an example, consider the following arrangement.

> At time 0, A pays to B an amount of £48.
> At time 1, B pays to A an amount of £140.
> Finally, at time 2, A pays to B an amount of £100.

We wish to know at what intrinsic rate we can say that the payments balance each other, taking account of the fact that the payments were made at different times, and of the appropriate discounts.

We solve the equation

$$48 - 140v + 100v^2 = 0$$

and obtain for v either 0·6 or 0·8 [see (2.7.9)], and hence for i either $\frac{2}{3}$ or $\frac{1}{4}$.

Indeed either rate of interest makes the respective payments balance each other. The payment by A of 48 increases, by 32 (or by 12) to 80 (or to 60). Now B pays A 140, i.e. 60 (or 80) more than he had received, with interest. This negative balance of his increases, with interest, to 100 (or again to 100), and this is made up by the payment of 100, which he receives from A.

In such cases there is no criterion which would decide which is the 'true' intrinsic rate of interest.

15.2. ACTUARIAL MATHEMATICS

15.2.1. Life Contingencies

We have considered, at the end of section 15.1.3, a case where an investor receives the capital sum after some number of years, and where the precise year when this happens is uncertain, each year having a given probability.

We shall now consider contracts where payments are again subject to probability, but where the latter are not constant throughout the years. They depend on the survival of an insured person.

As a basis for the computation of the probabilities we use a *life table*, i.e. a list of values

$$l_{x_0}, \ldots, l_x, l_{x+1}, \ldots$$

which gives the number of survivors to be expected at ages $x_0 + 1, \ldots, x$, $x + 1, \ldots$ out of a number of l_{x_0} of lives aged x_0. The probability of a person aged x to survive to age $x + 1$ is then $l_{x+1}/l_x = p_x$, say and to survive to age $x + n$ is l_{x+n}/l_x. The probability of a person aged x to die during the next year is $(l_x - l_{x+1})/l_x = 1 - p_x = q_x$, say.

Consider, then, an insurance policy providing for the payment of a sum assured 1 at the end of n years to a person aged now x, provided the person is then alive. With a rate of interest i the value of such an insurance is

$$\frac{l_{x+n}}{(1+i)^n} \frac{1}{l_x} = l_{x+n} v^n / l_x = {}_nE_x, \text{ say.} \tag{15.2.1}$$

This is the net single premium to be paid for the insurance. The actual premium prescribed by an insurer for the insurance will be higher, because a loading will be added for administrative costs, risk of fluctuations in mortality, profit of the insurance company, and sometimes for participation of the policy-holder in such profits. We shall ignore such loadings and describe only the computation of net premiums, for various types of insurance.

A *Whole-Life policy* provides for the payment of a capital at the end of the year of death of the insured person. The present value of such an insurance, for a unit sum assured, on a person aged x now is

$$\sum_{t=0}^{\infty} \frac{l_{x+t} - l_{x+t+1}}{l_x} v^{t+1} = A_x, \text{ say.} \tag{15.2.2}$$

If the sum assured is S, the value of such an assurance is SA_x.

An *Endowment Assurance* for n years provides for the payment of the sum assured at the end of the year of death, if death occurs within n years, or at the end of n years, if the insured person is then alive. Its present value for an insured aged x, and a sum assured 1, is

$$\sum_{t=0}^{n-1} \frac{l_{x+t} - l_{x+t+1}}{l_x} v^{t+1} + {}_nE_x = A_{x:\overline{n}|}, \text{ say.} \tag{15.2.3}$$

If the sum assured is S, the value of such an assurance is $SA_{x:\overline{n}|}$.

The values ${}_nE_x$, A_x, and $A_{x:\overline{n}|}$ are single premiums. If the policy-holder pays instead a constant annual premium while the insurance is in force, then we must know the value of these payments.

To obtain this value, we observe that the present value of 1 p.a. paid, by a person now aged x, at the beginning of each year while this person is alive, called a *life annuity*, equals

$$\sum_{t=0}^{\infty} \frac{l_{x+t}}{l_x} v^t = \ddot{a}_x, \text{ say.} \tag{15.2.4}$$

(Do not confuse this notation with $\ddot{a}_{\overline{n}|}$.) If payments are made at most n times, then the value of such a *temporary annuity* is

$$\sum_{t=0}^{n-1} \frac{l_{x+t}}{l_x} v^t = \ddot{a}_{x:\overline{n}|}, \text{ say.} \tag{15.2.5}$$

Simple algebraic transformation shows that

$$A_x = 1 - d\ddot{a}_x \quad \text{and} \quad A_{x:\overline{n}|} = 1 - d\ddot{a}_{x:\overline{n}|}$$

where $d = iv = 1 - v$, as defined earlier.

If the annuity is payable in amounts of $1/m$ at the beginning of each mth part of a year, as long as a person aged x is now alive, then $\ddot{a}_x - (m - 1/2m)$ is an acceptable approximation.

The yearly premium for a Whole-Life Assurance of unit sum assured, payable while the policy is in force, is

$$\frac{A_x}{\ddot{a}_x} = \frac{1}{\ddot{a}_x} - d = P_x \tag{15.2.6}$$

and that of an Endowment Assurance, term n, is

$$\frac{A_{x:\overline{n}|}}{\ddot{a}_{x:\overline{n}|}} = \frac{1}{\ddot{a}_{x:\overline{n}|}} - d = P_{x:\overline{n}|}. \tag{15.2.7}$$

All these values depend, of course, on the rate of interest used. The value of a *life annuity*, or of a temporary annuity, decreases with increasing interest rate. If the rate is zero, then $\ddot{a}_x - 1$ is simply the expected future survival time of a person aged x.

For the purpose of computing premiums and other actuarial values it is convenient to have tables available which contain, for rates of interest which are in use, values for successive ages of

$$D_x = l_x v^x, \; D_{x+1} = l_{x+1} v^{x+1}, \ldots$$

and also of

$$N_x = D_x + D_{x+1} + \ldots.$$

Using these so-called *commutation columns*, we can write

$$_nE_x = D_{x+n}/D_x, \quad \ddot{a}_x = N_x/D_x, \quad \ddot{a}_{x:\overline{n}|} = (N_x - N_{x+n})/D_x.$$

For computing Whole-Life and Endowment Insurance premiums, the following commutation columns are also used:

$$C_x = (l_x - l_{x+1})v^{x+1} = vD_x - D_{x+1} \tag{15.2.8}$$

and

$$M_x = C_x + C_{x+1} + \ldots + C_{x+t} + \ldots = vN_x - N_{x+1}. \tag{15.2.9}$$

With this notation we have

$$A_x = M_x/D_x, \quad \text{and} \quad A_{x:\overline{n}|} = (M_x - M_{x+n} + D_{x+n})/D_x.$$

15.2.2. Policy Values

If an insurance office underwrites contracts for a number l_x of persons, aged x, each for a sum assured 1, and if their rate of survival to age $x + t$ $(t = 1, 2, \ldots)$ is precisely l_{x+t}/l_x, then at the rate of interest which is used, there will be no loss or gain from the net premiums when all contracts of this group have terminated.

Consider, on the other hand, the situation during the course of the insurance contract, for example that of a whole-life or of an endowment assurance, for a constant sum assured (i.e. independent of the time of death or survival).

At ages which are usually relevant for such assurances, the rate of mortality increases with age. Therefore, if the policy-holder paid each year a premium to cover the risk of that year, viz. that of death during that year, or in the case of an endowment assurance in its last year the certainty of the sum assured becoming due at the end of the year, then he would have to pay increasing annual premiums.

However, as a rule the annual premiums remain constant during the time of the assurance. It follows that if we consider a group of l_x persons of age x, who take out the same type of insurance, contracting for the same annual premium, then the payments of the survivors of these persons during the early years of the contract will be more than sufficient to pay for the sums assured which become due, and the insurer can accumulate a fund from the surplus. Later the total of amounts becoming due will be higher than that of the premiums paid during the same year, and the insurer will draw on the accumulated fund. At the termination of all contracts his receipts, with interest, will just balance his payments, again with interest—provided, of course, that everything happens according to the assumptions of survival and death.

Dividing the fund by the number of those insured persons whose policy is still in force, we obtain the premium reserve, or *policy value*, of an insurance. Remembering that the payments of persons $l_x, l_{x+1}, \ldots$ who were alive at ages $x, x+1, \ldots$ were accumulated with compound interest i, while the sum assured was paid at the end of the respective year of death, and dividing the fund among the l_{x+k} persons alive after k years, we have the policy value of a Whole-Life Assurance in its *retrospective form*

$$_kV_x = P_x \sum_{t=0}^{k-1} \frac{D_{x+t}}{D_{x+k}} - \sum_{t=0}^{k-1} \frac{C_{x+t}}{D_{x+k}}. \qquad (15.2.10)$$

This can be transformed into the *prospective form*

$$_kV_x = A_{x+k} - P_x\ddot{a}_{x+k} = 1 - \frac{\ddot{a}_{x+k}}{\ddot{a}_x}. \qquad (15.2.11)$$

This expression is also understandable from first principles. The value of the expected future payments, A_{x+k}, must be provided for, taking into account the expected future income $P_x\ddot{a}_{x+k}$.

Similarly, for an Endowment Assurance, we obtain the policy value

$$_kV_{x:\overline{n}|} = A_{x+k:\overline{n-k}|} - P_{x:\overline{n}|}\ddot{a}_{x+k:\overline{n-k}|}$$

$$= 1 - \frac{\ddot{a}_{x+k:\overline{n-k}|}}{\ddot{a}_{x:\overline{n}|}}. \qquad (15.2.12)$$

For $k = 0$ this is zero, and for $k = n$ it is 1, as it must be.

The relationship between successive policy values, for a Whole-Life Assurance, is

$$_{k+1}V_x = \frac{_kV_x + P_x}{vp_{x+k}} - \frac{q_{x+k}}{p_{x+k}} \qquad (15.2.13)$$

and the same relationship holds for an Endowment Assurance if we replace P_x by $P_{x:\overline{n}|}$.

EXAMPLE 15.2.1. To give a numerical example, we take as our basis for mortality assumptions a table constructed from 'The Mortality experience of life insurance companies collected by the Institute of Actuaries', London 1869, which lists the experience of 20 British Offices. We use the table concerning healthy males (referred to in the professional literature as table H^M). As rate of interest we assume $3\frac{1}{2}\%$, and consider an Endowment Assurance, term 10, year of entry 50. We then use the following extract from the table:

Age x	l_x	D_x	$l_x - l_{x+1}$
50	72795	13034	1144
51	71651	12395	1193
52	70458	11777	1243
53	69215	11178	1296
54	67919	10598	1353
55	66566	10035	1414
56	65152	9490	1475
57	63677	8961	1541
58	62136	8449	1612
59	60524	7952	1682
		103869	

This gives a premium of

$$\frac{1}{\ddot{a}_{x:\overline{n}|}} - d = \frac{13034}{103869} - 0.03382 = 0.09167.$$

The development of the fund for initially 72795 persons of age 50, all of them paying a premium of 0.09167 for a sum assured 1, proceeds as follows:

Start of year	Total premium paid	Fund after receipt of premiums	With $3\frac{1}{2}\%$ interest at end of year	Sums assured paid	Fund end of year
1	6673	6673	6907	1144	5763
2	6568	12331	12763	1193	11570
3	6459	18029	18659	1243	17416
4	6345	23761	24593	1296	23297
5	6226	29523	30556	1353	29203
6	6102	35305	36541	1414	35727
7	5972	41099	42530	1475	41055
8	5837	46892	48533	1541	46992
9	5696	52688	54533	1612	52921
10	5548	58469	60516	60524	—

(The difference between 60524 and 60516, $1\frac{1}{3}$ in 10000, is due to rounding.)

15.2.3.   Other Assurance Types

The types of insurance which we have mentioned are by no means all possible ones. We mention just two more.

Let an annuity of 1 p.a. be payable at the start of year 1, 2, . . . while both of two persons, aged x and y respectively, are alive. The discounted present value at the start of the insurance, in other words its *single premium*, is

$$\ddot{a}_{xy} = \frac{l_x l_y + v l_{x+1} l_{y+1} + \cdots}{l_x l_y} = \frac{D_{xy} + D_{x+1,y+1} + \cdots}{D_{xy}} \qquad (15.2.14)$$

where D_{xy} is defined as $v^{(1/2)(x+y)} l_x l_y$.

A *reversionary annuity* is an annuity on the life of a person now aged x, starting after the death of a person now aged y. Its single premium is

$$\ddot{a}_x - \ddot{a}_{xy}, \qquad (15.2.15)$$

where $\ddot{a}_x$ is given by (15.2.4).

We remark that we have made use of the symbols of the International Actuarial Notation, as given, for instance, in the Journal of the Institute of Actuaries, vol. 75 (1949), pp. 121–129.

15.2.4.   Continuous Mortality

If the sum assured is payable immediately after death, or if the interval at which an annuity is payable converges to zero, then we introduce the *force of mortality*

$$\mu_x = -d \log_e l_x / dx \qquad (15.2.16)$$

[see IV, § 2.11 and IV, § 3.1.1 for definitions of $\log_e$ and of the derivative] and the *force of discount*

$$\delta = \log_e (1 + i) = -\log_e (1 - d). \qquad (15.2.17)$$

We use also the notation

$$\bar{D}_x = e^{-\delta x} l_x, \qquad \bar{N}_x = \int_0^\infty \bar{D}_{x+t} dt,$$

$$\bar{M}_x = \int_0^\infty e^{-\delta(x+t)} l_{x+t} l_{x+t} dt.$$

The single premium for a Whole-Life Assurance is then

$$\bar{A}_x = \bar{M}_x / \bar{D}_x \qquad (15.2.18)$$

and that of an Endowment Assurance, term n, is

$$\bar{A}_{x:\overline{n}|} = (\bar{M}_x - \bar{M}_{x+n} + \bar{D}_{x+n}) / \bar{D}_x. \qquad (15.2.19)$$

If an annuity is payable at intervals of $1 + m$ of a year in instalments of $1 + m$, while a person now aged x is alive, and if we let the interval $1/m$ converge to zero, then we obtain a *continuous*, or *momently annuity*. Its value is

$$\bar{N}_x/\bar{D}_x = \bar{a}_x \tag{15.2.20}$$

and a satisfactory approximation is

$$\breve{a}_x - \tfrac{1}{2} - (\delta_x + \mu_x)/12. \tag{15.2.21}$$

If the annuity terminates at the latest after n years, then its value is

$$\bar{a}_{x:\overline{n}|} = (\bar{N}_x - \bar{N}_{x+n})/\bar{D}_x. \tag{15.2.22}$$

The premium payable continuously, while an Endowment Assurance, term n, is in force, will be

$$\bar{P}_{x:\overline{n}|} = \frac{1}{\bar{a}_{x:n}} - \delta. \tag{15.2.23}$$

For the policy values of an Endowment Assurance, term n, we have after k years

$$_kV_x = \bar{A}_{x+k:\overline{n-k}|} - \bar{P}_{x:\overline{n}|}\bar{a}_{x+k:\overline{n-k}|}$$
$$= 1 - \bar{a}_{x+k:\overline{n-k}|}/\bar{a}_{x:\overline{n}|}. \tag{15.2.24}$$

Such formulae depend on a knowledge of l_x for continuous values of x, and for this purpose various formulae have been assumed, such as for instance the formula of *Gompertz-Makeham*

$$l_x = ks^x g^{c^x} \tag{15.2.25}$$

where s, g and c are suitable parameters, and k depends on l_0. Then [by IV (2.11.7)]

$$\log_e l_x = \log_e k + x \log_e s + c^x \log_e g,$$

so that

$$\mu_x = -\log_e s - \log_e c . \log_e g . c^x$$
$$= a + bc^x, \text{ say.} \tag{15.2.26}$$

This formula is convenient when calculating values of *joint life assurances*. The probability of both a person aged now x, and a person aged now y to be alive after n years is

$$l_{x+n}l_{y+n}/l_x l_y = \exp\left[-\int_0^n (\mu_{x+t} + \mu_{y+t})dt\right]. \tag{15.2.27}$$

Now if $\mu_{x+t} = a + bc^{x+t}$, $\mu_{y+t} = a + bc^{y+t}$, then

$$l_{x+n}l_{y+n}/l_x l_y = \exp\left[-\int_0^n \{2a + b(c^{x+t} + c^{y+t})\}dt\right].$$

We define w by $c^x + c^y = 2c^w$, and hence

$$c^{x+t} + c^{y+t} = 2c^{w+t}.$$

Hence

$$l_{x+n}l_{y+n}/l_xl_y = \exp\left[-\int_0^n (2a + 2bc^{w+t})dt\right]$$

$$= \exp\left[-2\int_0^n \mu_{w+t}dt\right] = l_{w+t}l_{w+t}/l_wl_w. \qquad (15.2.28)$$

In this manner the probability of survival of a pair aged, respectively, x and y, is reduced to the probability of survival of a pair of persons of equal age. Similar transformations are valid for groups of more than 2 persons.

S.V.

CHAPTER 16

Boolean Algebra

16.1. LAWS OF A BOOLEAN ALGEBRA

A Boolean Algebra B is a mathematical system consisting of a collection of elements with two binary operations [see § 1.5] called *union*, written $\vee$, and *intersection*, written $\wedge$, which obey the following laws of combination:

1. (i) For every two equal or distinct elements a and b in B, $a \vee b$ is a single element in B.

   (ii) For every two equal or distinct elements a and b in B, $a \wedge b$ is a single element in B.

These are called the *closure* laws.

2. For all elements a and b in B

   (i) $a \vee b = b \vee a$ and

   (ii) $a \wedge b = b \wedge a$.

These are called the *commutative* laws.

3. For all elements a, b and c in B,

   (i) $(a \vee b) \vee c = a \vee (b \vee c)$ and

   (ii) $(a \wedge b) \wedge c = a \wedge (b \wedge c)$.

These are called the *associative* laws.

4. For all elements a, b and c in B,

   (i) $a \wedge (b \vee c) = (a \wedge b) \vee (a \wedge c)$ and

   (ii) $a \vee (b \wedge c) = (a \vee b) \wedge (a \vee c)$.

These are called the *distributive* laws.

5. For all elements a in B,

   (i) $a \vee a = a$ and

   (ii) $a \wedge a = a$.

These are called the *idempotent* laws.

6. (i) B contains a unique element 0 with the properties that

       $a \vee 0 = a$ and $a \wedge 0 = 0$

       for all elements a in B.

   (ii) B contains a unique element 1 with the properties that

       $a \vee 1 = 1$ and $a \wedge 1 = a$

       for all elements a in B.

We call 0 and 1 the *identity elements* with respect to union and intersection respectively.

7. For every element a in B there exists a unique element $\bar{a}$ in B such that

(i) $a \vee \bar{a} = 1$ and

(ii) $a \wedge \bar{a} = 0$.

We call $\bar{a}$ the *complement* of a.

For all elements a and b in B, complements obey the laws

(iii) $\overline{a \vee b} = \bar{a} \wedge \bar{b}$ and

(iv) $\overline{a \wedge b} = \bar{a} \vee \bar{b}$.

These are called *De Morgan's* Laws.

For all elements a in B

(v) $\overline{(\bar{a})} = a$.

This is called the law of *involution*.

These seven sets of laws may appear at first glance to be rather artificial and a bit much to expect of any system, but they are obeyed by many systems which occur frequently both in mathematics and its applications. The most important example, and indeed the system which originally inspired this collection of laws, is the algebra of classes or sets [see § 1.1], with the familiar operations of set union and intersection [see § 1.2.1]. In this algebra the empty set plays the part of 0 and the universal set U plays the part of 1 [see § 1.2.4]. The complement of an element is the usual set complement [see § 1.2.4].

The particular collection of laws we have presented are not the only ones that can be used to define a Boolean Algebra and indeed some of our laws are redundant in that they can be deduced from some of the other laws. However, once we adopt a certain set of laws we must use these and only these laws to deduce further results about the Boolean Algebra.

We note that many of the laws of a Boolean Algebra are satisfied by numbers, with addition (+) playing the part of union and multiplication (.) playing the part of intersection. There are important exceptions however which deserve careful examination.

EXAMPLE 16.1.1.   Law 4(ii) is not always true for numbers, since for example

$$3 + (2.4) = 11, \quad \text{whereas} \quad (3 + 2).(3 + 4) = 35.$$

EXAMPLE 16.1.2.   Laws 5(i) and 5(ii) are not true in general for numbers, since for example

$$2 + 2 \neq 2 \quad \text{and} \quad 3.3 \neq 3.$$

In fact, the only numbers for which law 5(i) is true is the number 0 since $0 + 0 = 0$, and the only numbers for which 5(ii) is true are 0 and 1 since

$$0.0 = 0 \quad \text{and} \quad 1.1 = 1.$$

We emphasize that as yet we attach no particular meaning or interpretation to either the elements of the Boolean Algebra B or the operations whereby we combine them. We have nothing to go on except the seven laws we have assumed, but it is surprising just how much we can prove as a consequence of these laws. We start with a few simple results.

EXAMPLE 16.1.3. Prove that $\bar{0} = 1$ and $\bar{1} = 0$.

Proof
$$1 = 0 \lor \bar{0}, \quad \text{by 7(i)}$$
$$= \bar{0} \lor 0, \quad \text{by 2(i)}$$
$$= \bar{0}, \quad \text{by 6(i)}.$$

Similarly,
$$0 = 1 \land \bar{1}, \quad \text{by 7(ii)}$$
$$= \bar{1} \land 1, \quad \text{by 2(ii)}$$
$$= \bar{1}, \quad \text{by 6(ii)}.$$

Note that $\bar{1} = 0$ is in fact a consequence of $\bar{0} = 1$, because $\bar{0} = 1 \Rightarrow \overline{(\bar{0})} = \bar{1}$, by virtue of the fact that the complement of an element is unique, which implies that $0 = \bar{1}$ by 7(v).

It is useful to look upon the laws of a Boolean Algebra as rules which can help us to simplify a wide range of expressions involving elements which have been combined by means of the operations $\lor$ and $\land$.

EXAMPLE 16.1.4. For elements a and b in a Boolean Algebra $\{B, \lor, \land\}$, show that
$$(a \land b) \lor (a \land \bar{b}) = a.$$

Proof. Now,
$$(a \land b) \lor (a \land \bar{b}) = a \land (b \lor \bar{b}), \quad \text{by 4(i)}$$
$$= a \land 1, \quad \text{by 7(i)}$$
$$= a, \quad \text{by 6(ii)}.$$

EXAMPLE 16.1.5. For elements a, b, c and d in a Boolean Algebra $\{B, \lor, \land\}$ prove that
$$[a \land (b \land c)] \lor \{[(a \land b) \land \bar{c}] \lor [a \land (b \land d)]\} = a \land b.$$

Proof. The expression on the left-hand side is equal to

$$\{[a \land (b \land c)] \lor [(a \land b) \land \bar{c}]\} \lor [a \land (b \land d)], \quad \text{by 3(i)}$$
$$= \{[(a \land b) \land c)] \lor [(a \land b) \land \bar{c}]\} \lor [a \land (b \land d)], \quad \text{by 3(ii)}$$
$$= \{(a \land b) \land (c \lor \bar{c})\} \lor [a \land (b \land d)], \quad \text{by 4(i)}$$
$$= \{(a \land b) \land 1\} \lor \{a \land (b \land d)\}, \quad \text{by 7(i)}$$
$$= \{(a \land b) \land 1\} \lor \{(a \land b) \land d)\}, \quad \text{by 3(ii)}$$
$$= \{(a \land b) \land \{1 \lor d\}, \quad \text{by 4(i)}$$
$$= (a \land b) \land 1, \quad \text{by 6(ii)}$$
$$= a \land b, \quad \text{by 6(ii)}.$$

The laws of a Boolean Algebra $\{B, \vee, \wedge\}$ have been given in pairs to emphasize the symmetry that exists between the operations $\vee$ and $\wedge$ and the elements 0 and 1. This symmetry leads to what is called the *principle of duality*, i.e. if in any valid equation in a Boolean Algebra we interchange the operations $\vee$ and $\wedge$ throughout, and also interchange the elements 0 and 1 throughout, we obtain another valid equation. For example, we have shown that

$$(a \wedge b) \vee (a \wedge \bar{b}) = a.$$

The principle of duality tells us that

$$(a \vee b) \wedge (a \vee \bar{b}) = a$$

is also a valid equation for all elements a and b. Similarly $0 \wedge 1 = 0$ gives rise to the equation

$$1 \vee 0 = 1.$$

We note that the complement of a set remains the same when we apply the principle of duality. This is because, in a sense, law 7(v) is its own dual.

16.2.   THE INCLUSION RELATION

On the algebra of sets, we frequently use the relation "X is a subset of Y" (written $X \subset Y$) to denote the fact that every element of the set X is also an element of the set Y [see § 1.2.3]. Since it is not immediately obvious what the analogue of this useful concept should be in a Boolean Algebra, we must first fish around a little in the algebra of sets. The device we employ is to express the inclusion relation on sets in terms of set intersection only [see § 1.2.4], and then the appropriate definition in a Boolean Algebra will become obvious. A little thought will convince us that for sets X and Y, $X \subset Y$ if and only if $X \cap Y = X$, so with this in mind we make the following definition.

If a and b are elements of a Boolean Algebra $\{B, \vee, \wedge\}$ we say that a is included in b if

$$a \wedge b = a.$$

We write this relation as $a < b$ and say that a is included in b, or b includes a.

Many of the properties of the inclusion relation in set theory hold in an arbitrary Boolean Algebra and we proceed to prove a number of these, using only our seven basic axioms.

EXAMPLE 16.2.1.   In a Boolean Algebra $\{B, \vee, \wedge\}$ for any element a, we have

$$0 < a \quad \text{and} \quad a < 1.$$

Proof. $0 \wedge a = a \wedge 0, \quad$ by 2(ii)

$$= 0, \qquad \text{by 6(i)}$$

so, by definition, $0 < a$.

Also

$$a \wedge 1 = a, \quad \text{by 6(ii)},$$

so, by definition, $a < 1$.

EXAMPLE 16.2.2.   If a, b and c are elements of a Boolean Algebra $\{B, \vee, \wedge\}$ such that $a < b$ and $b < c$, then $a < c$.

Proof.   We know that

$$a \wedge b = a \quad \text{and} \quad b \wedge c = b,$$

and we wish to show that

$$a \wedge c = a.$$

Now,

$$
\begin{aligned}
a &= a \wedge b, & &\text{by hypothesis} \\
&= a \wedge (b \wedge c), & &\text{by hypothesis} \\
&= (a \wedge b) \wedge c, & &\text{by 3(ii)} \\
&= a \wedge c, & &\text{by hypothesis,}
\end{aligned}
$$

so $a < c$.

EXAMPLE 16.2.3.   For a Boolean Algebra $\{B, \vee, \wedge\}$

$$(a \wedge b) < (a \vee b),$$

for all elements a and b in B.

Proof.

$$
\begin{aligned}
(a \wedge b) \wedge (a \vee b) &= [(a \wedge b) \wedge a] \vee [(a \wedge b) \wedge b], & &\text{by 4(i)} \\
&= [a \wedge (b \wedge a)] \vee [a \wedge (b \wedge b),] & &\text{by 3(ii)} \\
&= [a \wedge (a \wedge b)] \vee [a \wedge (b \wedge b),] & &\text{by 2(ii)} \\
&= [a \wedge (a \wedge b)] \vee [a \wedge b], & &\text{by 5(ii)} \\
&= [(a \wedge a) \wedge b] \vee [a \wedge b], & &\text{by 3(ii)} \\
&= (a \wedge b) \vee (a \wedge b), & &\text{by 5(ii)} \\
&= a \wedge b, & &\text{by 5(i).}
\end{aligned}
$$

Thus $(a \wedge b) < (a \vee b)$.

Note that there are alternative ways of defining the inclusion relation which are equivalent to the one we have given. For example, we have the following: For elements a and b in a Boolean Algebra $\{B, \vee, \wedge\}$
  (1) $a < b$ if and only if $a \vee b = b$
  (2) $a < b$ if and only if $a \wedge \bar{b} = 0$.
However, (2) above is a little more difficult to work with in general than either (1) or the definition we have given.

16.3.   TRUTH TABLES

We shall see later that in certain contexts it is important to be able to decide whether two expressions involving elements of a Boolean Algebra are equal or not. The simple but ingenious device of a truth table gives us an effective, almost mechanical, method of making this decision in any given case. Suppose, for example, that we wish to investigate whether the expressions

$$a \vee (\bar{a} \wedge b) \quad \text{and} \quad a \vee b,$$

where a and b are elements of a Boolean Algebra $\{B, \vee, \wedge\}$ are equal or not. We form Table 16.3.1. Under the columns headed a and b, we list all the different possibilities (in this case four) that can be achieved by assigning the *truth values* 0 and 1 to the elements a and b. We then proceed to fill in the columns headed $a \vee b$, $\bar{a}$, $\bar{a} \wedge b$ and $a \vee (\bar{a} \wedge b)$ using the following rules:

$$\bar{0} = 1, \quad \bar{1} = 0, \quad 0 \vee 0 = 0 = 0 \wedge 0, \quad 1 \wedge 1 = 1 \vee 1 = 1;$$
$$1 \wedge 0 = 0 \wedge 1 = 0, \quad 1 \vee 0 = 0 \vee 1 = 1.$$

For example, since the element $\bar{a}$, heading column 4, is the complement of the element a, heading column 1, column 4 is formed by 0 for each 1 and 1 for each 0 in column 1. Also, column 5 is formed by combining columns 4 and 2 using $1 \wedge 0 = 0, 0 \wedge 0 = 0, 1 \wedge 1$ and $0 \wedge 1 = 0$.

1	2	3	4	5	6
a	b	$a \vee b$	$\bar{a}$	$\bar{a} \wedge b$	$a \vee (\bar{a} \wedge b)$
1	1	1	0	0	1
1	0	1	0	0′	1
0	1	1	1	1	1
0	0	0	1	0	0
		↑			↑

Table 16.3.1

The fact that the entries in columns 3 and 6 are identical (i.e. the same entries in the same order) tells us that the elements $a \vee b$ and $a \vee (\bar{a} \wedge b)$ have the same truth value for all possibilities of a and b and, therefore,

$$a \vee (\bar{a} \wedge b) = a \vee b.$$

Truth Table 16.3.2 tells us that in general $\overline{(a \wedge b)}$ is not the same as $\bar{a} \wedge \bar{b}$.

Incidentally, truth tables can also be used to check inclusions $x < y$, since

a	b	$\bar{a}$	$\bar{b}$	$a \wedge b$	$\overline{(a \wedge b)}$	$\bar{a} \wedge \bar{b}$	$(\bar{a} \wedge \bar{b}) \wedge \overline{(a \wedge b)}$
1	1	0	0	1	0	0	0
1	0	0	1	0	1	0	0
0	1	1	0	0	1	0	0
0	0	1	1	0	1	1	1
					↑	↑	

Table 16.3.2

we need only verify by a truth table that $x \wedge y = x$. For example, we see from Table 16.3.2 that

$$(\bar{a} \wedge \bar{b}) < \overline{(a \wedge b)}$$

since the truth table tells us that

$$(\bar{a} \wedge \bar{b}) \wedge \overline{(a \wedge b)} = (\bar{a} \wedge \bar{b}),$$

by comparing columns 7 and 8 of the table.

As the number of elements involved in an expression increases, the size of

a	b	c	$b \vee c$	$a \wedge (b \vee c)$	$a \wedge b$	$a \wedge c$	$(a \wedge b) \vee (a \wedge c)$
1	1	1	1	1	1	1	1
1	1	0	1	1	1	0	1
1	0	1	1	1	0	1	1
1	0	0	0	0	0	0	0
0	1	1	1	0	0	0	0
0	1	0	1	0	0	0	0
0	0	1	1	0	0	0	0
0	0	0	0	0	0	0	0
				↑			↑

Table 16.3.3

the related truth table also increases, but the basic principles remain the same. For three elements a, b and c of a Boolean Algebra $\{B, \vee, \wedge\}$ we have $8(=2^3)$ distinct possibilities to consider, and in general for n elements we have 2^n cases to examine. In this example a comparison of the fifth and eighth columns of the truth table (Table 16.3.3) shows that

$$a \wedge (b \vee c) = (a \wedge b) \vee (a \wedge c),$$

for all elements a, b and c of a Boolean Algebra $\{B, \vee, \wedge\}$.

16.4. SWITCHING CIRCUITS

Boolean Algebra has an interesting and totally unexpected application to switching circuits, discovered by Shannon in 1938.

Those with even an elementary knowledge of physics will know that there are two fundamental methods of connecting electric circuits to form new circuits. If a and b are circuits we can connect them in series, denoted as follows:

or we can connect them in parallel, denoted as follows:

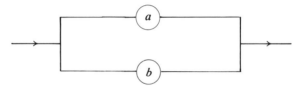

We write the circuit formed by connecting a and b in series $a \wedge b$, and the circuit formed by connecting a and b in parallel by $a \vee b$. Since $a \wedge b$ and $a \vee b$ are again circuits, we have the closure laws of section 16.1 and a little thought, or indeed a few elementary experiments, will show that the collection of circuits with the binary operations of placing circuits in series and parallel form a Boolean Algebra.

A little care is needed in defining what we mean, for example, by saying that,

$$a \vee b = b \vee a$$

in this context. The circuits corresponding to these expressions in the Boolean Algebra are given by

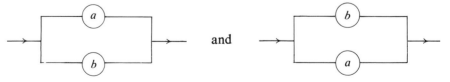

and

respectively, and in saying that the two circuits are 'equal' we mean the following: If current enters a circuit at the left-hand side and leaves by the right-hand side to return to the battery or power source, we claim that the circuits $a \lor b$ and $b \lor a$ have exactly the same transmittance, i.e., the two circuits behave in exactly the same way as far as transmission of current is concerned.

If we use the symbol 1 to denote the fact that current is flowing in a circuit, i.e. the circuit is switched on, and the symbol 0 to denote the fact that current is not flowing in a circuit, i.e. the circuit is switched off, we can analyse any circuit conveniently in terms of truth tables (see § 16.3).

(i) Series connection

For example, the second row of Table 16.4.1 simply says that if a is switched on and b is switched off then the series circuit $a \land b$ is switched off. In fact, the circuit $a \land b$ is switched on if and only if both a and b are switched on. This type of circuit is, therefore, referred to as an AND circuit.

a	b	$a \land b$
1	1	1
1	0	0
0	0	0
0	0	0

Table 16.4.1

(ii) Parallel connection

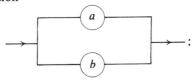

In this case we see that the parallel circuit $a \lor b$ is switched on if and only if either a or b is switched on. This type of circuit is, therefore, referred to as an OR circuit (Table 16.4.2).

a	b	$a \lor b$
1	1	1
1	0	1
0	1	1
0	0	0

Table 16.4.2

Note that in the example referred to earlier, we can see that the circuits corresponding to $a \lor b$ and $b \lor a$ are equal by showing that they have the same truth table.

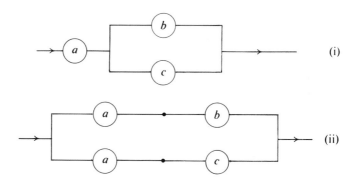

Figure 16.4.1

Moving on to slightly more complicated circuits, one of the distributive laws in a Boolean Algebra (see § 16.1).

$$a \wedge (b \vee c) = (a \wedge b) \vee (a \wedge c)$$

tells us that the transmittances of the two circuits shown in Figure 16.4.1 are the same.

In a practical situation, it is likely that circuit (i) would be a better choice than circuit (ii). Similarly, the other distributive law

$$a \vee (b \wedge c) = (a \vee b) \wedge (a \vee c)$$

tells us that the two circuits shown in Figure 16.4.2 have the same transmittance.

We define the complement of a circuit a to be a circuit which is always on when a is off and off when a is on, and we denote this circuit by $\bar{a}$. The circuit 0 is a circuit which is always off and the circuit 1 is a circuit which is always on.

A good example of the type of circuit which can be analyzed very easily using Boolean Algebra is the 'two-way switch' which is to be found in many houses. This type of circuit enables say an upstairs light to be controlled from either of two switches, one upstairs and the other downstairs. We describe this circuit as shown in Figure 16.4.3. The entire circuit, controlled by the two switches S_1 and S_2, can be written as

$$c = (a \wedge b) \vee (\bar{a} \wedge \bar{b}).$$

The circuit has the truth table shown in Table 16.4.3. The switches at S_1 and S_2, denoted by arrows, can be either on or off, and we wind up with the table of

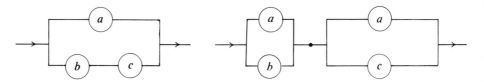

Figure 16.4.2

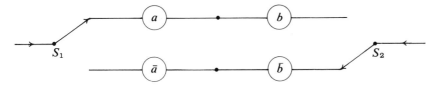

Figure 16.4.3

possibilities shown in Table 16.4.3. Regarding a switch (arbitrarily) as on when the arrow in Figure 16.4.3 is pointing upwards and off when pointing downwards, we see that, no matter what the initial state of the circuit, a single change in either S_1 or S_2 will reverse the state of the circuit, and this of course is exactly what we require. Figure 16.4.3 corresponds to row two of Table 16.4.4 and it is clear that the current will not flow in the circuit c in this case.

In the theory of switching circuits, Boolean Algebra is most useful in the following type of situation. Suppose we have a rather complicated circuit and we wish to see if a simpler circuit, perhaps costing less, taking up less space and having fewer component parts, can perform the same function. We first construct the expression in a Boolean Algebra which corresponds to the given circuit. Then, we simplify the expression as far as possible using the laws of the Boolean Algebra. Finally, we construct a new circuit corresponding to the simplified expression, and the new circuit will have the same transmittance as the original.

The following is a good example of this procedure. Consider the circuit given by Figure 16.4.4.

The Boolean Algebra expression for this circuit is

$$[(a \lor b) \land c] \lor [\{\bar{a} \land (\bar{b} \lor c) \lor \bar{b}] =$$
$$[(a \land c) \lor (b \land c)] \lor [(\bar{a} \land \bar{b}) \lor (\bar{a} \land c) \lor \bar{b}] =$$
$$[(a \land c) \lor (\bar{a} \land c)] \lor (b \land c) \lor \bar{b} \lor (\bar{a} \land \bar{b}) =$$
$$[(a \lor \bar{a}) \land c)] \lor (b \land c) \lor [(1 \land \bar{b}) \lor (\bar{a} \land \bar{b})] =$$
$$[(1 \land c) \lor (b \land c)] \lor [(1 \lor \bar{a}) \land \bar{b}] =$$
$$[(1 \lor b) \land c] \lor (1 \land \bar{b}) =$$
$$[(1 \land c) \lor (1 \land \bar{b}) =$$
$$c \lor \bar{b}.$$

a	b	$\bar{a}$	$\bar{b}$	$a \land b$	$\bar{a} \land \bar{b}$	c
1	1	0	0	1	0	1
1	0	0	1	0	0	0
0	1	1	0	0	0	0
0	0	1	1	0	1	1

Table 16.4.3

S_1	S_2	c
1	1	1
1	0	0
0	1	0
0	0	1

Table 16.4.4

Thus the original circuit has the same transmittance as the following simpler circuit.

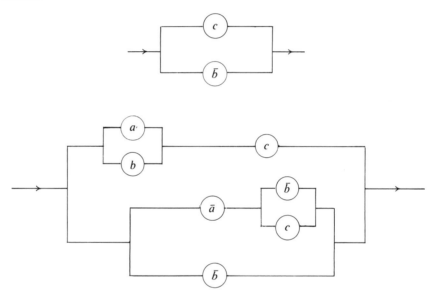

Figure 16.4.4

16.5.   MATHEMATICAL LOGIC

One of the most useful and interesting applications of Boolean Algebra is to the analysis of classical logic and the development of mathematical logic. This application depends on the fact that the algebra of classes is a Boolean Algebra (see § 16.1) and the device most commonly used is to express statements in classical logic as relations in the Boolean Algebra by means of the inclusion relation of section 16.2.

Consider the following argument and conclusion:
1. All people are mortal.
2. All women are people.
3. Therefore all women are mortal.

In the algebra of classes, let M denote the class of all mortal beings, W the class of all women and P the class of all people. The above statements then become

1. $P < M$.
2. $W < P$

and the conclusion becomes

3. $W < M$.

Sometimes we may have to make slight adjustments in language before we can apply the Algebra of classes. Thus the statements 'x is a fat person' and 'y is a woman' must become 'x is a member of F, the class of all fat people' and 'y is a member of W, the class of all women'. Of course $F \wedge W$ is the class of all fat women and $F \vee W$ is the class of all people who are either fat or a woman or both. The statement 'all women are fat' can be written as $W < F$ or, equivalently, $W \wedge F = W$.

Notice that we do not claim that any of our statements such as 'all women are fat' are true in any absolute sense. We are merely interested in the conclusions that can be drawn if we make the assumption that a given collection of statements are true.

Frequently, Boolean Algebra can help us sort out rather tricky logical situations which would be very difficult to analyse without symbols. Consider for example the following question:

'If all boiled red lobsters are dead, and all boiled dead lobsters are red, does it follow that all dead red lobsters are boiled?'

In the class of all lobsters $L (=1)$ let B denote the subclass of all boiled lobsters, R the class of all red lobsters and D the class of all dead lobsters. The above statements become

$$B \wedge R < D \quad \text{and} \quad B \wedge D < R$$

or

(i) $(B \wedge R) \wedge D = B \wedge R$ and (ii) $(B \wedge D) \wedge R = B \wedge D$.

We ask, does it necessarily follow that

$$R \wedge D < B \quad \text{or equivalently} \quad \text{(iii)} \ (R \wedge D) \wedge B = R \wedge D?$$

If it is true that (iii) is always a consequence of (i) and (ii), then (iii) will be true regardless of what the classes B, R and D are, provided of course they satisfy (i) and (ii).

Suppose $D = R = L = 1$, the class of all lobsters, while $B = 0$, the empty class. Now, $(B \wedge R) \wedge D = (0 \wedge 1) \wedge 1 = 0 \wedge 1 = B \wedge R$, so (i) is satisfied, and $(B \wedge D) \wedge R = (0 \wedge 1) \wedge 1 = 0 \wedge 1 = B \wedge D$, so (ii) is satisfied also. Finally, $(R \wedge D) \wedge B = (1 \wedge 1) \wedge 0 = 1 \wedge 0 = 0$, while $(R \wedge D) = 1 \wedge 1 = 1$. Since $1 \neq 0$, it is not always true that (iii) is a consequence of (i) and (ii), so not all dead red lobsters are boiled.

If we widen the scope of our discussion a little, we can consider simple propositions such as 'London is a city' or 'Snow is not black' directly, rather than translate them into the language of classes. We use letters p, q, r, etc. to denote these simple propositions. Let p be the proposition 'London is a city' and q the proposition 'Snow is not black'. The negation of a proposition x is written $\bar{x}$, and, for example, $\bar{p}$ is the proposition 'London is not a city', and $\bar{q}$ is the proposition 'Snow is black'.

The proposition 'London is a city and snow is not black' is denoted by $p \wedge q$, and $p \vee q$ is the proposition 'London is a city or snow is not black', where the word 'or' is used in the inclusive rather than the exclusive sense.

The notion of *implication* is written as $p \rightarrow q$, read 'p implies q', which means that if the proposition p is true, then the proposition q is also true. If we consider only simple propositions which have exactly two truth values, i.e. a given proposition is either true (denoted by 1) or false (denoted by 0), then we can use a truth table to cover the various possibilities that can arise.

p	q	$p \wedge q$	$p \vee q$	$\bar{p}$	$p \rightarrow q$	$p \leftrightarrow q$
1	1	1	1	0	1	1
1	0	0	1	0	0	0
0	1	0	1	1	1	0
0	0	0	0	1	1	1

$p \leftrightarrow q$ means $p \rightarrow q$ and $q \rightarrow p$ i.e. p and q are equivalent.

Table 16.5.1

The rather unexpected fact that if p and q both have truth value 0 then $p \rightarrow q$ and $p \leftrightarrow q$ both have truth value 1, can be understood by remembering common phrases such as 'If he is a mathematician, then I'm a Martian'.

Propositions form a Boolean Algebra under the laws of composition that we have outlined in section 16.1 and we can use our knowledge of Boolean Algebra to determine the logical conclusion of a series of simple propositions as in the following example.

> If this kettle is made of brass, then it contains copper and it contains zinc, and if this kettle is made of bronze, then it contains copper and it contains tin.
> This kettle is made of brass or this kettle is made of bronze.

What conclusion can be drawn?

> Let p be the proposition 'This kettle is made of brass'.
> Let q be the proposition 'This kettle is made of bronze'.
> Let r be the proposition 'This kettle contains copper'.
> Let s be the proposition 'This kettle contains zinc'.
> Let t be the proposition 'This kettle contains tin'.

In the language of the Boolean Algebra of propositions, the premises can be written

$$p \rightarrow r, \quad p \rightarrow s, \quad q \rightarrow r, \quad q \rightarrow t, \quad p \vee q, \quad \text{or, more simply,}$$
$$p \rightarrow r \wedge s, \quad q \rightarrow r \wedge t, \quad p \vee q.$$

We ask, what is the implication of $p \vee q$, given the implications of the propositions p and q individually? It is not difficult to verify, using a truth table or otherwise, that

$$p \vee q \rightarrow (r \wedge s) \vee (r \wedge t) = r \wedge (s \vee t).$$

The conclusion we can draw, therefore, is

'This kettle contains copper and this kettle contains either zinc or tin'.

BIBLIOGRAPHY

Gardner, M. (1968). *Logic Machines, Diagrams and Boolean Algebra*, Dover.
Hohn, F. E. (1966). *Applied Boolean Algebra*, MacMillan.
South, G. F. (1974). *Boolean Algebra and Its Uses*, Van Nostrand.

CHAPTER 17

Units of Measurement, Conversion Factors, and Dimensional Analysis

17.1. INTRODUCTION

17.1.1. Physical Quantities, Units, and Dimensions

A *physical quantity* is an entity or concept in nature to which we may assign a numerical magnitude by a suitable process of measurement. The measurement process involves following accepted procedures to express the quantity in terms of some physical standard unit. Thus we may write

$$Q = Q'[Q] \qquad (17.1.1)$$

where Q is a symbol to represent the quantity, Q' is the measured *magnitude* (a number, called the measure), and $[Q]$ is the *unit*—a quantity of the same type as Q defined to have unit magnitude. Examples would be length, mass, temperature, energy, electrical resistance, etc. When we write that a body has mass

$$m = 3 \cdot 7 \, \text{kg} \qquad (17.1.2)$$

kg indicates the unit [1 kilogram], the mass of an internationally agreed piece of platinum-iridium alloy kept in the cellars of the Bureau International des Poids et Mesures in Paris (BIPM).

Either Q or Q' may be used as physical variables in mathematical equations. Writing a length as

$$L = 20 \, \text{ft} = 6 \cdot 096 \, \text{m} \qquad (17.1.3)$$

shows however that Q' is dependent on the system of units chosen so that care must be taken about consistency and specification of units in equations using the Q'. When the symbol Q is used for the physical variable, it is independent of the units chosen, and therefore invariant under changes of systems of units.

Equation (17.1.1) shows that physical quantities have magnitude, and *dimension*, which is a generalization of the unit. Anything which can be measured in length units is said to have dimensions of length $[L]$; anything measured in velocity units has dimensions of length over time

$$[v] = \frac{[L]}{[T]} = [LT^{-1}].$$

493

17.1.2. Systems of Units

The relationships that exist between physical quantities as exhibited in physical phenomena and the laws of nature imply corresponding relationships between their units (and their dimensions). Thus, Newton's equation of motion

$$F = Kma$$

determines a relation between the units of force and those of mass and acceleration, so that it is possible to *derive* a unit for force in terms of the units for mass and acceleration, if these have already been defined. We conventionally take K to be unity and dimensionless and define $[F] = [m][a]$, i.e. the unit of force is that force which applied to unit mass gives it unit acceleration. Since

$$[a] = \frac{[v]}{[T]} = \frac{[LT^{-1}]}{[T]} = [LT^{-2}] \quad \text{and} \quad [m] = [M],$$

therefore $[F] = [MLT^{-2}]$. In the International System of Units (SI Units), force is measured in units of kg m s^{-2}, called the newton. Units derived in this way *without introducing numerical factors*, are said to form a *coherent* system.

From this it follows in the law of gravitation

$$F = \frac{Gm_1m_2}{r^2},$$

that the gravitational constant G has a magnitude which must be determined experimentally, and that it has the dimensions of $F/(m_1m_2/r^2)$, i.e.

$$\frac{[MLT^{-2}]}{[M^2L^{-2}]} = [M^{-1}L^3T^{-2}].$$

It would have been equally possible to treat G as unity and dimensionless to give an alternative definition of force in terms of gravitational attraction, with units of $[F] = [M^2L^{-2}]$. In this case K would have to be experimentally determined and would have dimensions $[ML^{-3}T^2]$. The dimensions of physical quantities therefore depend on the fundamental physical quantities chosen as the initial independent basic dimensional quantities (e.g. mass, length, time, etc.) and on the relationships used to derive the units and dimensions of other physical quantities.

Setting up a system of units requires
   (i) selection of those physical quantities to be used to specify the fundamental base units and dimensions of the system;
   (ii) definition of the *base units* and how they are to be realized in practical measurement;
   (iii) specification of relations to be used to obtain the *derived units* for other physical quantities.

The freedom of choice permitted in this way, and the fact that the differing fields of science—mechanics, optics, chemistry, electricity, magnetism, engineering, etc. developed independently, has produced a Babel of independent systems and measurement. Thus within a hundred years the following systems of units were in use among different groups of scientists and engineers: the metre-gramme-second-ohm units, foot-grain-second units, centimetre-gramme-second (cgs) units, cgs-electrostatic units, cgs-electromagnetic units, cgs-Gaussian units, cgs-International units, electrical units, metre-kilogramme-second-ohm (MKSΩ) units, MKS-ampère units, rationalized and unrationalized electromagnetic units, the foot-slug-second units, gravitational foot-slug-second units, the ill-defined British Imperial System of Units, US international units, and various temperature scales. Many inconvenient conversion factors were necessary, and they frequently had to be altered as new standard units were defined, or old ones adjusted in the wake of new knowledge and improvements in measurement technique. At the time of the First International Electrical Congress in 1881, there were in common use in different countries 12 different units of electric potential difference, 10 of electric current, and 15 of resistance.

At last, with SI units (Le Système International des Unités), a system has been accepted by most countries of the world as a practical system capable of meeting the needs not only of science, but of engineering and commerce also. International learned journals are now making use of SI units mandatory.

Intimately related to the choice of systems of units is the need to define the base standard units so that they may be realized in practice, and transferred with as much accuracy as possible, and be permanent and stable with time. Realization is now generally separated from definition, to allow definitions to remain independent of improvements in techniques of measurement and realization. Thus, the second was based for many years on the period of revolution of the earth and defined as the fraction 1/86400 of the mean solar day. However, the earth has been found to rotate with irregularities amounting to one or two seconds per year, and to be slowing down gradually, corresponding to a lengthening of the second by 3 parts in 10^8 per century. The definition has therefore been replaced by one defined as 9 192 631 770 periods (reciprocal of the frequency) of monochromatic electromagnetic radiation absorbed by the atomic caesium-133 atom. This definition is stable and unaffected by improvements in measurement technique. It is reproducible to about 1 part in 10^{13} in modern atomic clocks.

Except for the standard kilogram, which is the mass of a particular lump of platinum-iridium alloy, all other base units on the SI system are precisely defined, stable quantities separated from their means of realization, and most non-SI and non-metric units are now defined in terms of the SI standards, e.g. 1 inch $\equiv 2 \cdot 54 \times 10^{-2}$ metres exactly. The definitions of the SI base units are given in section 17.2.1.

17.2. SI UNITS—LE SYSTÈME INTERNATIONAL DES UNITÉS

SI units comprise *base*, *derived* and *supplementary* units. Seven physical quantities have been chosen to be dimensionally independent, and to be used to define the SI base units. They define the dimensions used in the system. The derived units (Tables 17.2.3 and 17.7.1) belong to a coherent system of simple combinations of the base units without numerical factors, e.g. N (newton) = kg × m × s^{-2}. There is only one SI unit for each physical quantity, but *prefixes* may be added to form decimal multiples and fractions of the units (see Table 17.2.4). The *symbols* for the SI units are international, but the spelling of the *names* of the symbols depends on the language being used.

The SI units are based on the metre, kilogram, second and ampère, because they form a coherent system with widely used pre-existing units (the joule, watt, volt, coulomb, ohm, farad, weber, henry, tesla) in which no numerical factors are required in obtaining the derived units for physical quantities in terms of the base units (see, for example, Tables 17.2.3 and 17.7.1); SI units are not coherent with the centimetre-gram-second or other base units. (Conversion factors between SI units and commonly used units in other systems are given in Tables 17.2.5, 17.2.6, 17.2.7 and 17.7.1.)

17.2.1.   The SI Base Units

The seven SI base units with their names and symbols are shown in Table 17.2.1.

Table 17.2.1.   SI base units

Physical quantity	Symbol for quantity	Name of SI unit	Symbol for SI unit	Dimension
Length	l	metre	m	*L*
Mass	m	kilogram	kg	*M*
Time	t	second	s	*T*
Electric current	I	ampère	A	*I*
Thermodynamic temperature	T	kelvin	K	Θ
Amount of substance	n	mole	mol	*N*
Luminous intensity	I	candela	cd	*J*

The following are the definitions of the base units:

Length	The metre is the length equal to 1 650 763·73 wavelengths in vacuum of the radiation corresponding to the transition between the levels $2p_{10}$ and $5d_5$ of the krypton-86 atom.
Mass	The kilogram is the unit of mass [and not of weight or of force]; it is equal to the mass of the international prototype of the kilogram.
Time	The second is the duration of 9 192 631 770 periods of the radiation corresponding to the transition between the two hyperfine levels of the ground state of the caesium-133 atom.

Electric current	The ampère is that constant current which, if maintained in two straight parallel conductors of infinite length, of negligible circular cross-section, and placed 1 metre apart in vacuum, would produce between these conductors a force equal to 2×10^{-7} newton per metre of length.
Thermodynamic temperature	The kelvin, unit of thermodynamic temperature, is the fraction $1/273 \cdot 16$ of the thermodynamic temperature of the triple point of water.
Amount of substance	1. The mole is the amount of substance of a system which contains as many elementary entities as there are atoms in $0 \cdot 012$ kilogram of carbon 12. 2. When the mole is used, the elementary entities must be specified and may be atoms, molecules, ions, electrons, other particles, or specified groups of such particles.
Luminous intensity	The candela is the luminous intensity, in the perpendicular direction, of a surface of $1/600\,000$ square metre of a black body at the temperature of freezing platinum under a pressure of 101 325 newtons per square metre.

17.2.2.   SI Supplementary Units

Two units have been called *supplementary* units, and not classified as base units or derived units. They are the geometric units for plane angle and solid angle.

Plane angle	The radian is the plane angle between two radii of a circle which cut off on the circumference an arc equal in length to the radius.
Solid angle	The steradian is the solid angle which, having its vertex in the centre of a sphere, cuts off an area of the surface of the sphere equal to that of a square with sides of length equal to the radius of the sphere.

They may be regarded as dimensionless (of dimension unity), being, respectively, the ratio of two lengths and two areas. They may, however, be used to

Table 17.2.2.   SI supplementary units

Physical quantity	Symbol for quantity	Name of SI unit	Symbol for SI unit
Plane angle	$\alpha, \beta, \gamma, \theta, \phi$	radian	rad
Solid angle	ω, Ω	steradian	sr

form derived units, e.g. radians per second (for angular velocity), watts per steradian (for radiant intensity) (see Table 17.7.1).

17.2.3.   SI Derived Units

Derived units are expressed by algebraic multiplication or division of the base units by each other. Some have been given special names and symbols, others are expressed in terms of base units and the named units. The specially named ones are defined in Table 17.2.3 and others in Table 17.7.1 which also gives conversion factors to relate them to non-SI units.

17.2.4.   SI Prefixes

Table 17.2.4 shows the SI prefixes which may be attached to units to form decimal multiples and decimal fractions.

Table 17.2.3.   Derived SI units with special names

| Physical quantity | SI unit | | | |
	Name	Symbol	Expression in terms of other units	Expression in terms of SI base units
Frequency	hertz	Hz		s^{-1}
Force	newton	N		$m\,kg\,s^{-2}$
Pressure, stress	pascal	Pa	$N\,m^{-2}$	$m^{-1}\,kg\,s^{-2}$
Energy, work, quantity of heat	joule	J	$N\,m$	$m^2\,kg\,s^{-2}$
Power, radiant flux	watt	W	$J\,s^{-1}$	$m^2\,kg\,s^{-3}$
Quantity of electricity, electric charge	coulomb	C		$s\,A$
Electric potential, potential difference, electromotive force	volt	V	$W\,A^{-1}$	$m^2\,kg\,s^{-3}\,A^{-1}$
Capacitance	farad	F	$C\,V^{-1}$	$m^{-2}\,kg^{-1}\,s^4\,A^2$
Electric resistance	ohm	Ω	$V\,A^{-1}$	$m^2\,kg\,s^{-3}\,A^{-2}$
Conductance	siemens	S	$A\,V^{-1}$	$m^{-2}\,kg^{-1}\,s^3\,A^2$
Magnetic flux (ϕ)	weber	Wb	$V\,s$	$m^2\,kg\,s^{-2}\,A^{-1}$
Magnetic flux density, magnetic induction, magnetic field (B)	tesla	T	$Wb\,m^{-2}$	$kg\,s^{-2}\,A^{-1}$
Inductance	henry	H	$Wb\,A^{-1}$	$m^2\,kg\,s^{-2}\,A^{-2}$
Luminous flux	lumen	lm		$cd\,sr^*$
Illuminance	lux	lx	$lm\,m^{-2}$	$m^{-2}\,cd\,sr^*$
Activity (ionizing radiations)	becquerel	Bq		s^{-1}
Absorbed dose	gray	Gy	$J\,kg^{-1}$	$m^2\,s^{-2}$

* In this expression the steradian (sr) is treated as a base unit.

Table 17.2.4. SI prefixes

Factor	Prefix	Symbol	Factor	Prefix	Symbol
10^{18}	exa	E	10^{-1}	deci	d
10^{15}	peta	P	10^{-2}	centi	c
10^{12}	tera	T	10^{-3}	milli	m
10^{9}	giga	G	10^{-6}	micro	μ
10^{6}	mega	M	10^{-9}	nano	n
10^{3}	kilo	k	10^{-12}	pico	p
10^{2}	hecto	h	10^{-15}	femto	f
10^{1}	deca	da	10^{-18}	atto	a

17.2.5. Rules for the Printing of SI Symbols

1. Symbols for units are always printed in roman (upright) type, regardless of the type used for the text in which they occur.

2. Unit symbols are usually lower case letters, unless the symbol is derived from a proper name, when the first letter of the symbol is capitalized. It should be noted that the *names* of units are not capitalized. Examples:

> m   metre
> cd   candela
> K   kelvin
> Pa   pascal.

3. Symbols do not change in the plural, and are not followed by a full stop, unless they end a sentence:

> cm   centimetre (not cm.)
> 5 km   5 kilometres (not 5 kms.).

4. SI prefixes immediately precede the symbol, with no intervening space or punctuation. Compound prefixes should not be used. Examples:

> μA   microampère
> GW   gigawatt
> kHz   kilohertz
> mg   milligram (not μkg).

5. A symbol with an attached prefix is regarded as a single symbol and may be raised to a power without the use of parentheses:

> cm^2 means square centimetre $(0.01\ m)^2$ and not $0.01\ m^2$
> dm^3 means cubic decimetre $(0.1\ m)^3$ and not $0.1\ m^3$.

6. Multiplication and division of units. A product may be represented in any of the ways:

> N m   *or*   N·m   *or*   N.m.

The space must be left and the ordering used to avoid confusion with mN, the symbol for millinewton.

Simple quotients may be represented in any of the ways:

$$\mathrm{m\ s^{-2}} \quad or \quad \mathrm{m/s^2} \quad or \quad \mathrm{m.s^{-2}} \quad (never\ \mathrm{m/s/s}).$$

Complicated expressions should use negative exponents or parentheses:

$$\mathrm{J\ s\ k^{-1}\ mol^{-1}} \quad or \quad \mathrm{j\ s/(k\ mol)}.$$

17.2.6. Units Outside the SI System

Certain widely used units outside the SI have been accepted for use with it. They are tabulated below.

It is considered preferable with the International System *to avoid use of the named CGS units erg, dyne, poise, stokes, gauss, œrsted, maxwell, stiles, phot* (see Table 17.7.1).

Table 17.2.5.   Units in use with the International System

Name	Symbol	Value in SI unit
minute	min	$1\ \mathrm{min} = 60\ \mathrm{s}$
hour	h	$1\ \mathrm{h} = 60\ \mathrm{min} = 3600\ \mathrm{s}$
day	d	$1\ \mathrm{d} = 24\ \mathrm{h} = 86\ 400\ \mathrm{s}$
degree	°	$1° = (\pi/180)\ \mathrm{rad}$
minute	′	$1' = (1/60)° = (\pi/10\ 800)\ \mathrm{rad}$
second	″	$1'' = (1/60)' = (\pi/648\ 000)\ \mathrm{rad}$
litre	l	$1\ \mathrm{l} = 1\ \mathrm{dm^3} = 10^{-3}\ \mathrm{m^3}$
tonne	t	$1\ \mathrm{t} = 10^3\ \mathrm{kg}$

Table 17.2.6.   Units to be used with the International System for a limited time only

Name	Symbol	Value in SI units
nautical mile		$1\ \text{nautical mile} = 1852\ \mathrm{m}$
knot		$1\ \text{nautical mile per hour} = (1852/3600)\mathrm{m/s}$
ångström	Å	$1\ \text{Å} = 0{\cdot}1\ \mathrm{nm} = 10^{-10}\ \mathrm{m}$
are	a	$1\ \mathrm{a} = 1\ \mathrm{dam^2} = 10^2\ \mathrm{m^2}$
hectare	ha	$1\ \mathrm{ha} = 1\ \mathrm{hm^2} = 10^4\ \mathrm{m^2}$
barn	b	$1\ \mathrm{b} = 100\ \mathrm{fm^2} = 10^{-28}\ \mathrm{m^2}$
bar	bar	$1\ \mathrm{bar} = 0{\cdot}1\ \mathrm{MPa} = 10^5\ \mathrm{Pa}$
standard atmosphere	atm	$1\ \mathrm{atm} = 101325\ \mathrm{Pa}$
gal	Gal	$1\ \mathrm{Gal} = 1\ \mathrm{cm/s^2} = 10^{-2}\ \mathrm{m/s^2}$
curie	Ci	$1\ \mathrm{Ci} = 3{\cdot}7 \times 10^{10}\ \mathrm{s^{-1}}$
röntgen	R	$1\ \mathrm{R} = 2{\cdot}58 \times 10^{-4}\ \mathrm{C/kg}$
rad	rad	$1\ \mathrm{rad} = 10^{-2}\ \mathrm{J/kg}$

Table 17.2.7.   Other units generally deprecated

Name	Value in SI units
fermi	1 fermi = 1 fm = 10^{-15} m
metric carat	1 metric carat = 200 mg = 2×10^{-4} kg
torr	1 torr = (101 325/760) Pa
kilogram-force (kgf)	1 kgf = 9·806 65 N
calorie (cal)	1 cal = 4·1868 J
micron (μ)	$1\,\mu = 1\,\mu$m = 10^{-6} m
X unit	
stere (st)	1 st = 1 m^3
gamma (γ)	$1\,\gamma = 1$ nT = 10^{-9} T
γ	$1\,\gamma = 1\,\mu$g = 10^{-9} kg
λ	$1\,\lambda = 1\,\mu$l = 10^{-6} l

17.3.   OTHER SYSTEMS OF UNITS

The growth in applications of electricity and magnetism lead to the widespread use amongst engineers of the so-called 'practical' electrical units (the ohm, volt, ampere, coulomb, farad, joule, watt, henry). Though metric they are coherent (see § 17.1.2) with metre-kilogram-second (mks) base units, but not with centimetre-gram-second (cgs) units which were in use by physicists from about 1870.

Physicists consequently preferred the 'electromagnetic' or 'electrostatic' systems for magnetic and electrical phenomena, which were coherent with cgs units, or the mixed form—'Gaussian' units. All units for electromagnetic quantities were *derived* units, defined through the equations of electromagnetism, with dimensions in terms of mass, length and time.

Electrical engineers stuck to their 'practical' electrical units, plus a miscellany of metric, and non-metric (e.g. imperial) units used by mechanical and heat engineers.

All these systems are now obsolescent, but Table 17.7.1, which gives SI units for many physical quantities, gives numerical conversion factors between the SI and other systems.

17.4.   CHECKING BY DIMENSIONS

Physical quantities have *magnitude* and *dimensions* and equations involving physical variables express an equality of both the magnitude and dimensions of each side of the equation. The separate terms in the equation must therefore have the same dimension, and the equation is said to be *dimensionally homogeneous*. This is true whatever consistent set of units is being used.

An equation may therefore be checked by comparing the dimensions of each separate term for the required homogeneity.

The SI dimensions are mass $[M]$, length $[L]$, time $[T]$, current $[I]$, temperature $[\Theta]$, amount of substance $[N]$, and luminous intensity $[J]$ (see Table 17.2.1).

EXAMPLE.   The oscillatory frequency f of a tuned circuit is $f = 1/(2\pi\sqrt{LC})$ where L is the inductance, C the capacitance of the circuit. The terms have the following dimensions (from Table 17.7.1):

$$f = [T^{-1}]$$
$$L = [L^2 M T^{-2} I^{-2}]$$
$$C = [L^{-2} M^{-1} T^4 I^2]$$
$$1/\sqrt{LC} = [L^2 M T^{-2} I^{-2} . L^{-2} M^{-1} T^4 I^2]^{-1/2} = [T^2]^{-1/2} = [T^{-1}].$$

This matches the dimensions of f as required (2π is a number, without dimensions).

17.5.   CONVERSION OF UNITS

One often requires to convert the value for a physical quantity expressed in one set of units to the value in another set, when a direct conversion factor is not available. A simple, reliable, and self-checking method is multiplication by appropriate dimensionless conversion factors. For example, to convert 36 inches to metres, one would write

$$36 \text{ in} = 36 \text{ in} \left(\frac{2\cdot54 \text{ cm}}{1 \text{ in}}\right)\left(\frac{1 \text{ m}}{100 \text{ cm}}\right) = 0\cdot9144 \text{ m}.$$

Each factor is a ratio of equal quantities (e.g. $2\cdot54$ cm $= 1$ in), so that the method is equivalent to multiplication by *unity factors*. The units must match throughout, and so are easily checked by cancellation. As a more complicated example:

$$1 \text{ BTU}/(\text{hr.ft.}^\circ\text{F}) = \left(\frac{1 \text{ BTU}}{\text{hr.ft.}^\circ\text{F}}\right)\left(\frac{1055 \text{ J}}{1 \text{ BTU}}\right)\left(\frac{1 \text{ hr}}{3600 \text{ s}}\right)\left(\frac{1 \text{ ft}}{12 \text{ in}}\right)\left(\frac{1 \text{ in}}{0\cdot0254 \text{ m}}\right)\left(\frac{1^\circ\text{F}}{5/9 \text{ K}}\right)$$

$$= 1\cdot73 \text{ J}/(\text{s.m.K}) = 1\cdot73 \text{ W m}^{-1} \text{ K}^{-1}.$$

Basic conversion factors between SI and non-SI units are given in Table 17.7.1.

17.6.   FORMULATION OF EQUATIONS WITH INCONSISTENT UNITS

For repetitive calculations, it is often convenient to formulate an equation which will accommodate the numerical values of various physical quantities given in convenient but inconsistent units. For example, the volume flow rate Q, cross-sectional area A of a tube, and mean flow velocity v of a fluid are related by

$$Q = Av. \tag{17.6.1}$$

Data on these quantities may be given in m³/hr, in², and m/s respectively. To convert (17.6.1) to accept data in these units we write, in the notation of section 17.1.1

$$Q = Q' \text{ m}^3/\text{hr}$$
$$A = A' \text{ in}^2$$
$$v = v' \text{ m/s}$$

where Q', A', v' are the numerical values of the quantities in the units given. Substituting in (17.6.1) we get

$$Q'(\text{m}^3/\text{hr}) = A'(\text{in}^2)v'(\text{m/s})$$

or

$$Q' = A'v' \frac{\text{in}^2 . \text{m} . \text{hr}}{\text{m}^3 . \text{s}}.$$

Multiplying by appropriate unity factors will now eliminate the inconsistent units to yield an appropriate numerical (dimensionless) equation:

$$Q' = A'v' \frac{\text{in}^2 . \text{m} . \text{hr}}{\text{m}^3 . \text{s}} \left(\frac{(0.0254)^2 \text{ m}^2}{1 \text{ in}^2} \right) \left(\frac{3600 \text{ s}}{1 \text{ hr}} \right) = 2 \cdot 32 A'v'.$$

The process is checked by the cancellation of all the units. Thus, for the units given,

$$Q' = 2 \cdot 32 A'v'.$$

17.7.   TABLES OF CONVERSION FACTORS BETWEEN SI AND NON-SI UNITS

Tables 17.7.1 and 17.7.2, shown on pages 504–509 give the conversion factors between SI and non-SI units.

17.8.   DIMENSIONAL ANALYSIS

It was noted in section 17.4 that physical equations are dimensionally homogeneous, i.e. all the terms are expressible as the same combination of physical dimensions. This requirement imposes a restriction on the functional form of relations involving physical variables. This is used in dimensional analysis to predict and investigate the possible functional relations between variables for a physical situation purely from considerations of the dimensions of the variables involved. It is not necessary to solve the fundamental equations governing the situation, or even to write them down. It is, however, necessary to have sufficient insight into the situation to be able to recognize the relevant physical variables involved. Dimensional analysis cannot reveal whether variables have been omitted, still less indicate what they might be; nor can it confirm the validity of the set of variables fed in. Nor does it distinguish between the effect of different quantities having the same dimensions, though it will treat these correctly.

Table 17.7.1.   Physical quantities, SI units, and conversion factors

Physical quantity	Recommended symbol	SI unit	Symbol for SI unit	Value of non-SI unit in SI units and international symbol for unit
SPACE AND TIME				
angle (plane angle)	$\alpha, \beta, \gamma, \theta, \phi$	radian	rad	$1° = \pi/180$ rad $1' = \pi/(180 \times 60)$ rad $1'' = \pi/(180 \times 3600)$ rad
solid angle	ω, Ω	steradian	sr	
length	l	metre	m	$1\ \text{Å (angström)} = 0\cdot1\ \text{nm} = 10^{-10}$ m $1\ \text{in} = 2\cdot54\ \text{cm} = 2\cdot54 \times 10^{-2}$ m $1\ \text{mile} = 1\cdot609344\ \text{km} = 1609\cdot344$ m $1\ \text{nautical mile} = 1852$ m $1\ \text{astronomical unit} = 149\,600 \times 10^{6}$ m $1\ \text{pc (parsec)} \approx 3\cdot0857 \times 10^{16}$ m
area	$A, (S)$	square metre	m²	$1\ \text{ha (hectare)} = 1\ \text{hm}^2 = 10^4\ \text{m}^2$ $1\ \text{b (barn)} = 100\ \text{fm}^2 = 10^{-28}\ \text{m}^2$ $1\ \text{acre} \approx 4\cdot046\ 86 \times 10^3\ \text{m}^2$
volume	$V, (v)$	cubic metre	m³	$1\ \text{l (litre)} = 1\ \text{dm}^3 = 10^{-3}\ \text{m}^3$ $1\ \text{UK gallon} \approx 4\cdot546\ 09 \times 10^{-3}\ \text{m}^3$ $1\ \text{US gallon} \approx 3\cdot785\ 41 \times 10^{-3}\ \text{m}^3$
time	t	second	s	$1\ \text{min} = 60\ \text{s}; 1\ \text{hr} = 60\ \text{min}$ $1\ \text{d} = 24\ \text{hr}$
frequency	f, ν	hertz	$\text{Hz} = \text{s}^{-1}$	
angular frequency, $\omega = 2\pi f$	ω	reciprocal second	s^{-1}	
angular velocity, $\omega = d\theta/dt$	ω, Ω	radian per second	rad/s	
period, $T = 1/f$	T	second	s	
speed, velocity	u, v, w, c	metre per second	m/s	$1\ \text{mile/hr} \approx 0\cdot447\ 040\ \text{m/s}$

Quantity	Symbol	SI unit		Conversion factors
acceleration (linear)	a	metre per second squared	m/s^2	$1\ ft/s^2 \approx 0.3048\ m/s^2$
angular acceleration	α	radian per second squared	rad/s^2	
wavenumber, $\sigma = 1/\lambda$	$\sigma,\ \tilde{\nu}$	reciprocal metre	m^{-1}	
MECHANICS				
mass	m	kilogram	kg	$1\ lb$ (pound, avoirdupois) $= 0.453\ 592\ 37\ kg$; $1\ t$ (tonne) $= 10^3\ kg$; $1\ u$ (atomic mass unit, unified) $= m(^{12}C)/12 \approx 1.660\ 57 \times 10^{-27}\ kg$
density	ρ	kilogram per cubic metre	kg/m^3	$1\ lb/ft^3 \approx 16.018\ 5\ kg/m^3$
force	F	newton	$N = m\ kg/s^2$	$1\ dyn$ (dyne) $= 10^{-5}\ N$; $1\ pdl$ (poundal) $\approx 0.138\ 255\ N$; $1\ lbf$ (pound force) $\approx 4.448\ 22\ N$
weight, mg	$G,\ (P,\ W)$	newton	$N = m\ kg/s^2$	
moment of force	M	newton metre	$N\ m = m^2\ kg/s^2$	
moment of inertia	I	kilogram metre squared	$kg\ m^2$	
angular momentum	L	kilogram metre squared per second	$kg\ m^2/s$	
pressure	p	pascal	$Pa = N/m^2 = m^{-1}\ kg\ s^{-2}$	$1\ atm$ (atmosphere) $= 1.013\ 25 \times 10^5\ Pa$; $1\ bar = 10^5\ Pa$; $1\ dyn/cm^2 = 10^{-1}\ Pa$; $1\ torr = 1\ mm\ Hg \approx 133.322\ Pa$
shear stress normal stress	$\left.\begin{array}{l}\tau\\\sigma\end{array}\right\}$	pascal	$Pa = N/m^2 = m^{-1}\ kg\ s^{-2}$	$1\ dyn/cm^2 = 10^{-1}\ Pa$
energy work	$\left.\begin{array}{l}E,\ W\\W\end{array}\right\}$	joule	$J = N\ m = m^2\ kg/s^2$	$1\ erg = 10^{-7}\ J$; $1\ eV$ (electron volt) $\approx 1.602\ 19 \times 10^{-19}\ J$; $1\ Btu$ (British thermal unit) $\approx 1055.06\ J$; $1\ therm = 10^5\ Btu \approx 1.055\ 06 \times 10^8\ J$; $1\ kWh$ (kilowatt hour) $= 3.6 \times 10^6\ J$

Table 17.7.1.—*Continued*

Physical quantity	Recommended symbol	SI unit	Symbol for SI unit	Value of non-SI unit in SI units and international symbol for unit
power	P	watt	$W = N m/s = m^2 kg/s^3$	1 Btu/h $\approx$ 0.293 072 W 1 erg/s = 10^{-7} W
dynamic viscosity, $\tau_{xz} = \eta \dfrac{du}{dz} x$	η, μ	pascal second	$Pa\ s = m^{-1} kg\ s^{-1}$	1 P (poise) = 0·1 Pa s
kinematic viscosity, $\nu = \eta/\rho$	ν	square metre per second	m^2/s	1 St (stokes) = 1 cm²/s = 10^{-4} m²/s
surface tension	σ, γ	newton per metre	$N/m = kg/s^2$	1 dyn/cm = 10^{-3} N/m
modulus of elasticity	E	pascal	Pa	1 dyn/cm² = 0·1 Pa
THERMODYNAMICS thermodynamic temperature	T, Θ	kelvin	K (not °K or deg K)	$\theta/°C = T/K - 273.15$ $\theta/°F = \frac{9}{5}\theta/°C + 32$ 1 °C = 1 K (temperature interval) 1 °F = $\frac{5}{9}$ K (temperature interval)
heat	Q	joule	$J = m^2 kg/s^2$	1 cal (calorie) = 4·186 8 J
thermal conductivity	λ, K	watt per metre kelvin	$W\ m^{-1} K^{-1} = m\ kg\ s^{-3} K^{-1}$	1 cal/(cm s K) = 4·186 8 W m⁻¹ K⁻¹ 1 Btu/(h ft °F) $\approx$ 22·430 3 × 10⁶ W m⁻¹ K⁻¹
heat flux density	$q, (\phi)$	watt per square metre	$W/m^2 = kg\ s^{-3}$	1 cal/(cm² s) = 4·186 8 × 10⁴ W/m² 1 Btu/(h ft²) $\approx$ 3·15 460 W/m²
specific heat capacity	c	joule per kilogram kelvin	$J\,kg^{-1} K^{-1} = m^2 s^{-2} K^{-1}$	1 cal gm⁻¹ K⁻¹ = 4·186 8 × 10³ J kg⁻¹ K⁻¹ 1 Btu lb⁻¹ °F⁻¹ = 4·186 8 × 10³ J kg⁻¹ K⁻¹
heat capacity	C	joule per kelvin	$J/K = m^2 kg\ s^{-2} K^{-1}$	1 cal/K = 4·186 8 J/K
latent heat	L	joule per kilogram	$J/kg = m^2 s^{-2}$	1 cal/gm = 4·186 8 × 10³ J/kg 1 Btu/lb = 2·326 × 10³ J/kg

Quantity	Symbol	Unit name	Unit symbol / definition	Conversion factors
entropy	S	joule per kelvin	$J/K = m^2\,kg\,s^{-2}\,K^{-1}$	
enthalpy, $H = U + pV$	H	joule	J	
LIGHT				
radiant power, radiant flux	Φ, P	watt	$W = m^2\,kg/s^3$	
radiant flux density	ϕ, ψ	watt per square metre	$W/m^2 = kg/s^3$	
radiant intensity	I	watt per steradian	$W/sr = m^2\,kg/s^3$	
radiance	L	watt per steradian per square metre	$W\,sr^{-1}\,m^{-2} = kg/s^3$	
irradiance	E	watt per square metre	$W/m^2 = kg/s^3$	
luminous power, luminous flux	Φ, Φ_v	lumen	$lm = cd\,sr$	
luminous intensity, $I_v = \partial\Phi_v/\partial\omega$	I, I_v	candela	cd	
luminance $L_v = \dfrac{\partial^2\Phi_v}{\partial S\,\partial\omega\,\cos\theta}$	L, L_v	candela per square metre	cd/m^2	1 lambert $= \frac{1}{\pi}$ cd/cm² $= 3183 \cdot 10$ cd/m² 1 foot lambert $= \frac{1}{\pi}$ cd/ft² $\approx 3 \cdot 426\ 26$ cd/m² 1 sb (stilb) $= 1$ cd/cm² $= 10^4$ cd/m²
illuminance $E_v = \partial\Phi_v/\partial s$	E, E_v	lux	$lx = lm/m^2 = cd\,sr/m$	1 foot candle $= 1$ lm/ft² $\approx 10 \cdot 763\ 9$ lx 1 ph (phot) $= 10^4$ lx
AMOUNT OF SUBSTANCE AND MOLAR QUANTITIES				
amount of substance	n, ν	mole	mol	1 gram-molecule $= 1$ mole
amount of substance concentration, n/V		mole per cubic metre	mol/m^3	
molar gas constant	R	joule per mole kelvin	$J/(mol\,K)$	
molar heat capacity	C_m, C	joule per mole kelvin	$J/(mol\,K)$	1 cal/(gm-mole K) $= 4 \cdot 186\ 8$ J mol⁻¹ K⁻¹

Table 17.7.2   Electricity and Magnetism

The values of the electric and magnetic constants ε_0 and μ_0 are given below for the cgs-electrostatic, and cgs-electromagnetic systems. The constant c in the conversion factors has the value $2.997\,924\,58 \times 10^{10}$ (equal to the speed of light in cm/s).

Physical quantity	Symbol	SI value	cgs-e.s.	cgs-e.m.
permittivity of vacuum	ε_0	$10^{11}/(4\pi c^2)$ F/m	$1/c^2$ cm^{-2} s^2	1
permeability of vacuum	μ_0	$4\pi \times 10^{-7}$ H/m	1	$1/c^2$ cm^{-2} s^2

Physical quantity	Recommended symbol	SI unit	Symbol for SI unit	Values of unrationalized cgs unit in SI units (name of cgs unit in brackets)	
				1 e.s.u. =	1 e.m.u. =
electric current	I	ampère	A	$10/c$ A	10 A (1 abamp)
electric charge	Q	coulomb	C = A s	$10/c$ C (1 statcoulomb)	10 C (1 abcoulomb)
electric potential, potential difference	V, ϕ	volt	V = m^2 kg s^{-3} A^{-1}	$c/10^8$ V (1 statvolt)	10^{-8} V (1 abvolt)
electric field strength	E	volt per metre	V/m = m kg s^{-3} A^{-1}	$c/10^6$ V/m	10^{-6} V/m
electric displacement, electric flux density	D	coulomb per square metre	C/m^2 = m^{-2} s A	$10^5/4\pi c$ C/m^2	$10^5/4\pi$ C/m^2

quantity	symbol	SI unit		
electric dipole moment	$\mathbf{p}, \boldsymbol{\mu}$	$Cm = m\,s\,A$	$1/10c\ C\,m$	$1/10\ C\,m$
electric polarization	$\mathbf{P}$	$C/m^2 = m^{-2}\,s\,A$	$10^5/c\ C/m^2$	$10^5\ C/m^2$
permittivity, $D = \varepsilon E$	ε	$F/m = m^{-3}\,kg^{-1}\,s^4\,A^2$	$10^{11}/4\pi c^2\ F/m$	$10^{11}/4\pi\ F/m$
capacitance	C	$F = m^{-2}\,kg^{-1}\,s^4\,A^2$	$10^9/c^2\ F$	$10^9\ F$
magnetic flux density; magnetic induction	$\mathbf{B}$	$T = Wb/m^2 = kg\,s^{-2}\,A^{-1}$	$c/10^4\ T$	$10^{-4}\ T\ (1\ gauss)$
magnetic flux	Φ	$Wb = Vs = m^2\,kg\,s^{-2}\,A^{-1}$	$c/10^8\ Wb$	$10^{-8}\ Wb$ (1 maxwell)
magnetic field strength	H	$A/m = m^{-1}\,A$	$10^3/4\pi c\ A/m$	$10^3/4\pi\ A/m$ (1 oersted)
magnetic moment	$\mathbf{m}$	$A\,m^2$	$1/10^3 c\ A\,m^2$	$10^{-3}\ A\,m^2$
magnetization, $\mathbf{M} = \mathbf{B}/\mu_0 - \mathbf{H}$	$\mathbf{M}$	A/m	$10^3/c\ A/m$	$10^3\ A/m$
permeability, $\mu = \mathbf{B}/\mathbf{H}$	μ	$H/m = m\,kg\,s^{-2}\,A^{-2}$	$4\pi c^2/10^7\ H/m$	$4\pi/10^7\ H/m$
inductance	L, M	$H = Wb/A = m^2\,kg\,s^{-2}\,A^{-2}$	$c^2/10^9\ H$	$10^{-9}\ H$
electric charge density, Q/V	ρ	$C/m^3 = m^{-3}\,s\,A$	$10^7/c\ C/m^3$	$10^7\ C/m^3$
surface charge density, Q/A	σ	$C/m^2 = m^{-2}\,s\,A$	$10^5/c\ C/m^2$	$10^5\ C/m^2$
resistance	R	$\Omega = V/A = m^2\,kg\,s^{-3}\,A^{-2}$	$c^2/10^9\ \Omega$	$10^{-9}\ \Omega$
electric conductance	G	$S = \Omega^{-1} = m^{-2}\,kg^{-1}\,s^3\,A^2$	$10^9/c^2\ S$	$10^9\ S$
resistivity	ρ	$\Omega m = m^3\,kg\,s^{-3}\,A^{-2}$	$c^2/10^{11}\ \Omega\,m$	$10^{-11}\ \Omega\,m$
conductivity	σ, γ	$S/m = m^{-3}\,kg^{-1}\,s^3\,A^2$	$10^{11}/c^2\ S/m$	$10^{11}\ S/m$

17.8.1. Method of Indices

In general, we seek a relationship between one physical variable Q_1 and the other $n - 1$ variables $Q_2, Q_3, \ldots, Q_n$ involved in a physical system. We look for relationships of the form

$$Q_1 = Q_2{}^{a_2} Q_3{}^{a_3} \ldots Q_n{}^{a_n}$$

where the indices a_2, a_3, etc. are to be determined. The dimensions of each side must be identical, i.e.

$$[Q_1] = [Q_2{}^{a_2} Q_3{}^{a_3} \ldots Q_n{}^{a_n}].$$

We substitute for each of $[Q_1]$, $[Q_2]$, ..., the appropriate dimensional product of fundamental quantities, e.g.

$$[Q_1] = [M^{p_1} L^{q_1} T^{r_1} \ldots]$$

and then in turn equate the indices of each of the fundamental dimensions M, L, T on the left- and right-hand sides of the equation. If j dimensions are involved, this yields j simultaneous equations. If k of these are independent ($k \leq j$), we can solve for k of the indices $a_2, a_3, \ldots, a_n$ in terms of the other $n - k$ which remain indeterminate. The number of variables n in the problem is thus effectively reduced by k.

As an example, consider the volume flow rate Q of a fluid of viscosity η through a capillary tube of radius r, when a pressure difference P is applied across a length l of tube. Let

$$[Q] = [P^{a_2} r^{a_3} \eta^{a_4} l^{a_5}].$$

Substituting the known dimensional combination for each of Q, P, r, η, l in terms of $[M]$, $[L]$, and $[T]$ we obtain

$$[L^3 T^{-1}] = [(ML^{-1}T^{-2})^{a_2}(L)^{a_3}(ML^{-1}T^{-1})^{a_4}(L)^{a_5}],$$

that is

$$[L^3 T^{-1}] = [M^{a_2 + a_4} L^{-a_2 + a_3 - a_4 + a_5} T^{-2a_2 - a_4}].$$

Equating powers of $[M]$, $[L]$, and $[T]$ on each side of this dimensional equation then gives

$$\begin{aligned} 0 &= a_2 + a_4 & \text{(from } M) \\ 3 &= -a_2 + a_3 - a_4 + a_5 & \text{(from } L) \\ -1 &= -2a_2 - a_4 & \text{(from } T). \end{aligned}$$

Solving gives $a_4 = 1$, $a_3 = -1$, $a_3 + a_5 = 3$, that is

$$[Q] = \left[\frac{P r^{3 - a_5} l^{a_5}}{\eta} \right] = \left[\frac{P r^3}{\eta} \left(\frac{l}{r} \right)^{a_5} \right].$$

a_5 is indeterminate, so we may therefore write the solution as a summation over different possible values of a_5:

$$Q = \frac{P r^3}{\eta} \sum_{a_5} C_{a_5} \left(\frac{l}{r} \right)^{a_5} = \frac{P r^3}{\eta} \cdot f\left(\frac{l}{r} \right).$$

This is consistent with Poiseuille's formula $Q = \pi r^4 P / 8 l \eta$, but the dimensional method has been unable to distinguish precisely between the effects of the lengths

r and l. It does, however, reveal that for tubes with the same ratio l/r, $Q \propto Pr^3/\eta$. The result may be written in the form

$$\left(\frac{Q\eta}{Pr^3}\right) = f\left(\frac{l}{r}\right) \quad \text{or} \quad F\left\{\frac{Q\eta}{Pr^3}, \frac{l}{r}\right\} = 0. \tag{17.8.1}$$

Both expressions in round brackets ( ) are *dimensionless groups*, i.e. products of variables with no dimensions, so that *the result of the dimensional analysis is a functional relation between dimensionless groups* $\Pi_1 = Q\eta/Pr$ and $\Pi_2 = l/r$, i.e. $F(\Pi_1, \Pi_2) = 0$. In general, $n - k$ independent dimensionless groups occur. [They may be combined by multiplication or division to yield simpler or more physically significant groups if necessary.] Dimensionless groups may be found very straightforwardly by the method of the following section.

17.8.2. Method of Dimensionless Groups

The relation between the physical variables $Q_1, \ldots, Q_n$ of the physical system may be expressed in the form

$$f(Q_1, Q_2, \ldots, Q_n) = 0.$$

The requirements of dimensional homogeneity enable this to be reduced to the form

$$F(\Pi_1, \Pi_2, \ldots, \Pi_{n-k}) = 0,$$

where Π_1, Π_2, etc. are dimensionless groups formed from $Q_1, \ldots, Q_n$, and k is the number of independent indicial equations (see § 17.8.1).

The following method shows how the dimensionless groups may be found in a simple step by step method from the variables $Q_1, \ldots, Q_n$. We consider an elaboration of the example of section 17.8.1, where the volume flow rate Q of fluid through a pipe is taken to be a function not only of the radius r, fluid viscosity η, length l and pressure difference P across the pipe, but also of the mean fluid velocity v and density ρ. Thus, the variables and their dimensions are as shown in the first column:

$$Q = L^3T^{-1} \quad\bigg|\quad Q = L^3T^{-1} \quad\bigg|\quad Q/r^3 = T^{-1} \quad\bigg|\quad \boxed{\frac{Q}{r^3}\cdot\frac{r}{v}} = 1$$

$$P = ML^{-1}T^{-2} \quad\bigg|\quad P/\rho = L^2T^{-2} \quad\bigg|\quad P/\rho r^2 = T^{-2} \quad\bigg|\quad \boxed{\frac{P}{\rho r^2}\cdot\frac{r^2}{v^2}} = 1$$

$$r = L \quad\bigg|\quad r = L \quad\bigg|\quad r/r = 1 \quad\bigg|$$

$$\eta = ML^{-1}T^{-1} \quad\bigg|\quad \eta/\rho = L^2T^{-1} \quad\bigg|\quad \eta/\rho r^2 = T^{-1} \quad\bigg|\quad \boxed{\frac{\eta}{\rho r^2}\cdot\frac{r}{v}} = 1$$

$$l = L \quad\bigg|\quad l = L \quad\bigg|\quad \boxed{L/r} = 1 \quad\bigg|$$

$$v = LT^{-1} \quad\bigg|\quad v = LT^{-1} \quad\bigg|\quad v/r = T^{-1} \quad\bigg|\quad \frac{v}{r}\bigg/\frac{v}{r} = 1$$

$$\rho = ML^{-3} \quad\bigg|\quad \rho/\rho = 1 \quad\bigg|$$

Table 17.8.1.   Values of fundamental physical constants

Units in the International System of Units (SI) have been used throughout. Users of cgs units should omit the portion of the formulae given in the square brackets [ ], and any remaining electronic charge terms will then be in esu. The figures in parentheses ( ) in the 'value' column represent the best estimates of the standard deviation uncertainties in the last two digits quoted. Values are taken from the 'recommended consistent values of fundamental constants' recommended by the Committee on Data for Science and Technology, of the International Council of Scientific Unions (see *J. Phys. Chem. Ref. Data*, **2**, No. 4, 663 (1973)).

Quantity	Symbol	Value	Units	Uncertainty, parts in 10^6
General				
speed of light in vacuum	c	2.997 924 580 (12)	10^8 m s^{-1}	0·004
permeability of vacuum	μ_0	4π exactly	10^{-7} H M^{-1}	—
permittivity of vacuum $1/\mu_0 c^2$	ε_0	8.854 187 818 (71)	10^{-12} F m^{-1}	0·008
Planck constant	h	6.626 176 (36)	10^{-34} J Hz^{-1}	5·4
		4.135 701 (11)	10^{-15} eV Hz^{-1}	2·6
$h/2\pi$	$\hbar$	1.054 588 7 (57)	10^{-34} J s	5·4
		6.582 173 (17)	10^{-16} eV s	2·6
elementary charge	e	1.602 189 2 (46)	10^{-19} C	2·9
Avogadro constant	N_A	6.022 045 (31)	10^{23} mol^{-1}	5·1
atomic mass unit, 10^{-3} kg mol^{-1} $N_A{}^{-1}$	u	1.660 565 5 (86)	10^{-27} kg	5·1
		931·501 6 (26)	10^6 eV	2·8
Faraday constant of electrolysis $N_A e$	F	9.648 456 (27)	10^4 C mol^{-1}	2·8
gravitational constant	G	6.672 0 (41)	10^{-11} N m^2 kg^{-2}	615
Bohr magneton [c] ($e\hbar/2m_ec$)	μ_B	9.274 078 (36)	10^{-24} J T^{-1}	3·9
		5.788 378 5 (95)	10^{-5} eV T^{-1}	1·6
nuclear magneton [c] ($e\hbar/2m_pc$)	μ_N	5.050 824 (20)	10^{-27} J T^{-1}	3·9
		3.152 451 5 (53)	10^{-8} eV T^{-1}	1·7

Category		Symbol	Value	Unit	%
Particles	Bohr radius $[\mu_0 c^2/4\pi]^{-1}(h^2/m_e e^3) = \alpha/4\pi R_\infty$	a_0	5.291 770 6 (44)	10^{-11} m	0.82
	charge to mass ratio	e/m_e	1.758 804 7 (49)	10^{11} C kg^{-1}	2.8
	magnetic moment of the electron	μ_e	9.284 832 (36)	10^{-24} J T^{-1}	3.9
	mass of the electron at rest	m_e	9.109 534 (47)	10^{-31} kg	5.1
			5.485 802 6 (21)	10^{-4} u	0.38
			0.511 003 4 (14)	MeV	2.8
	1 electron volt	eV	1.602 189 2 (46)	10^{-19} J	2.9
	in frequency units e/h		2.417 969 6 (63)	10^{14} Hz	2.6
	in wavenumber units e/hc		8.065 479 (21)	10^5 m^{-1}	2.6
	in temperature units e/k		1.160 450 (36)	10^4 K	31
Proton	magnetic moment of free proton	μ_p	1.410 617 1 (55)	10^{-26} J T^{-1}	3.9
	mass of proton at rest	m_p	1.672 648 5 (86)	10^{-27} kg	5.1
			1.007 276 470 (11)	u	0.011
			938.279 6 (27)	MeV	2.8
Neutron	ratio of proton mass to electron mass	m_p/m_e	1836.151 52 (70)	—	0.38
	mass of neutron at rest	m_n	1.674 954 3 (86)	10^{-27} kg	5.1
			1.008 665 012 (37)	u	0.037
			939.573 1 (27)	MeV	2.8
Spectroscopic	fine structure constant $[\mu_0 c^3/4\pi](e^2/hc)$	α	7.297 350 6 (60)	10^{-3}	0.82
		α^{-1}	137.036 04 (11)	—	0.82
	Rydberg constant (fixed nucleus) $[\mu_0 c^2 4\pi]^2(m_e e^4/4\pi \hbar^3 c)$	R_∞	1.097 373 177 (83)	10^7 m^{-1}	0.075
			2.179 907 (12)	10^{-18} J	5.4
			13.605 804 (36)	eV	2.6
			3.289 842 00 (25)	10^{15} Hz	0.075
Thermal	Boltzmann constant R/N_A	k	1.380 662 (44)	10^{-23} J K^{-1}	32
	molar gas constant $p_0 V_m/T_0$	R	8.314 41 (26)	J mol^{-1} K^{-1}	31
	molar volume of ideal gas at stp	V_m	22.413 83 (70)	10^{-3} m^3 mol^{-1}	31
	($T_0 = 273.15$ K; $p_0 = 101\,325$ Pa $= 1$ atm)				
	Stefan-Boltzmann constant $2\pi^5 k^4/15 h^3 c^2$	σ	5.670 32 (71)	10^{-8} W m^{-2} K^{-4}	125

We eliminate the dimensions M, L and T, in turn by multiplying or dividing each variable by a suitable power of one of the other variables. Thus, to eliminate M, we can divide P, η, and ρ by ρ. The resulting dimensions do not involve M (see column 2). To eliminate the dimension L, we divide by appropriate powers of r. The first dimensionless group (l/r) appears (column 3). To eliminate T, we may divide appropriately by v/r. Three more dimensionless groups (Q/vr^2), ($P/\rho v^2$), ($\eta/\rho rv$) appear (see column 4). The $(n - k)$ independent dimensionless groups are found quickly by this method.

The groups may be recombined to form groups of more direct physical significance for the problem. For example, $(Q/vr^2).(\eta/\rho rv).(\rho v^2/P) = (Q\eta/Pr^3)$, and is the significant group in the Poiseuille flow problem in the preceding section. We may thus take our solution as

$$\frac{Q\eta}{Pr^3} = f\left(\frac{P}{\rho v^2}, \frac{l}{r}, \frac{\eta}{\rho r}\right). \tag{17.8.2}$$

Comparing with (17.8.1), we see that introducing the additional variables ρ, v has generated two extra dimensionless groups, $P/\rho v^2$, and $\eta/\rho rv$ (Reynold's number, of great importance in fluid flow in determining whether turbulence will occur). Equation (17.8.1) applies only for non-turbulent flow, where velocity is not a significant variable.

Dimensional analysis is rarely able to give a complete solution, but can give additional weight, and as with (17.8.2), could help identify important parameters (like $\eta/\rho rv$) which may be held constant or varied systematically in the experimental investigation of a problem.

REFERENCES

Armstrong Lowe, D. (1975). *A Guide to International Recommendations on Names and Symbols for Quantities and on Units of Measurement*, World Health Organisation, Geneva.

Page, C. H. and Vigoureux, P. (1973). *The International System of Units* (translation approved by International Bureau of Weights and Measures), National Physical Laboratory, (London H.M.S.O.).

Pankhurst, R. C. (1964). *Dimensional Analysis and Scale Factors*, Chapman & Hall, London; Rheinhold, New York.

Ipsen, D. C. (1960). *Units, Dimensions and Dimensionless Numbers*, McGraw-Hill.

Index

515